HUMAN PERFORMANCE ENGINEERING

HUMAN PERFORMANCE ENGINEERING

A Guide for System Designers

Robert W. Bailey
Bell Telephone Laboratories

Prentice-Hall, Inc. Englewood Cliffs, New Jersey 07632

Library of Congress Cataloging in Publication Data

Bailey, Robert W.
Human performance engineering.

Bibliography: p.
Includes index.
1. Human engineering. 2. System design. I. Title.
TA166.B33 620.8'2 81-21038
ISBN 0-13-445320-4 AACR2

Editorial/production supervision by Kathryn Gollin Marshak
Cover artwork by Susan Farris
Cover design by Fred Charles Ltd.
Manufacturing buyer: Gordon Osbourne

Printed in the United States of America

10 9 8 7 6

ISBN 0-13-445320-4

PRENTICE-HALL INTERNATIONAL, INC., *London*
PRENTICE-HALL OF AUSTRALIA PTY. LIMITED, *Sydney*
PRENTICE-HALL OF CANADA, LTD., *Toronto*
PRENTICE-HALL OF INDIA PRIVATE LIMITED, *New Delhi*
PRENTICE-HALL OF JAPAN, INC., *Tokyo*
PRENTICE-HALL OF SOUTHEAST ASIA PTE. LTD., *Singapore*
WHITEHALL BOOKS LIMITED, *Wellington, New Zealand*

To my wife Brenda

TABLE OF CONTENTS

THE HUMAN (USER)

THE ACTIVITY - BASIC DESIGN

THE ACTIVITY - INTERFACE DESIGN

PREFACE

When developing a new system, many of the most critical decisions designers must make are related to human performance. Informed decisions require that designers understand several human performance considerations, including how users sense, process information, and respond.

Even a series of good design decisions does not ensure acceptable human performance. Designers have control over only a portion of the many conditions that govern human performance (possibly less than fifty percent). For example designers can:

— "Build-in" conditions that are potentially motivating, but new system management must ensure that users are motivated.

— Specify the number and types of people that would perform best in the system, but others select those who do the work.

— Develop a complete set of training materials, but others must conduct quality training sessions.

— Prepare a clear and meaningful set of written instructions, but others must read and comply with them.

In addition, designers are restricted in the ways they can improve the performance of users. Generally, they are limited to decisions that fall within six major areas:

1. Provide work, interfaces, and facilitators that are *motivating*
2. Provide *selection criteria* that enable the best matching of user abilities, skills, and attitudes with the work
3. Optimize the basic *work design*
4. Optimize the design of critical *interfaces*
5. Develop effective *training* materials
6. Develop meaningful and useful *instructions* including performance aids.

People have been trying to improve human performance for several thousand years. Throughout human history various forms of motivation alone were used to improve performance. Other approaches such as choosing individuals who are suited for a given job, training workers, or designing work to account for the ways people sense, process information, and respond in certain situations were not considered until relatively recently.

By about 2100 B.C. some people were improving the performance of others by requiring them to serve apprenticeships. However, improving performance by using special training facilities did not begin for almost another 4000 years. Tests were being used to select people at least by 1100 B.C., but the first modern selection tests were not used until World War I.

In the early 1940s a major focus for improving human performance was to modify, in some performance-related way, the design of equipment. A few years later we had the formal beginning of a new discipline called "human engineering."

Taylor in a 1957 *American Psychologist* article stated that the aim of human engineering was to "apply the knowledge of human behavior...to the structuring of *machines*" (p. 251). In 1960 Van Cott published a book entitled *Human Factors Methods for System Design.* In the foreword to his book, he broadened the scope of human engineering, stating: that the "primary goal of human factors engineering is to help design an optimal *system* ... to help design not only the equipment, but also jobs, and to some extent the characteristics of the personnel in the system." One of the most influential human performance books in the 1970s, *Human Engineering Guide to Equipment Design,* points out that human engineering is concerned "with conceptually synthesizing a *total system,* with determining the role that human performance will play in such a system, and then designing the environment and man-machine interface to make that performance possible" (p. 3).

There is no question that designers ought to be concerned with using all viable methods to improve human performance. This includes motivating; matching (testing); human-oriented designing of work and interfaces; and preparing effective training, instructions, and performance aids. *The primary purpose of this book is to provide designers, particularly those with limited background in psychology, with some knowledge of how people sense, process information, and respond; as well as to introduce data, principles, and methods that are useful in eliciting an acceptable level of human performance in systems.* This book is intended to introduce the concept of human performance engineering. To present all available information on the topics covered would require much more than a single volume.

Since its beginning in 1925, Bell Laboratories has been concerned with improving human performance. At that time Harvey Fletcher and his colleagues were conducting research on the nature of speech and hearing. These studies included precise measurements (for the first time) of the sensitivity of the ear and the human ability to perceive complex sounds under varying conditions. Their work enabled Bell Laboratories designers to more closely match telephone system products with human abilities. In 1955 Robert N. Ford combined the work on psychoacoustics with work being conducted on other behavioral topics into the Social Science Research Laboratory. Now called the Acoustical and Behavioral Research Center, this Bell Laboratories group is directed by Max V. Mathews, and has over 100 researchers, including about 60 psychologists.

The first human factors group at Bell Laboratories was established in 1946. Within six years, the original group of three people had grown into a 26-member department, headed by John E. Karlin. This department consulted with designers on human factors issues. Based on their studies and recommendations, for example, weight of telephone handsets was reduced from 18 to 6 ounces, the controversy over whether to use letters and numbers or just numbers for telephone numbers was resolved, and the best layout of keys on the TOUCH-TONE* telephone pad for push-button telephones was determined.

In 1967 H. O. Holt organized a second human factors group at Bell Laboratories—the Human Performance Technology Center. Within a few months the new Center consisted of the Human Performance Technology

* Trademark of AT&T.

Department headed by W. F. Fox, the Training Technology Department headed by F. L. Stevenson, and the Personnel Subsystem Applications Department headed by R. Marion. This Center was established to help ensure that the human performance of Bell System *employees,* particularly those working with computer-based systems, was considered along with the performance of Bell System customers.

In 1975 much of the human factors effort was decentralized with small groups placed in individual project organizations, while still retaining two major groups, i.e., J. E. Karlin's Human Factors Department and W. F. Fox's Human Performance Engineering Department. In addition to Mathew's research group, Bell Laboratories presently has about 35 supervisory groups with over 300 behavioral scientists (primarily psychologists) helping to ensure acceptable levels of human performance for Bell System customers and employees.

At the initiative of J. R. Harris, chairman of the Behavioral Science Directors committee, a committee composed of select Bell Laboratories human factors supervisors was organized in 1978. Original members on this Advisory Committee for Applied Behavioral Science were M. S. Schoeffler (chairman), M. J. Katz, E. T. Klemmer, W. S. Peavler, J. C. Vassallo, and J. E. Zielinski. One of their first acts was to recommend that a textbook on human factors be prepared for design engineers. They recommended an author, secured management approvals, and worked out guidelines on the scope and content of the book. This volume is the result of that effort. Final review of the textbook was undertaken by the original committee members, plus new members: C. S. Harris, A. S. Kamlet, B. L. Lively, and M. K. Seagren.

While the book was being prepared, comments were requested from over 100 Bell Laboratories design engineers who participated in company sponsored courses entitled "Human Performance Considerations in System Design." Comments also were received from students who were enrolled in the author's "Human Performance Engineering" courses at Columbia University and Stevens Institute of Technology.

Helpful suggestions were received from several colleagues including: A. Ackerman, R. W. Bennett, K. M. Cohen, T. C. B. Davison, M. C. Day, J. J. Dever, D. J. Dooling, S. H. Ellis, F. S. Frome, G. W. Furnas, R. J. Glushko, M. G. Grisham, C. Hoffman, V. F. Iuliano, M. D. Jackson, J. A. Kadlac, A. J. Kames, J. Klem, J. L. Knight, Jr., T. K. Landauer, K. McKeithen, K. L. Medsker, F. Miller, M. F. Poller, A. Rieck, C. A. Riley, L. Sager, H. R. Silbiger, M. W. Soth, S. J. Starr, J. E. Tscirgi, G. T. VeSonder, J. T. Walker, and C. G. Wolf.

Special assistance was provided to the author by specialists in certain topical areas. The following people surveyed the literature, organized available information, and prepared initial drafts of chapters: J. L. Collymore, Performance Aids; J. D. Cox, Conducting Comparison Studies; S. Fagan,

Training; D. M. Gilfoil (C. L. Mauro Associates), Motivation and Environment; and T. H. Gross, Controls, Displays and Workspace Design. In addition, J. D. Cox, M. A. Craig and P. Marchese ensured completeness and accuracy of the References, and C. L. Barnett and E. M. Schaffer provided chapter summaries.

The original draft of the book was edited by M. H. Roycraft, with assistance from N. M. Burgas, S. Gazsi and C. Margolis. The original art work was provided by S. M. Farris, D. Pisacreta, and M. Safran. Audio-visual support was provided by R. Osborne and J. R. Shepheard. Clerical support and typing was provided by G. P. Brown (Coordinator), S. M. Bush, S. L. Cuba, V. Ho, K. A. Lewis, J. Lorincz, A. M. Mishak, S. A. Napolitano, K. Papsin and E. Yu.

The final version was edited by D. Barker with assistance from M. A. Craig. V. L. D'Andrea, M. J. Reedy and B. M. Castagna did the final typing. W. Czok, V. Kulihin and P. Mahoney provided additional art work.

Management support and assistance in the preparation of the book came from T. H. Crowley, Executive Director of the Computing Technology and Design Engineering Division; P. A. Turner, Director of the Human Performance and Support Center; H. O. Holt, past Director of the Human Performance Technology Center; V. G. Stetter, Head of the Human Performance Engineering Department; W. F. Fox, past Head of the Human Performance Engineering Department; C. B. Rubinstein, Head of the Human Factors Department; D. P. Clayton, Head of the Systems Training Department; J. H. Shoemake, past Head of the Systems Training Department; R. L. Layton, Head of the Continuing Education Department; W. E. Vreeland, Group Supervisor of the In-hours Continuing Education Program and Instructional Support; R. A. Meacham, Group Supervisor of Technical Publications; B. L. Wattenbarger, Supervisor of the Human Performance Studies Group; R. G. Keesing, Supervisor of the Training Support Group; J. H. Ten Eyck, Supervisor of Technical Publications; and B. D. Richardson, Supervisor of Word Processing and Technical Services.

Robert W. Bailey
Bell Laboratories

HUMAN PERFORMANCE ENGINEERING

1

PSYCHOLOGY AND SYSTEMS

INTRODUCTION

People performing in systems have in common the fact that they are each *somebody* doing *something, someplace*. The possibilities range from a telephone installer connecting a cable high atop a pole in the middle of winter, to a chauffeur driving an automobile on a curvy mountain road, or a craftsperson building a violin in a small workshop on a sultry afternoon. In each case, there is somebody doing something, someplace. A *human* is performing some *activity* in some *context*.

PERFORMANCE OR BEHAVIOR

Performance easily can be confused with behavior. Performance is meeting your objective—a result. The actions leading to this result are behavior. This story from Thomas Gilbert's *Human Competence* illustrates the difference.

The Capitol Grill slumped in an ill-smelling corner of Columbia, halfway between the elegance of the Old Campus and the grand old Capitol building.... Few of the 50 T-shirted students had food in mind as they slouched around the front of the Grill in the mortal lock of the morning sun. They had gathered there to await the truck that would carry them out into the piney woods, where they would begin a new job—for many of them, their first.

There ... they would dig for spent bullets, the refuse of soldier training on a Fort Jackson firing range... In 9 years a million GIs had enfiladed these beach-white sands, leaving a half-billion pieces of lead....

Barton Hogg had achieved a dream he thought might be worthy of Midas... There must be $100,000 lying there for him, just for the sifting....

But he was worried. The 60 laborers he had found by scraping the countryside weren't getting the lead out fast enough.... But he would have to admit that his overalled regiment looked busy enough, bent over their shovels and sieves in a long line, just as he had deployed them. He had them working in cadence: a shovel of bleached sand into the hardware-cloth box, a sifting of the box, and then the thudding dump into the milk pails he had bought from army salvage. He had worked out the cadence himself, and was quite pleased with it. Now, if the 50 college students he had had the inspiration to hire could work as well, perhaps he could [become]...a rich man.

The truck arrived annoyingly late, and the platoon of students poured off in shouting disarray. Hogg's heart sank as he watched this undisciplined crew.... A few of them even carried portable radios, and some had newspapers under their arms; not a pair of overalls among them...

They listened to Hogg's instructions with the same blank inattention they had learned to give their professors, and they followed his instructions just as poorly. Straggling off into groups (the radios seemed to form the social nucleus), they proceeded to work completely out of cadence. Most were soon on their haunches and shouting blasphemies, radios blaring, with no hint of order. Soon the shovels were discarded, and they were scraping the sieves directly into the sand. Hogg ran from one student to another, shouting each to his feet and inserting the shovel back into his hands. This went on all morning, and to no avail. Shaping these guys up was like sculpting in mercury. Derisive hoots chased him into retreat.

Defeated, Hogg spent the afternoon in the shade of a truck, visions of Midas shattered. That evening he called them together once more; and in anger and tears, he fired them, every last one of them, with a bitter diatribe on the lack of morals of a lost generation.

The next morning... buckets of lead, left at odd angles in the sand, attested to the rout of the incompetent students. Hogg found that ***the unruly gang had sifted out three times as much lead per labor-hour as the cadenced crew!*** (adapted from *Human Competence*, by Thomas Gilbert. Copyright © 1978 by McGraw-Hill, pp. 13-15. Used with the permission of the McGraw-Hill Book Company.)

FOCUSING ON PERFORMANCE

Like Barton Hogg, many people confuse performance with behavior. Consider someone using a rifle for target practice. We watch as the person lifts the gun, sights down the barrel, and pulls the trigger. We observe a set of behaviors that can be measured. For example, we can time how fast the person is able to raise the rifle, sight, and fire (or each behavior separately). We can measure the width of the person's stance, the steadiness of the aimed gun and the pressure exerted on the trigger. And we can interview the person about the planning and thinking involved in shooting a rifle. But no matter how thorough these measurements are, we still do not have any information on the performance level of the person shooting the rifle, i.e. how many shots are on targets.

Frequently we find a designer working hard to ensure that a system measures certain carefully selected behaviors rather than performance. Consider, for example, companies that have elaborate systems for reporting tardiness or absenteeism. They may accurately measure some interesting *behaviors,* but end up providing little, if any, useful information about *performance.* One good way to measure the performance of our rifle shooter is to inspect the bullet holes in the target. By evaluating the location and pattern of bullet holes, we know something about the performance level of the shooter.

Directly measuring the observable behaviors and indirectly measuring the nonobservable intellectual process does not provide information on performance. However, if we find degraded performance through performance measures, these behavioral measurements may provide clues about the cause. For example, to find out why the shooter missed the target on half the shots, we may use a videotape to carefully examine behavioral measures. Perhaps we see the trigger severely jerked on some shots and gently squeezed on others. We now know why some shots missed the target, and can make behavior changes that will improve performance.

Golfers frequently give each other advice on their observable golfing behavior (e.g., "stand closer to the ball," "use an open stance," or "keep your head down.") This (free) advice is usually given, however, only *after* the golfer's performance (e.g., a dribbled shot, bad slice, ball hit in the water) is degraded sufficiently to warrant the comments. The focus on behaviors usually comes about only after an evaluation of performance.

However, even a careful analysis of a person's behavior may not always reveal the reasons for degraded performance. Frequently, the reasons are related to nonobservable human conditions (e.g., poor eyesight), or to difficulties with the *activity,* and/or the *context.* For example, the difficulties most closely related to degraded performance may be that the resistance of the trigger is too stiff, the gunsight is moved slightly between shots, or the target shooting takes place in a gusting wind. Performance could then be improved by correcting these conditions.

PERFORMANCE DEFINED

Performance then is defined as the *result* of a pattern of actions carried out to satisfy an objective according to some standard. The actions may include observable behavior or nonobservable intellectual processing (e.g., problem solving, decision making, planning, reasoning). Things change when people perform.

ESTABLISHING STANDARDS

Any performance objective must be met according to some *standard.* The two most common standards are quality and quantity. To key data into a computer using a terminal suggests an action ("keying") aimed at fulfilling an objective (converting data to computer-readable form). What is missing is some indication of how accurate and fast the keying must be—a standard. Without such a standard there is no way to measure the performance, at least at a level where suggestions for improvement can be made. When a good set of standards exists, designers can compare them with the outcomes of user actions, and evaluate any differences. In the target shooting example, finding where the bullets hit or how rapidly the person fired the shots is only meaningful if the shooter knew what was expected before shooting. The standards must be known to the user, and be meaningful and measurable.

Unfortunately, in some systems the designers allow standards to simply evolve. For example, they do not set requirements for accuracy or the rate at which actions must take place. Under these conditions, there is no way to determine if performance is acceptable; virtually all performance is acceptable. Often it is mistakenly assumed that designers, users, and user management all have a common standard or expectation and that any deviation from this "common standard" will be quickly recognized and corrected. This näive approach frequently results in considerable disappointment with the human performance levels of new systems. At least four standards should be made a part of every system. The two most common standards are accuracy and user processing time.

Accuracy

Accuracy standards can be established in almost any system. In keying data for input to a computer system, for example, the standard can be an accuracy level of 99 percent, or an error rate of 1 percent. An error is defined as a deviation from an expected outcome; in this case, an error could be counted if the operator keys a "B" instead of a "V," or writes the code "IFR" when "IDK" should be written.

Measuring the relative proportion of errors made in a given time gives an indication of how closely a user meets the standard set for that work. In addition, inspecting the errors themselves may help explain why a person may not be achieving the required level of performance. In the target shooting example, assume that the person shot ten bullets at the target and the resulting holes are arranged as in Figure 1-1.

If the person aimed for the center and ended up with a pattern of holes like those shown, we have good reason to suspect that the performance degradation may be related to the gunsight. The gunsight may need realigning. In this case, the errors themselves suggest a solution to the performance problem. Another possible pattern of errors is shown in Figure 1-2.

The pattern of shots in Figure 1-2 suggests a totally different solution. Rather than a gunsight problem, it appears to be a "steadiness" problem. The person may be jerking the trigger on some shots, or the barrel of the gun may be so heavy that it sways, or perhaps the person does not sight in the same manner each time a shot is made. Whatever the problem, the number and pattern of errors suggest solutions.

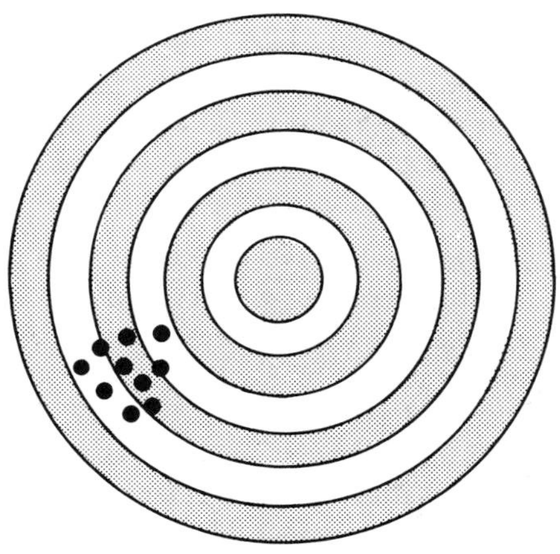

Figure 1-1. Performance Degradation Possibly Related to Gunsight

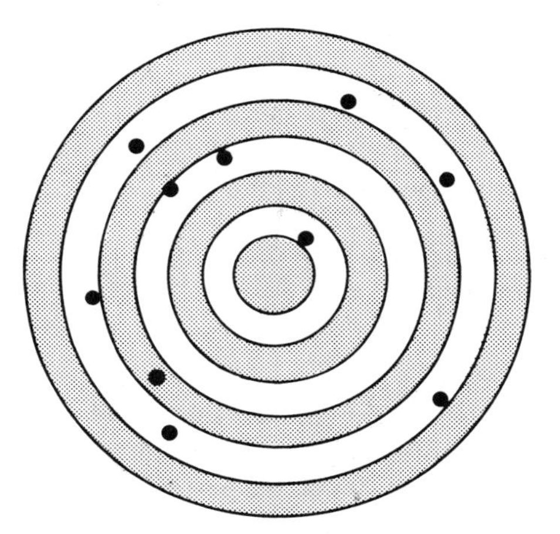

Figure 1-2. Performance Degradation Possibly Related to Steadiness

If the accuracy shooter has no idea what the accuracy standards are (i.e., none have been levied), virtually any performance will be acceptable. This is like having the person shoot at a blank wall and then adding the target (see Figure 1-3). The shooter is always a winner, and as long as the bullets strike the target at all (i.e., performance is not totally degraded) the performance is "acceptable." Many systems where designers have not concerned themselves with accuracy standards reflect this state of affairs.

User Speed of Performance

The quantity standard—that is, the rate at which a person works, or the speed of performance—may be considered in terms such as keystrokes per day, number of shots fired per minute, or crossword puzzles solved per hour. Without a time standard, it may take two, three, or more times longer to reach

Figure 1-3. Performance Degradation Possibly Related to Accuracy Standards

an objective and there is no way to evaluate a person's speed. Usually, if a designer does not set a rate standard, the users will. Some users, recognizing that their early speed (i.e., production rate) will set the standard for years to come, may set a standard so slow as to have the viability of the new system seriously questioned.

Skill Development Time

A third important standard is the time necessary to develop the skill that is unique to performing a new activity. We refer to this as the *time to develop a unique skill.* This may be a few minutes, as in the case of Barton Hogg and his lead collectors, or it may be several months as for pilots of small airplanes. For some skills, such as those of a surgeon, the time may be measured in years. If no such standard is ever set, the time taken to develop a skill may, and frequently does, exceed reasonable limits.

Human performance in any system is very much dependent on efficiently developing the necessary skills. To let these skills develop unsystematically and to not know when they have matured is a very costly way to operate a system. In the past, the time needed to develop a unique skill has been kept short largely by simplifying work. As we will discuss later, this solution has probably created as many human performance problems as it has solved.

The standards set for quality and quantity depend on the user developing the required skill. Measuring the quality and quantity of work before the unique basic skills are developed is misleading. The standards for quality and quantity should be for *skilled* users. For the target shooter, for example, the expectation of having eighty percent of all shots hit the bullseye, while shooting at a rate of one bullet per second should be evaluated only *after* the unique skills are developed. In fact, the quality and quantity standards could be used to help tell when the necessary skills exist.

A designer should also keep in mind that as skills continue to develop, the original quality and quantity standards may no longer apply. To encourage the continuous development of skills, a designer may want to establish quality and quantity standards that become stricter as the user gains experience with the system.

User Satisfaction

The fourth basic standard is *user satisfaction*. Creating satisfying work should be a goal of all designers. Unfortunately, too much of the work developed for system users is boring and trivial. It is unlikely that a user will complete this tedious work for the sheer satisfaction of doing it rather than for the pay or the chance to be with friends all day. Even though pay and friends are forms of satisfaction, the work itself should be satisfying. Satisfaction is usually measured indirectly using interviews or questionnaires.

Designers usually have control over these four standards. A designer should strive to ensure that work is done with the greatest possible accuracy within the shortest time. In addition, the performance itself should be satisfying, and the designer should ensure that the time to develop the necessary, unique skills is as short as possible. We could include other standards, but if a designer works hard to ensure that the design of a system reflects these four, there is a good chance that the system will achieve an acceptable level of human performance.

PERFORMANCE EXAMPLES

Recall that the definition of performance is the *result* of action or actions carried out to satisfy an *objective* according to some *standard*. Table 1-1 illustrates and compares the action, objective, and standards of performance for several different activities, and emphasizes again the major differences between behavior and performance. These examples will help a designer understand the nature of performance covered in the remainder of this book.

Table 1-1. Performance Examples

General Activity	*Actions*	*Objectives*	*Standards*
	Observable		
Dig for lead in sand (Barton Hogg)	(a) Manipulate shovel (b) Shovel sand into box (c) Sift sand in box (d) Dump lead into bucket	(a) Find lead (b) Accumulate lead in buckets	Accuracy: keep only lead, not rocks Rate: five pounds per hour Skill Development Time: 15 minutes Satisfaction: none specified

General Activity	*Actions*	*Objectives*	*Standards*
	Nonobservable (a) Decide next spot to dig (b) Identify lead		
Target shooting with a rifle	*Observable* (a) Pick up rifle (b) Take proper stance (c) Put rifle to shoulder (d) Sight down barrel (e) Pull trigger	Hit target with bullets	Accuracy: 80% in bullseye Rate: one shot per second Skill Development Time: three months Satisfaction: high
	Nonobservable (a) Plan to pick up rifle (d) Determine level of concentration (c) Remember to squeeze and not jerk trigger		
Terminal Use	*Observable* (a) Sit at keyboard (b) Strike keys with finger (c) Look at source data sheet	Enter data into computer	Accuracy: 99.8% Rate: 30,000 keystrokes a day Skill Development Time: three months Satisfaction: high
	Nonobservable (a) Emphasize speed over accuracy (b) Try to ignore distractions		
Driving a car	*Observable* (a) Put key in ignition (b) Start car (c) Look for obstructions (d) Press accelerator	Move car from point A to point B	Accuracy: no accidents Rate: within posted speed limit Skill Development Time: three months

General Activity	Actions	Objectives	Standards
	(e) Drive onto highway (f) Watch for other cars (g) Shift gears		Satisfaction: high
	Nonobservable (a) Consider destination (b) Think about directions (c) Interpret road signs		
Problem solving during a nuclear reactor problem	*Observable* (a) Observing displays (b) Observing warnings (c) Talking/swearing/ angry remarks (d) Flushed faces (e) Walking back and forth (f) Waving arms in air (g) Nodding head (h) Pushing buttons (i) Moving levers (j) Listening	(a) Determine what is wrong (b) Take appropriate action	Accuracy: do not make one wrong decision or take one wrong action Time: solve the problem before the core is damaged and/or radiation is released Skill Development Time: one month Satisfaction: moderate do not want discouragement)
	Nonobservable (a) Considering the meaning of display readings (b) Thinking of possible alternative solutions		

MEASURING HUMAN PERFORMANCE

To measure the height of a table, we might refer to the number of inches on a tape measure. To measure the weight of an object, we use pounds recorded on scales. Human performance is commonly measured using the standards just discussed—accuracy, speed, training time, and satisfaction.

Measuring human performance is essential to the successful operation of a system. A more detailed discussion of the characteristics of these standards and how we can use them as performance measures will follow. Later in the book we will discuss how to convert these standards to *human performance requirements,* and then how to use them to guide the development of the user-related portions of a system.

Measuring Accuracy

Human performance is frequently measured in terms of *accuracy*—performing an action with the fewest errors. In sports emphasis is placed on accuracy as, for example, in rifle marksmanship or executing the coordinated body movements necessary in gymnastics. In the work situation, particularly in computer-based systems, emphasis is placed on inputting large amounts of data in a short time with the fewest errors. In both situations, those who make few errors are superior performers.

One of the main goals in almost any system is fewer errors. In fact, many believe that the very essence of acceptable human performance is to have activities performed in a reasonable time with few or no errors. Because faulty design decisions lead to errors, designers can control errors by making informed decisions. Chapanis (1965) made this insightful statement on the subject:

> Human factors engineers are the first to grant that people make mistakes. But they raise these important questions also: Is some of the blame to be found in the design of the equipment that people use? Do people make more mistakes with some kinds of equipment or vehicles than with others? Is it possible to redesign machines so that human errors are reduced or even eliminated? Research over the past few decades provides us with a resounding "YES" to all these questions. This then is the rationale behind the approach of the human factors engineer: he starts with the certain knowledge and conviction that people are fallible and careless, and that they have human limitations, but he then turns to the machine and the job to see whether he can eliminate their *error provocative features. Courtesy of Brooks/Cole Books.*

Measuring Speed of Performance

Improving efficiency in a system often means reducing the speed of human performance to a minimum. The prediction of efficiency is based on the study of variables affecting the speed of performance in many different activities, such as reading, computing, checking. Frequently, the designer's ultimate goal is to have activities performed in the shortest possible time. This results in the most work being performed per person in a given time, and a need for the fewest people.

Measuring Training Time

Another important performance measure is the total *training time* required to bring system users to a desired level of performance. One of the main goals of a designer is to find ways of designing activities so that training time is reduced to the minimum and whatever level of proficiency is obtained in training is maintained after training has been completed.

Performance aids can substantially reduce the time required to train a person. Instructions, if well done, also will minimize the need for training. It makes good sense not to spend a good deal of time training someone to perform an activity when the activity only will be required twice a year. In these cases, a set of instructions that can lead someone through the activity step by step should be used in lieu of training.

Generally the less time it takes to train people, the lower the cost of operating the system. Reducing errors and shortening processing time produce the same result. Fewer errors require fewer corrections. Shorter processing times require fewer people. Both lower the cost of reaching the system objective. Reducing errors, processing time, and training time all contribute to a system that costs less to operate.

Measuring User Satisfaction

The final standard may turn out to be the most critical. It concerns whether or not the human performing an activity in a particular context receives *satisfaction*: Is the activity/context situation rewarding? Some work activities are satisfying for a large number of people, unfortunately, large numbers of work activities are only satisfying for a few people.

A designer should strive to build a system that will allow work to be done in the shortest time, with few errors *and at the same time satisfy the worker*. There is an irony in expecting people to want to work (versus being on welfare) when we are still designing jobs that are not satisfying.

It is curious that people will pay to perform such activities as playing pinball and electronic games. Places exist where people can go and spend a good deal of money to perform these activities. People also work without pay and do so reliably and efficiently for many churches and charitable organizations. That people are willing to spend many hours in these endeavors without pay suggests that the work is satisfying.

Evaluating Tradeoffs

By attempting to perfect the results of all performance measures, the designer constantly faces tradeoff decisions. In some cases decisions to increase satisfaction may lead to more errors or slower processing time. On the other hand, by always making decisions that reduce errors and processing and training times, a designer may develop a system that is not at all satisfying to work in.

In another situation, if a designer places primary importance on few errors, then he or she will make design decisions to slow down the manual processing of information. In some computer systems designers may find that slower computer response times lead to more time for a person to review their input and thus detect and correct errors. A decision to minimize training time may also lead to longer processing times because the individual must read and digest instructions rather than automatically perform certain activities. Relying on printed instructions and/or performance aids in lieu of training may also lead to more errors in a system. Thus, improving the results of one performance measure may often mean lessening the importance of others. The skill of the designer becomes one of making tradeoffs that lead to the best overall human performance. An acceptable level of human performance helps ensure an acceptable level of system performance.

We achieve acceptable human performance, then, by attempting to perfect the results of these performance measures within constraints (e.g., economic or technical). But we need to keep in mind that a designer cannot seek to eliminate errors while ignoring the fact that people in the system dislike showing up for work, or seek to reduce training when 50 percent of all transactions are in error. The secret is *balance,* and the best balance requires a good understanding of the relative contributions to human performance made by the human, the activity, and the context.

Human performance, then, should be measured using a clear and consistent set of performance measures. The measurements should focus on a particular group of people performing the same activity in a similar context. When measuring accuracy, the data must be collected over a period of time long enough to get reasonably reliable results. These results can then be compared with a standard. Ideally, the standard is determined ahead of time and documented as human performance requirements. Thus, the acceptable accuracy level for these activities should be known, as well as the acceptable processing times, the acceptable time period for training, and the acceptable criteria for scoring a questionnaire on preference, comfort, and satisfaction for the activity in the context. When measurements taken may indicate problems, a designer can make human performance adjustments through changes to the people, activity, or context.

Adequate human performance in a particular system can be measured using any one or all of these performance measures. In fact, no matter what other performance measures are used, one could argue that human performance should *always* be measured in terms of how well a designer deals with *errors, manual processing time, skill development time* and *user satisfaction.*

HUMAN PERFORMANCE VS. SYSTEM PERFORMANCE

Many designers seeking to measure *human* performance actually measure *system* performance. In many systems human performance is only one consideration. The adequacy of equipment, computers, or even people outside the system boundaries may all partially determine the success of a system. When trying to understand why a system may have problems, the various components must be evaluated *separately*. Too often people equate poor system performance with poor human performance without taking into account other system components.

One of the most interesting examples of this is the consistency with which human error is considered the cause of so many airline accidents. Within two or three days of any major airline accident the newspapers usually begin to report that human error is responsible. More investigation frequently indicates that other components were equally at fault. Even with many automobile accidents, investigators attempt to attribute the mishap to degraded human performance, frequently ignoring the mechanical adequacy of the automobile and the conditions under which the automobile was driven (including misleading road signs, rough road surfacing, sharp corners, or unusual weather conditions).

This book is about *human* performance, not *system* performance. The human is the most complex of all components in any system and rightly deserves to be singled out as the most likely reason an accident occurs or a system falters. However, human performance is often degraded because of poor design decisions pertaining to the activity being performed including the tools being used or even the context in which an activity is performed.

Taylor (1957) provides a good example of problems associated with measuring human performance rather than system performance. Comparing the performance of a boy on a bicycle with that of a boy on a pogo stick helps to illustrate the difficulties. In this case the main performance measurement is the speed at which they travel a quarter of a mile. After several trials we find that the boy on the bicycle consistently traveled the distance in a shorter time. What can we conclude about human performance in this situation? The answer is: very little. The performance measure we selected is not a measure of human performance but a measure of system performance—boy/pogo-stick versus boy/bicycle. It is apparent that one system is better than the other but this does not mean that human performance in one system is better than human performance in the other. The boy on the pogo stick actually may have been doing a better job of pogo-stick jumping than the bicycle-rider was doing of bicycle-riding. As long as we are dealing with a system-level performance measure, human performance can only be inferred, and the inference in this case could easily be misleading.

This is one reason it is so difficult to judge the adequacy of human performance in a large system. It is much better to measure human performance separately. Unfortunately, in many situations there is no meaningful, uncontaminated way of evaluating human performance within a system. In our example it is not possible to measure hopping independently of the physical characteristics of the pogo stick, or pedaling in the absence of pedals. One cannot pedal a pogo stick or effectively hop through the air on a bicycle.

In a system, human performance is interrelated with all other components working to satisfy the objective of the system. A single component does not act independently of the other components. Even in relatively simple systems—such as the pogo stick and bicycle examples—it is difficult to measure human performance. In more complex systems it is even more difficult. But it can be done if a designer makes an effort to do so beginning *early* in the design process.

Once the designer has successfully separated the human component from other system components, he or she must evaluate the different elements of human performance. Any measurement of human performance must take into account human characteristics, the activity being performed, and the context in which the activity is performed. For example, we cannot even study the human alone and expect to understand a great deal about performance. We cannot study something as simple as a person walking without having some idea of where the walking is being done (i.e., a muddy field, asphalt track, or two feet of water).

For example, consider a system that was designed to produce White Pages telephone directories. Each directory contains about 50,000 listings (names, addresses, and telephone numbers) and is published once a year. The directory has been published for the last 15 years and about 25 percent of the listings requires some change from one issue to the next.

There are several people involved in the production of the directory, ranging from those who deal with the customer (taking requests to install new phones or to change names or addresses) to those who actually deliver directories to customers. In between these two groups are those who are responsible for ensuring that a printer makes all the necessary deletions, additions, or changes to listings. These users must delete listings when people have their service discontinued, add listings for new service, and make changes to listings when requested by a customer or when an error is detected.

How can the performance of these people be evaluated independently of the performance of this system? System performance is generally evaluated in terms of producing a directory on time and with few errors. One way of evaluating human performance, then, is to have each person in the system be responsible for all aspects of the production of a single book. In this way, system performance can be equated to human performance. However, with a large book this is not practical. Nevertheless, if certain individuals are responsible for a certain range of alphabetical listings there still remains the possibility of separating human performance from system performance.

Degraded system performance may or may not be directly related to the human performance of this group of people. For example, the directories may not be produced on schedule because changes are not received in a timely manner, the printer has many other jobs and puts this one off until last, or the people who deliver the directory store them in a warehouse for six to eight weeks waiting for additional help.

HUMAN PERFORMANCE MODEL

Predicting human performance in any situation requires an understanding of the *human,* the *activity* being performed, and the *context* in which it is performed (see Figure 1-4). This model of human performance is general enough to serve as model for many, if not all, performance situations.

HUMAN PERFORMANCE MODEL

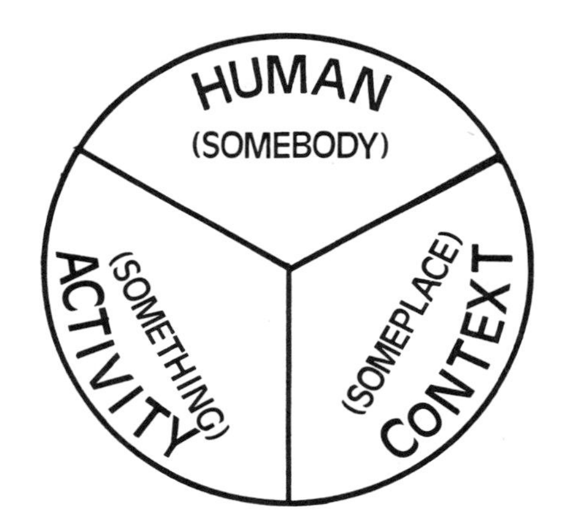

SOMEBODY DOES SOMETHING SOMEPLACE

Figure 1-4. Human Performance Model

People sometimes seek to attain *super* (near perfect) performance. For example, a professional golfer who demonstrates super performance during a golf tournament can make several thousand dollars; a track star who sets a new world record is in a position to make a great deal of money. A premium is paid for super performance in dance, painting, and in many of the crafts, for example, the making of violins or guitars. Near perfect performances usually have one thing in common: In each, a highly skilled individual performs a familiar (and usually satisfying) activity in a favorable context.

Some other situations require *optimal* performance. Life itself may be on the line in such situations as the Apollo lunar landings, a pilot's communication while landing a commercial aircraft, or people working with high-voltage electricity, large mechanical apparatus, or nuclear reactors.

Few designers have the requirement, resources, or know-how to design for super or optimal performance. Designers usually aim for an *acceptable* level of human performance, particularly where *degraded* performance may cause financial losses. This includes systems where excessive errors or excessive training time lead to unmanageable costs, or where boring activities cause high turnover, thereby increasing the costs of recruiting and training.

Nevertheless, to have a super, optimal or acceptable level of human performance (or simply to avoid degraded performance), a designer must take into account each of the following elements (Figure 1-5):

- The general state or condition of the *human*
- The *activity*, including any required tools or equipment
- The *context* in which an activity is performed.

The designer has different degrees of control over each of these three elements. Usually designers have the most control over the *activity*. Designers usually have less control over the *people* selected for the performance and the *context* in which the performance takes place. Not always, however. The Apollo lunar landing project is unique in that designers had control over all three major elements and could determine to a large extent the final level of human performance.

In most situations, however, the designer is severely limited in his or her control over the elements that affect human performance. A good example is automobile design. The automobile designer has control over the activity itself, but has little control over the potential user or where the activity will be performed. The driver may be inexperienced, or may be intoxicated; he or she may be driving on icy roads at a speed of 60 to 70 miles per hour. Designers in this case usually must assume an "average" driver performing in a normal context.

Even a good understanding of each of the elements separately (human, activity, and context) is not sufficient to predict human performance. The interaction between elements is also critical. For example, in computer-based activities the interaction between the human and the computer (i.e., *human* and *activity)* must be carefully designed. Some interactions between the *activity* and the *context* may present design problems. A user would find that the act of opening a door by turning a large wheel with both feet firmly planted on the earth much different from the same act performed weightless two hundred miles above the earth, or under the pressure of three hundred feet of water.

The interactions between the *human* and the *context* also may vary. A very noisy room may have a different effect on a well-rested person than it has on a person in need of sleep.

As illustrated in Figure 1-5, in all performance situations we have somebody doing something someplace. The designer should consider these three elements separately and in combination to help reach the desired level of human performance. This helps to simplify ensuring that the critical elements and their interaction are recognized and dealt with. Each of the elements will be briefly discussed.

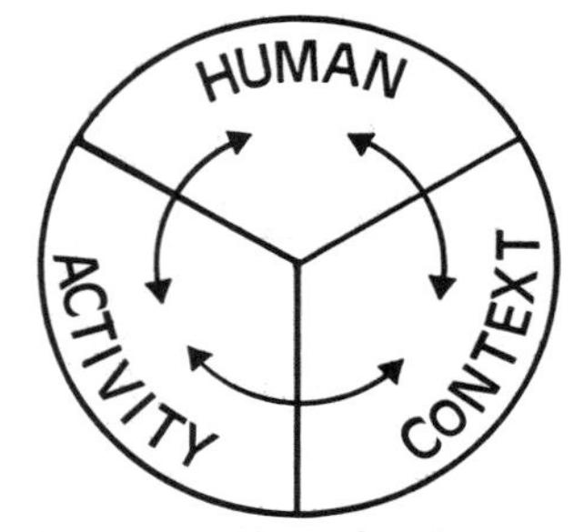

ACCEPTABLE HUMAN PERFORMANCE DEPENDS ON:

(a) THE ADEQUACY OF EACH MAJOR COMPONENT

(b) THE ADEQUACY OF THE INTERFACE BETWEEN AND AMONG MAJOR COMPONENTS

Figure 1-5. Human Performance Model Showing Interactions Between Major Elements

The Human

The human is the most complex of the three elements.

Human performance can be affected either positively or negatively by a wide range of conditions or influences that exist within a user, even without considering the nature of the activity or the context. The designer should understand the possible sources of deficiencies in people and take them into account when making decisions. For example, color coding of displays is now common in many systems. However, color coding should not be the sole form of coding because many people are color blind.

The major considerations of the human element of a system are the sensors, brain (cognitive) processing, and responders (see Figure 1-6). People bring a wide range of basic abilities to an activity. These include good vision and adequate hearing (sensors); arms, fingers and a mouth that function properly (responders); and the ability to think reason and make decisions (brain processing). To attempt to design a system without having a good understanding of how people sense, respond, and process information is like attempting to wire a house without understanding the principles of basic electricity. In both cases it can be done, but the results may leave much to be desired.

Degraded performance could result if any of the basic capabilities required to perform an activity are lacking or reduced. In one system, a telephone operator was having difficulty performing at a switchboard where she was taking calls through headphones. For weeks she had been retrained and encouraged in many different ways by her supervisor without any improvement in performance. One day in an interview it became apparent that she was having difficulty hearing. Her hearing was tested and found severely impaired which interfered greatly with her ability to hear a voice through earphones. The solution to her degraded performance was to install an amplifier in her earphones.

Alluisi and Morgan (1976) have suggested that it is also necessary to consider the temporal influences (e.g., biological rhythms, sleep, fatigue) and organismic influences (e.g., illness, drug reactions) when considering human performance.

Human deficiencies, for whatever reason, may rule out optimal human performance. Thus, we come to expect degraded human performance in certain activities if, for example, a person has not had adequate sleep, has less than perfect vision, has not learned certain basic skills, or does not desire to perform the activity requested.

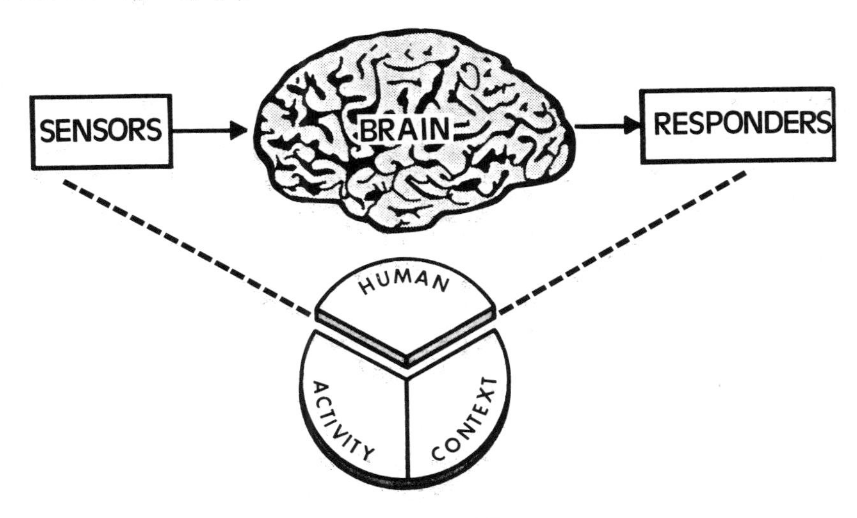

Figure 1-6. The Human Element

System designers usually assume that potential users of a system will have certain skills such as speaking, listening, writing, and in some cases typing. The cognitive skills of perception, decision making, problem solving, and movement control are also assumed. If any of these skills are lacking, a designer needs to provide an efficient way for the user to learn them. This, of course, presupposes that a user is able to learn new skills in a reasonable time. Designers should be cautioned: These assumptions may not always be true. Some potential users may not have the necessary skills to do a job; furthermore, there are users who are incapable of learning the needed skills in an amount of time that is practical.

Another characteristic associated with the human is the "desire to perform," or motivation. Even if the person has the basic abilities and has acquired the necessary skills, the questions remains of whether or not the person *will* perform. Thus, we must consider not only what the person *can* do through basic abilities and acquired skills, but also what the person *will* do.

In addition, we assume that people are mentally healthy. For example, they are not severely depressed, or do not have a chronic high anxiety; do not have a performance-affecting phobia; are not compulsive in their reactions; and do not have perverse or obsessed ideas relating to the performance of an activity. To complicate matters further, we tend to assume that all people are the same, and that one individual does not change over even a short period of time. However, a depressed or anxious person may "level out," a phobia may be overcome, or irrational thinking may be replaced with clear, rational thought.

Designers of new systems do not usually have much control over the actual people selected to perform in their system. Ideally, a designer could design for a specific individual, as in the Apollo lunar landing missions where there was a small select group of astronauts known to the designers long before critical human performance design decisions were made. Unfortunately, this is rarely the case. The next best situation is for the designer to have a good idea of the characteristics of a potential user population so that design decisions can best accommodate this target population. Because most systems are designed for groups of people, it becomes necessary to deal with strengths and weaknesses expected in the potential user population. The designer should then make sure his or her assumptions about the user population are well documented for those who *do* select system users.

The Activity

The second major consideration in understanding human performance is the activity performed. Major characteristics of this element are shown in Figure 1-7.

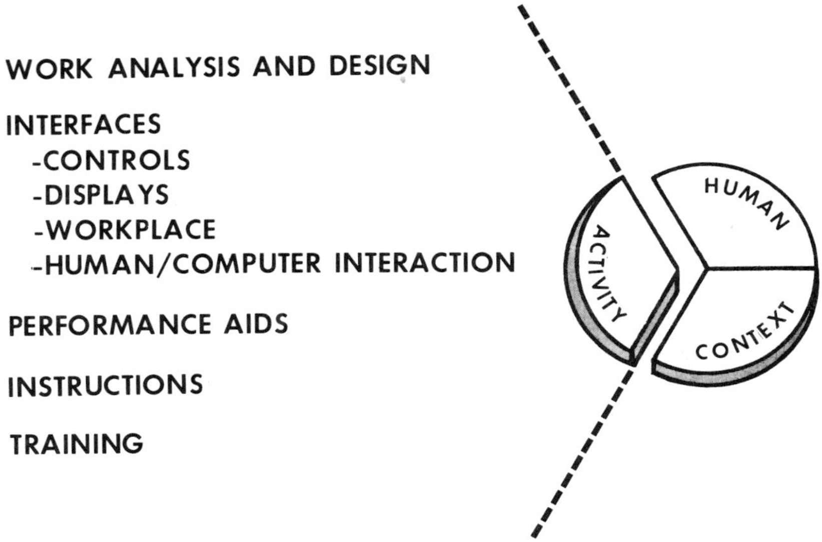

WORK ANALYSIS AND DESIGN

INTERFACES
- -CONTROLS
- -DISPLAYS
- -WORKPLACE
- -HUMAN/COMPUTER INTERACTION

PERFORMANCE AIDS

INSTRUCTIONS

TRAINING

ACTIVITY
HUMAN
CONTEXT

Figure 1-7. The Activity

One wonders if a hundred years ago anyone might have thought that thousands of people would someday be employed to create jobs for thousands of other people. Cazamian (1970) writes:

> For a period measured in millennia, work was carried out in the form of crafts. The craftsman or artist was at one and the same time the organizer and executor of his own works. With a look to agriculture one sees a very slow, from generation to generation, refining of hand tools brought about by the users themselves. Today, we live in a situation where the craftsman's function has, so to speak, split into two parts: those of the organizer (system designer) and those of the executor (system users). In many, many cases the designer is no longer a user and a user has limited or no input to a designer.

On the one hand, we see the designer concerned with economic and technical aspects. On the other side, we have the user whose thinking is more oriented toward just getting through the day. For many users, work is merely a means of existence. In a large number of cases, however, success at work may determine a person's other successes and failures in life. Thus, the activity to be performed should be thoughtfully created by an innovative designer, rather than simply being left to evolve.

Because the designer can control certain conditions relating to the performance of an activity, he or she must know which factors lead to better performance and which tend to degraded performance. For example, designers should know what kinds of work are best done by people; what tasks should be combined into a module of work; and what training is required to build sufficient skills for an acceptable level of human performance. If a system uses tools such as a computer, some special interface considerations must be taken into account.

The Context

The final consideration in this human performance model concerns the context in which a human performs a particular activity (see Figure 1-8). There are actually two different considerations, the *physical context* and the *social context*. It can make a considerable difference if an individual is attempting to connect a cable on top of a telephone pole in shirt-sleeves in Florida as opposed to performing the same activity in Minnesota during the winter, wearing a heavy coat, hood, and gloves. In both cases it could be the same person performing the same activity; the major difference is the context.

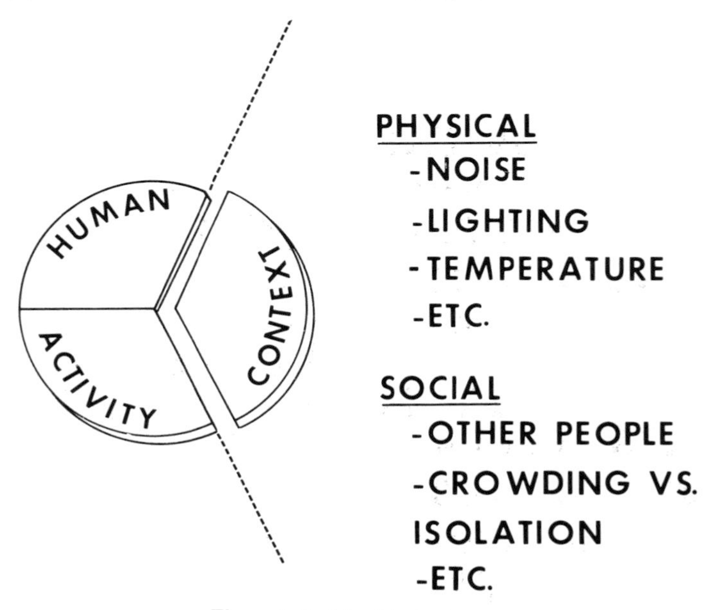

Figure 1-8. The Context

The performance of a person attempting to communicate over the telephone can be degraded if noise interferes with the perception of speech by one or both parties. Noise, in fact, is probably the single most studied physical context factor. Other physical context-related conditions of interest to researchers in recent years include weightlessness, vibration, and insufficient oxygen.

Conditions in the social context that may affect human performance include the effects of other people, crowding, and isolation. The effects of the social and physical context are well demonstrated by considering a football player who performs the same task over and over during practice and in a series of games. The main difference is the context in which the performance takes place. During a football season, the player may perform in snow, rain, sunshine, 20-degree temperatures, 95-degree temperatures, with a noisy partisan crowd or a noisy nonpartisan crowd, or in a practice game with no observers. As the so-called "home court advantage" seems to suggest, the context may make a big difference in the human performance that takes place.

PEOPLE IN SYSTEMS

Systems can be defined as groups of people working with the necessary tools, to meet some goal. Systems designers work to determine what must be done and the best way to do it in order to reach the goal. One or more clerical workers calculating, printing, keying, and doing other tasks to ensure that a payroll will be produced weekly or monthly could be defined as a payroll system. If these people use tools, ranging from pencils to computers, they are still working toward some common goal and the tools merely enable them to reach this goal more accurately, more quickly, or perhaps in a more satisfying way.

There was a time when it was convenient to refer to "man/machine" systems. Historically, the man/machine concept was a compromise in an attempt to get designers to pay attention to *people* as well as machines during the early stages of a system's development. In the early development of military aircraft systems, for example, the aircraft was designed, developed, and manufactured with considerable, almost exclusive, emphasis on getting the aircraft airborne, getting it to its destination, enabling it to carry and drop its payload or fire its guns, and returning home again. It was not usually seen as a human/machine problem, but as a machine problem with people added. Not too many years ago it would have been a difficult problem indeed to encourage aircraft designers to adopt a different approach to system design. These early engineers faced tough technical problems and often they erroneously assumed that people could perform any activity assigned.

Singleton (1974) feels that the best systems are those developed with the characteristics of users as the main frame of reference. This suggests that the human be placed in a central role, with effective system design being anthropocentric. By focusing on the human, the designer has focused on most complex component in any system. If the ability to predict performance is an indication of a component's complexity, then surely the most difficult performance to predict is that of the human. Once the human has been placed in the central role, design decisions concerning the activity (including all tools) and the context can be made to enable users to meet the goals of the system.

A *Total System* approach to system design is proposed and discussed later in the book. With this approach, a designer gives equal attention to all major components.

HUMAN PERFORMANCE ENGINEERING

Human performance engineering includes the *scientific study* of performance-related processes and functions; the *translation* of research results into meaningful human performance data, design principles, methodologies and techniques; and the appropriate *application* of this information in systems. As we see in Figure 1-9, the people associated most with this field include system designers (engineers), human performance specialists, and experimental psychologists.

Scientific Study

Some types of human performance research have been going on for well over one hundred years. This research can be divided into three types. The first is *basic general* research, in which scientists try to better understand *human characteristics* as they relate to performance. This includes research in such areas as memory, sensory capacities and limitation, decision-making, and problem-solving abilities. This research is primarily concerned with the functioning of people in laboratory situations. It tends to be general; for example, findings about the memory of individuals in a particular study are usually considered representative of memory for other people in similar situations.

A second type of research, also general in nature, focuses more on either *activities* or *context*. This is *applied general* research, and the emphasis is usually on measuring a group of "normal" or representative people performing an activity in a given context. Examples of research in this area are: studies of typing errors; studies of hand printing speed; the effects of noise on different activities; the effects of different-sized push buttons; work motivation; the effects of different instructional techniques; and the content and layout of instructions. These findings generally can be applied to other situations with a similar activity and/or context.

The third type of research—*specific* research—is usually conducted to answer a particular question. Results are difficult to generalize beyond the human/activity/context situation in which the research was done. For example, what is the best configuration for a keyboard for directory assistance; what is the best shape for the hand-held portion of a telephone set; how should information be presented to a person trying to detect problems on a telephone line; or where should the dial be placed for a telephone installed in a car.

Translation

The second major aspect of human performance engineering then, deals with translating research results, as well as other relevant information on methodologies and techniques, into a form where it can be easily and readily used by designers.

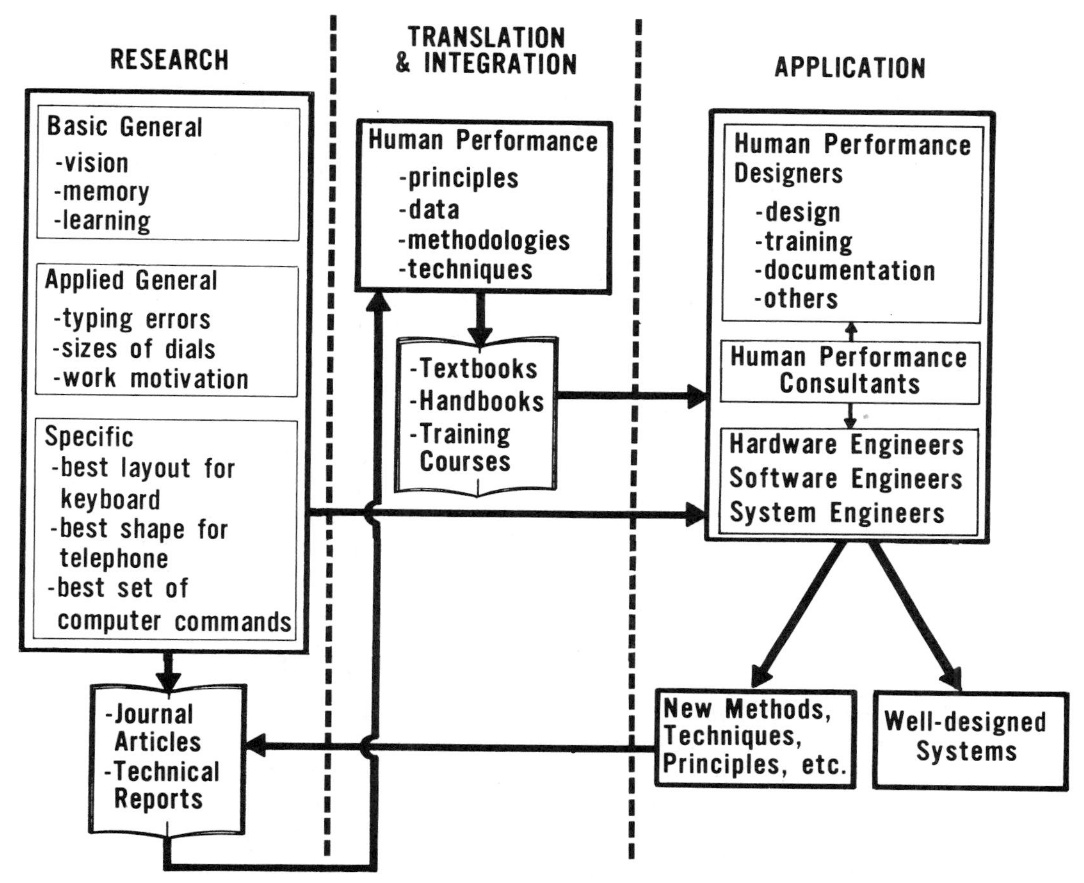

Figure 1-9. Human Performance Engineering

Alluisi and Morgan (1976) have pointed out that despite the commonly held notion that general research findings automatically lead to useful implementation, human performance research must first undergo a difficult *translation* if it is ever to be used in an application. Designers should be aware that there continues to be considerable difficulty in translating research results and application experience into a form that can be useful in the design of new systems.

There are actually two problems here. The first is not having the interest and/or necessary resources to convert or translate research results and other materials into meaningful human performance data, design principles, methodologies and techniques. The second is that much research that goes on in the name of improving or gaining insights into human performance actually has little or no relevance and will never find its way into any meaningful application. In addition, much of the research on the effects of noise, heat and cold, or light-level on human performance has been done in such artificial and poorly controlled situations that it is not useful.

Garner (1972) has listed and discussed many areas of research that were prompted by people trying to resolve practical problems. These include research in selective attention, space perception, speech perception, pattern recognition, and absolute judgment. Where the research has been prompted by specific design problems, it tends to be more immediately useful to designers.

In spite of some problems, over the years there has been sufficient good research as well as practical experience to establish a scientific foundation in the area of human performance engineering.

Application

The field of human performance engineering, then, consists of doing meaningful *research*, and *translating* research findings into appropriate and meaningful data and principles. But perhaps most importantly, human performance engineering consists of *applying* knowledge to the design of human performance in systems. This book introduces and discusses how designers can apply human performance information to system design.

MAKING INFORMED DESIGN DECISIONS

Designing systems for people is serious business. It is not fair to hold a designer responsible for human performance problems that are beyond his or her control. But it *is* fair to hold designers responsible for *poor* design decisions that lead to degraded performance—particularly if these decisions result from ignorance of human performance technology. Consider an excerpt from the Code of Hammurabi written in 2150 B.C.

> If a builder has built a house for a man and his work is not strong and the house falls in and kills the householder, that builder shall be slain.

In a similar fashion, perhaps with a slightly reduced penalty, designers should be held responsible for their decisions that lead to less than adequate human performance.

FOR MORE INFORMATION

Chapanis, A., *Man-Machine Engineering,* Monterey, California: Brooks/Cole Publishing Company, 1965.

DeGreene, K. B., (Ed.) *Systems Psychology,* New York: McGraw-Hill, 1970.

2

HISTORY OF HUMAN PERFORMANCE ENGINEERING

INTRODUCTION

We do not know when people first became concerned with improving human performance. The first humans most likely found securing food and protecting themselves against the weather, wild animals, and other people a great challenge. One can easily imagine a group's leader standing, back to the fire, trying to encourage others in the group to run faster and aim their spears more accurately. Better performance meant the survival of both individuals and groups.

Even the earliest tools reflect degrees of improvement over the years—improvements that most likely resulted in better human performance. These early tools, primarily sticks and stones, were used for hunting and protection. Today's most sophisticated tool, the computer, is designed to meet current needs, one of which is still survival. We don't know if the skill required to use a stone tool is more or less than the skill required to use a computer, or if the absolute level of individual human performance improved over the past 50,000 years.

Athletic contests provide the earliest record of improvements in human performance. The Olympic games began in Greece at least 3,500 years ago. The first recorded Olympic champion was Coroebus of Elis, a cook, who won the sprint race in 776 B.C. These early sporting events, as those today, emphasized improving the human element, with the activity (event) and context (weather, number of spectators, etc.) assumed as givens. In 708 B.C. participants began using standard size tools, such as the javelin and discus, in some events. And in 680 B.C. a chariot race was introduced. Possibly for the first time, the emphasis shifted from preparing only the human element to also preparing the vehicle and horses, items necessary to increase human speed.

From a human performance point of view, there is a difference in complexity between running a footrace and attempting to control two, four, or six running horses. In the latter case, consideration must extend beyond the human to improve the activity itself as well as the human/activity interface. In addition to athletic contests, early Greece also had competition in trumpeting, heralding, singing, and recitation. Thus, attempts to improve the human performance of others, and demonstrate that improvement, have a relatively long recorded history.

MOTIVATION

The earliest and probably still most widely used method of improving the performance of another is "motivation," usually through persuasion or coercion. Typically, one person tries to improve the performance of others by convincing them that improved performance is somehow better. The threat of pain or reward of pleasure could accompany the discussion. Parents frequently use this approach with their children, as do management people with their subordinates.

Sigmund Freud probably made the first systematic attempt to understand motivation as related to human performance. In 1901 Freud published *The Psychopathology of Everyday Life*. This book is essentially a theoretical statement on motivation-related causes of degraded human performance. The chapter titles give an indication of the material covered: "The Forgetting of Proper Names," "The Forgetting of Sets of Words," "Misreadings and Slips of the Pen," etc.

Freud presented about 100 examples, many analyzed in great detail. Here is one characteristic example, without Freud's analysis.

> A Herr Y fell in love with a lady; but he met with no success, and shortly afterwards she married a Herr X. Thereafter, Herr Y, in spite of having known Herr X for a long time and even having business dealings with him, forgot his name over and over again, so that several times he had to enquire what it was from other people when he wanted to correspond with him. (p. 25)

Also in the early 1900s Frederick W. Taylor introduced his methods of "Scientific Management." He attempted to improve performance by improving motivation. One of his best-known studies is of a crew of pig iron handlers, loading lengths of iron each weighing 92 pounds. Each man loaded an average of 12 1/2 tons per day. Taylor assumed that all workers could be motivated by money. An excerpt from his report of a worker named Schmidt illustrates this:

> "Schmidt, are you a high-priced man?"
>
> "Vell, I don't know vat you mean."
>
> "Oh yes, you do. What I want to know is whether you are a high-priced man or not."
>
> "Vell, I don't know what you mean."
>
> "Oh, come now, you answer my questions. What I want to find out is whether you are a high-priced man or one of these cheap fellows here. What I want to find out is whether you want to earn $1.85 a day or whether you are satisfied with $1.15, just the same as all those cheap fellows are getting."
>
> "Did I vant $1.85 a day? Vas dot a high-priced man? Vell, yes, I vas a high-priced man."
>
> "Oh, you're aggravating me. Of course you want $1.85 a day--everyone wants it! You know perfectly well that has very little to do with your being a high-priced man. For goodness sake, answer my questions, and don't waste any more of my time. Now come over here. You see that pile of pig iron?
>
> "Yes."
>
> "You see that car?"
>
> "Yes."

"Well, if you are high-priced man, you will load that pig iron on that car tomorrow for $1.85. Now, do wake up and answer my question. Tell me whether you are a high-priced man or not."

"Vell--did I got $1.85 for loading dot pig iron on dot car tomorrow?" "Yes, of course you do, and you get $1.85 for loading a pile like that every day right through the year. That is what a high-priced man does, and you know it just as well as I do." *Courtesy Harper and Row Publishing Company.*

Although Taylor's treatment of Schmidt may seem unfair, much of Taylor's work led to improved worker performance.

TRAINING

Advances in tools, weapons, and shelter by early human inhabitants suggest that these early people passed on to others the knowledge and skill gained in mastering their circumstances. We could call the process "training" in cases where this was done by deliberate example, or the use of words. When another person received the message successfully, learning took place—knowledge and skill were transferred from one person to another.

As larger groups of people formed, certain skilled specialties appeared. Direct instruction and experience transmitted the skills and knowledge of these crafts. We know this type of apprenticeship training was being used at least by 2100 B.C. because the code of Hammurabi contains rules and procedures for governing apprentice relationships. This form of training is still used to improve human performance in fields such as masonry, carpentry, and plumbing. Only since the early 1900s has training in the areas of dentistry, medicine, and law moved away from a strong reliance on apprenticeship.

In America, an early form of vocational education appeared in 1745 at the Moravian settlement in Bethlehem, Pennsylvania, with a course in carpentry (Steinmetz, 1967). By the early 1800s training schools for certain crafts were becoming more popular. For example, the Masonic Grand Lodge of New York, in 1809, established vocational training facilities.

Factory schools for training workers began to appear in the late 1800s. Hoe and Company established one of the first in New York City in 1872. According to Steinmetz, the large volume of business at this manufacturer of printing presses made it necessary to establish a factory school to train machinists. The company found the old-style apprentice approach inadequate for improving performance of new workers.

Silvern (1970) noted that industrial training increased in popularity in the early 1900s because the public schools were not preparing their graduates for immediate employment. School classrooms presented academic courses while skills needed for adequate work performance were ignored. Academic and skill instruction ("shop" classes) were finally incorporated in schools around 1930.

SELECTION (TESTING)

Selecting those people who can best perform an activity also improves performance. This requires that each prospective worker be evaluated. One good way to evaluate large numbers of people is to use tests.

The first evidence of using tests to select people for a particular performance comes from a Biblical account dated about 1100 B.C. (Judges 7:5-6). The Bible recounts that Gideon was instructed to select a small group of warriors to deliver the Israelites from oppression by the Midianites. Approximately 22,000 people answered his first call for volunteers. The battle strategy called for far fewer people, so he told all those who were "fearful and afraid" to return to their tents. Twelve thousand returned, leaving about 10,000—still too many. To both reduce the number and retain those specifically needed, Gideon devised a selection test. He had each person who remained go down individually to the nearest water (possibly a stream or small lake) and take a drink. This simple test divided the remaining people into two groups: (a) those "that lappeth of the water with his tongue, as a dog lappeth" and (b) those "that boweth down upon his knees to drink." Only 300 lapped the water, while the other 9700 bowed down on their knees. Gideon was only interested in those that lapped and sent the others home. It is not clear why. Possibly because their stiff-necked refusal to bow down even to drink made them the stuff of martyrs. With his select group of 300, the Bible tells us, he went on to conquer the Midianites.

The use of testing also has been traced to the ancient Chinese (around 1100 B.C.) by DuBois (1965), where applicants were screened by aptitude tests for higher positions in the civil service. The early Greek philosophers also contributed to the development of psychological testing. Plato, in the *Republic*, proposed an aptitude test to select persons who would be suited to a military career.

Modern testing primarily owes its start to Sir Francis Galton, an English biologist. In 1882 he established an anthropometric laboratory in South Kensington Museum, London, where, for a small fee, individuals could have certain traits measured, including vision and hearing, muscular strength, reaction time, and other simple functions.

Galton believed that tests of sensory discrimination could gauge a person's intellect. In his *Inquiries into Human Faculty and Its Development* (1883), Galton wrote: "The only information that reaches us concerning outward events appears to pass through the avenue of our senses; and the more perceptive the senses are of difference, the larger is the field upon which our judgment and intelligence can act." Galton had also noted that the mentally retarded tend to be defective in the ability to discriminate heat, cold, and pain—an observation that further strengthened his conviction that sensory discrimination "would on the whole be highest among the intellectually ablest."

James Cattell, one of America's earliest psychologists, used the term "mental test" for the first time in an article written in 1890 (Anastasi, 1963). This article described a series of tests administered annually to college students to determine their intellectual level. The tests, administered individually, included measures of muscular strength, speed of movement, sensitivity to pain, keenness of vision and of hearing, weight discrimination, reaction time, and memory. In his choice of tests, Cattell shared Galton's view that a measure of intellectual functions could be obtained through tests of sensory discrimination and reaction time.

The tests developed by Galton, Cattell, and others (including the famous Binet intelligence tests) were all designed to be administered individually. *Group testing* began shortly after the United States entered World War I. There was a need for rapid classification of a million and a half recruits according to general intellectual level. Administrative decisions depended on such information, including assignment to different types of service and admission to officer training camps. For the first time in modern history, tests were used to help match people with work.

The trend toward using group intelligence tests as rough, preliminary screening instruments continued at a rapid rate during the 1930s and 1940s in both the military and private industry. Those who "passed" the intelligence test (i.e., were selected) received more detailed measures of special aptitudes. Among the latter were tests of mechanical, clerical, and managerial aptitudes. During World War II, test psychologists developed specialized test "batteries," or combinations of tests. Special test batteries were constructed for pilots, bombardiers, radio operators, range finders, and several other military specialists.

HUMAN-ORIENTED DESIGN

Astronomers

One of the most fascinating and well-documented examples of the concern for human performance in work took place in England around 1800. At the Greenwich Observatory, the Astronomer Royal, Maskelyne, and Kinnebrook, his assistant were charged with observing the times of stellar transits. The observations were important since upon them depended the calibration and accuracy of the clock used to establish world standards of time. Maskelyne was convinced that through the year 1794 no discrepancy existed between the observations of the two of them. In August 1795, however, Kinnebrook was found to be recording times about a half-second later than Maskelyne. He was told of the error, and it seems that Kinnebrook tried to correct it. Nevertheless, it reportedly increased during the succeeding months until in January, 1796, it had become about eight-tenths of a second. So critical was accuracy that at this point, Maskelyne dismissed Kinnebrook (Boring, 1929).

It is interesting to consider the difficult human performance task that Maskelyne and Kinnebrook were performing. The accepted manner of observing stellar transits at that time, and for at least 50 years after, was the "eye and ear" method of Bradley. The field of the telescope was divided by parallel cross-wires in the reticle. The observer had to note, within one-tenth of a second, the time at which a given star crossed a given wire. The person proceeded as follows:

1. Looked at a clock, and noted the time to a second.
2. Began counting seconds with the heard beats of the clock.
3. While doing Steps 1 and 2, tracked the star across the field of the telescope.
4. Noted and "fixed in mind" the star's precise position at the beat of the clock just before the star came to the critical wire.
5. Noted the star's position at the next beat after it crossed the wire.
6. Estimated the place of the wire between the two positions in tenths of the total distance between the positions.
7. Added these tenths of a second to the time in seconds that had been counted for the beat before the wire was reached.

The Bradley method was accepted and regarded as accurate to one or at least two-tenths of a second. In the face of this belief, Kinnebrook's error of eight-tenths of a second was a large one and tended to justify his dismissal and Maskelyne's conclusion that he had fallen "into some irregular and confused method of his own." About 20 years later a history of Greenwich Observatory included the Maskelyne-Kinnebrook incident.

Bessel, a Konigsberg astronomer, sent to England for a copy of Maskelyne's complete observations, and, after studying them, determined to see whether this difference between astronomers, which seemed incredibly large in view of the supposed accuracy of the method, could be found among other observers. His first data were collected in 1820. For the next twenty years astronomers collected and analyzed human performance data in an attempt to find ways to *improve human performance*.

Finally, in about 1850, the chronograph was developed. This helped to reduce the negative effect of observer error. Like most error reduction programs, greater accuracy in stellar observation came about by automating an error prone task. Nevertheless, the results of numerous early researchers proved that human performance could be improved if the causes of the degradation could be identified and controlled.

Printers

Others were concerned about human performance in printing. Printers' errors had plagued both writers and readers from at least 1456—the time of the Gutenberg Bible. Originally, all detected mistakes were corrected with a pen in each copy. But in 1478 printed errata began to appear. Some of these errata lists were very lengthy. For example, one book published in 1507 had fifteen folio pages of errata; a much smaller book of only 172 pages, published in 1561, also contained fifteen pages of errata. At about the same time, another author had trouble getting his writings printed correctly, and was forced to publish an 88-page volume that contained only errata for his past publications (Wheatley, 1893).

Most printers' errors were considered unavoidable nuisances and were begrudgingly expected and accepted by both authors and readers. But some errors were totally unacceptable. These were errors in printed Bibles. The most severe errors were those that actually changed the intended meaning of scriptural messages. For example, in a Bible printed in 1634, the first verse of the 14th Psalm was printed as "The fool hath said in his heart there is God;" and in another Bible, I Corinthians, verse 9, was printed as "Know ye not that the unrighteous shall inherit the kingdom of God?" But probably the worst error of all appeared in a Bible published in 1631. In this Bible the word "not" was left out of the seventh commandment, thus leaving "Thou shalt commit adultery." The penalties paid by printers for Biblical errors (i.e., degraded human performance) ranged from heavy fines to excommunication.

Printers' errors being such a major concern, it is not surprising to find that another of the first documented, systematic attempts to improve human performance was with manual typesetters (Blades, 1872).

Contributions from 1900 to 1920

The early 1900s saw much attention directed toward improving people's performance (usually productivity) by altering the design of their work. Increasing productivity usually meant increasing people's speed. Much of the original impetus for this early design work came from Taylor (1911, 1947), as well as from influential publications by Munsterberg *(Psychology and Industrial Efficiency,* 1913), Gilbreth *(Brick Laying System,* 1911) and Frank and Lillian Gilbreth *(Applied Motion Study,* 1917).

The Gilbreths tended to focus on identifying and eliminating wasted motions. Gilbreth's first work, a study of the motions involved in laying bricks, enabled him to reduce the motions of the bricklayer from 18 to 5, thereby improving individual performance from 120 to 350 bricks per hour.

Frank Gilbreth was an engineer and his wife Lillian Gilbreth was a psychologist. They worked together for many years and their results demonstrated, early in this century, the advantages of having engineers and psychologists cooperating on projects involving human performance. One of the most interesting of the Gilbreths' contributions was their analysis and breakdown of tasks into basic elements of motion, which they called "therbligs" (i.e., "Gilbreth" spelled backwards, with t and h reversed).

The motions required to write one's name on a sheet of paper using a felt-tip pen kept with other pens in a shirt pocket illustrates therbligs (Blum, 1949). In therblig terms the person must:

1. search (the pocket)
2. find
3. select (if more than one)
4. remove cap
5. place cap on back
6. transport (to paper)
7. position
8. use
9. remove cap from back
10. replace cap
11. transfer (back to pocket)
12. search (for exact spot)
13. find
14. position.

After therblig identification, the following are asked to evaluate any human performance improvement:

1. Is each therblig necessary?
2. Can the task be made simpler by having fewer motions?
3. Can there be less motion in performance or degree?
4. Can the steps be combined?
5. Can the sequence be changed?
6. Can more than one be done at the same time?

Such an analysis makes it immediately obvious that in the above example, a desk-set pen in a fixed position and with no cap over the point would improve human performance (i.e., take less time) over a pocket model.

Contributions from about 1920 to 1940

Attempts to improve human performance are not new to the telephone industry. Around the same time the Gilbreths were carrying out their research, the New York Telephone Company published a report of efforts to improve customer performance by making telephone directories more legible (Baird, 1917).

Murrell (1965) observed that no organized effort to study the effect of working conditions on human performance was made until the end of World War I. At that time the Industrial Fatigue Research Board was set up in England. For the first time, a group of people trained in behavioral science entered industry to study people working. The Board's work differed from the contributions of the Gilbreths in that the Gilbreths' principles of motion study were based to a large extent on observation, whereas the Board relied on controlled experiments. Up to 1929 the Board produced 61 reports.

In 1927 the famous Hawthorne studies were begun by Elton Mayo at the Western Electric Hawthorne plant (Roethlisberger and Dickson, 1939). These studies extended over a period of 12 years and began with the seemingly simple and straightforward problem of determining the relation between changes in illumination intensity and production. The answer proved elusive.

Five experiments were conducted. After the first, they concluded that more experimental controls were needed and that they had to eliminate non-illumination factors which affected production output. In the second study both the control and experimental groups increased their production to an almost identical degree. In the third study the light level was decreased until the test subjects were working in very dim light (about three footcandles); even so, they maintained their level of efficiency.

The fourth study had two volunteers work in a room until the light intensity equaled that of ordinary moonlight; the subjects maintained their production level and reported no eyestrain and *less* fatigue than when working under bright lights. In the fifth and final experiment, the light was increased daily and the subjects reported they liked bright lights. When light bulbs were replaced with some that projected the same intensity, the subjects commented favorably on the "increased" illumination. When illumination was decreased the subjects said that less light was less pleasant. However, throughout this study there was no change in production.

Although they have received some criticism in the past few years (cf. Parsons, 1974; Franke and Kaul, 1978), the Hawthorne studies helped to focus on the effects of working conditions on human performance.

During the 1930s interest in human performance declined due to heavy unemployment. With so many people available for work, people interested in human performance tended to emphasize selection techniques, primarily screening tests.

Contributions since about 1940

World War II was a time of rapid scientific development. It brought about a large number of entirely new and sophisticated types of equipment for human use. Unfortunately, many of these devices were not designed to elicit acceptable levels of human performance. Degraded human performance resulted. Taylor (1957) has noted that "bombs and bullets often missed their mark, planes crashed, friendly ships were fired upon and sunk, and whales were depth charged." (p. 249)

Up until this time, the American, British, and German armed forces had attempted to handle most human performance problems with motivation, training techniques and selection tests. American and German psychologists had been involved in training and testing since at least 1920 (Fitts, 1946). But in the early 1940s the problems associated with operating many of the new machines increased. Taylor observed that:

> Regardless of how much he could be stretched by training or pared down through selection, there were still many military equipments which the man just could not be moulded to fit. They required of him too many hands, too many feet, or in the case of some of the more complex devices, too many heads. Sometimes they called for the operator to see targets which were close to invisible, or to understand speech in the presence of deafening noise, to track simultaneously in three coordinates with the two hands, to solve in analogue form complex differential equations, or to consider large amounts of information and to reach life-and-death decisions in split seconds and with no hope of another try. *Courtesy of the American Psychologist.*

Of course, people often failed in performing these activities. As a result, psychologists became more active in working with engineers to produce machines that required less of their users while at the same time taking full advantage of people's special abilities. Design began to focus more on the activity to be performed and the tools to be used, with particular attention given to the design of the human/machine interface.

As World War II continued both American and German psychologists began working on designing military equipment to more closely accommodate the capacities of users. German human performance studies included the shape and color of reticles for gun sights, amount of magnification for telescopic sights, the best positions of the body in relation to a control, the best type of movement for accurate adjustment, and the design of controls (Fitts, 1946).

In 1944, when the war was almost over, the first study primarily concerned with equipment design was conducted in the United States (Parsons, 1972). In that year a joint project of the Applied Psychology Panel of the National Defense Research Committee and the Armored Medical Research Laboratory investigated the sources of errors in Army field artillery. Like those of the Germans, many of these first studies dealt with evaluating new gun-sight scales designed to eliminate errors. Also in 1944, the Applied Psychology Panel established a large field laboratory in Texas to conduct research to improve the design of gun sights for B-29 aircraft artillery.

Shortly thereafter, some truly innovative pioneering studies on improving human performance in *combat information centers* were begun at the System Research Laboratory of Harvard University. Combat information centers were complexes where radar and other information was viewed on various display scopes, evaluated, and distributed for weapons and battle direction. Significant early military-sponsored work was also done at Oxford and Cambridge in England, and Johns Hopkins University in the United States.

No doubt as a result of much of the work begun during or shortly after the war, a relatively large group of psychologists developed an interest in human performance and continued to work in this field, supported primarily by government funds and working mainly on military problems. Literally thousands of studies on performance in military systems have been conducted and reported. Most are in the form of in-house technical reports, but many have been published in psychological journals.

In 1949 two of the first books in the area were published, *Applied Experimental Psychology* (Chapanis, Garner, and Morgan) and *Human Factors in Undersea Warfare* (Panel on Psychology and Physiology). Since that time, numerous books have been written on human performance. Also in 1949 an interdisciplinary group of people including psychologists, design engineers, work study engineers, physiologists, industrial medical officers, and others with a special interest in human performance, met at Oxford to form the Ergonomics Research Society. The new word "ergonomics" was created from the Greek *ergos,* work, and *nomos,* natural laws.

A second major society, the Human Factors Society, was founded in the United States in 1957. The Human Factors Society was organized to provide professional and personal interchange of ideas among workers concerned with human performance. In order to disseminate new knowledge and to promote the application of this knowledge to design, both the Ergonomics Society and the Human Factors Society publish journals.

3

HUMAN LIMITS AND DIFFERENCES

THE CONCEPT OF LIMITS

Good designers must know three things about the people for whom they are designing.

1. What they *can* do—their basic abilities and skills.
2. What they *cannot* do—their limitations.
3. What they *will* do—what they will be motivated to perform.

We tend to think that people can and will do anything, despite much evidence to the contrary. People's abilities and skills have very definite limits. For example, on Earth, people *cannot* run 100 miles an hour, jump 25 feet high or lift 10,000 pounds.

Many designers tend to focus only on people's strengths, and attempt to expand these abilities. Galileo's telescope improved our ability to see long distances, Ford's automobile enabled us to travel more quickly, the Wright brothers' flying machine enabled people to move even faster, and the computer enables people to make computations more quickly. New ways of improving human abilities will no doubt come to pass in the future.

However, designers must recognize people's *limitations,* and should not expect people to perform beyond their ability. For example, any high jumper will have great difficulty jumping higher than nine feet; certainly we do not have high jumpers that can jump ten, twelve, or fifteen feet. This type of physical limit seems inherent and fairly obvious. If a designer built a system that required people to jump unassisted over a nine-foot wall, the system would be inoperable and fail, and the failure even may be blamed on faulty human performance. We cannot expect people to reliably perform beyond their limits.

A designer should become familiar with human performance limits, particularly those of the group of people who will be using his or her new system. Commonly, designers assume their own characteristics are also those of the users. But being human does not make one typical or representative of the user population.

Figure 3-1 shows a hypothetical set of extreme limits, the widest possible boundary that could be placed around a system. If the system required people within the system to lift 6,200 pounds, send Morse code at 840 taps per minute, or speak at a rate of 500 words per minute, it would certainly fail. Very few people in the world can perform at the extreme limits in any one skill, and if we require a person to exercise more than one of these skills we probably would find that *no one* in the world has the capability to perform. In the whole world probably less than five people (in some cases only one) can perform any *one* of the four extreme items shown in Figure 3-1.

These extreme limits change from time to time and continue to be slightly extended. But any increase in capability, even for those on the extreme outer boundaries, is very slow.

Instead of extreme limits, most designers should deal with a set of limits more representative of the entire population. In fact, they usually must identify and deal with a set of more *specific limits* characteristic of their user population. For example, the specific limits of each astronaut involved in the Apollo lunar landing missions were known and well defined. This was practical because the user population was small. On the other hand, a designer attempting to construct a map of the New York subway system must consider a set of limits covering a much wider user population, one that would not be well defined at all. The latter case includes people who can read English and people who cannot; those who can see different colors and those who are color-blind; those who ride the subways daily and those with no experience. Even in this case, however, we can make certain assumptions about human limits.

Thus, when designing a system that requires typing, designers do not plan for people who can type five hundred words per minute, but for people who can type sixty words per minute. Systems are not designed for people who can lift five hundred pound boxes (without help) constantly for eight hours a day, but are designed for people who can consistently lift fifty pounds.

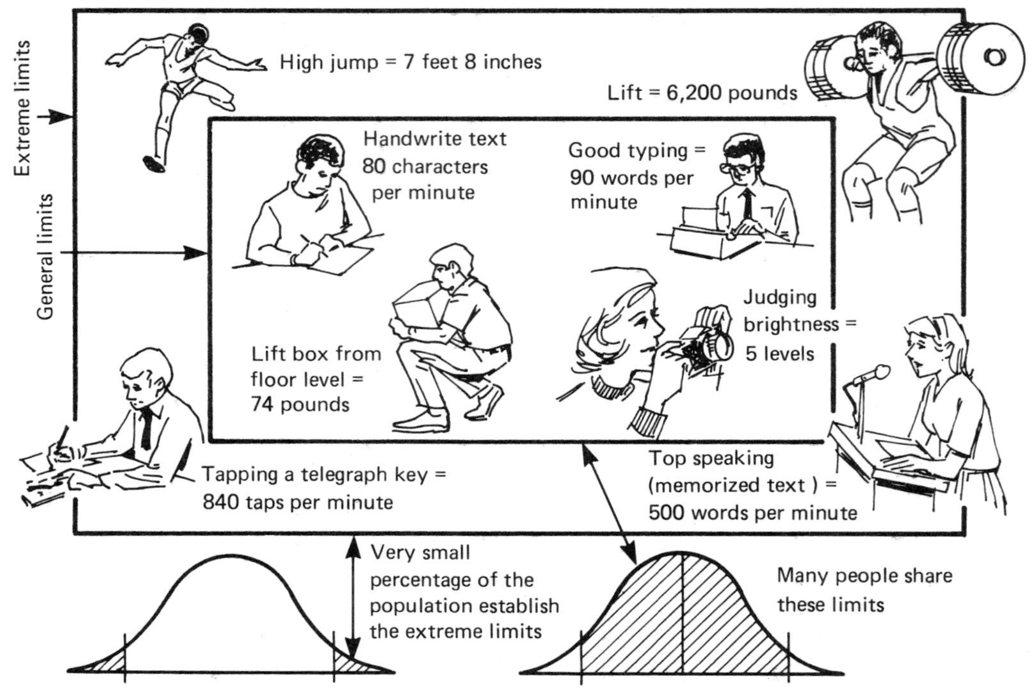

Figure 3-1. Human Performance Limits

It may seem obvious that systems are not designed to intentionally degrade human performance. It may also seem obvious that a designer would not intentionally make design decisions that would result in degraded human performance. Why, then, do we find system after system reflecting design decisions that seem to ignore human limits? Why would a designer require people to do the equivalent of leaping nine feet in the air by requiring a person to sense, process, and respond to information faster or with greater accuracy than humanly possible?

The real secret to making good design decisions, and the key to not compromising human limitations, is to understand the full range of human limits in the user population. This includes appreciating that some people have limits that others do not have. Human limits can be divided into three major areas: sensory, responder, and cognitive processing limits.

SENSORY LIMITS

Sensory limits include the basic sensory *thresholds,* as well as sensory *deficiencies.* Sensory thresholds include the least amount of light that can be perceived, the smallest lettering that can be reliably read, and the faintest noise that can be reliably heard. To be noticed, differences in stimuli—visual, auditory, touch, or other—must be large enough. We will discuss specific thresholds in Chapter 4. Now, it is sufficient to recognize that a set of basic stimulus characteristics must be present for the human sensory apparatus to function.

Not all people have a full complement of senses. Even those that can, for example, see, hear, touch, and smell, have these abilities in different amounts. Sensory deficiencies are common. Considering the number of people that have seeing defects of one type or another, probably best shows the varying levels of sensing. For example, almost 50 percent of the population in the United States wears glasses. Obviously if someone's deficient vision is not corrected, his or her performance on a vision-related activity could be degraded.

Another visual deficiency, color blindness, is present in some form in about 8 percent of the male and 1 percent of the female adult population. A designer who frequently uses color coding and who is not in a position to screen out people with defective color vision may find degraded performance.

RESPONDER LIMITS

Two common, performance-related responder limits are the user's reach and strength. Some automobiles today have a hood latch that only can be released by exceptionally strong people. The designer has not recognized the strength limit of the user population. Not being able to reach a car's headlight switch while strapped in by a seatbelt illustrates the overlooked reach limit.

Human sensory and responder limits, one would think, would be the most obvious. We can easily pick out the people who wear glasses or use crutches as well as those with obvious coordination problems or those with speech defects. Even so, these limits are frequently overlooked. Less obvious characteristics such as defective color vision can be fairly accurately determined through a quick test. Even if they have no better information than their own senses or a quick test, designers can always fall back on statistical estimates of the limits in a given population.

COGNITIVE PROCESSING LIMITS

Probably the most difficult limits to identify are those associated with the brain—cognitive processing limits. When considering human performance, two sets of limits are especially interesting: *response time* and *accuracy*.

Response Time

Response time can be conveniently separated into the time taken to recognize that a signal for a certain action has occurred and decide on the appropriate movement (reaction time), and the time it takes to move (movement time). For example, consider the time required for you to stop an automobile when a child runs after a ball into the street. Your response is broken down as follows.

You become aware of the situation and make the decision to move the foot—this is the *reaction* component. The *movement* component consists of actually moving your foot from the gas pedal to the brake pedal and pushing down.

If a new system were being developed and a designer determined that one user task needed to be completed in one-tenth of a second (100 milliseconds), would this result in degraded human performance? The "reaction" portion of response time (the time to initiate a movement) consists of a series of delays. Wargo (1967) has suggested that the delays look like those in Table 3-1. The total delay ranges from 113 to 528 milliseconds (i.e., about one-tenth of a second to about one-half of a second). In general, "fast" people under ideal circumstances can react to a visual stimulus in about two hundred milliseconds. To expect people to react in a shorter time will lead to disappointment when the new system is operational. This is a good example of a cognitive limit. It takes time to react. In answer to our question, a design requirement of 100 milliseconds will most certainly result in degraded human performance.

Table 3-1. Reaction Times

Delays	*Typical Times (msec)*
Sensory receptor	1-38
Neural transmission to brain	2-100
Cognitive-processing delays (brain)	70-300
Neural transmission to muscle	10-20
Muscle latency and activation time	30-70
Total	113-528

Reaction time varies with the sensor that is used. The different reaction times associated with different sensors have been known for well over a hundred years. For example, someone can *hear* a signal and make a simple response, on the average, in 150 milliseconds. However, to *see* a signal and respond, the average time will probably be closer to 200 milliseconds. To *smell* a stimulus and respond could take an average of 300 milliseconds. Sensing pain and responding could take as much as 700 milliseconds (Swink, 1966). However, people react fastest if they see, hear, and touch the stimulus all at the same time.

These reaction time limits can be shortened if the person is well practiced in the activity, is alerted shortly before the signal occurs, or if the stimulus is increased in size or intensity. The reaction time is lengthened when a person is fatigued, is using a depressant drug, or must make a very complex movement in response to the signal.

If a system requires reactions that are close to or exceed the cognitive limits, performance will be degraded. For example, it was once considered feasible that pilots of two supersonic planes flying in a head-on collision course could alter their respective courses after seeing the other aircraft. However, after calculating the time required for responding once the other plane was sighted, it was determined that neither of the pilots would have sufficient time to move. As a result, both the detection and movement processes were automated.

In communicating with one another, we have definite limits on the speed with which this can take place. Table 3-2 shows estimates of many of these limits.

Accuracy

Unlike reaction time, which is physiologically limited, accuracy seems more under the person's control. Certain critical activities, once learned, can be performed with near-perfect accuracy. It appears that people establish their own accuracy level on a task basis. For each type of activity, an individual sets an accuracy criterion and attempts to meet it. For example, the typical error rate for reaction-time experiments (a person sees and responds to a signal) is about 1 percent to 3 percent; when hand printing, the character-level error rate is about 0.5 percent and when keying, the error rate is much lower at .03 percent.

When driving a car people seem to set a more stringent accuracy criterion than they do when typing. And in telephone dialing, people generally have a somewhat lower accuracy level for using a TOUCH-TONE™ telephone pad than for using a dial telephone. On the average, people *dial* calls with 97 to 98 percent accuracy; when using TOUCH-TONE telephone pushbuttons, the accuracy level is only about 95 percent. Unfortunately, some activities are so poorly designed that they encourage errors to occur. In some cases people are actually "trapped" into committing errors.

Accuracy, then, is activity-related and may vary considerably with people performing the same activity, and even slightly for one person on the same activity. Do not assume that people consciously make a decision on their accuracy criteria. With experience, a person seems to arrive at a level that is the most comfortable in terms of achieving the activity's objective.

There seems to be a relationship between speed and accuracy. Often, the faster an activity, the higher the probability that errors will occur. In these cases, slowing the activity reduces errors. However, in some activities speed has little relationship to the number of errors. And in still other activities the

Table 3-2. A Comparison of Top Speeds for a Variety of Communication Tasks.

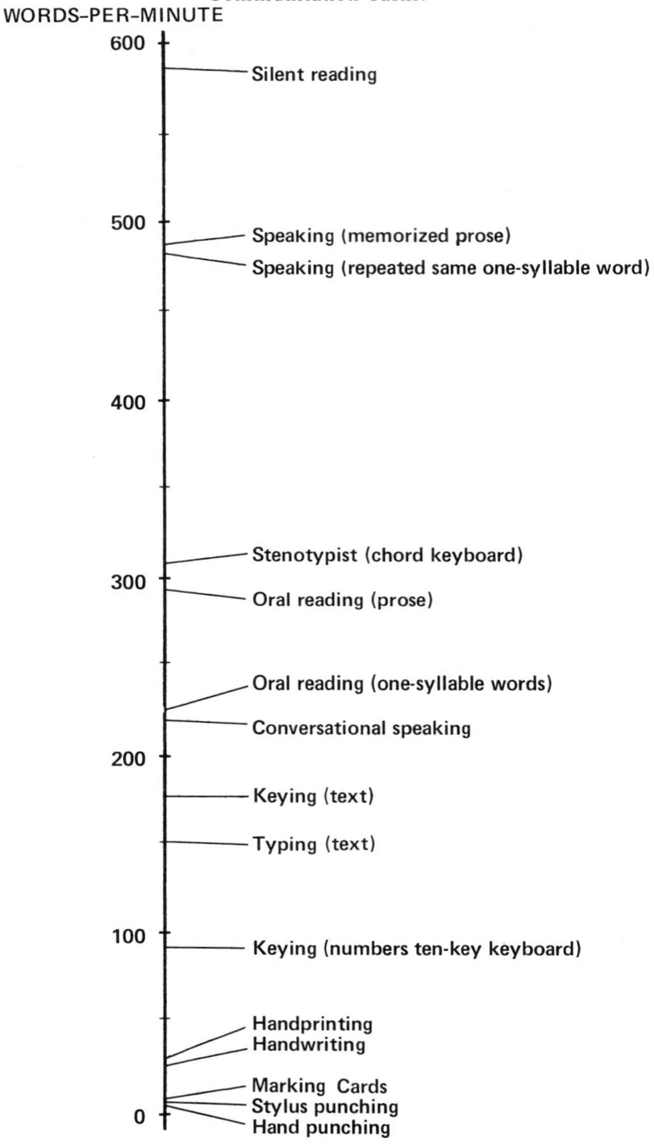

(Derived from Shackel, 1979; Turn, 1974; Newell, 1971; Deininger, 1967; Devoe, 1967; Hershman and Hillix, 1965; Seibel, 1964; Pierce and Karlin, 1956)

fastest performers actually have the lowest percentage of errors. One such activity is keying by experienced operators. Klemmer and Lockhead (1962) reported a difference among experienced key operators in both speed and accuracy. The fastest key operators were about *twice as fast* as the slower key operators, and the slower key operators tended to make *ten times more errors* than the fastest key operators.

Making discriminations also involves cognitive processing. People frequently have difficulties making accurate judgments even in rather straightforward, simple situations. Designers should know that people can discriminate among only a small number of different sizes, brightnesses, line lengths, etc. In fact, when stimuli are presented separately people can usually only discriminate between five to nine different categories. For example, if loudness is used as the signal, most people will only be able to discriminate between five different loudnesses; if brightness is used, most people will only be able to discriminate between five different light brightnesses.

There is even evidence that people cannot readily tell the difference between an Anthony dollar and a quarter. The Susan B. Anthony dollar was introduced in 1978 after government surveys showed that the U. S. public would not accept a coin larger than 28 mm in diameter. So they made the Anthony dollar only 26.4 mm in diameter. People complain, however, that the Anthony dollar is too easily mistaken for a quarter. U. S. Mint officials brush this aside as a matter of "perception" rather than "reality." They point out that the size difference between a quarter and Anthony dollar is the same as the difference between a nickel and a quarter. In addition, the dollar is 43 percent heavier than the quarter. Many people still find it difficult to tell the difference. Overlooking this human limitation may turn out to be very costly to the government.

The human ability to estimate has very definite limits. People do not do all things as well as most designers think they should. For example, people tend to overestimate time when they are passively involved and underestimate time when they are actively involved. Thus, if a person sitting in a chair is asked to say when five minutes have passed, he or she will tend to overestimate the time. The estimate will be that five minutes have elapsed when perhaps only four minutes have actually gone by. On the other hand, if a person is totally involved in building a model airplane and is asked to judge when five minutes have passed, he or she will tend to underestimate—when the person judges five minutes, seven minutes may have already gone by. People do not accurately represent time, and time estimations depend considerably on the activity being performed during the time interval.

When people estimate physical quantities, their judgments, though more predictable, still vary. People tend to underestimate distance. They tend to overestimate vertical height when looking down (e.g., from the Empire State Building) and underestimate when looking up (from the street to the top of the Empire State Building). People tend to overestimate temperature when it's hot and underestimate it when it's cold. Weight is usually overestimated if bulky and underestimated if compact. Someone asked to estimate a number of items without counting will consistently underestimate. These examples all show peoples' limits in performance requiring judgment.

Most designers assume that people can perform one task at a time. This may not be the case. It may only appear so because people usually *learn* to do one thing at a time. People can *do* two or more things at the same time. Take for example driving a car and talking to someone in the seat next to you, or walking and whistling a tune. Certainly as the skills are first being learned it is difficult to do two or more at the same time, but once a skill has been developed people can do amazing things at the same time. For example, Neisser (1976) reported on people who can listen and speak, and listen and read at the same time. Expecting users to perform only one activity at a time, and at the most two, seems to be more habit than reality. Nevertheless, when dealing with human limits, requiring people to perform more than one task at a time means that each activity must be well learned and practiced until a high skill level in both (or all three) has been attained. Otherwise, people are limited to the performance of one activity at a time. If forced to do more than one without sufficient skill, degraded human performance will result.

Most people cannot do everything well. Their ability to sense, process, and make appropriate responses in different situations has limits. All designers should appreciate these limits. It is the designers' responsibility to know and understand the extent of these limits prior to making design decisions that would be affected by limitations. If designers do this, it is much more likely their systems would elicit an acceptable level of human performance.

INDIVIDUAL DIFFERENCES

No two people are the same, and people continually change as well. Even a small group of potential users that appear to be the same, on closer inspection are different. There are no identical twins as far as human performance is concerned. The *differences between people* must be recognized as another way of considering limits. Some differences are inherent and long-term—people are born with a certain set of characteristics that last a lifetime. These long-term differences include sex, as well as certain basic physiological capabilities such as the existence of sensory organs and responders.

Other long-term differences include certain psychological capabilities that people are born with, including the ability to perceive, reason, and remember. These capabilities are not the same for all people, and even though they may vary slightly throughout a lifetime their basic existence or non-existence seems to remain fairly constant. Perceptual, reasoning and verbal skills are frequently measured and reported as intelligence quotients (IQ). It is doubtful whether all that is measured by an IQ test is inherent. In fact, much of what is measured is probably learned. The fact remains that certain basic psychological abilities are inherent, and these include the ability to perceive, reason, and remember.

As far as movement control is concerned, there are numerous examples of people who are born with an impaired ability. Again, it is very apparent when an impairment is extreme—as in cases of total paralysis. It is less apparent when the ability to control movement is only partially impaired. Some people are born with the potential for movement control that results in an experienced ballerina, professional athlete or eloquent speaker. With others, it seems that no matter how hard they try, they cannot become professional golfers, Olympic swimmers or world-class tennis stars.

Physiological changes also take place in the body over time. Many of these changes are due directly to the aging process. The body structure obviously becomes larger as an individual grows from infancy through childhood to adulthood. Later in life the physical structure actually begins to shrink slightly. Sensory capacities also change. This can be seen readily by observing the larger proportion of eyeglasses and hearing aids among older people.

People begin with basic capabilities that definitely differ. This means that some people are more limited than others. As changes take place, whether due to new learning, skill development or physiological changes, they tend to widen the differences among people. Most designers are concerned with systems that involve human performance by adults, and recognize that these adults can range in age from eighteen to eighty. The multitude of differences between people must be taken into account, and the better they are taken into account the better the likelihood of a successful system.

Design decisions should also reflect people's numerous short-term differences. These short-term differences include, fatigue, stress, illness, and drug effects. Recognizing the existence of these differences is important because the limits of people will change as they are put under additional stress, become more fatigued, or perhaps develop a cold.

Human limitations, then, are not stationary and are directly related to both long-term and short-term differences. Human limits not only change as a person grows older, but they may change on a day-to-day, hour-by-hour basis. Good designers take into account that these differences exist. Their designs reflect this knowledge.

4

SENSING

INTRODUCTION

According to the simplified model of human information processing shown in Figure 4-1, the sensors (e.g., eye, ear, and nose) ensure that information is picked up and forwarded to the brain. Information comes to the sensor as a "stimulus." If strong enough (but not too strong) it is received and passed to the brain.

STIMULI

A stimulus is a physical event, or a change in physical energy, that causes physiological activity in a sense organ. The stimulus for the ear is sound; for the finger tips, pressure; and for the nose, odor. A stimulus may activate a sensor and yet not be the appropriate stimulus for the specific sense organ involved. For example, if you press hard on your eyeball, you will experience a visual sensation, but not the sensation that is characteristic of normal vision. Thus, in the eye, pressure works as a stimulus but it is an inappropriate stimulus; the appropriate stimulus is light.

When the sensors are properly functioning, they are flooded with stimuli. And while they continually receive a tremendous amount of information, much

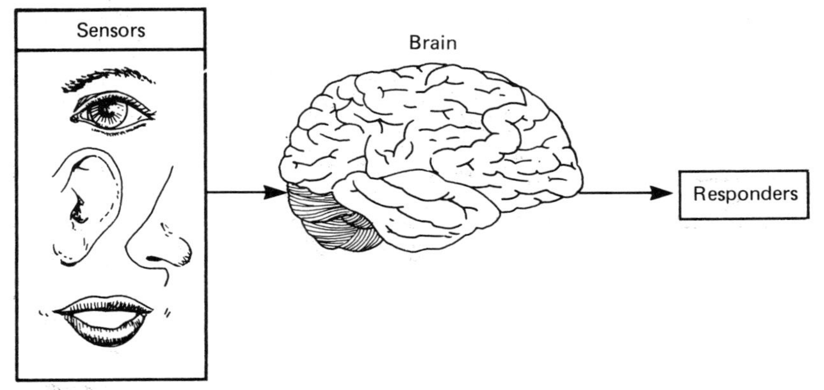

Figure 4-1. Simplified Model of Human Information Processing with the Sensors Emphasized

of it is filtered out by cognitive processes in the brain. Except when we close our eyes or hold our nose or breath, the senses themselves have no way to avoid the stimuli that are bombarding them. The brain seems to provide this protection at several levels. For example, a person can be told what stimulation to attend to and what to ignore. Through the psychological processes of set and attention a person learns to "filter out" certain types of irrelevant information. Other cognitive processes enable people to establish priorities among the different senses when selecting information.

SENSORS AND RECEPTORS

Most people can sense a wide range of stimuli from both external and internal sources. Each sensor is designed to respond to a specific type of stimulus. The major function of sensors is to receive and transform stimulus energy into a form that the brain can recognize and process.

One of the ways sensors can be classified is based on the source of the stimulus and the location of the sensor. According to this method of classification, there are four types:

1. Those located in the eyes, ears, and nose, that give us information concerning changes that take place at a distance from the body
2. Those located in the skin that give us information concerning changes immediately adjacent to the body
3. Those located in the visceral organs that tell us about changes in our internal organs
4. Those found in muscles, tendons, joints, and the labyrinth (ear), that give us information concerning movements of the body and the position of the body in space.

The senses also may be classified according to the type of energy that is the proper stimulus. See Table 4-1.

Table 4-1. Various Forms of Energy and the Senses Each Stimulates

Stimuli	*Sense*
Electromagnetic	Vision
Mechanical	Hearing
	Touch
	Pain
	Vestibular
	Kinesthetic
Thermal	Cold
	Warmth
Chemical	Taste
	Smell

(adapted from *Human Behavior: A Systems Approach*, by N. W. Heimstra and V. S. Ellingstad. Copyright © 1972 by Wadsworth Publishing Company, Inc. Reprinted by permission of the publisher, Brooks/Cole Publishing Company, Monterey, California.)

All the senses are selectively receptive to certain types of stimuli. For example, vision responds to radiant energy-but only to wavelengths between 400 nm (nanometers, or one millionth of a millimeter) and 700 nm. The temperature sensors respond to infrared wavelengths. The tactile sense may respond to relatively slow pressure changes, while the ear responds to very rapid pressure changes. The sense of smell is especially sensitive to chemical stimuli in gaseous form; while the sense of taste is most sensitive to chemicals in liquid form.

Sometimes the sensitivities overlap. For example, both the tactile and auditory senses respond to pressure changes (oscillations) in the 20 Hz to 1000 Hz range. Even so, under most circumstances there is a restricted set of stimuli that activate each of the different sensors.

SENSORY LIMITS

A light may be so dim that we cannot see it or a sound so quiet that we cannot hear it. The simplest definition of a threshold is that it is a point or a region on an intensity scale below which we do not detect the stimulus and above which we do. Both thresholds and upper limits are important to human performance. The threshold represents the smallest amount of stimulation necessary to produce a sensation. Table 4-2 shows some approximate thresholds for five senses. Galanter (1962) used familiar stimuli to help in understanding how sensitive the senses are.

Table 4-2. Some Approximate Sensory Thresholds

Sense	*Detection Threshold*
Sight	Candle flame seen at 30 miles on a dark clear night
Hearing	Tick of a watch under quiet conditions at 20 feet
Taste	Teaspoon of sugar in 2 gallons of water
Smell	Drop of perfume diffused into the entire volume of a three-room apartment
Touch	Wing of a bee falling on your cheek from a distance of 1 centimeter

(adapted from Galanter, 1962)

But each sense has, in addition to a threshold, an upper limit-a point on the physical stimulus—energy continuum above which the sensation will not become more intense no matter how much the stimulus increases. For some sensations this limit is not known because stimuli intense enough to reach this point would damage the sensors. For example, sound pressure intense enough to establish an upper limit for loudness would probably cause deafness.

Table 4-3 lists in a more technical form the thresholds and highest practical limits for various sensors. It gives the smallest detectable amount of energy and the largest amount permissible before pain or permanent damage occurs. Stimuli near either of these limits may lead to unreliable sensing.

Another type of threshold is the *difference threshold*. This threshold is the minimum physical difference in the amount of stimulation that produces a perceptible (just noticeable) difference in the intensity of a sensation. For example, if a person must rely on a change in light or noise intensity as a warning signal, the amount of decrease or increase in intensity necessary for an observer to detect a change in the brightness of a light or loudness of a sound is the difference threshold. A problem with difference thresholds is that they vary with the overall intensity. In general, the size of the difference threshold increases as stimulus values increase, i.e., it is harder to tell the difference between 100 and 105 pounds than between 10 and 15.

Table 4-3. Minimum Thresholds and Maximum Practical Intensities for Several Senses

Sense	*Smallest Detectable (Threshold)*	*Largest Tolerable or Practical*
Sight	10^{-6} mL	10^{4} mL
Hearing	$2x\,10^{-4}$ *dynes/cm*^{2}	10^{3} *dynes/cm*^{2}
Touch (pressure)	Fingertips, 0.05 to 1.1 erg (One erg approx. = kinetic energy of 1 mg dropped 1 cm).	Unknown
Smell	Very sensitive for some substances, e.g., $4x\,10^{-7}$ molar concentration of quinine sulfate.	Unknown
Temperature	$15x10^{-5}$ gm—cal/cm^{2}/sec. for 3 sec. exposure of 200 cm^{2} skin.	$22x10^{-2}$ gm—cal/cm^{2}/sec. for 3 sec. exposure of 200 *cm*^{2} skin.
Position and movement	0.2-0.7 deg. at 10 deg./min. for joint movement.	Unknown
Acceleration	0.02g for linear acceleration; 0.08g for linear deceleration.	5 to 8g positive 3 to 4g negative.

(adapted from Van Cott and Kinkade, 1972)

SENSING AND PERFORMANCE

People have many senses, including vision, hearing, cold, warmth, pain, touch, smell, taste, kinesthesis, and the vestibular sense. We will discuss the senses most closely related to human performance.

VISION

General

More is known about vision than any of the other senses. The light that stimulates the eye is a form of electromagnetic radiation. As shown in Figure 4-2, it belongs to the same class of physical phenomena as radio waves, radar waves, and X-rays. The eye converts light into a form that can be used by the brain. For human performance, vision may be the most important sense.

As shown in Figure 4-2, visual stimuli come only from a narrow band in the electromagnetic spectrum—a band that covers wavelengths ranging

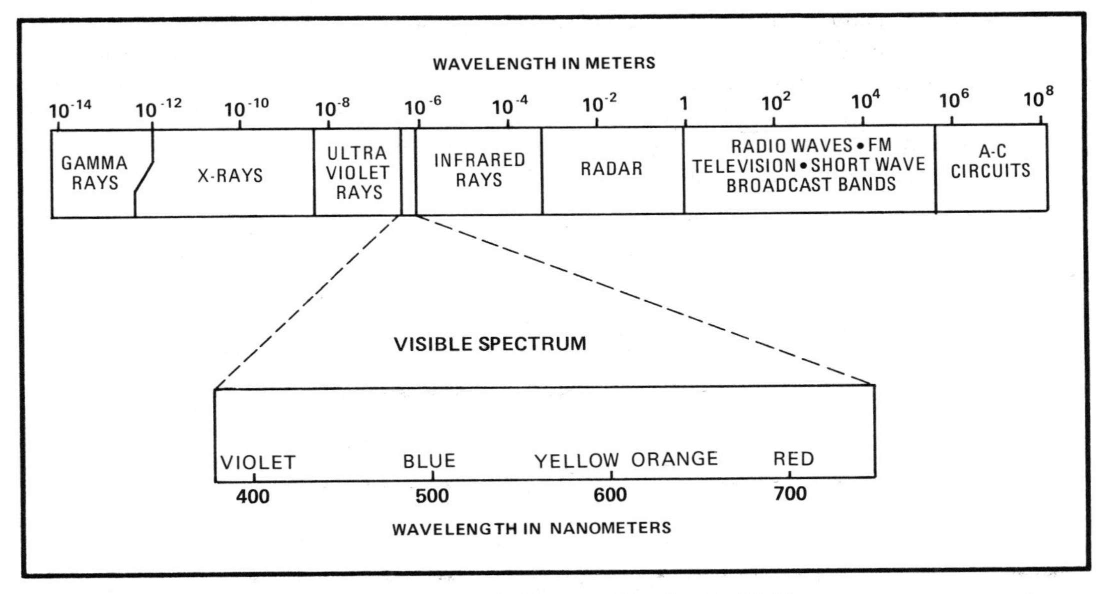

Figure 4-2. The Electromagnetic Spectrum Showing the Visible Wavelengths (adapted from Riggs, 1971)

approximately from 400 nanometers to 700 nanometers (Riggs, 1971). A nanometer is one-billionth of a meter.

Light sensitivity at a given moment depends on many things, including the time that the eye has been exposed to a certain level of illumination; the region of the retina stimulated; the nature of the stimulus (such as the intensity and duration of light); physiological and psychological conditions of the individual; the background against which something is seen; and the visual angle, or size, of the object being observed. Several of the most important considerations are discussed in the following pages.

Size of Stimulus

The measurement of size is the "visual angle." This is the angle formed at the eye by the viewed object (see Figure 4-3). Usually, this is given in degrees of arc [1° (degree) = 60' (minutes of arc); 1' = 60" (seconds of arc)].

Use this formula:

$$\text{Visual angle (minutes of arc)} = \frac{(57.3)(60)L}{D}$$

where L=the size of the object measured perpendicular to the line of sight, and D=the distance from the front of the eye to the object. The visual receptors are actually about 7 mm behind the foremost point of the eye, but this distance is so small it has little effect on most calculations. This formula is for angles less than 10 degrees, and assumes that the line of regard bisects L. When the visual angle is greater than 10 degrees or if the object is not perpendicular to the line of sight, or the line of regard does not bisect L, other formulas would be more appropriate (see Graham, 1965).

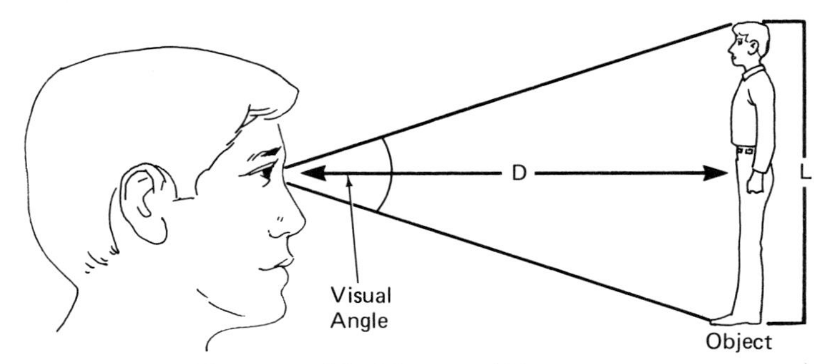

Figure 4-3. The Meaning of Visual Angle

Generally, the minimum perceptible visual angle is approximately 1 second of an arc (e.g., a thin wire against a bright sky). Some comparative visual angles are shown in Table 4-4. It is usually recommended that under good viewing conditions the visual angle should subtend at least 15 minutes of arc. And under degraded viewing conditions (e.g., low light levels) the angle should subtend at least 21 minutes of arc (Buckler, 1977).

Table 4-4. Comparisons of Visual Angles for Several Familiar Objects

Object	*Distance*	*Visual Angle*
Sun	93,000,000 mi	30'
Moon	240,000 mi	30'
Quarter	Arm's length (70 cm)	2°
Quarter	90 yd	1'
Quarter	3 mi	1"
Lowercase pica type letter	Reading distance (40 cm)	13'

(adapted from Cornsweet, 1970) *Courtesy Academic Press*

Brightness

A second important consideration is the brightness of the stimulus. Strictly speaking, brightness is the subjective sensation when viewing an object. Luminance is the amount of light per unit area that is emitted by a surface. Unfortunately, luminance can be expressed in terms of millilamberts, footlamberts, candelas per meter squared or nits. Table 4-5 shows some luminance conversion factors.

Table 4-5. Luminance Conversion Factors

	Candela per meter squared (cd/m^2)	Footlambert (ftL)	Millilambert (mL)	Nit (nt)
1 Candela per meter squared =		0.2919	0.3142	1
1 Footlambert =	3.426		1.076	3.426
1 Millilambet =	3.183	0.9290		3.183
1 Nit =	1	0.2919	0.3142	

The approximate luminance values for a variety of commonly experienced conditions are shown in Figure 4-4.

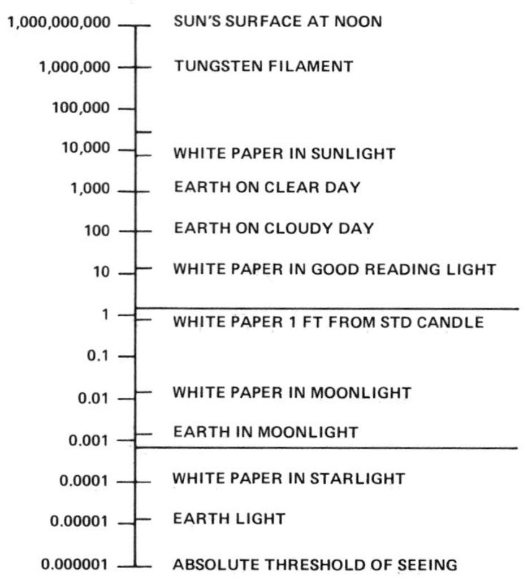

Figure 4-4. Example of Various Luminance Levels in Millilamberts (adapted from Van Cott and Kinkade, 1972; and Riggs, 1971)

Visual Field

The visual field is that area that can be seen when the head and the eyes are motionless. Figure 4-5 shows that the monocular field (for one-eyed vision) extends from 104° from the line of sight on the temporal side to some 60°-70° on the nasal side. The visual field can be enlarged by rotating the eyes: the horizontal field for each eye alone is approximately 166°, and for both eyes, about 208°. The visual field obviously can be further extended by movements of the head or body through 360°.

The Human Eye

Four of the most important features of the human eye are shown in Figure 4-6. Light enters the eye and passes through the *cornea*, *pupil* and *lens*, on its way to the ***retina***. The cornea and lens help to bend and focus the light on the retina. Unlike the cornea, however, the shape of the lens is continually being modified to change the focus. The curvature of the lens is increased for distant vision and decreased for near vision. The size of the pupil helps to control the amount of light that enters the eye and reaches the retina. The pupil widens (dilates) in dim light and contricts in bright light. The retina is the part of the eye that contains light-sensitive receptors.

Peripheral Color Vision

Not all receptors in the retina are equally sensitive to color. Toward the periphery, objects can still be distinguished while their colors cannot. Figure

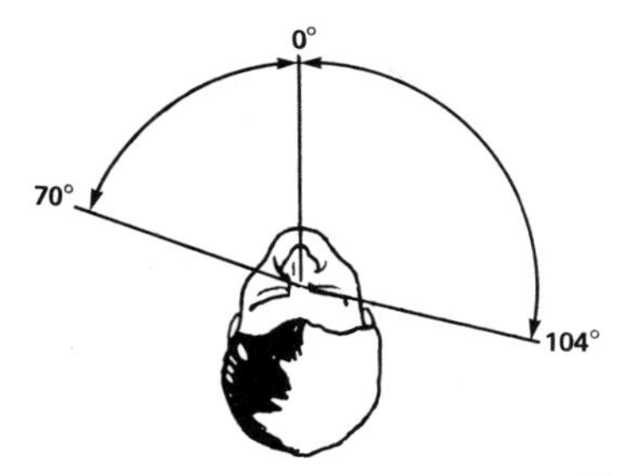

The visual field (one-eyed vision with the eye motionless) extends from 70° on the nasal side to 104° on the temporal side

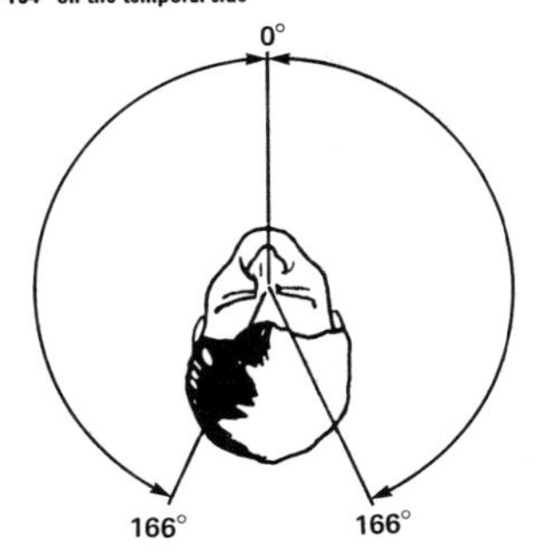

The visual field for each eye when the eyes are allowed to rotate, but not the head, is about 166°

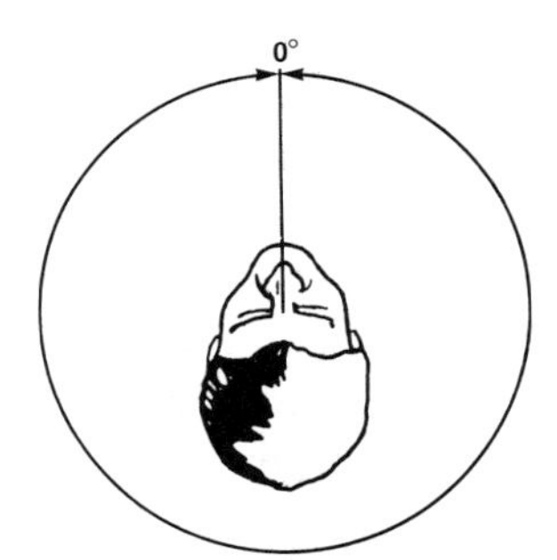

The visual field when the eye, head, and body are allowed to rotate is 360°

Figure 4-5. The Visual Field (adapted from Woodson and Conover, 1964)

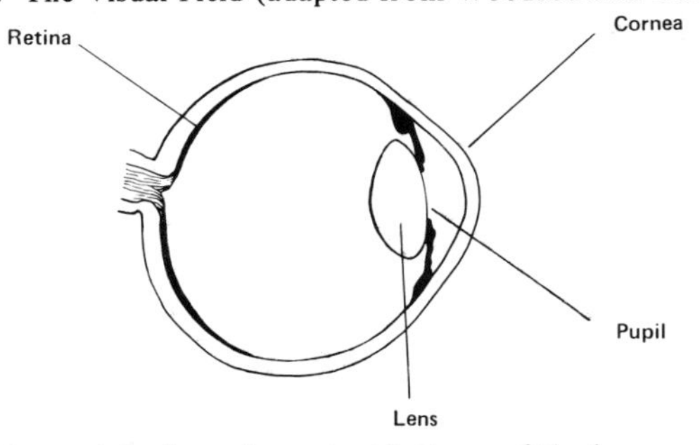

Figure 4-6. Some important features of the human eye.

4-7 shows the visual regions in which the various colors can, under normal illumination, be correctly recognized. Figure 4-7 shows that the area where color is perceived extends for about 60° on either side of a fixed point when the eyes are motionless. Color perception occurs from 30° above to 40° below the horizontal line of vision. With the eyes motionless, blue can be recognized at about 60°, with colors yellow, red, and green recognizable as we move progressively closer to the fixed point.

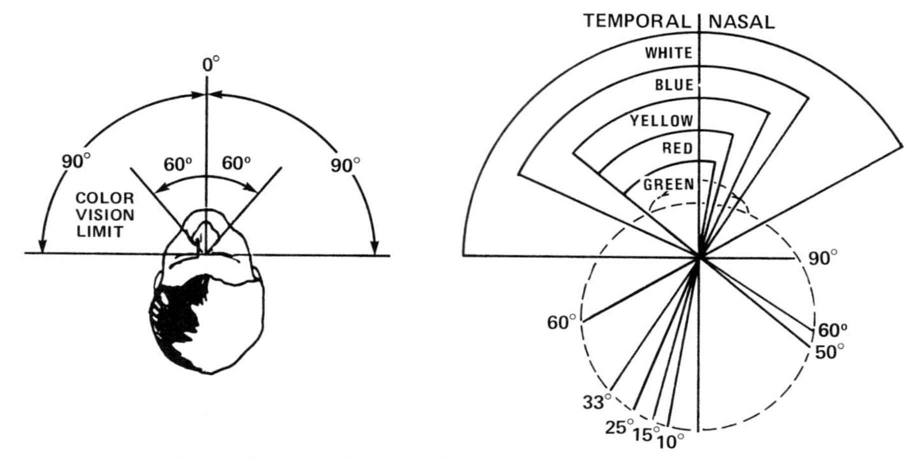

Figure 4-7. The Visual Field for Color with Eye Motionless (adapted from Woodson and Conover, 1964)

Visual Acuity

Visual acuity is the eye's ability to perceive fine details. Devices have been developed to measure visual acuity. In the most common, an individual stands 20 feet from a standard Snellen eye chart that shows rows of letters in different sizes. Acuity is indicated by the smallest letters a person can read. A person with impaired vision recognizes letters at 20 feet. A person with normal vision would recognize letters at 60 feet; he or she is then said to have 20/60 vision. In this case corrective lens are usually prescribed. A person with 20/10 vision, on the other hand, sees letters at 20 feet away that a person with normal vision must view from 10 feet away.

In the past, before printing presses or electric lights, the uncorrected eye may have satisfied the necessary seeing requirements. If the individual was a hunter, herdsman, or farmer, he or she probably had sufficient visual acuity. In modern countries, at least 50 percent of people 21 and over require eyeglasses or contact lenses to successfully perform daily activities.

Visual Defects

Some problems of visual sensors can be corrected and others cannot. Many people wear glasses or contact lenses to help correct deficiencies. A designer must realize that these people do not always wear their glasses or contact lenses. This may lead to degraded human performance.

Hyperopia and Myopia

Two of the most common visual defects, hyperopia and myopia (nearsightedness) are usually due to shape abnormalities in the eyeballs. Figure 4-8 illustrates the structure of normal, hyperopic and myopic eyes.

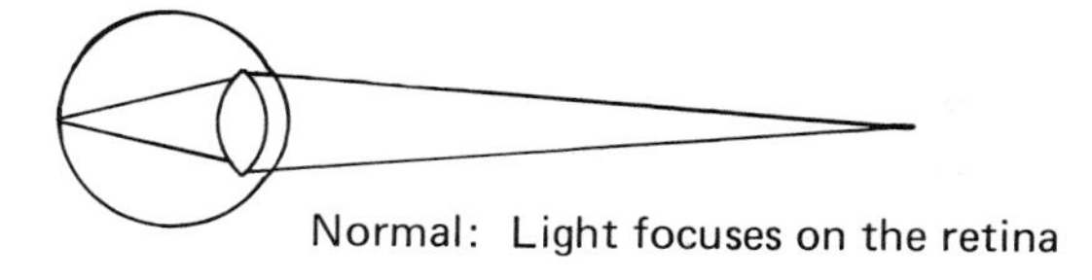
Normal: Light focuses on the retina

Hyperopia: Light focuses in back of the retina for near objects

Myopia: Light focuses in front of the retina for far objects

Figure 4-8. Structure of Normal, Hyperopic, and Myopic Eyes

In hyperopic people, the eyeball is shorter than normal, so that the refracted light rays from near objects focus somewhere beyond the sensory surface at the back of the eye. The surface is stimulated adequately by images of distant objects.

Hardening of the lens with age often brings on increasing hyperopia. As the lens loses elasticity, the individual finds it more difficult to focus clearly, because the lens cannot contract to allow nearby objects to come into proper focus. Generally, hyperopic people experience deterioration of vision for this reason in the years following age 40. By about 60, the hyperopic person must hold an object at least 39 inches away in order to see it clearly.

The reverse is true for myopic persons. In nearsightedness, the abnormally long eyeball causes light rays to focus best at a point somewhere just short of the sensory surface, so that the surface is not sufficiently stimulated. The myopic person usually has little trouble in seeing close objects clearly.

Drugs

Drugs may temporarily affect vision. Most people are familiar with the warning against driving if certain prescribed drugs are used. Although driving is probably the most hazardous activity affected, on-the-job performance can also suffer. Green and Spencer (1969) have listed about 200 commonly prescribed drugs known to affect vision.

Age

Age usually causes physiological changes that bring about a decline in vision. For instance, the pupil shrinks with age. Consequently, older people's vision can be improved by increased illumination. Visual acuity is relatively poor in young children, but improves as a person reaches young adulthood. From about the middle twenties to the fifties there is a slight decline in visual acuity, and a somewhat accelerated decline thereafter. The use of glasses or contacts usually compensates for this decline.

Night Blindness

Another visual impairment that can be corrected is night blindness—a condition where a person has less than normal vision in dim light. It may be due to a vitamin A deficiency which can be corrected in as short a time as 30 minutes with an intravenous injection of vitamin A. An excess of vitamin A, however, will not enhance sight. In fact, too much vitamin A can cause a toxic condition leading to nausea, fatigue, and *blurred vision,* all of which could degrade human performance.

We have discussed only a few of several visual deficiencies that could exist and that are correctable if recognized. The objective is to make a designer aware of the implications of not taking into account these and other problems when considering human performance.

Blindness

Two types of visual deficiencies usually not correctable are blindness (partial or total) and color defectiveness. A person is defined as blind if vision (after correction) is 20/200 in his or her better eye. This degree of visual acuity is generally inadequate for meeting day-to-day visual demands. There would certainly be human performance problems if a blind person was expected to perform in a system where the possibility of blindness had not been taken into account when it was designed. A person is defined as being "partially sighted" if the vision in his or her better eye (after correction) is less than 20/70 but better than 20/200. Such people need special considerations if expected to perform adequately in vision-oriented systems. In some cases, the existing design may be adequate, but in other cases, special material, equipment, or procedures may be required to produce a satisfactory level of human performance.

Color Defectiveness

There are two major defects associated with seeing colors: color weakness and color blindness. The most common defect is *color weakness.* These people are capable of seeing all colors, but tend to confuse some of them, *especially under dim light.* Their ability to distinguish different colors tends to be less acute than for people with normal color vision.

The second most common defect is *color blindness.* Most people who are color blind tend to confuse red, green and gray.

Finally, it should be noted that only a very small percentage of people (about 3 in 100,000) see no color or only one color. Some of these people see everything as gray; others see everything as the same color.

Many more men than women have defective color vision. Approximately 8 percent of men are color defective, while less than 1 percent of women (Hsia and Graham, 1965). Other studies have shown the incidence among Negro and Indian males to range from 2 percent to 4 percent and among Chinese from 5 percent to 7 percent (Kherumian and Pickford, 1959). Goldschmidt (1963) lists percentages that range from about 4 percent to 10 percent across 11 groups of different national origins. Thus, in some places in the world, it is possible to have an all male user population in which one out of every ten has defective vision. You can easily see the implications for using color in systems. Color coding should not be used without some form of backup coding.

AUDITION

General

The sense of hearing is probably second only to vision as the most important sense associated with human performance. Hearing is particularly important in spoken communication. In addition, it provides sound information ranging from loud warning signals to soft, soothing music. Our ears probably allow the greatest amount of "annoying" stimulation. People can close their eyes if they wish to eliminate visual sensations, but cannot as easily avoid auditory stimulation.

Sound is the result of pressure fluctuation generated by vibrations from some source. Hearing is the phenomenon of sensing these vibrations. Before any auditory sensations are perceived, acoustic energy must set off a series of mechanical, and neural events beginning with sound waves moving through the air and striking the eardrum, which transmits them to receptors in the inner ear. Nerve impulses are then transmitted to the brain.

Frequency and Intensity

Sound is measured in terms of frequency and intensity. *Frequency* refers to the number of cycles per unit of time in a periodic vibration. The psychological correlate of frequency (that is, the auditory sensation we experience) is *pitch.* The rate at which the eardrum vibrates determines what one hears, and the rate of movement is determined by the frequency of the sound wave. Familiar frequencies associated with singing and playing the piano are shown in Figure 4-9.

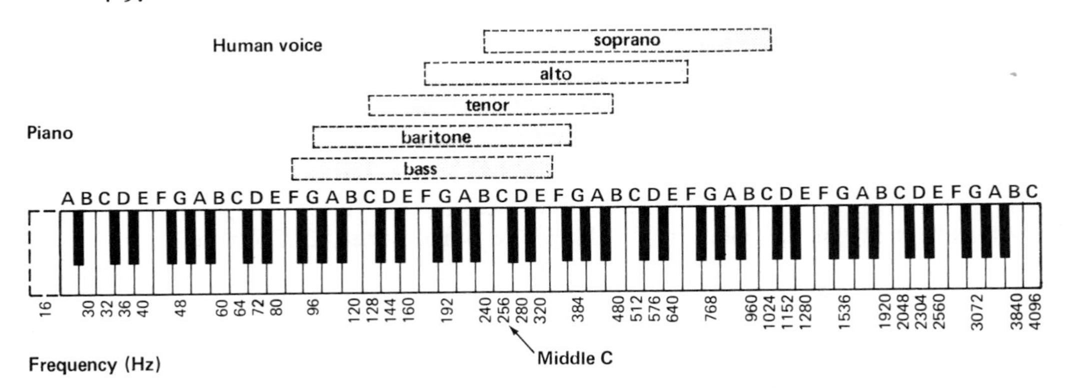

Figure 4-9. Familiar Frequencies Associated with Singing and Piano Playing (adapted from Gilmer, 1970)

We generally accept that the human ear responds to frequencies from about 20 to 20,000 Hz. However, hearing has been reported for sounds up to 100,000 Hz and down to about 5 Hz. The normal person is most sensitive to frequencies in the region from about 1000 to 4000 Hz. After age 50, the ability to perceive tones at higher frequencies gradually declines. Few people over the age of 65 can hear tones with a frequency over 10,000 Hz. This loss of perception of high frequencies interferes with identifying others by their voices and with understanding conversation in a group.

The *intensity* of the sound depends on the pressure of the sound wave that strikes the eardrum. The psychological correlate of intensity is *loudness.* The range of sound intensities to which the human ear responds is so wide that sound intensity is measured on a very large scale. The loudest sound that people can hear without experiencing discomfort has a pressure about one million times greater than the weakest.

The *decibel* (dB) is the unit used to measure the intensity of sound. For most purposes, we can regard a decibel scale as a set of numbers like those read on a thermometer: certain numbers correspond to certain loudnesses. To get an idea of how decibel levels are associated with some familiar sounds, see Figure 4-10. The decibel concept is discussed further in Chapter 13.

Extended exposure to high sound—pressure levels (over 100 dB) may result in permanent damage to the ear. Prolonged exposure to high-intensity sound can result in the diminishing of the ear's sensitivity to all frequencies, but especially to the higher frequencies.

Below the level of 20 dB, the ear rapidly loses its ability to detect frequency changes. Above this level, however, the ear will fairly consistently detect a difference as small as 3 Hz in a tone of 1,000 Hz or less. Also, above 20 dB, an intensity increment of about 0.5 to 1 dB is detectable.

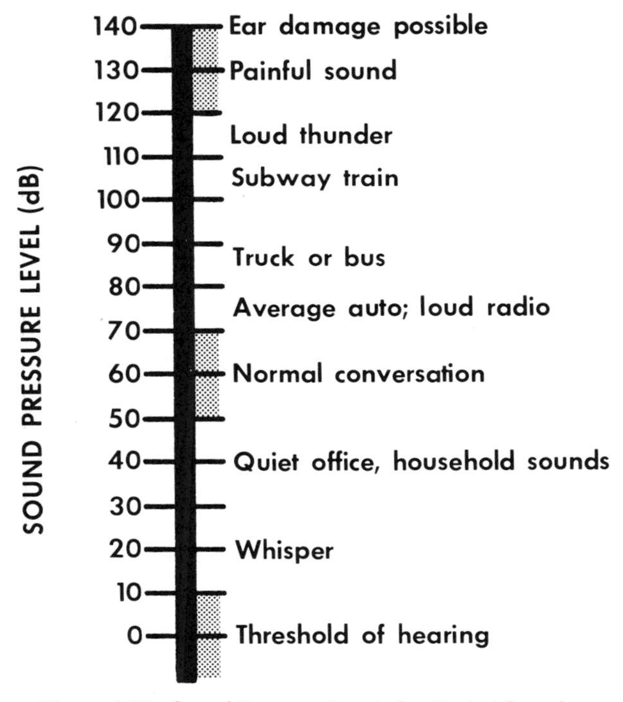

Figure 4-10. Sound Pressure Levels for Typical Sounds

Hearing sensitivity varies greatly among individuals, within a normal group by as much as 20 dB or more. Even an individual's hearing sensitivity may vary 5 dB within a very short period of time. Hearing sensitivity differences are most frequently associated with age (older people tend to have more difficulty) and past exposure to loud noises (the more exposure, the greater the probability of having a hearing difficulty).

At the lower frequencies a tone must be much louder to be heard than a tone in the mid-frequencies. Frequencies between 1000 Hz and 8000 Hz require the least intensity to be heard.

Hearing Speech

One aspect of sound that relates to numerous human performance situations has to do with speech perception. Communication among people is most frequently by means of speech. Average speech at a distance of one meter corresponds roughly to sound falling in the range of 60 and 75 dB. The speech spectrum of a normal voice lies almost entirely within the frequency from 100 to 8000 Hz. Over half is expended in frequencies below 1000 Hz. When noise is present, particularly noise in the frequencies from 100 to 8000 Hz, speech is frequently masked so as to make interpretation by the listener difficult or impossible (see Chapter 13 for a more detailed discussion).

Under normal circumstances, when speech intensity reaches a level of about 40 dB, maximum intelligibility is attained. Above 100 dB, intelligibility decreases slightly. When the intensity of speech and noise combined is in excess of about 80 dB, distortion within the ear itself lowers speech intelligibility. In this case, ear plugs, which bring the intensity of speech and noise down by as much as 20 dB, reduce distortion and make these sounds more intelligible. For satisfactory communication of most voice messages in noise, the speech level should exceed the noise level by at least 6 dB.

Auditory Defects

Frequently, designers erroneously assume that all people working in a new system will have adequate hearing, and because of this, place much weight on spoken communication, and other auditory signals. Actually, hearing impairments are relatively common, and the chances of a user having difficulty in hearing are fairly high. Designers should take into account the basic abilities of *all* potential users.

Deafness is a two-way problem. It may not only handicap a person as a listener, but also may handicap a person as a speaker. Because totally deaf people cannot hear their own voices, many have great difficulty in learning to speak so that they can be easily understood.

There are two kinds of deafness that may figure in human performance problems. *Conduction deafness* involves roughly the same hearing loss at all frequencies, that is, the person suffering from it has about the same difficulty in hearing at one frequency as at another. The term *conduction deafness* is used because the deafness originates in deficiencies of mechanical conduction in the ear. The ear may be stopped up, the eardrum may be broken, or organs of the middle ear may be damaged. The effect of conduction deafness is much the same as putting earplugs in one's ear.

The second kind of hearing loss is *nerve deafness.* As the name suggests, in this type of deafness something is wrong with the auditory nervous system.

Either the nerves themselves have been damaged, or damage has been done in the inner ear. It is characteristic of nerve deafness that hearing loss is much greater at higher frequencies. This means that the nerve-deaf person can hear low-pitched sounds reasonably well but hears high-pitched sounds rather poorly or not at all. Such a person has a great deal of trouble understanding speech because some of the higher frequencies are very important in speech comprehension. They are unable to distinguish easily between word sounds. The curves for each type of deafness are shown in Figure 4-11.

Loss of hearing is very common in older people. In fact, nearly all people can expect to have at least mild hearing problems by the time they reach age 60.

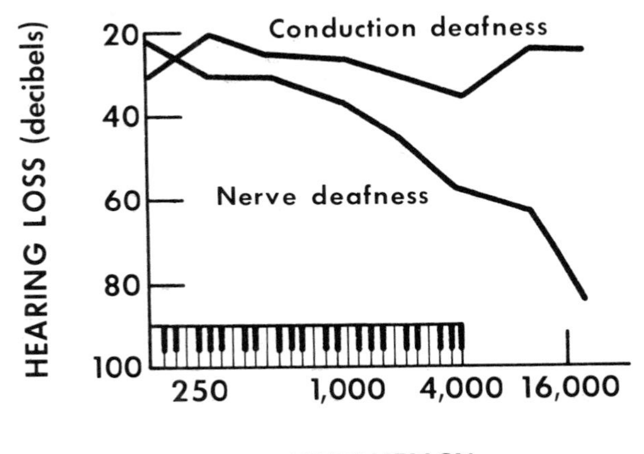

Figure 4-11. A Graphic Comparison of Hearing Loss Due to Conduction vs. Nerve Deafness

People with hearing impairments should be accommodated as users in a system by making use of their other senses. On the other hand, the "hard of hearing" usually have sufficient hearing to benefit from most of the regularly used decisions for enhancing human performance. This of course assumes that those people who are hard of hearing are equipped with the necessary amplification devices such as hearing aids. If they are not so equipped, or if they have adjusted the volume to reduce the noise level around them, then the hard of hearing represent a user group that requires special attention. For example, auditory alarms as the sole source of alerting people will not work well for some people who are hard of hearing and are of no use for the deaf. A designer should build redundant means of alerting people.

One of the most difficult problems for a designer is temporarily impaired hearing. Being exposed to loud noises for extended periods of time, whether from aircraft noise or loud music, can temporarily impair a person's hearing. A much more subtle effect on hearing comes from the use of certain drugs. By the latter part of the ninteenth century it was known that particular drugs, then in common use and in more frequent use today, could produce temporary or even permanent impairment of hearing.

For example, with the discovery of the antibiotic streptomycin, came the finding that the drug could cause impairment of hearing, as well as vestibular (balance) disturbances. A large number of other drugs, some of which are very commonly used such as aspirin, can temporarily or permanently impair hearing. Gignoux, Martin and Cajgfinger (1966) describe a case involving a sixteen year old girl with normal hearing who became totally deaf after taking forty aspirin tablets in a suicide attempt. She partially recovered her hearing but for reasons unknown, a loss of 65-80 dB persisted in one ear as compared to a 20 dB loss in the other. Jarvis (1966) reports an example of partial deafness in a patient who took two or three aspirin tablets every two hours for three days after a tooth extraction.

Designers must realize that systems designed to operate adequately only with people who have good hearing are destined to exhibit human performance problems. People will continue to expose themselves to loud noises (e.g., rock music) and to take drugs, prescribed or otherwise, both of which are known to affect hearing. Most designers are not in a position to restrict the outside activities or the drug intake of people who will be working in their systems, and consequently should design a system that functions well even if the hearing of some users is impaired.

CUTANEOUS SENSES

The surface of the body can be classified as a sense organ in the same way as the eyes or the ears. Because the skin covers the entire surface of the body, it provides widespread contact with the immediate environment. The skin's several kinds of nerve endings respond to mechanical, thermal, electrical, and chemical stimuli. The sensations produced by these different stimuli take the form of pressure, pain, cold and warmth.

Touch (Pressure)

Touch or pressure is experienced when a depression is formed upon the skin by some mechanical stimulus. The skin sensor indicates that an object is touching the body, and within certain limits, where it is touching, what size and what shape it is, and whether it is moving, still, or vibrating.

The threshold for pressure varies with the area of the skin, and depends to a large measure on the concentration of nerve fibers and skin thickness at each location. Among the most sensitive areas of the body are the tip of the tongue and the fingertips, among the least sensitive are the kneecaps.

The great sensitivity of certain areas of the skin, including the fingers and hands, has enabled designers to develop numerous different types of controls that can be identified by touch. In many aircraft, for example, the control used for lowering the wheels when landing is in the form of a small wheel. The use of touch can also provide some advantage when the visual and auditory channels are overloaded. The touch sense is able to receive and respond to stimuli every bit as quickly as the auditory sense and in many cases faster than the visual sense. In high noise areas or in cases where visual and auditory detection may be impaired (for example, in early stages of hypoxia), touch warning signals appear to offer significant advantages.

The possibility of using the touch sense on areas of the body other than fingers and hands has been considered as a way of enhancing human performance in systems. There are now available a variety of systems designed to provide optical information to skin areas using a matrix of artificial photoreceptors and a corresponding matrix of skin stimulators. These aid the blind with reading, recognition of pictorial information, or improve mobility in their surroundings.

Geldard (1960) developed a touch-communications technique by which subjects were able to learn a touch code for letters and numbers. He used vibrators located at different positions on the chest to send signals consisting of changes in duration and amplitude of vibrations. His best subject was able to learn the touch code and receive communications at the rate of 38 words-per-minute.

Linvill designed a reading aid for the blind with the stimulus array on the skin of the fingertip. With an improved model of this device, blind subjects can read at rates of over 60 words-per-minute (Linvill and Bliss, 1966; Bliss, 1971). Collins and Bach-y-Rita (1973) have developed a technique for transmitting pictorial information via the skin. This scanning technique uses both mechanical and electrical stimulation of the skin. It uses a camera with a zoom lens controlled by a subject. The person aims the camera in any desired direction and the video signal is converted to a touch stimulus that is picked up by touch receptors on the back or the trunk.

Perhaps the greatest disadvantage of using the touch sense is that pressure sensations remain only as long as the rate of movement into or out from the skin continues. With constant pressure and no movement, sensation soon ceases.

Pain

People's sensitivity to pain has received little recognition by system designers. Although overstimulation of virtually any sensor will produce the sensation of pain, its use as a warning or in other ways in systems has been minimal. Pain can occur with either mechanical, thermal, electrical, or chemical stimuli. Receptors are spread through almost all tissues of the body. There are actually three kinds of pain: superficial or cutaneous pain; deep pain from muscles, tendons, and joints; and visceral pain. The threshold for pain varies with the place stimulated. Probably the most sensitive place of the body is the cornea of the eye and the area that is least sensitive to pain is the sole of the foot.

People have a considerable ability for adapting to pain. If for example a needle is held to the skin with a steady, unvarying force, the pain aroused gradually loses its intensity and eventually disappears altogether, leaving a sensation of pressure, which then gradually declines and also disappears. As with touch, if pain is used in a system, the effect of the pain must be capitalized on quickly. Unfortunately, people react relatively slowly to pain stimuli. Subjects in one study, for example, were able to respond to visual stimuli on the average of about 200 ms, whereas they required about 700 ms to respond to pain stimuli.

Pain may be aroused by stimuli such as radiant heat or cold that are not in direct contact with the skin. The sensation of heat or cold is aroused first, then a sensation of pain, followed by adaptation and a return of the sensation of heat or cold. This confusion may lead to ambiguous sets of responses to the stimuli which would lead to unpredictable and degraded performance.

Temperature

The skin also has receptors that are sensitive to temperature changes. Thermal receptors are stimulated by raising and lowering the temperature of the skin. Normal *skin* temperature is about 91.4^o F. (32 or 33° C) and stimuli at this temperature elicit no thermal sensation.

Note that the normal skin temperature of 91.4^o F is considerably below the average deep body temperature of 98.6^o F. At any one time, the different regions of the skin can be at different temperatures. And although different regions of the skin may be at different temperatures, adaptation can exist simultaneously at all regions. As shown in Figure 4-12, thermal adaptation will only occur when the skin temperature is between about 60° and 105° F. Below this range the skin will transmit a cold feeling, and above it, a feeling of warmth.

Again, we have the problem of relatively quick adaptation to temperature changes. This suggests that the temperature receptors may be used as sources of information in a system, but it would require a relatively quick interpretation because the receptors tend to adapt so quickly.

TASTE

A person can normally identify four distinct qualities of taste: sweet, sour, salty, and bitter. All taste sensations are combinations of these four primary taste qualities. The most sensitive to stimuli is the bitter taste sense.

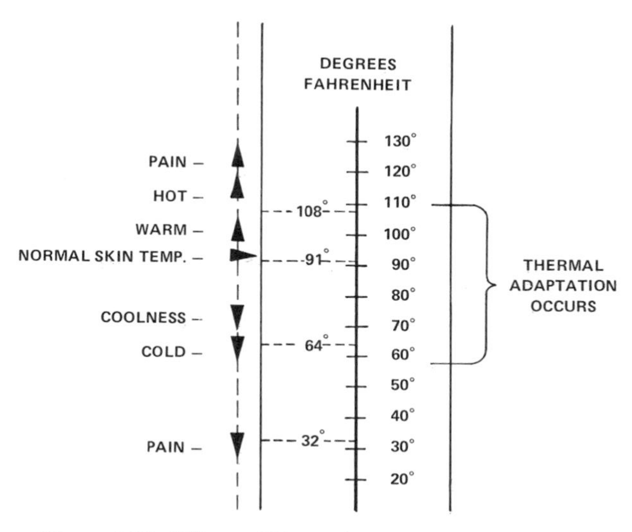

Figure 4-12. Effects of Temperature on Skin Receptors

Actually, in many cases taste is the result of the interaction of several senses. Pepper tastes as it does because it stimulates both taste and pain receptors. The sense of smell is particularly important in determining what we taste. Food can be almost tasteless to someone suffering from a head cold mainly because of a loss of the sense of smell. Thus, much of what we call taste is actually smell. In fact odors from food may pass upward into the nasal area, stimulating the olfactory system as much as thousands of times as strongly as the taste system. The taste of food also may be affected by its temperature and its appearance.

Reduced taste sensitivity is associated with the atrophy and loss of taste buds from the tongue as people approach the age of 70 and beyond. Sensitivity may also be affected by temperature, internal factors such as a salt deficiency, adaptation, and individual differences. In addition, many people are taste-blind for certain substances. A substance used frequently by psychologists for demonstrating taste blindness is phenylthiocarbamide, for which 15 to 30 percent of the subjects exhibit taste blindness.

This leads us to conclude that there would be great difficulty in using taste as a source of information in many, if not most, systems. However, a creative designer may find ways of capitalizing on the unique characteristics of the taste sense for providing information to help achieve an adequate level of human performance. In one system, a telephone company employee touched the end of wires to his tongue. If he tasted salt he rightly concluded that the wire was "live."

SMELL

The sense of smell is activated when an odorous stimulus, in the form of vaporized chemical substances, comes into contact with specialized areas in the upper region of each nostril. Some smells also may involve mechanical and chemical stimuli.

The primary qualities of smell are not as well known or understood as those of vision, hearing, or taste. There are probably at least six qualities of odor: spicy, fruity, burnt, resinous, flowery, and putrid, although there may be as many as 50 or more primary sensations of smell. Smell stimuli are almost always mixtures of basic smells, resulting in complex odors.

From a designer's point of view, one of the most important characteristics of smell is the minute quantity of the stimulating agent required to effect a smell sensation. It is estimated that smell is 10,000 times as sensitive as taste. For instance, the substance methyl mercaptan can be smelled when only 1/25,000,000,000 mg is present in each milliliter of air. Because of this low threshold, this substance is mixed with natural gas to give the gas an odor that can be detected when it leaks from a gas pipe. Even though the threshold concentrations of substances that evoke smell are extremely slight, concentrations only 10 to 15 times above the threshold values evoke maximum intensity of smell. This is in contrast to most sensory systems of the body, in which the ranges of detection are much larger.

In some systems it may be useful to use odor as a warning rather than bells or flashing lights. For example, most of us have been involved in fire drills while at school and have noted the lack of urgency in these drills as people left the building. One wonders if the drills are actually helpful since they represent such a low-fidelity simulation. If smoke and flames were actually part of the drill (making a higher fidelity situation) would the people perform in the same manner? It is possible they would not. Using a noxious odor that is poured into the air rather than a bell that rings, would greatly improve the motivation of the people to leave the building. This would make the fire drills more like an actual situation where a fire exists. Thus, noxious odors could be used in situations where people are expected to leave an area quickly, as tear gas is already used. In these cases the sense of smell would become every bit as important as sight or hearing and would be very useful to help achieve a particular system objective.

One of the main problems with using the smell sense is that smell discrimination is influenced by any odor to which a person is already adapted. Also, the existence of *odor blindness* in many people should be recognized by designers who may want to use certain odors as warning stimuli in a system. The simple blocking of the nasal passages with the common cold can eliminate the ability to smell. Finally, like many other senses, receptors in the nose adapt very quickly and cease to respond to stimuli. Thus, even a strong odor will gradually become imperceptible.

There are some substances that elicit differences in the sense of smell between men and women. The most extraordinary compound tested is exaltolide. Most men and most children cannot smell this substance at all, but most women find it very strong and the strength seems to vary with the menstrual cycle. In pregnancy, the ability to smell exaltolide diminishes during the first two months. At the end of the third month it regains its normal level but continues to increase until the baby is born. Gurrier, et al. (1969) describe six women who lost their sensitivity to exaltolide following hysterectomy and ovariectomy. In all cases, these patients regained their abilities to smell these substances with hormonal replacement therapy.

KINESTHETIC SENSE

The kinesthetic sense is used when people are positioning, making movements, controlling forces, judging weight, etc. In order for people to control actions, they need to know the position of body parts both before and after a movement. The primary source of this information is the kinesthetic or muscle sense. For example, people can close their eyes and still walk, sit down, and get up. Kinesthesis provides information on the position of the limbs, how far they have moved, and the posture of the body as a whole. A unique feature of kinesthesis is that stimulation comes from within the organism itself rather than from the outside world. Kinesthetic stimuli are always present, although most of the time we are not aware of their presence and sensitivity varies with different parts of the body.

One of the main functions, perhaps the most important function, of the kinesthetic sense is to enable people to control their voluntary muscular activities without the aid of vision. When the eyes are used in executing positioning movements (for example, shifting gears in a car and keying on a typewriter) kinesthetic cues are not used too much. However, this type of visual positioning requires more time to accomplish than when more reliance is placed on the kinesthetic sense (i.e., "feeling" where to shift the gears or which key to depress on a keyboard).

When considering the large number of movements made daily by people performing different activities, the kinesthetic sense may be as important to successful performance as the visual or auditory senses. Design decisions should be made that expedite the shifting of dependence from visual cues to kinesthetic cues as quickly as possible when new skills are being learned. In addition, designers need to develop activities that provide clear and unambiguous opportunities for the kinesthetic sense to function properly.

SENSORY ADAPTATION

All senses adapt, but as we have seen, some adapt much more quickly than others. The sensitivity of each sensor is modified by the continuous presentation of stimuli to that receptor or adaptation. With adaptation the receptors become less effective. Few people realize that there is one process of adaptation that *heightens* the performance of the receptors. When the eye adapts to dark places, certain visual receptors become *more sensitive* during the course of adaptation.

SENSORY INTERACTION

The senses have been discussed as though they function independently in the information-gathering process. Normally, the senses combine to produce an integrated experience. When eating, for example, the sight of food arranged on the table, the conversational tones and background music, and the tactile sensations, aromas and taste of the food all combine to enhance the experience. So when considering the adequacy of human senses, designers need to be aware that they are not used individually but in combination.

FOR MORE INFORMATION

Kling, J. W. and Riggs, L. A., (Ed.) *Experimental Psychology,* 3rd Edition, New York: Holt, Rinehart and Winston, Inc., 1971.

Van Cott, H. P. and Kinkade, R. G., *Human Engineering Guide to Equipment Design,* Washington, D. C.: U. S. Government Printing Office, 1972.

5

THE BODY AND PERFORMANCE

INTRODUCTION

In designing systems for optimal human performance, we need to understand not only the senses but also the capabilities of the body that enable an individual to respond (see Figure 5-1). This understanding is especially important as a designer envisions the potential user population of a system.

BODY DIMENSIONS

A designer must consider the human body's relationship to performance in at least two ways. First, certain body dimensions seem to be related to optimal performance in some activities. It seems that the taller professional basketball players are, the more success they have on the court. Second, body dimensions must be considered in designing work areas so that a user will "fit" into the space provided and will be able to make the required *reaches* or *movements*. An automobile, for example, should comfortably accommodate larger drivers, yet have the controls readily accessible to smaller drivers.

Because they vary greatly, human body dimensions can pose many problems for designers. Fortunately, all people fall within certain limits. If these extremes are defined, they can be considered in any design for human

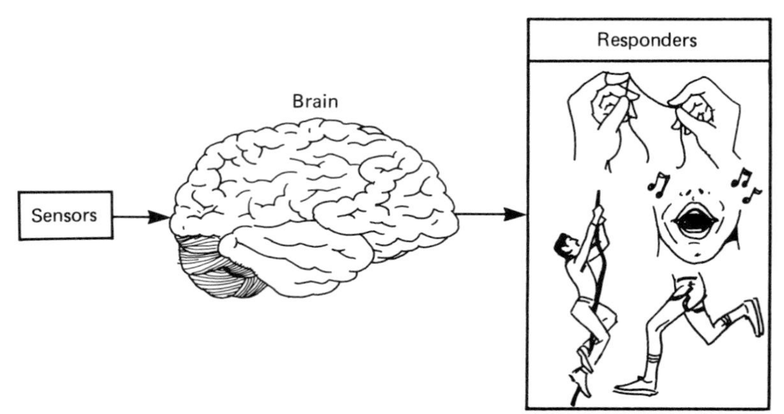

Figure 5-1. Simplified Model of Human Information Processing with the Responses Emphasized

performance. For example, the smallest group of people on record are the Pygmy people of central Africa. Gusinde (1948) found a mean stature of 4'8" (143.8 cm) for adult males and 4'6" (137.2 cm) for females. Conversely, the tallest group of people are the Nilotes of the Sudan, where the mean was 5'11" (182.9 cm) for males and 5'6" (168.9 cm) for females (Roberts, 1975). The tallest human reliably recorded was Robert Wadlow (1918-1940), 8' 11"; while the shortest was Pauline Musters (1876-1895), at 1' 11" tall. This range, from about 2 feet to 9 feet, establishes the extreme outer limits for human adults.

Outer limits have also been established for other body dimensions. Studies of body dimensions, or *anthropometry,* have been conducted throughout history. In ancient Egypt, the length of the middle finger was considered to be a third of the height of the head and neck, and a nineteenth of the whole body. Leonardo da Vinci followed the rule that the head, from crown to chin, is contained eight times in the stature. The systematic study of individual differences in size probably derives from the pioneer work of the Belgian astronomer Quetelet who published his *Anthropometrie* in 1870. Since that time much anthropometric literature has been written, particularly in the last few years, and details of body measurements are available for over 10,000 population samples throughout the world. Several such studies were conducted in the 1920s and 1930s, but it wasn't until World War II that well-controlled studies of body measurements were begun.

Based on the findings of these studies, we now have information concerning many body dimensions. One interesting fact that comes from these results is that there seems to be a world-wide increase in height. American soldiers of World War II, for example, were an average of seven-tenths of an inch taller than those of World War I (Davenport and Love, 1921; Newman and White, 1951). Dreyfuss (1967) estimates that the average height of men in the United States from 1900 to 2000 could increase as much as three inches.

These increases cannot be totally accounted for by evolution or even better nutrition. One interesting explanation for this difference has been suggested by Landauer and Whiting (1979). After reviewing several studies concerning consequences of stress in infancy, they found that there is a significant two and a half inch average difference in adult male stature in societies that had *stressful infant care practices* than in those that did not. The data suggest that the stress (e.g., piercing, scarification, separation of infant from mother, and smallpox vaccination) must occur before the age of two in order to have an enhancing effect on growth. Perhaps modern society is providing more of these infant stressors than societies of the past.

Full growth in males occurs at about age twenty and in females at about age seventeen. Around age sixty, height begins to decrease. Old age, however, is not the only time humans lose height. In fact, although temporary, the height of an individual usually decreases as the day progresses because of the compression of the intervertebral discs. Body heights are greatest immediately after rising and least before retiring. The difference averages about 0.95 inches in adult males (Backman, 1924).

The design-related problems associated with the variations of body dimensions in different populations have begun to emerge as a result of the increase in world trade and the attempt to establish world-wide standards. International studies have focused on the extent of differences in body dimensions and their importance for design, particularly in systems that will have international use.

AVERAGE PERSON FALLACY

Designing for the body dimensions of the so-called *average person* is usually a mistake. The result of such a design is that the smaller 50 percent of the users may be unable to reach the controls comfortably or read the displays, and the larger 50 percent will not have sufficient room to move about comfortably. No one is average in all dimensions, and few people are average in even a few dimensions. (See Figure 5-2.)

To test the concept of the average person, Daniels and Churchill (1952) categorized 4063 men according to ten measurements used in clothing design. For each of the ten dimensions, the men making up about the middle 30 percent were considered average.

The results are as follows:

		Percent of original sample remaining
(1)	Of the original 4063 men, 1055 were of approximately average stature.	25.9
(2)	Of these 1055 men, 302 were of approximately average chest circumference.	7.4
(3)	Of these 302 men, 143 were of approximately average sleeve length.	3.5
(4)	Of these 143 men, 73 were of approximately average crotch height.	1.8
(5)	Of these 73 men, 28 were of approximately average torso circumference.	0.69
(6)	Of these 28 men, 12 were of approximately average hip circumference.	0.29
(7)	Of these 12 men, 6 were of approximately average neck circumference.	0.14
(8)	Of these 6 men, 3 were of approximately average waist circumference.	0.07
(9)	Of these 3 men, 2 were of approximately average thigh circumference.	0.04
(10)	Of these 2 men, 0 were of approximately average inseam length.	0.00

Of the original 4063 men, not one was average in all ten dimensions, and less than 4 percent were average in even the first three. The average person simply does not exist.

AVERAGE PERSON FALLACY

Original Population

Average Height

Average Height and Weight

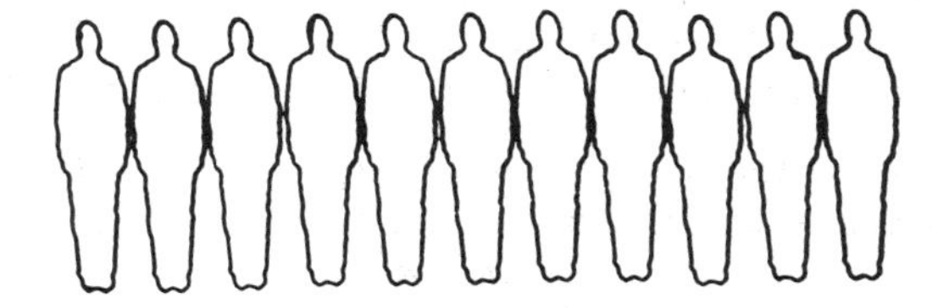

Average Height, Weight, and Chest Circumference

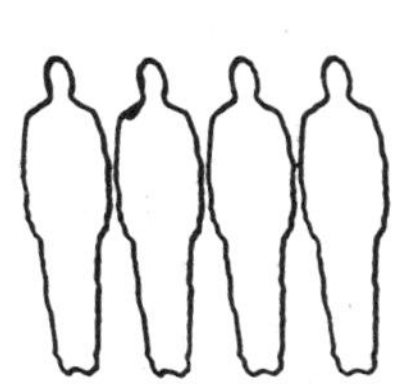

Average Height, Weight, Chest Circumference, and Knee Height

Figure 5-2. Average Person Fallacy

ANTHROPOMETRIC MEASURES

Instead of using an average person as a design model, designers are encouraged to use tables of anthropometric measurements. These measurements provide information on body dimensions that are of value in a wide range of design situations. Good design practice uses this information to accommodate at least 95 percent of the potential users in a system. Designing for 95 percent of the potential user population eliminates only a few people, and at the same time avoids costly problems encountered in dealing with extreme dimensions. For example, in decisions associated with the distance a person can reach when secured by a seat belt, a designer should use the dimensions of a person at the 5th percentile. This would mean that the headlight switch would be easily reached by 95 percent of the people who would drive the automobile. On the other hand, decisions associated with head, shoulder, and leg clearance should be based on the height and other dimensions of the largest users, i.e., those at the 95th percentile. Percentile values are needed for each critical dimension. Thus, the designer must first determine which dimensions directly relate to human performance in a particular activity.

Percentile Ranks

In any distribution the percentile rank of any specific value is the percent of cases out of the total that falls *at or below* the specific value. For example, out of a group of 345 people, we wish to find the percentile rank that 115 fall at or below. This would be figured as follows: (100) (115/345) = (100)(.33) = 33. Thus, the percentile rank is 33, or the 115th score is at the 33rd percentile.

The maximum score must have a percentile rank of 100, and the minimum score must have a percentile rank of 100(1/N) where N equals the total number of people in the group under study. The percentile rank of the median must be 50, since 50 percent of all observed values in a frequency distribution must lie at or below the median.

Static and Functional Measures

Two basic kinds of measurements are of value to a designer.

1. *Static* measurements are those taken when an individual is in a non-moving position. For example, standing back against a wall for measuring height.

2. *Functional* measurements are those taken with people assuming common movement positions where the position itself helps to determine the dimension of the body parts involved.

Functional measurements are used, for example, for determining clearance in a crawl position or reach from a seated position. When dealing with either static or functional dimensions the type and amount of clothing that a person is wearing, including the height of the shoes, headgear (e.g., helmet), and the type of equipment that an individual might be carrying or wearing (e.g., a parachute, weedsprayer, gun) must be taken into account.

Human Dimensions

The measurements presented in Tables 5-1, 5-2, 5-3 and 5-4 are for adults, with men and women shown separately. It would be best to have a set of *combined* measurements. Unfortunately, available measurements were taken when people working in one or another occupation were typically men *or* women. However, these measurements illustrate the dimensions that are useful for most applications. For "clearance" purposes a designer should use the 95th percentile on Table 5-1 for men, and for "reach" purposes the 5th percentile on Table 5-2 for women. There are extensive measurements available in other sources for specialized applications (cf. Damon, Stoudt and McFarland, 1966; Garrett and Kennedy, 1971).

Table 5-1. Body Dimensions for Male Workers in the Age Range 18 to 45

		Dimension Name	Dimension (in inches except as noted)				Added Increments	
			5th Percentile	50th Percentile	95th Percentile	Standard Deviation	Light Clothing	Heavy Clothing
		Weight (LBS)	120.5	159.1	197.5	23.36	5.0	10.0
A	1.	Vertical Reach	77.0	82.5	88.0	3.33	1.0	1.0
	2.	Stature	64.4	68.7	73.0	2.60	4.0	4.0
	3.	Eye Height	60.0	64.3	68.9	--	1.0	1.0
	4.	Crotch to Floor	29.7	32.7	35.7	1.84	1.0	1.0
	5.	Waist Breadth	9.0	10.7	12.2	0.94	1.0	3.0
	6.	Hip Breadth	11.8	13.1	14.4	0.79	1.0	3.0
B	1.	Head Width	5.6	6.0	6.4	0.23	4.0	4.0
	2.	Interpupillary Distance	2.2	2.4	2.7	0.16	--	--
	3.	Head Circumference	21.1	22.1	23.1	0.63	--	--
	4.	Neck Circumference	13.4	14.7	16.1	0.82	--	--
	5.	Head Length	7.2	7.7	8.1	0.29	4.5	4.5
	6.	Head Height	8.0	8.7	9.8	--	2.5	2.5
	7.	Ear to Tip of Lip Length	3.69	3.7	3.71	--	--	--
	8.	Ear to Top of Head	4.7	5.2	5.7	0.31	2.5	2.5
	9.	Ear Breadth	1.3	1.4	1.6	0.11	--	--
		Ear Length	2.2	2.5	2.7	0.16	--	--
C	1.	Thumb Tip Reach	29.4	32.5	35.7	1.91	1.0	1.0
	2.	Chest Circumference	32.6	36.9	41.3	2.63	--	--
		Chest Depth	7.8	9.1	10.4	0.79	1.0	3.0
	3.	Waist Circumference	26.3	31.6	36.9	3.22	--	--
		Waist Height to Floor	38.4	41.9	45.3	2.11	1.0	1.0
		Waist Depth	6.7	7.9	9.5	0.9	1.0	2.0
	4.	Hip Circumference	33.0	37.1	41.2	2.46	--	--
		Hip Depth	7.6	8.8	10.2	0.8	1.0	2.0
	5.	Upper Thigh Circumference	18.7	21.8	24.9	1.89	--	--
		Gluteal Furrow Height	28.9	31.6	34.2	1.62	1.0	1.0
	6.	Calf Circumference	12.7	14.4	16.1	1.05	--	--
	7.	Ankle Circumference	9.4	10.4	11.3	0.57	--	--
		Ankle Height	2.3	2.7	3.2	--	1.0	1.0
	8.	Foot Length	9.8	10.5	11.2	0.45	1.5	1.5
		Foot Width	3.5	3.8	4.1	0.19	1.0	1.0
	9.	Shoulder Height	52.5	56.6	60.6	2.45	1.5	1.5
	10.	Forearm Circumference (Flexed)	10.2	11.6	13.0	0.85	--	--
	11.	Biceps Circumference (Flexed)	10.9	12.7	14.5	1.08	--	--
D	1.	Head to Seat Height	33.3	35.7	38.1	1.44	3.5	3.5
	2.	Eye to Seat Height	28.7	31.0	33.3	1.41	1.0	1.5
	3.	Shoulder Breadth	16.2	17.9	19.5	1.00	1.0	3.0
	4.	Hip Breadth Sitting	11.9	13.4	15.0	0.94	1.0	3.0
E	1.	Hand Length	6.7	7.5	8.1	0.38	--	1.0
	2.	Hand Breadth	3.2	3.5	3.8	0.19	--	1.5
	3.	Wrist Circumference	6.2	6.7	7.3	0.34	--	--
	4.	Hand Circumference	7.8	8.5	9.3	0.45	--	--
	5.	Hand Thickness	1.2	1.3	1.4	0.08	--	--
F	1.	Knee Height	19.5	21.3	23.1	1.08	1.5	1.5
	2.	Popliteal Height	15.6	17.2	18.8	0.98	1.5	1.5
	3.	Buttock to Popliteal Length	17.6	19.2	20.9	0.99	0.5	1.0
	4.	Buttock to Knee Length	21.6	23.4	25.3	1.12	1.0	1.5
	5.	Elbow to Wrist Length	9.9	11.3	12.7	0.84	0.5	1.0
	6.	Thigh Clearance	5.9	6.6	7.1	--	0.5	1.0
	7.	Shoulder to Elbow Length	13.3	14.5	15.7	0.73	1.0	1.5
	8.	Elbow Rest Height	7.4	9.1	10.8	1.04	0.5	0.5
	9.	Mid-Shoulder to Seat Height	22.5	24.6	26.6	1.25	1.0	1.5

Adapted from Israelski, 1977;
Garrett and Kennedy, 1971;
Dreyfus, 1967;
Woodson and Conover, 1964;
Hertzberg, 1954.

STRUCTURAL BODY DIMENSIONS – MALE AND FEMALE

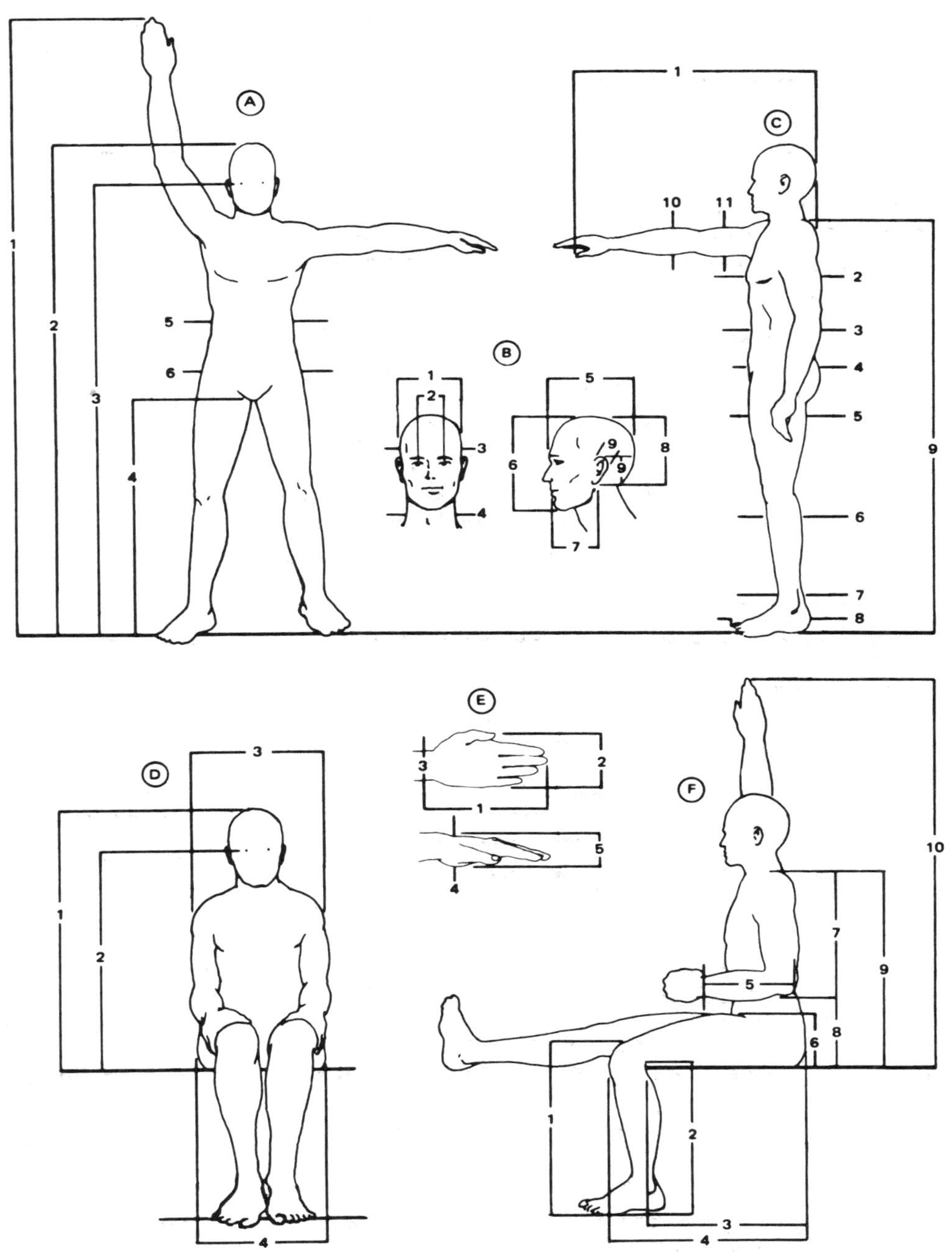

Woodson and Conover, 1964

Table 5-2. Body Dimensions for Female Workers in the Age Range 18 to 45

		Dimension (in inches except as noted)				Added Increments	
	Dimension Name	5th Percentile	50th Percentile	95th Percentile	Standard Deviation	Light Clothing	Heavy Clothing
	Weight (LBS)	102.3	126.1	156.4	16.6	3.5	7.0
A	1. Vertical Reach	72.9	78.4	84.0	3.4	1.0	1.0
	2. Stature	60.0	63.8	67.8	2.4	4.0	4.0
	3. Eye Height	56.0	59.0	62.5	2.5	1.0	1.0
	4. Crotch to Floor	26.8	29.3	32.0	1.6	1.0	1.0
	5. Waist Breadth	8.4	9.4	10.9	0.8	1.0	3.0
	6. Hip Breadth	12.4	13.7	15.3	0.9	1.0	3.0
B	1. Head Width	5.3	5.7	6.1	0.2	4.0	4.0
	2. Interpupillary Distance	2.1	2.5	2.8	--	--	--
	3. Head Circumference	20.6	21.6	22.7	0.6	--	--
	4. Neck Circumference	12.2	13.3	14.4	0.7	--	--
	5. Head Length	6.8	7.3	7.7	0.3	4.5	4.5
	6. Head Height	8.0	8.6	9.4	0.5	2.5	2.5
	7. Ear to Tip of Lip Length	3.5	3.6	3.7	0.7	--	--
	8. Ear to Top of Head	4.6	5.0	5.5	0.3	2.5	2.5
	9. Ear Breadth	1.0	1.2	1.4	0.1	--	--
	Ear Length	1.8	2.0	2.3	0.2	--	--
C	1. Thumb Tip Reach	26.7	29.1	31.7	1.5	1.0	1.0
	2. Chest Circumference (Bust)	32.1	35.0	39.5	2.2	--	--
	Bust Height to Floor	43.3	46.5	50.1	2.0	1.0	1.0
	Chest Depth	8.2	9.2	10.7	0.8	1.0	3.0
	3. Waist Circumference	23.4	26.1	30.4	2.1	--	--
	Waist Height to Floor	36.7	39.4	42.5	1.8	1.0	1.0
	Waist Depth	5.8	6.6	8.0	0.7	1.0	2.0
	4. Hip Circumference	33.8	37.4	41.6	2.3	--	--
	Hip Height to Floor	29.8	32.5	35.4	1.7	1.0	1.0
	Hip Depth	7.3	8.3	9.6	0.7	1.0	2.0
	5. Upper Thigh Circumference	19.1	21.8	24.7	1.7	--	--
	Gluteal Furrow to Floor	26.1	28.6	31.3	1.6	1.0	1.0
	6. Calf Circumference	12.0	13.5	15.0	0.9	--	--
	7. Ankle Circumference	7.5	8.3	9.2	0.5	--	--
	Ankle Height	2.3	2.7	3.1	0.2	1.0	1.0
	8. Foot Length	8.7	9.5	10.2	0.4	1.5	1.5
	Foot Width	3.2	3.5	3.8	0.2	1.0	1.0
	9. Shoulder Height	48.4	51.9	55.6	2.2	1.5	1.5
	10. Forearm Circumference (Flexed)	8.9	9.8	10.8	0.6	--	--
	11. Biceps Circumference (Flexed)	9.1	10.4	12.1	1.0	--	--
D	1. Head to Seat Height	31.7	33.7	35.8	1.25	3.5	3.5
	2. Eye to Seat Height	27.1	29.0	31.0	1.20	1.0	1.5
	3. Shoulder Breadth	15.1	16.4	18.1	0.90	1.0	3.0
	4. Hip Breadth Sitting	13.3	15.0	17.0	1.13	1.0	3.0
E	1. Hand Length	6.7	7.2	7.9	0.4	--	1.0
	2. Hand Breadth	2.7	3.0	3.2	0.2	--	1.5
	3. Wrist Circumference	5.4	5.9	6.4	0.3	--	--
	4. Hand Circumference	6.7	7.2	7.8	0.4	--	--
	5. Hand Thickness	0.8	1.0	1.1	0.1	--	--
F	1. Knee Height	17.8	19.6	21.4	0.9	1.5	1.5
	2. Popliteal Height	15.0	16.2	17.4	0.7	1.5	1.5
	3. Buttock to Popliteal Length	17.1	18.7	20.7	1.1	0.5	1.0
	4. Buttock to Knee Length	21.0	22.6	24.4	1.0	1.0	1.5
	5. Elbow to Wrist Length	8.3	9.2	10.1	0.5	0.5	1.0
	6. Thigh Clearance	4.1	4.9	5.7	0.5	0.5	1.0
	7. Shoulder to Elbow Length	11.2	12.2	13.3	0.6	1.0	1.5
	8. Elbow Rest Height	7.4	9.0	10.6	1.0	0.5	0.5
	9. Mid-Shoulder to Seat Height	21.2	22.8	24.6	1.1	1.0	1.5

Adapted from Israelski, 1977;
Clauser et al., 1972;
Garrett, 1971;
Dreyfuss, 1966;
Woodson and Conover, 1964;
Daniels et al., 1953.

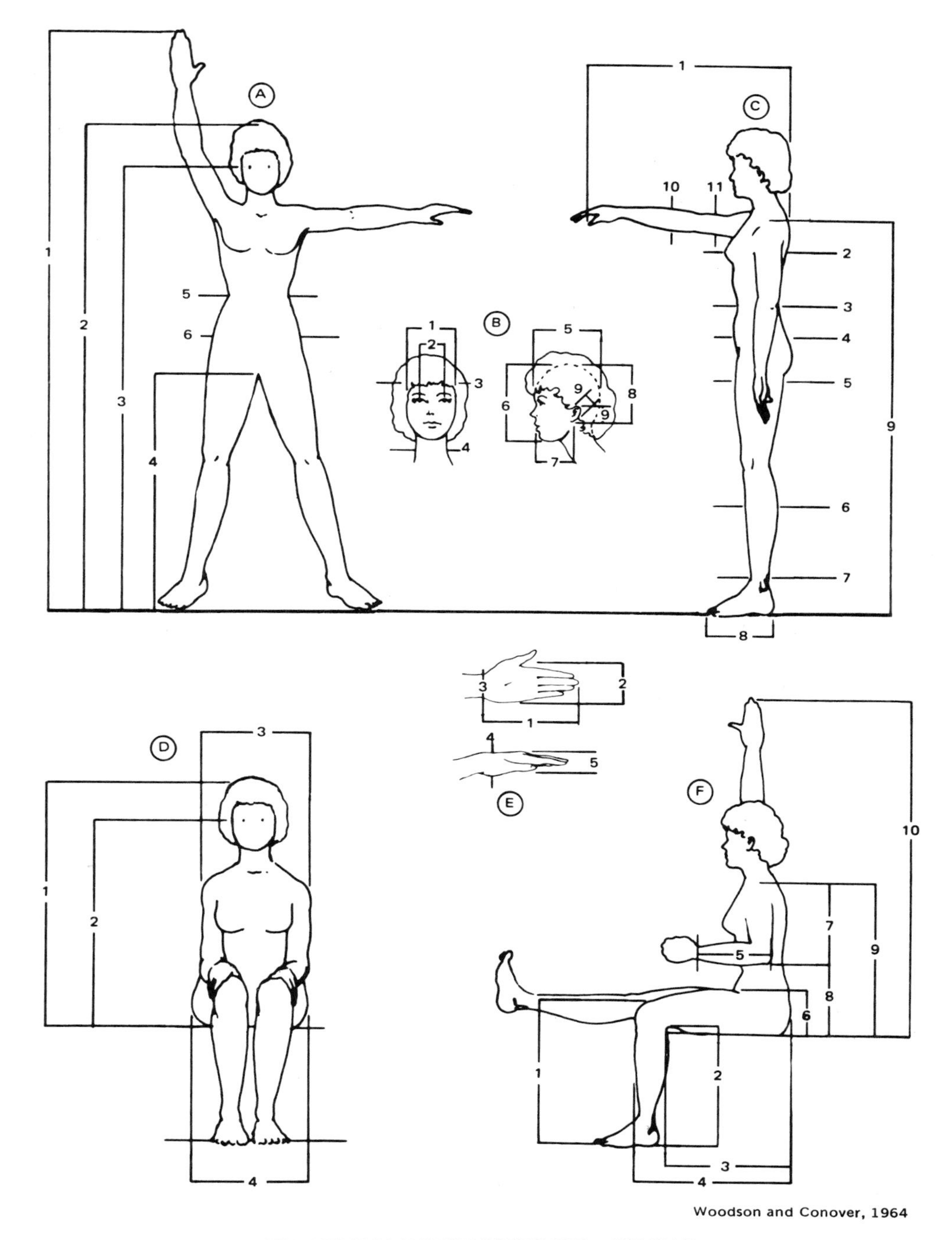

STRUCTURAL BODY DIMENSIONS – FEMALE

Table 5-3. Male Body Dimensions for Common Working Positions.

Figure	Dimension (Inches)	Percentile 95th	Standard Deviation	Added Increments Light Clothing	Added Increments Heavy Clothing
A	Prone Stretched Length	95.8	3.4	1.5	2.5
	Prone Height	16.4	1.3	4.0	4.0
B	Supine Length	73.9	2.4	4.0	4.0
	Supine Fist Reach	32.2	1.5	1.5	2.5
C	Crawling Length	58.2	2.6	4.0	4.0
	Crawling Height	30.5	1.3	4.0	4.0
D	Kneeling Height, Upright	54.4	1.8	4.0	4.0
	Kneeling Leg Length	28.7	1.3	1.5	1.5
E	Squatting Height	47.0	1.9	4.0	4.0
	Squatting Breadth	25.7	2.1	1.0	2.0
F	Bent Torso Height	55.9	2.8	4.0	4.0
	Bent Torso Breadth	19.1	0.9	1.0	3.0
G	Arm Span	76	— —	— —	2.0

Adapted from Israelski, 1977;
Diffrient, Tilly and Bardayjy, 1974;
Dreyfuss, 1967;
Hertzberg, Emanuel and Alexander, 1956.

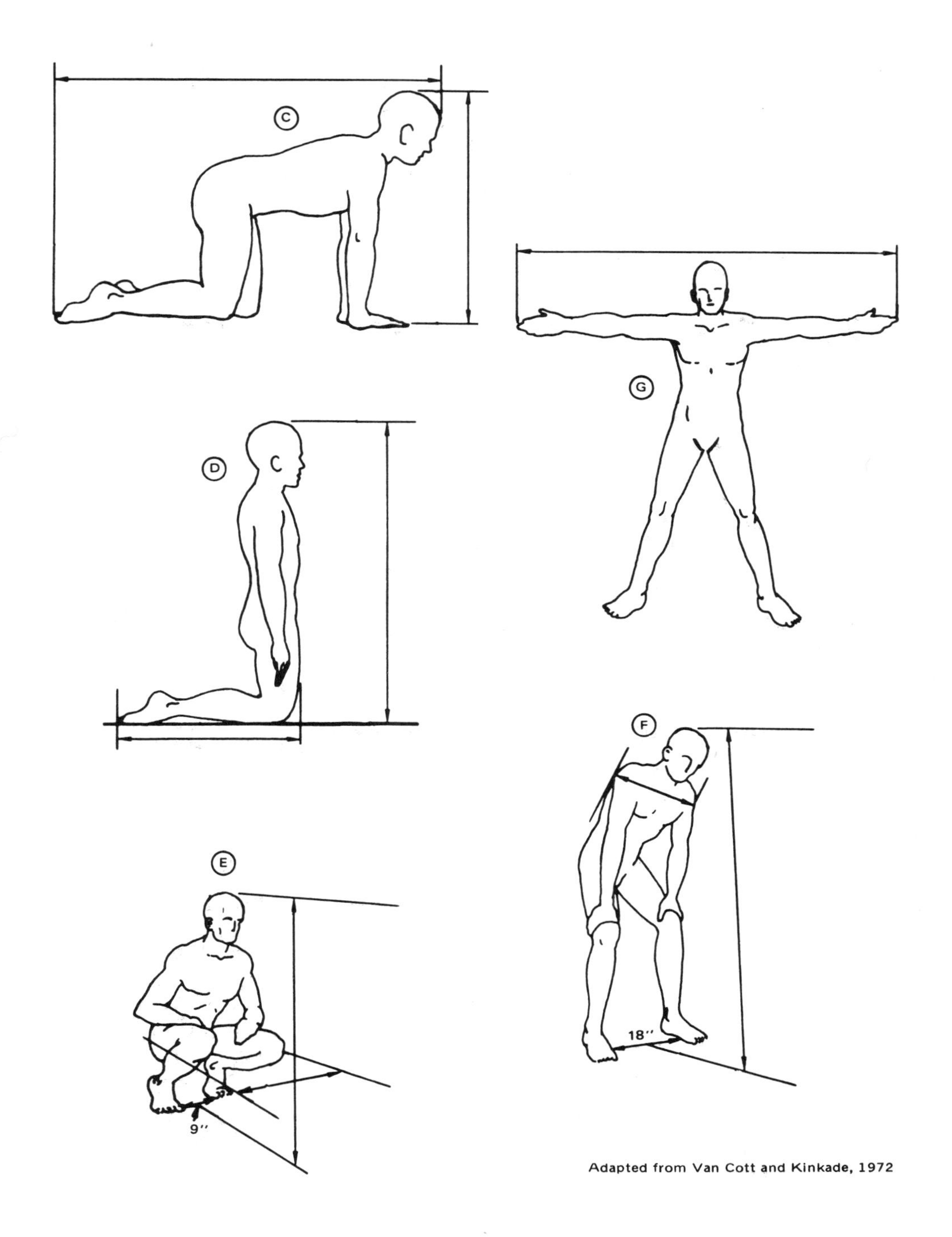

Adapted from Van Cott and Kinkade, 1972

FUNCTIONAL BODY DIMENSIONS (CONT)

Table 5-4. Functional Reach Dimensions for 5th Percentile

Reach contours in this table are representative of 5th percentile male workers with light clothing using the right arm. Correction factors are 0.90 for females and 0.80 for heavy clothing, i.e., multiply the grasping reach by these correction factors (Israelski, 1977).

	HEIGHT ABOVE SEAT REFERENCE POINT (INCHES)							
Angle to Left or Right	10	15	20	25	30	35	40	45
L 165								10.50
L 150								8.75
L 135								7.75
L 120						10.75	11.25	7.50
L 105						12.25	11.75	7.25
L 90						13.75	12.25	7.25
L 75						15.00	12.50	7.50
L 60			17.50	18.25	17.25	16.00	13.25	7.75
L 45		19.00	19.50	20.00	19.00	17.25	14.00	8.50
L 30		21.75	21.50	22.50	21.50	19.25	15.50	9.50
L 15		23.25	23.50	24.00	23.75	21.00	17.00	11.00
0		24.75	25.50	26.25	25.50	22.25	19.00	12.75
R 15		26.50	28.00	28.25	27.25	24.75	21.00	15.50
R 30	27.00	28.50	30.00	30.25	29.00	26.75	22.75	17.50
R 45	28.25	30.00	31.00	31.00	30.25	28.25	24.75	19.00
R 60	29.00	31.00	32.00	31.50	31.00	29.00	25.50	20.50
R 75	29.25	31.50	32.25	32.00	31.25	29.50	26.00	20.50
R 90	29.25	31.00	32.25	32.25	31.25	29.75	26.25	21.00
R 105	28.75	30.75	31.75	31.50	31.00	29.75	26.75	21.50
R 120	27.75	29.50	30.50	30.50	30.25	29.00	26.25	21.25
R 135	26.25							20.00

Illustration of Height above Seat Reference Point (use with Table 5-4)

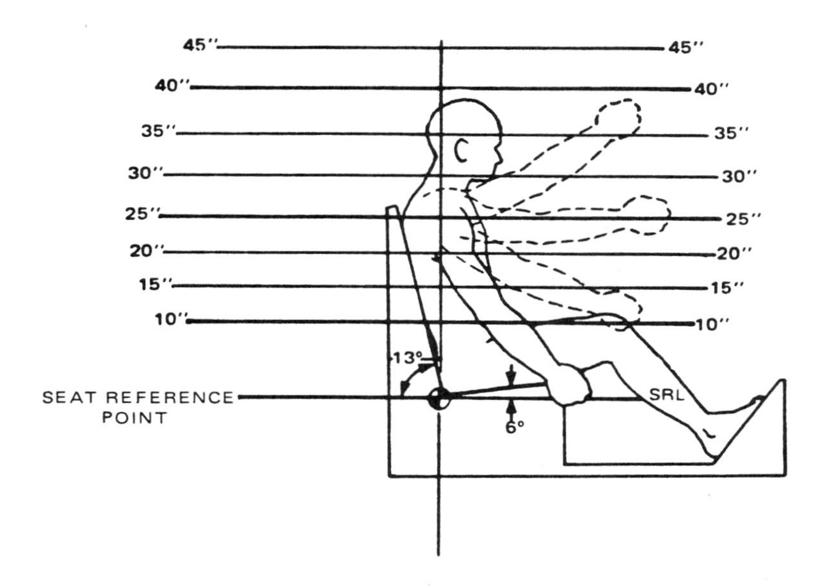

Illustration of Angle to Left or Right (use with Table 5-4)

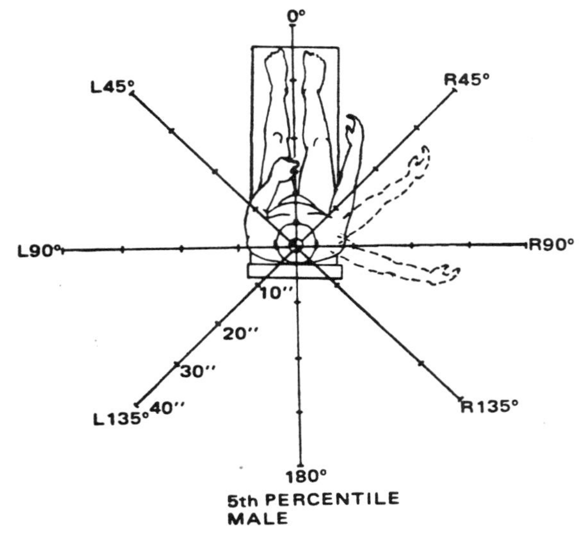

Adapted from Kennedy, 1964

Performance and Body Dimensions

Efficiently performing certain tasks seems to be easier for persons with certain anthropometric combinations (cf. Roberts, 1975). Conversely, Tanner (1964) has suggested that the lack of the proper dimensions can make it impossible for some people to reach an acceptable level of performance. Maas (1974) argues that a set of "ideal" measurements exists for many different activities, particularly those activities, such as sports, that emphasize movement. Indeed, the ancient Greeks believed that a given physique was a prerequisite for success in a given sport. For example in his book *Peri Gymnastikis,* Philostratos Flavius (170-250 A.D.) describes the physiques of winning Olympic participates.

There have been numerous studies conducted in an effort to associate body measurements with human performance. After reviewing the available literature, Carter (1970) concluded that champion performers of a particular sport exhibit similar patterns of body size, with patterns tending to become narrower as levels of performance increase.

Differences in performance levels also have been observed between male and female athletes. For example, in most sporting events the top man will out-perform the top woman. This disparity in performance, however, does not seem to be a result of gender-related *sensory* or *cognitive* abilities. Rather, because their bodies are different—men are usually stronger than women.

As the performance of an individual in a particular activity becomes more cognitive, that is, depends more on decision-making and reasoning and less on movement, the importance of anthropometric measurements diminishes (as does the differences between male and female). For example, one study determined that the speed with which an untrained individual can voluntarily react to a visual stimulus has little relationship to body size (Pierson, 1962). Another investigation measured the speed of voluntary movements of the hand, forearm, and arm, and the length and strength of these segments. No significant correlations between anthropometric measurements and speed of movement were found (Rasch, 1954). For these activities, the speed of performance seems to be more closely related to cognitive processing than the size of body part.

RANGE OF MOVEMENT

Movement is considered to be a composite, an integration of both structure and function. The *structure* of human movement involves the complex interaction of the skeleton and the muscles. The skeleton supports the human body and provides the system of links and hinges that form body levers. Muscles are the source of power.

There are over 200 bones in the skeleton. Most are connected by joints that permit movement. With the exception of speech, the long bones of the legs and arms and the miniature long bones of the fingers and toes most directly affect human performance. The bones of the joints are held together by ligaments and muscles. The ligaments are inelastic, and in general *limit movement* as they tighten at the end of a movement. In joints where movement is restricted, ligaments are found on each side of the joint, remaining taut throughout the entire range of movement. The muscles operating the joint are paired, with one set flexing the joint and the other extending it.

The moveable joints of the body are of several different types. The three most important are: hinged joints (such as in the finger), pivot joints (such as in the elbow), and ball-and-socket joints (such as in the shoulder and hip).

Designers tend to assume that the range of motion for joints is fairly uniform within a user group. In reality, however, ranges of motion vary considerably and are usually determined by the joint's bony configurations, the attached muscles, tendons, and ligaments, and the amount of surrounding tissue. Joint mobility tends to decrease only slightly in *healthy* people between the ages 20 and 60. The incidence of arthritis, however, increases so markedly beyond age 45 that any older population usually reflects considerably decreased average joint mobility (Morgan et al., 1963).

On the average, women exceed men in the range of movement at all joints but the knee. Slender men and women have the widest range of joint movement, while obese individuals have the smallest. Average and muscular body builds have intermediate ranges. Physical exercise can increase the range of motion of a joint, but excessive exercise can result in the so-called "muscle-bound" condition where the range of motion decreases. The range of movement of one part of the body is affected by the position or movement of neighboring parts. In addition, movements made while prone are not necessarily the same as those made while standing. Numerous range-of-movement limitations can be found in Damon, Stoudt and McFarland (1966).

STRENGTH

In some activities the strength of an individual is as important as body dimensions and range of motion (cf. Kroemer, 1970, 1974, 1975). Strength increases rapidly in the teens, more slowly in the early twenties, reaches a maximum by about the middle to late twenties, remains at this level for five to ten years, and thereafter declines slowly but continuously. By about the age of forty muscle strength is approximately 90-95 percent of the maximum attained in the late twenties, this reduces to about 85 percent by age fifty, and about 80 percent by sixty.

However, not all muscle strength declines with age at the same rate. For example, hand grip seems to remain relatively strong in later years. Conversely, the strength of the back muscles decreases more rapidly with age than either the hands or arms. As one would expect, body build is related to strength. In general, the larger the muscle, the stronger a person is. Body position is an important factor affecting strength. When large force must be exerted, people usually assume the position in which they can best exert their maximum strength.

Designers should ensure that the maximum resistance for controls and the limits for lifting or carrying are based on the strength of the *weakest* potential user. In addition, resistance levels for controls should not require the application of maximum strength by any operator. These resistance levels should be low enough to prevent fatigue or discomfort but high enough to prevent inadvertent operation of the controls. They should also provide clear and concise kinesthetic cues.

It is interesting that the apparent maximum strength exerted by most people seems to be far less than their potential. In certain situations, one's capability can be astounding. Several years ago, a 123-pound woman was reported to have lifted a 3600-pound station wagon off her son who was trapped beneath it. More recently, news accounts told of a 12-year-old boy who rescued his father in the same manner (Morehouse, 1977). The enormous strength summoned by such life-and-death emergencies suggests that we normally use only a portion of our actual capacity.

Chaffin (1972) has suggested that the position of the hands relative to the feet is the major job-related determinant of strength. He simulated numerous lifting, pushing, and pulling tasks with the hands located at different positions in front of the body. The curves in Figures 5-3, 5-4, and 5-5 depict lifting, pushing, and pulling capabilities where (a) both hands are equally loaded in front of the body, (b) the object is slowly being moved, (c) the person is free to select the best body position, and (d) there is no accumulated muscle fatigue. One additional consideration must be given to pushing and pulling that is not usually given to lifting: friction between shoes and floor must be great enough to prevent slipping. The curves show the force delivered by both hands. Forces for only one hand were not measured directly but can be assumed to be slightly greater than one-half the two-handed value. All forces delivered are without supporting structures for bracing any part of the body.

Chaffin recommends that jobs performed exclusively by men be designed with regard to the lifting, pushing, and pulling capabilities of men at the 5th percentile. These are shown in Figures 5-3a, 5-4a, and 5-5a. For jobs in which both men and women are working, or just women, the curves in Figures 5-3b, 5-4b, and 5-5b are recommended.

The Figures 5-3, 5-4, and 5-5 are read according to the following examples:

1. The top curve in Figure 5-3 suggests, for example, that a 5th percentile male can lift a weight of 75 pounds when the object is about 40 to 55

inches above his ankles, and about 5 to 13 inches in front of his ankles. The same 5th percentile man can lift an object only weighing 25 pounds if it is the same 40 inches above his ankles, but located about 25 inches in front of his ankles. And this same man can only lift a one pound object that is 40 inches above his ankles, but over 35 inches in front of him.

Chaffin (1972) suggests that in the best job design, objects for lifting should be stationed no lower than 20 inches, as close to the body as possible, and where hands can easily grasp them. The design would also allow a person to carry the load at mid-thigh height and release it where the hands are no lower than 20 inches or higher than 30 inches.

2. The bottom curve in Figure 5-5 suggests that a 5th percentile woman can exert a pull force of about 50 pounds when the object is about 20 to 30 inches above the ankles and within 10 inches in front of the ankles. To obtain maximum pulling force, Chaffin (1972) suggests that most people position their hands about 20 to 30 inches above their ankles, and put one leg behind the other to stop themselves from falling backwards.

a. Lifting Capability for 5th percentile males

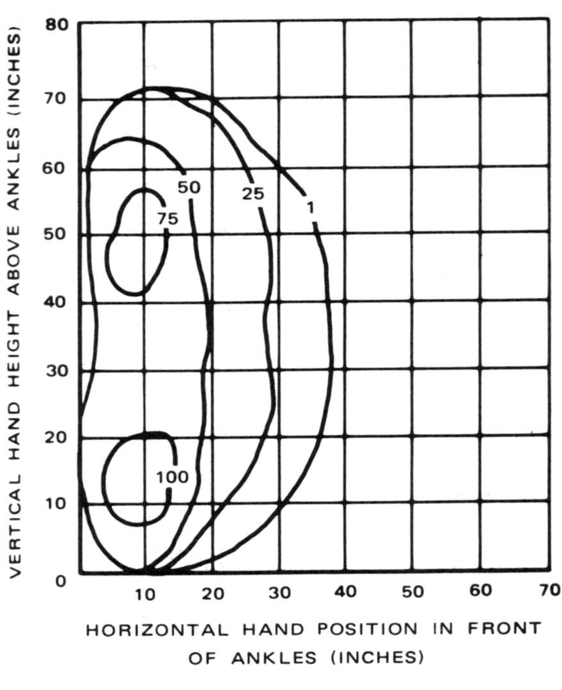

b. Lifting Capability for 5th percentile females

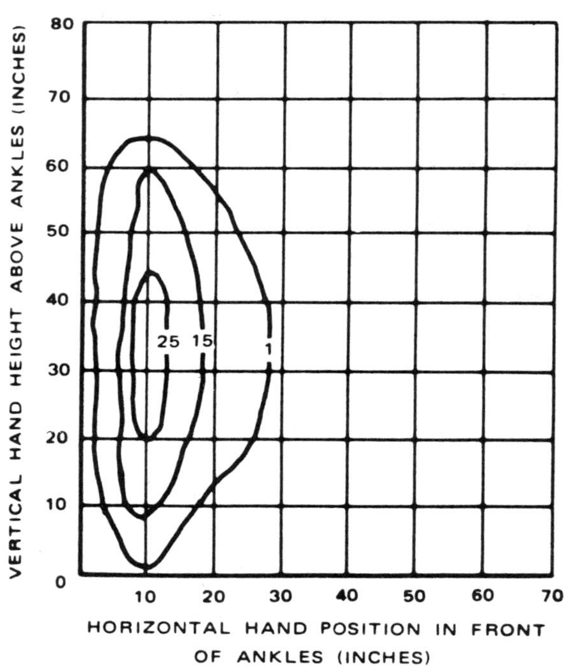

Figure 5-3. Lifting Capabilities - Male and Female

a. Pushing Capability for 5th percentile males

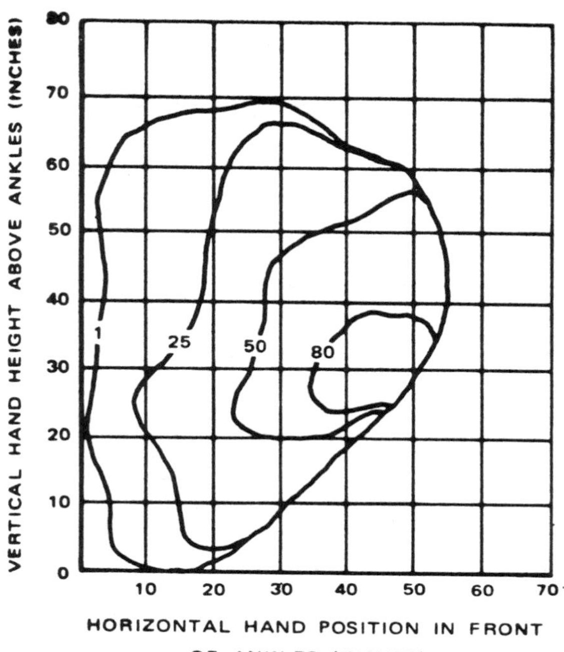

b. Pushing Capability for 5th percentile females

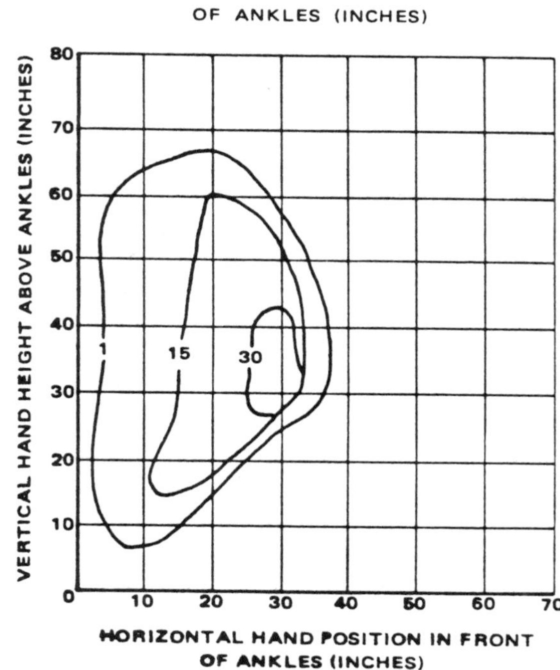

Figure 5-4. Pushing Capabilities - Male and Female

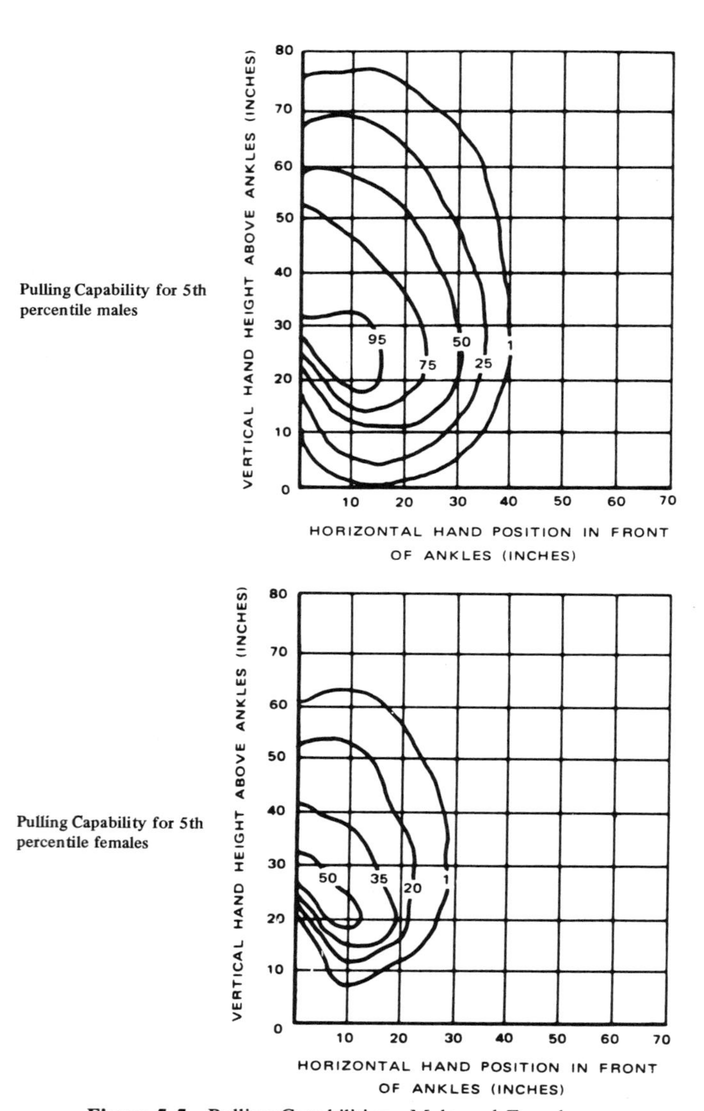

Figure 5-5. Pulling Capabilities - Male and Female

FOR MORE INFORMATION

Damon, A., Stoudt, H. W. and McFarland, R. A., *The Human Body in Equipment Design,* Cambridge, Mass.: Harvard University Press, 1966.

Dreyfuss, M., *The Measure of Man: Human Factors in Design,* Second Edition, New York: Whitney Library of Design, 1967.

Tichauer, E. R., *The Mechanical Basis of Ergonomics,* New York: John Wiley and Sons, 1978.

6

COGNITIVE PROCESSING AND PERFORMANCE

INTRODUCTION

The sense organs present the brain with information about the world. The brain interprets this information and may send appropriate messages to responders (e.g., fingers, legs, or mouth). The processes that take place between sensing and movement, called *cognitive processes* or *cognition,* are important to human performance. Cognition refers to all the processes by which sensory input is transformed, reduced, elaborated, stored, recovered, and used. The cognitive processes may also operate in the absence of external stimulation, as in internally produced reasoning, decision-making, and problem solving. Cognition is involved in all human performance.

Understanding human cognitive processes is similar to discovering how a computer is programmed. If a program stores and reuses information, by what routines or procedures is it done? It does not matter whether the computer stores information on magnetic tape or electronic cores. Designers are usually more interested in the program (psychology), than the hardware (physiology). Computer programs have much in common with human cognition; they are both descriptions of how information is manipulated. The computer program analogy helps us to understand cognitive processing in people. A program is only a flow of symbols, but it has reality enough to control the operation of the tangible machinery that executes physical operations.

STAGE PROCESSING

In the past few years, it has become popular to consider the processing of information by the brain in terms of *stages*. Information is believed to come from the sense organs to a *perceptual* stage. After processing, the information passes to a *translation* stage where it is translated from perception to action. The response is then selected and passes to a *movement control stage*. Each of these stages is thought to have access to a memory store. The important components of cognitive processing are shown in Figure 6-1. The characteristics of human memory will be discussed in Chapter 8.

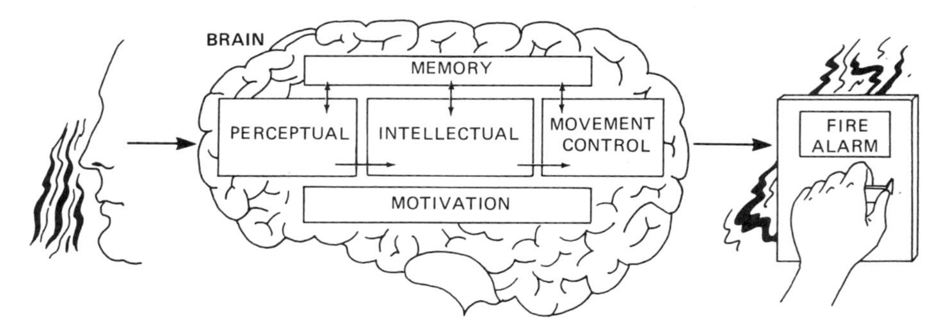

Figure 6-1. Major Components in Human Cognition

Donders' Experiments

F. C. Donders was the first person to study brain functioning in terms of stages of processing. Donders (1869) noted that the prevailing attitude was that the time required for a stimulated nerve to carry its message to the brain, and for the brain to activate the muscles was "infinitely short," and could not be measured. Donders felt that this brain processing time *could* be measured and decided to experiment.

In one experiment, an electrode was placed on each foot of a subject so that Donders could stimulate either foot. There were two conditions. In the first, Donders told the subject that he was going to stimulate the left (or the right) foot. The subject was then asked to make a response as rapidly as possible with the left (or the right) hand, depending on which foot was stimulated. The subject was to detect that a stimulus occurred and give the correct response as fast as possible. In this condition, the subject knew which foot would be stimulated and was prepared to respond with the correct hand.

In the second condition, the subject was told that the stimulus might be given to *either* foot. Again, the subject was instructed to respond with the left hand if the left foot was stimulated, and with the right hand if the right foot was stimulated. Donders would stimulate each foot randomly. In this situation the subject did not know in advance which foot would be stimulated and, therefore, which hand would be correct for the response. Thus, two additional operations in the mental processing were required in the second task. The subject had to first identify which foot was stimulated and then select the appropriate hand for the response. On the average, the second condition took about 67 msec. longer than the first.

Donders reasoned that if all other aspects of the experimental situation had been held constant in the two conditions, then the additional time necessary for completion of the second task could only be explained by the presence of the two additional mental processes of "stimulus recognition" and "response selection." He concluded that 67 msec. was "the time required for deciding which side had been stimulated and for establishing the action of the will on the right or left side." Donders' experiments uncovered evidence suggesting that there were different processing stages in the brain.

Modern Stage Processing Models

Donders' stage model came under much criticism and remained dormant along with his techniques until the 1960s. It was revived by Saul Sternberg at Bell Laboratories (cf. Welford, 1960; Sternberg, 1964, 1969a, 1969b; Smith, 1968; Taylor, 1976). Since that time, considerable attention has been given to identifying and further describing stages in cognitive processing. It is interesting that much of our present knowledge on cognitive stages still comes from experiments that measure reaction time.

Modern analyses of stages continue to decompose the total time required for a response into stages, and then further decompose these stages into even smaller units. These reaction time stage models make several assumptions, some of which have been found to be difficult to prove. For example, it is generally assumed that the stages are executed in a strictly serial manner, and that the stages are independent of each other. Given these assumptions, total reaction time is considered to represent the sum of the component stage times.

Human Information Processing Model

This book will assume that at least three stages of processing take place in the brain—*perception, intellection,* and *movement control* (Welford, 1976). The stage model shown in Figure 6-2 represents some of the cognitive processes of most interest to system designers. This general approach to relating cognition to human performance originated with Broadbent's (1958) model of attention.

The human information processing model is useful for studying and understanding cognitive processing. However, the stage model carries with it numerous difficulties. For example, when the three hypothetical stages are examined closely, it becomes very difficult to distinguish one stage from the next. Also, processing often does not seem to be strictly serial. Nevertheless, this stage model approach is a useful heuristic and will be used in our discussion of cognitive processing. It is used to give designers an insight into the processing that goes on in the brain to aid them in making design decisions that facilitate cognitive processing.

The stage model in Figure 6-2 also suggests that each of the three stages of processing make use of *memory*. Human memory is discussed in Chapter 8. In addition, Figure 6-2 shows *motivation* to be a major consideration related to human performance. Performance motivation is presented in Chapter 9.

The *sensors* shown at the extreme left of Figure 6-2, fall into two groups. The "external" group consists of sensors that receive data from sources such as the eyes, ears, nose, mouth, and skin. The "internal" group can be further divided into two categories. First, there are internal sensors in muscles, tendons, and joints that supply data concerning the control of movement. Second, there are a number of less understood internal sensors that measure the state of the blood, body dehydration, and other bodily conditions.

The *responders*, shown at the extreme right of Figure 6-2, also fall into two groups. One includes the hands, feet, vocal organs, and other voluntary muscles, considerably important in human performance. The other group consists of various reactors in the autonomic system such as the action of glands, heart, and lungs, that have less obvious effects on performance.

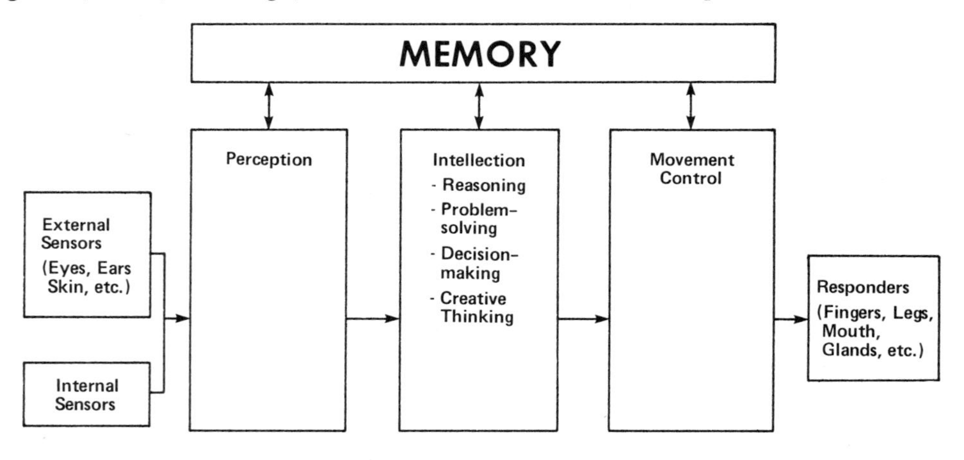

Figure 6-2. Stage Model of Major Human Performance Processes

If you were to pull open your desk drawer and see a snake coiled and ready to strike, it would be one thing to *recognize* it, quite another to *decide* what to do about it, and still another to *carry out* the required action. Thus, the process begins with data from the various sense organs flowing into the perceptual mechanism, then to the intellecting mechanism where a decision is made. Finally, after an appropriate response is determined, the orders are passed on to the movement control mechanism, which carries out and monitors the response to ensure proper execution.

We have considered the above stages in terms of the *time* taken for processing to occur in each, but they can also be seen in terms of *errors* that occur in each stage. An incorrect response—an error—could be made if there were a failure in any one of these three cognitive stages. For example, if the snake were perceived as a piece of old rope the response would be to throw it away. But if this were a poisonous snake, that response would obviously be incorrect. The error could be attributed to *faulty perception*. Assume, however, that the snake was perceived as a snake but the decision was made to reach in with the hand and remove the snake. If the snake were poisonous and did strike as the person was reaching for it, this would also be considered as an incorrect response. This error could be attributed to an *inappropriate decision*. Finally, assuming that the snake was appropriately perceived and that a decision to remove the snake with a stick was made, the execution of this decision may still be carried out, but in a jerky, awkward manner. If the person's movements are not smooth and precise, the snake might still be able to strike. In the latter case the incorrect response could be associated with *faulty movement control*.

Using another example, consider a typist looking at an "X" on a sheet of paper and then hitting the "X" key on a keyboard. The "X" is first perceived, a decision is made on how to translate what is perceived into an appropriate action, and then the orders for striking the "X" key are carried out by the movement control mechanism. Again, it is fairly easy to see how errors could occur at each point in the process.

As noted earlier, different amounts of time could be spent in each one of the three cognitive stages. To use an extreme example, typists can perceive an "X", decide the appropriate response, and then strike an "X" on a keyboard in less than a second. In this case the decision was made within milliseconds because it was well practiced and there were few alternatives. It is also possible that a business executive when asked for a decision on an important policy matter could ponder the question, consider a large number of alternative decisions, and after two or three hours write down a response. In this case the question was perceived by listening, and responded to (hours later) by writing. The interim cognitive processing took a considerable amount of time, measured in hours rather than milliseconds. Even though the situations were different and the time involved was different, it appears that the order of cognitive processing was the same (perception to intellecting to movement control).

PROCESSING STAGES AND SKILLED PERFORMANCE

Ability and Skill

Fleishman (1972) provides a clear distinction between ability and skill. According to his definition, *ability* refers to a basic trait of an individual. Many of these abilities are the product of previous learning. Abilities are what an individual brings to a new situation. These abilities become the basis for learning a new activity. It is assumed that the individual who has many highly developed abilities can more readily become proficient in performing a wide variety of other (similar) activities.

The term *skill* refers to a level of proficiency obtained in a specific activity. When we talk about proficiency in wiring a house, driving a car or keying at a terminal, we are referring to specific skills. Designers are usually interested in taking full advantage of existing abilities for developing specific user skills as quickly as possible.

Skilled Performance

The development of the specialized concept of skill took place during and shortly after World War II, and was stimulated by the study of highly skilled people performing real activities. Bartlett (1943) observed that skilled responses were not a series of unrelated simple movements but *coordinated actions*. The smooth sequence of movements needed to kick a ball or the integration of information required for playing a game of chess constitute skilled responses. Skill refers to the seemingly automatic execution of those cognitive processes that produce rapid and accurate performance. This definition may be applied to all skills, even though different skills may be dominated by one of the three stages (perceptual, intellectual or movement control). Welford (1976) argues that all skills involve all three of the major cognitive mechanisms but that different types of activities emphasize different processes. When we speak of "perceptual" or "intellectual" or "movement" skills, we are classifying in relative rather than in absolute terms.

Perceptual Skill

Welford suggests that *perceptual skill* consists of giving coherence to the sensory data that pour in through the sense organs, and linking that data to material already stored in memory. Individuals may differ in both their basic perceptual ability and the level of perceptual skill they have developed. Some people are able to recognize and readily identify large numbers of different people, rocks, or birds. Others have great difficulty in recognizing even a small number of different objects. The development of perceptual skill seems to require continuous exercising of the perceptual mechanism. Neisser (1976) has observed that perceptual skill differs from movement control skill (e.g., sculpting or playing tennis) in that the perceiver's effects on the world are usually negligible. A person does not change objects by looking at them or events by listening to them. Although perceiving does not change the world, it does change the perceiver (this idea will be discussed in more detail in Chapter 7).

Intellectual Skill

Many skills in the arts and trades probably have less to do with the ability to execute particular responses, such as making a brush stroke or connecting a pipe, than in deciding what colors to use in the painting, and how best to install plumbing in a house. A highly skilled musician transcends the mere playing of an instrument and makes ingenious new interpretations of the music. These are all forms of *intellectual skill* that are analogous to the skills of an administrator, manager, politician, or military officer (Welford, 1976). For all of these individuals the input data may be perfectly clear, and the actions needed to effect their decisions (e.g., writing or speaking) may be straightforward, due to extensive experience in perceiving and in movement control. Intellectual skill lies in the efficient linking of perception to an appropriate action based on reasoning, decision-making, or problem-solving.

Movement Control Skills

Once data have been perceived and an appropriate action chosen, the emphasis shifts to making a response, in many cases a movement. These are called *movement skills* and include such things as riding a bicycle, eating food with a fork or chopsticks, pronouncing words clearly, hitting a golf ball, and typing. As these skills develop the movements become highly coordinated. Well-practiced movements are noted for their *lack* of intellectual control. For example, after the skill of bike riding is acquired, we no longer have to think about moving the handle bars to maintain balance.

Fitts and Posner (1967) observed that the organization or patterning of movement control skill involves both spatial and temporal factors. They suggest:

The simple act of picking up a pencil involves skill in that the movement must be precise in amplitude and the fingers, in order to grasp the pencil, must move in a coordinated way at the right time in the reaching sequence. Similarly, speaking one's name requires the modulation of amplitude and pitch of the voice in a complex temporal pattern. The writing of a name may involve the coordinated activity of as many as twenty different small-muscle groups in the arm and hand. These simple acts of reaching, speaking, and writing become so highly overlearned and automatic in an adult that it is easy to forget the laborious way in which they were originally learned as a child.

We marvel at the execution of the soloist and the timing of the supporting symphony orchestra, or at the control of the quarterback as he throws a pass to the end who is running at full speed down the field. No less remarkable, however, are the linguistic skills that we employ every day in communicating with other people.

Many people feel that movement control skills are relatively simple in comparison to perceptual and intellectual skills. However, Fitts and Posner have rightly pointed out that programming a computer to hit a pitched baseball is as complex as programming it to play chess. In both cases all three cognitive processes are involved, although in the baseball example the emphasis is on the movement control stage and in the chess example the emphasis is on the intellectual stage.

Developing Skills

The acquisition of a skill may not increase basic abilities, but does result in improved efficiency of the cognitive processes. For example, the exertion of force is not a skill in itself, although the controlling and appropriate application of force does require skill. Similarly, visual and auditory reception of stimuli is not skilled until it is organized by the perceptual process. Thus, the reception of stimuli by our senses and the carrying out of actions by muscles are not included in what is being defined here as skill. *Skill is exclusively cognitive*. That is, skill refers to changes that take place in the brain; even though skilled execution relies to a large measure on the condition of the sensors and responders.

Senses do not seem to improve the more they are used (i.e., the constant use of the eyes does not improve their ability to see). The cognitive processes do improve with use. The responder's size, strength, and extent of flexibility are finite and for most adults change very little. Of course, in order to increase strength a person can build up the fibers in a muscle by exercise. However, new fibers are not added, and the building up of existing fibers has limits that are soon attained. In addition, improved strength is not necessarily related to improved skill. A person may become strong without becoming more skilled.

Certain body structures no doubt limit the extent to which some skills can be developed. As we will discuss later in the book, training is often used to help develop skills. No amount of training, however, can bring a severely brain-damaged person to perform activities requiring difficult decision making, creative problem solving, or smooth coordinated movements. In the same sense, it would be impossible for a person who is physically handicapped to be trained to run 100 meters in less than ten seconds.

Skills appear to improve indefinitely as long as they are practiced; however, the rate of improvement slows considerably after a person has had extensive experience. Skills that are developed and not practiced tend to deteriorate.

Improving Performance

We have discussed how skill can be related to the three processing stages. A designer should consider each activity being performed and decide which of the three processing stages dominates. The designer can then concentrate his or her efforts on making decisions that will lead to improved performance.

There are some cautions related to skilled performance that are of interest to designers. First, it is possible to design an activity that is impossible to perform no matter how skilled the individual. Secondly, although it is possible for people to become skilled in many activities, they frequently do not perform in one activity long enough to acquire all the necessary skills. As a result, many users never develop a complete set of even the minimum skills necessary to reach an acceptable level of performance.

By considering the dominant stage in each activity, the designer can make decisions concerning the best way to conduct training to develop the appropriate skills. For example, for movement–control dominated activities, the potential user should be provided considerable practice making the necessary movements. In an intellectual dominated activity, practice should minimize the perceptual and movement–control processes and place emphasis on the appropriate intellectual processing (e.g., practicing making critical decisions, solving problems).

The designer can expect different types of errors from each stage, and the cognitive processing time in each stage can be affected by certain design decisions. For example, barely perceptible stimuli, such as illegible print, can require substantially more perceptual processing time per item. Over a large number of items this "slowness" of processing can mean that more work time and possibly more people will be required to perform a given amount of work.

PROCESSING LEVELS

The designer should also consider that people can process performance-related information in at least two levels—the automatic and the conscious. See Figure 6-3. For example, driving home from work in a car pool with two other people, an experienced driver (i.e., a driver *skilled* in operating an automobile) is able to find the proper roads, and respond appropriately to road conditions, signs and other cars. All of this is done *at the same time* the driver is participating in a discussion with others in the car. In this example, both processing levels are being used. The *automatic* level controls performance related to the skill of driving. At the same time, the *conscious* level controls much of the performance related to listening, thinking, and speaking. Both types of performance can occur at the same time as long as one of the activities is performed automatically.

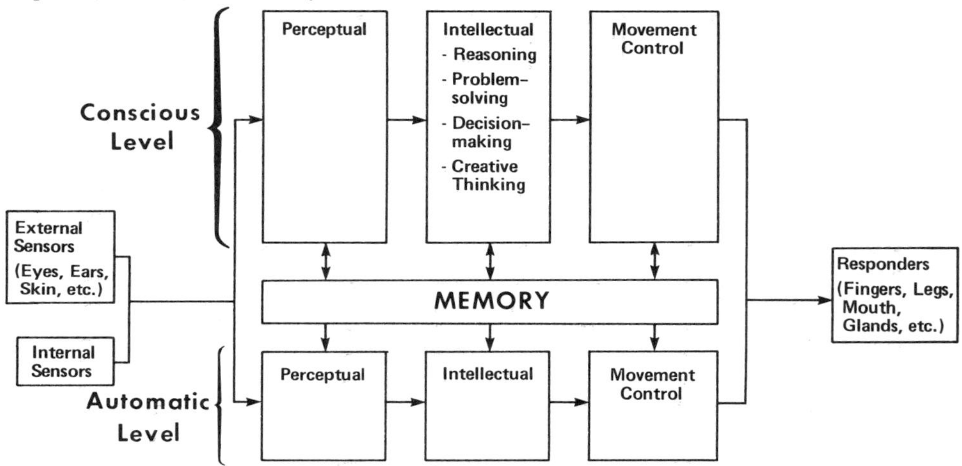

Figure 6-3. Two-Level Model of Human Performance Processes as Related to the Stage Processing Model

Automatic performance is usually considered skilled performance and can only take place after the skill for that particular activity has been developed. Skilled performance is made up of actions that were originally under conscious control, but that have become automatic and relatively inflexible.

Miller, Galanter, and Pribram (1960) reported an interesting comparison of automatic performance in people and machines. Consider a stereo that has a system for playing and changing records. Whenever the appropriate stimulus conditions are present—for example, when the arm is near enough to the spindle or when a particular button is pushed—the routine for changing the record is executed. There are "sensors" that discriminate between different size records, and there are "responders" that push the next record into place and lower the tone arm gently into the groove of the record. The entire performance is obviously automatic. This routine for changing records is built into the machine and requires only the proper stimulus to elicit the performance. For our discussion, the main difference between this machine and people is that people develop their own routines, which we frequently refer to as skills.

Keele (1968) suggests that a shift from conscious to automatic control of performance provides at least three advantages to a user. First, the degree of conscious attention required to perform an activity is reduced. Second, successive stimuli are anticipated so that appropriate movements can be planned in advance to coincide with the stimuli rather than lag behind. And third, it is possible for movements to be made at a faster rate—much faster than when conscious thought processes are required.

Most psychologists agree (cf. Woodworth, 1899; Lashley, 1951; Fitts, 1964; Pew, 1966; Welford, 1976) that skilled performance generally begins with strict conscious control, eventually reaching a level of automatic actions with occasional conscious monitoring, and finally reaching a point where it is almost exclusively automatic.

At the turn of the century, Book (1908) wired a typewriter to record the time of occurrence of successive key-strokes and then collected data while people learned the "touch method" of typing. His study represents an interesting example of the shifting from conscious to automatic control in a real-world activity. People first tried to consciously memorize the locations of different letters on the keyboard. Then they went through several discrete steps: look at the letter in the material that was to be typed; locate this letter on the keyboard; strike the key; and look to see if it was correct. Each character was typed with considerable conscious effort. After a few hours of practice these conscious actions began to fit together into small routines of automatic actions centered around individual letters. At this point we would say that the learner had acquired a *letter skill.*

With practice people began to anticipate the next letter and to develop new automatic routines for dealing with familiar sequences like "the" and "ing". The letter routines were combined into word routines and a *word skill* was developed. Eventually, the experienced typist was able to read the text several words ahead of the letters being typed, thus developing a *phrase skill.* The end result of the skill—development exercise was to have almost totally automatic typing performance with little conscious thought involved.

Conscious processing seems to rely more heavily on peripheral sensory feedback than does automatic processing. As the transition from conscious to automatic control continues, a person becomes less dependent upon peripheral feedback for performance and needs less time to check on the progress of the activity. In some activities the emphasis can be observed to shift from feedback-dependent movements to smooth, feedback free, execution. MacNeilage and MacNeilage (1973) observed that "the need for peripheral sensory feedback can be thought of as inversely proportional to the ability of the central nervous system to determine every essential aspect of the following acts." (p. 424)

As conditions change, or as novel circumstances occur, people must change their automatic performance or errors could occur. Unfortunately, some designers do not recognize that changes in people do not come about as readily as changes in software or in hardware. People have a tendency to deal with new situations in "old" automatic ways. For example, some radios in new automobiles have the tuning knob on the left and the volume control on the right (with one labeled TUNE and the other labeled TONE). This layout is just the opposite of how radios have been designed for well over 30 years. It is interesting to observe users as they "stop and think" before adjusting the volume or changing stations. Quick automatic motions without thought frequently result in turning the wrong knob, even after several months of experience with the new radio.

Conscious Awareness

When people attain a high level of skill, they can perform one or more actions without *conscious awareness*. In fact the concept of performing *automatically* suggests that the performance is taking place without conscious control, and in many cases without conscious awareness (Welford, 1976). Highly skilled people that are put in situations where considerable demands are placed on the efficient execution of the skill, experience moments of such intense involvement that there is no awareness of the passage of time. Only after the difficult situation has passed is there awareness of what has taken place—and even then the details may be fuzzy (Dreyfus and Dreyfus, 1979).

Conscious awareness is lost and automatic performance is gained when it is possible to dispense with monitoring. This is likely when the predictability of a situation is such that actions are always, or virtually always, accurate. In such cases, the performance does not require much checking for possible modification. It is just this kind of accuracy that seems likely to be attained and maintained by considerable practice on the same activity, especially the intensive practice that is an inevitable feature of highly repetitive tasks such as speaking, typing, keypunching, or driving an automobile. In addition, when performing automatically, with little or no monitoring, a person is free to consciously carry out actions related to other activities.

Performance that takes place automatically is not usually remembered. Automatic performance seems to make use of memory (obviously, or the driver would not know where to drive, which roads to use, or what the signs mean), but does not appear to feed new information into memory. At least not in a form that can be readily retrieved.

This may be why when a person drives home after work there are few, if any, memories associated with the trip itself. The driver may remember a conversation that took place, and a child that ran into the road (as well as the sick feeling associated with the "near miss"), but may have difficulty remembering other cars, taking the usual curves in the road, the landscape, or even the time spent. In one study, when asked to think back and estimate the time it took to perform a work activity, people had a tendency to believe that performance of a well-learned activity took considerably longer (up to 40 percent) than it really did (Butler, Domangue and Felfoldy, 1979).

Errors

The two-level model of human performance can be of help in making design decisions that will reduce human errors in a system. We have already discussed that different types of errors could be made in each of the three processing stages. In addition, different types of errors are characteristic of conscious and automatic performance.

People are frequently the *victims* of unwanted releases of automatic performance, just as animals are for their instinctive behaviors. James (1914) observed, "Who is there that has never wound up his watch on taking off his waistcoat in the daytime, or taken his latch-key out on arriving at the door-step of a friend?" Norman (1980) has suggested that these kinds of errors involve the principle: pass too near a well-formed habit and it will capture your behavior. In other words, if a habit is strong enough, even cues that only partially match the situation can activate the incorrect program that leads to incorrect performance and thus errors. Norman gives an example of automatic performance as being triggered and carried out without conscious awareness of why the performance was even needed. A colleague reported to him that before starting work at his desk at home he headed for his bedroom. After getting there he realized that he had forgotten why he had gone there. "I kept going," he reported, "hoping that something in the bedroom would remind me." Nothing did. He finally went to his desk, realized that his glasses were dirty, and returned to the bedroom for the handkerchief he needed to clean them.

A designer should seek to have few errors made as users develop skill on an activity. Kay (1951) and Von Wright (1957) have shown that errors made in the first few trials of performing a new activity tend to become ingrained. In fact, Kay has argued that one of the major problems in achieving an acceptable level of performance lies in getting rid of, or in other words *unlearning,* these early errors. It seems that for those activities that eventually will be automatic, a designer would do well to seek for *errorless learning* (cf. Terrace, 1963).

Speed and Errors

Pew (1969) suggested that there is a "robust" *speed-accuracy tradeoff* that is very much apparent in most choice reaction-time studies. Swensson (1972), on the other hand, found the speed-accuracy tradeoff to be very elusive. It is probably fair to say that the precise relationship of speed and accuracy is still not well understood. Consider, for example, two studies where subjects were given instruction to perform either (a) as fast, or (b) as accurately, or (c) as fast and accurately as possible (Howell and Kreidler, 1963; Fitts, 1966). A comparison of the accuracy levels for the three conditions in each study is shown below:

	Instructions		
Study	*Speed-Accuracy*	*Accuracy*	*Speed*
Howell and Kreidler	95.9%	97.8%	87.3%
Fitts	86.4%	89.7%	77.4%

Both studies found little difference between any of the groups with respect to *speed* of responding. The *speed instruction* resulted in no significant increase in speed, but a very large reduction in accuracy.

On the other hand Bahk (1980), using a tracking task, found that speed of performance under an *accuracy* instruction was much slower than under a *speed* instruction. The number of errors was the same with both instructions.

There is a common misconception that as people gain experience on an activity they always tend to make fewer errors. Designers should note that in activities where performance is primarily automatic, the proportion of errors will remain fairly constant, but the *speed* with which the activity is performed will increase with practice. Thus, experience gained on automatic activities affects speed of performance more than accuracy of performance. For example, with handprinting, people initially learn to write faster and make fewer errors. Once the basic skill is developed, the speed continues to increase while the proportion of errors remains about the same (and may even increase). When learning to type, as typists gain several months and eventually years of experience, they type more words-per-minute with the same proportion of errors to key strokes. People reach an acceptable error rate relatively soon after beginning to perform many activities, but continue to increase their speed of performance for a longer period of time. In fact, Crossman (1959) has shown that the time taken by workers to make a certain product was still decreasing after seven years.

A study by Howell and Kreidler (1963) illustrates this phenomenon. They had subjects perform a choice-reaction time activity over a period of four blocks, with each block containing five trials, for a total of 20 trials. As subjects gained experience their response speed significantly improved. But, as far as accuracy was concerned, they reported "no significant trend of improvement ... over the complete 20 trials" (p. 42). Accuracy remained at a fairly constant level throughout the experiment, and nearly all improvement in performance came about as a result of increased speed.

A second example comes from a field study (Bailey and Koch, 1976). In this study, the performance of several newly trained clerks was monitored daily for four consecutive weeks. For comparative purposes, the performance of experienced workers was also monitored. It was observed that the average time new clerks spent per customer became shorter as they performed over the four week period. That is, they worked *faster* with practice. Although the new clerks were not able to process customers as quickly as experienced clerks at the end of four weeks, the trend was certainly in that direction. But, for the same period of time, the error rate after only one week's experience for the new clerks was already at the same level as experienced clerks, and it remained at that level for the remainder of the study.

It is interesting to note that the majority of errors occurring during automatic processing in systems are likely contributed by the *most experienced* people. For example, as key operators gain experience with a computer system they learn to perform faster, but their *rate* of making errors remains fairly stable. Because they process more data, they have more opportunities to make errors; and because they continue to make about the same proportion of errors, they may actually have a greater number of errors.

The number of self-detected errors increases slightly as people become more experienced (cf., Conrad and Longman, 1965; West, 1967; Schaffer and Hardwick, 1969). The increased capability for self-detection does not appear to keep pace with faster performance. Even with more errors being self-detected, the absolute number of undetected errors also continues to increase as people gain experience (i.e., work faster). Thus, the term "experience" often reflects the ability of a person to do things faster. People seem to reach an accuracy level as soon as the basic skill begins to develop, and for the most part stay with that accuracy level over a long period of time.

Speed

In most well-learned activities, the fastest performance comes from performing totally automatically. Disruptions to this automatic processing will slow down the overall processing time. A designer should guard against unwanted interruptions in order to have activities that are under automatic control performed in the shortest time possible. This may mean that a set of activities be designed so that the automatic processing can run its course, with consciously controlled processing taking place before or after the activity. For example, a form that requires filling-in information familiar to the user, except for 2 or 3 unfamiliar items, should have the unfamiliar items grouped at the end of the form. In this way the automatic performance is allowed to run with a minimum of interference.

The fastest performance is only possible if conscious thought can be avoided. Consider an example about tennis by Gallwey (1974):

> I asked students to stand at net in the volley position, and then set the machine to shoot balls at three-quarter speed .... At first the balls seem too fast for them, but soon their responses quicken. Gradually I turn the machine to faster and faster speeds, and the volleyers become more concentrated. When they are responding quickly enough to hit the top-speed balls and believe they are at the peak of their concentration, I move the machine fifteen feet closer than before. At this point students will often lose some concentration as a degree of fear intrudes .... Soon they are again able to meet the ball in front of them with the center of their rackets. There is no smile of self-satisfaction, merely total absorption in each moment. Afterward some players say that the ball seemed to slow down; others remark how weird it is to hit balls when you ***don't have time to think about it.***(adapted from *The Inner Game of Tennis*, by W.Timothy Gallwey. Random House, Inc. Copyright © 1974.)

Morehouse (1977) reports on his successes as a psychologist in achieving super performance from world class athletes. One of his major points is to "*think* when you need to, and *do* when you need to, but make it a rule to keep thinking and doing separate" (p. 65). A person cannot effectively plan and act at the same time. Planning requires analysis. Consciously analyzing performance while performing detracts from quickly accomplishing the activity. Movements are slowed.

PROCESSING LEVELS AND SKILL DEVELOPMENT

Initially, almost all performance may require considerable conscious thought. When learning to drive, practically every move is consciously considered before it is initiated. For many activities, one of the main objectives of a designer is to realize a shift of as many tasks as possible from conscious to automatic control in the shortest time possible.

In relatively simple tasks a user is not told what to do, but is shown and then guided through the proper actions until the skill begins to be developed. In these cases, knowledge is gained as the skill develops, not before. But for complex activities a considerable amount of knowledge must be accumulated *before* the skill-building exercise begins. A designer should determine the type of activity to be performed, consider all reasonable alternatives for building the skill in the shortest time, and then decide the best way to proceed. For each task or series of tasks, the "best way" may be different. In some cases the knowledge needed can be obtained by selecting people who already have it. In other cases the needed knowledge is so minimal that a person can begin developing skills with some brief verbal instructions.

FOR MORE INFORMATION

Miller, C. A., Galanter, E. and Pribram, K. H., *Plans and the Structure of Behavior,* New York: Henry Holt and Company, 1960.

Singer R. N. (Ed.), *The Psychomotor Domain: Movement Behavior,* Philadelphia: Lea and Febiger, 1972.

Welford, A. T., *Skilled Performance: Perceptual and Motor Skills,* Glenview, Illinois: Scott, Foresman and Company, 1976.

7

PERCEPTION, PROBLEM SOLVING AND DECISION MAKING

PERCEPTION AND PERCEPTUAL SKILLS

The word "perceive" is often used as a synonym for "see;" however, perception may involve any sense, including hearing, tasting, smelling, or feeling. Another person's perception cannot be observed directly, and is usually inferred from observations of performance.

The Perceptual Process

Perception involves the interaction of two sources of information available to the perceiver. The first is the information available through our senses, and the second is the accumulated knowledge of the perceiver stored in memory. The whole process of perception hinges on being able to relate new experiences with old experiences in some meaningful way. Perceptual skill, in fact, may be defined as developing ways of quickly and efficiently combining new experiences with old.

Perceptual Flexibility

The perceptual process as a whole seems quite flexible. For example, Kohler (1962, 1964) reported a series of experiments where subjects wore special prism goggles that reversed the image on the eye, transforming the entire visual world into a mirror image. The subjects wore these prisms for several days or weeks. At the beginning, the subjects functioned clumsily, and would, for example, see someone apparently on their left, move to the right, and bump squarely into the person. After a while, however, the subjects adapted to this new way of looking at the world and were able to function quite well. One subject learned to ride a bicycle with ease while wearing the prisms. Then, when the prisms were removed, subjects had to go through an adaptation period to readjust their perceptions.

In another study, people were asked to reach for a target while looking through similar prism goggles that displaced their visual world several inches to the left or right. After watching their reaching hand for several minutes through the prisms, they were soon able to recorrelate vision with touch, and reestablished normal performance despite the optical distortion. When the prism goggles were removed, the people would reach for the same targets and miss in the opposite direction. Recent research has shown in these types of situations that, contrary to all appearances, *visual perception* does not change at all. It is *kinesthetic perception* that is affected (Harris 1980a). Instead of correcting the distorted visual input, people automatically adjust their kinesthetic perceptions to match the visual input. It is the persistence of the altered kinesthetic perception that requires the readjustment period once the prism goggles are removed. This readjustment period can last for quite a long time as shown by another study (Harris and Harris, in press; see Harris, 1980a). People spent fifteen minutes a day for four days drawing pictures and doodling while watching their hand through mirror-reversing prism goggles. Afterwards, when they tried to write letters and numbers correctly while blindfolded, they often wrote them backwards without being aware of their errors. It is interesting to note that only the hand that is observed through the prisms makes the kinesthetic adjustment, the other hand is unaffected by the visual distortions.

These and other similar findings have practical application in interactive computing. For example, if a visual movement is linked to a hand movement in another location (as when one uses a Rand Tablet or "mouse" to control a CRT), the arm's kinesthetic sense may become distorted. The distortion could happen without the user's awareness, and could induce errors in movements made without visual guidance, such as reaching for familiar control buttons without looking.

Designers should proceed cautiously when developing systems that use spatially separated controls and displays. To help avoid these kinds of perceptual difficulties, designers should develop systems with controls and displays coincident; for example, a light pen in computer graphics, or perhaps a device like the Knowlton keyboard (Knowlton, 1977), which works with either graphics or alphanumeric input. The Knowlton keyboard (Figure 7-1) makes use of computer-generated labels that are optically superimposed on an array of keys on a standard keyboard. This arrangement permits great flexibility of labeling and other human performance advantages. But in addition, it helps to ensure that visual feedback is seen in the same location as the finger that produces it (Harris, 1980b).

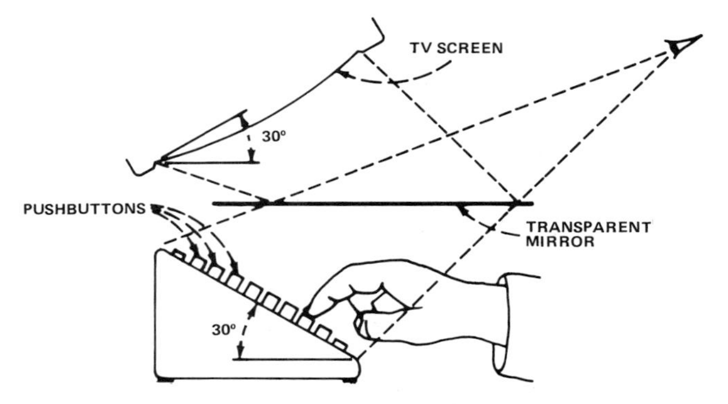

Figure 7-1. A Knowlton Keyboard with Computer-Generated Labels Optically Superimposed on an Array of Buttons. This arrangement not only permits great flexibility of labeling, but also prevents kinesthetic illusions by ensuring that visual feedback is seen in the same location as the finger that produces it.

Matching Patterns

One of the most remarkable features of the perceptual processes is the capacity to respond to a wide range of differing patterns. People usually make accurate responses irrespective of the size of the stimuli. For example, the size of type on a page has little effect, over a considerable range, on the speed or accuracy with which the page is read. We can view the printed page, the face of a friend, or any object from various distances and yet perceive it as constant in size. The ability to recognize spoken words is even more remarkable. Within limits, a speaker can vary his or her rate of speech or loudness; another voice with very different frequency characteristics can even take over and the listener would continue to understand. In fact, speech is quite intelligible if all frequencies above 1900 Hz are omitted, or if all below 1900 Hz are omitted instead.

The following example illustrates the complexities involved in recognizing patterns.

> In November, 1966, a 25-year-old soldier home on leave accidentally suffered carbon monoxide poisoning from leaking gas fumes.

> Seven months after the accident the soldier was admitted to a hospital for extensive tests. Most of his cognitive abilities, such as language use and memory, appeared normal. Most of his perceptual system was also intact. He could readily identify and name things through their feel, smell, or sound. In addition, his most elementary visual abilities were also preserved. Nevertheless, the soldier's visual perception was severely impaired. He was unable to recognize objects, letters, or people when he saw them. His impairment was so severe that on one occasion he identified his own reflection in a mirror as the face of his doctor! (Glass, Holyoak, and Santa, 1979).

The soldier was unable to recognize shapes or forms. At one point he was given a test in which he had to select which two of several patterns were the same. (See Figure 7-2.)

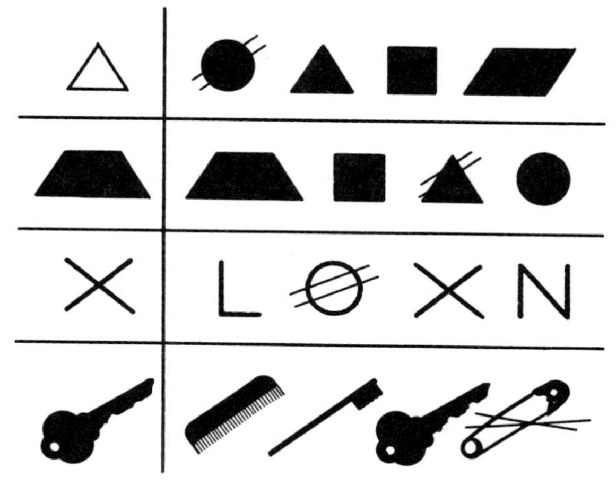

Figure 7-2. A Perceptual Test for Distinguishing Visual Shapes. The person is expected to match the shape on the left with one of the shapes on the right. (from Benson and Greenberg, 1969), *Courtesy Archives of Neurology*

The man was unable to match any of them correctly. The soldier's disorder dramatically illustrates the negative effects associated with losing even a small part of the perceptual process.

Perception is more than just sensing isolated lines or patches of color. The overall arrangement is crucial. For example, the line segment shown at the top of Figure 7-3 is identified more accurately when it is part of a picture that looks like a three-dimensional object (item a) than when it is within a less meaningful context (items b-f) (Weisstein and Harris, 1974).

In proofreading, we sometimes may fail to detect errors (distortions) because the context leads us to expect a certain pattern. This is illustrated in

LINE SEGMENTS

CONTEXTS

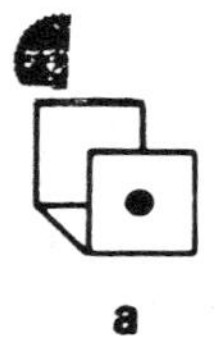 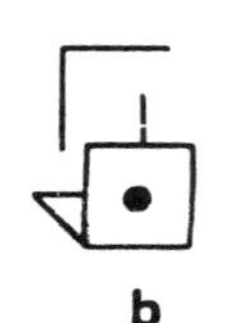 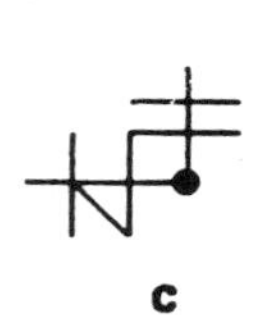 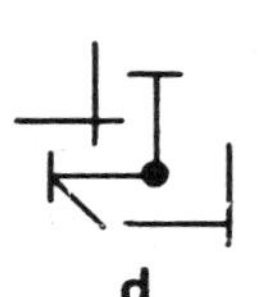 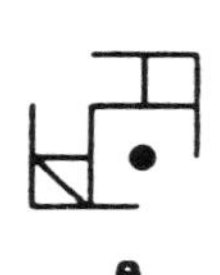 

a b c d e f

Figure 7-3. Visual Detection of Line Segments.

Figure 7-4. Quickly read the statement and determine if it is correct.

Jack and Jill went

went up the

hill to fetch a

a pail of milk

Figure 7-4. What's Wrong with This Sentence?

If someone reads Figure 7-4 quickly and is asked to point out the error, he or she would likely suggest that Jack and Jill fetched a pail of *water*, not milk. This is true. However, the extent of errors in this short passage goes far beyond the substitution of milk for water. The words "went," and "a" are repeated. People frequently overlook this because we learn to ignore certain types of errors in favor of recognizing a consistent and meaningful whole.

Perception can be thought of as having the immediate past and the remote past brought to bear on the present in such a way that the present makes sense. The *skilled* perceiver detects features and structures of which a naive viewer or listener is not even aware because he or she lacks past experience (or knowledge of what is task relevant). A younger child, for example, sometimes ignores information that older children and adults recognize quite effortlessly.

Schema

Neisser (1976) proposed a perceptual cycle, which is illustrated in Figure 7-5. The cycle begins with a *schema* which directs exploration and samples the environment; this in turn modifies the schema, which directs new exploration and so on. The term, used in slightly different ways by Bartlett (1932), Woodworth (1938), Piaget (1952) and Posner (1973), is defined here as the portion of the perceptual cycle that is internal to the perceiver, modifiable by experience, and somehow specific to what is being perceived. The schema accepts information from the senses and is changed by that information. When viewed as an information acceptor, a schema is like a *format* in computer programming language. Formats specify that information must have certain characteristics to be interpreted coherently. Other information could be ignored or lead to meaningless (perhaps confusing) results. A schema also functions like a *plan* (cf. Miller, Galanter and Pribram, 1960) for finding out about objects and events, and for obtaining more information to meet the format requirements.

Perceptual Cycle

We see perception as an active, constructive process; at each moment the perceiver constructs anticipations of certain kinds of information that enable him or her to accept it as it becomes available. The *perceptual explorations* are directed by the schema. Because we perceive best what we know how to look for, it is the schema (together with the information actually available in the world) that determines what will be perceived. Anticipatory schemas then are plans for perceptual action. These anticipations enable a person's eyes to move, head to turn, hands to explore, all to be in a better position to sample the world. The outcome of the explorations, the information picked up, modifies the original schema. Thus modified, it directs further exploration and becomes ready for more information.

The perceptual cycle is probably best illustrated with an example by Neisser (1976):

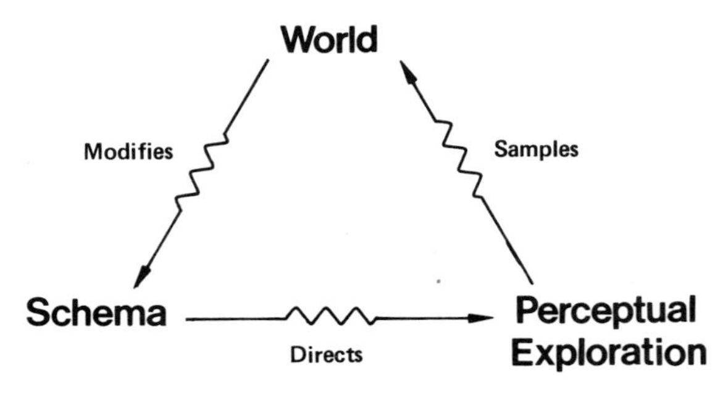

Figure 7-5. The Perceptual Cycle (adapted from *Cognition and Reality: Principles and Implications of Cognitive Psychology*, by Ulric Neisser, W.H. Freeman and Company. Copyright © 1976.)

> It is no evolutionary accident that babies are born with a tendency to look toward the sources of sounds, nor that the outer parts of the retina are sensitive to motion and change although they are poorly endowed for pattern vision. The sound of a footstep, like the first peripheral glimpse of a movement, is an effective guide to further exploration. In its own right, it indicates only that somebody is moving in a certain region of the environment. Nevertheless, it allows the perceiver to anticipate what a glance in that direction might reveal. This "anticipation" is not a deliberate and conscious hypothesis, of course, but a general readiness for information of a particular kind. If the perceiver actually executes the exploratory glance, he embarks on the perceptual cycle; otherwise, he fails to perceive the object. In the latter case the initiating information may still have some effect on him (he may startle or blink or return with renewed determination to what he was doing before) but it will be minor and transient.
>
> To see my visitor properly, then, I must swivel my head and eyes around for a better look. In that better look, the visitor's face will probably be imaged on the central fovea of my eye. But perception is not complete in that moment either; during the next few seconds I will shift my gaze repeatedly as I look at him. Each exploratory eye movement will be made as a consequence of information already picked up, in anticipation of obtaining more. I will not be aware of the fixations or their sequence; only of the visitor himself. Courtesy of W. M. Freeman.

Seeing, listening, and feeling are all perceptual skills that develop over time. *Expectations* appear to direct perception. For example, the old joke that the optimist sees the doughnut while the pessimist sees the hole does not imply that either is mistaken. Each will be confirmed by what he or she has seen.

People can identify a picture or word far more easily when it is anticipated or plausible (i.e., matches expectations) than when it is rare or out of context. People can identify a letter better when it is part of a word than when it occurs in a meaningless string of letters, even when more than one letter is equally plausible at that position in the word (Reicher, 1969; Johnston and McClelland, 1974).

Bruner and Minturn (1955) report a study where people were briefly shown either a letter (e,g., L, M, Y or A), or two numbers (e.g., 16, 17, 10, 12). Afterwards they were briefly shown a broken-B in which the curved part of the figure was separated from the vertical line (see Figure 7-6). The subjects indicated whether the broken-B was a B or a 13. As expected, the subjects showed a tendency to perceive a B when expecting a letter, and to perceive a 13 when expecting numbers.

Figure 7-6. This figure could be perceived as a B or 13, depending on your expectations

The time taken to perceive each of the items in Figure 7-7 is due in large measure to appropriate schema *already* existing in memory. To understand the message in some problems requires the development of new schema. However, once an answer is given or a rule is established and the same type of situation occurs, existing schema can be used and recognition occurs much faster. Most people can eventually sort out, organize, and arrive at a meaningful solution to each of the problems by making the necessary schema changes. By contrast, solving these problems is very difficult for computers.

Combining Sensory Input

We must not believe that people deal with each bit of sensory information that comes to the perceptual mechanism separately. People perceive total situations, not isolated sensory inputs. The perceptual process "puts it all together" in a way that helps people understand what is occurring in the world. For example, we see someone walk and also hear their footsteps; we hear someone talk and also observe movements of the facial features. When eating, we touch, taste, and smell as well as chew. When viewing an argument, we see the gestures and attitudes of the participants and hear their words and their tones of voice.

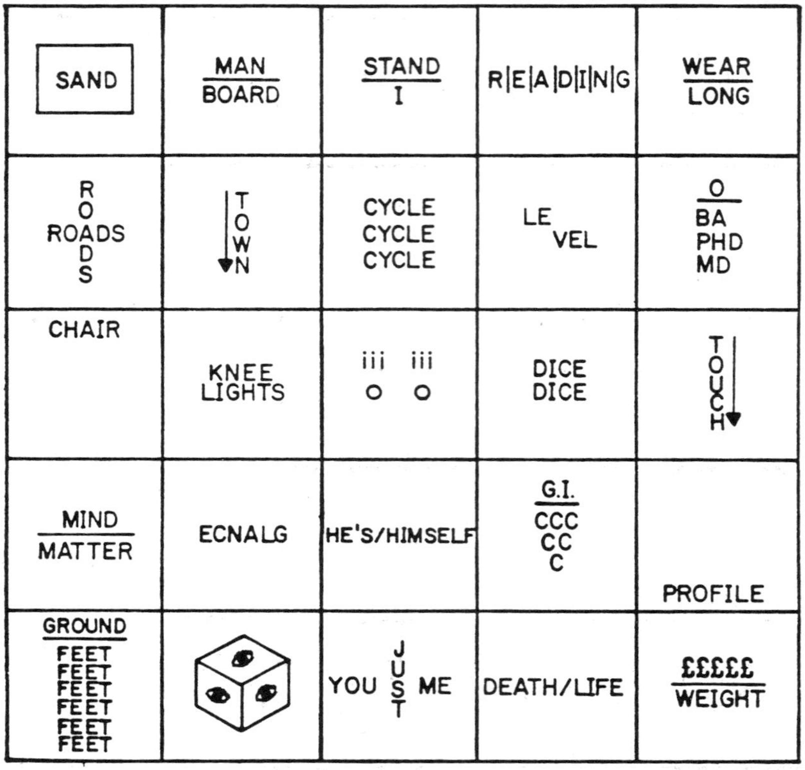

SAND	MAN / BOARD	STAND / I	R|E|A|D|I|N|G	WEAR / LONG
R O ROADS D S	T O W N	CYCLE CYCLE CYCLE	LE VEL	O / BA PHD MD
CHAIR	KNEE LIGHTS	iii iii O O	DICE DICE	T O U C H
MIND / MATTER	ECNALG	HE'S/HIMSELF	G.I. / CCC CC C	PROFILE
GROUND / FEET FEET FEET FEET FEET FEET		YOU J U S T ME	DEATH/LIFE	££££ / WEIGHT

Figure 7-7. What Is the Message in Each Box?

The act of perceiving is a composite of experiences, and it is this composite that is stored and becomes the schema to which new experiences are related.

Developing Perceptual Skills

Gregory (1966) reports the experiences of a man who, blind from the age of ten months, had his sight restored at the age of 52. When he was shown a simple lathe, he could not recognize it or see it clearly although he knew what a lathe was and how it functioned. When he was allowed to touch it, he closed his eyes and ran his hands over the parts of the lathe. Afterwards, when he stood back and observed it, he said, "Now that I've felt it, I can see it." This man had learned a perceptual skill which allowed him to see the world better by touch than by sight. Without knowing what to look for, or how to look for it, the object when first seen could not be perceived. Even the simplest perceptual skills must be learned.

Each time we reread a good book we inevitably gain new insights and new information. Each reading of a good book provides new information not because the words have changed but because the reader has changed. The schema used for the second reading is much different from the schema used in the first reading. Our understanding of the world builds line upon line and precept upon precept.

Assume that we are interested in teaching people to use a CRT display. In the language of cognition, one of the designer's tasks is to create a set of experiences that will enable each user to build schema consistent with the quick and accurate use of the CRT display. One way to approach this problem is to give each person free access to a CRT display, as well as whatever time is necessary to become proficient (this may take some users days, others months). A second approach, and one that is usually more efficient, is to develop a means of assisting the person in developing appropriate schema.

The latter approach is best known as *training* and is discussed in more detail in Chapter 21. Training has the advantage of guiding a potential user through the important conditions of an activity in some controlled manner, building schema as the user has each new experience. Leaving a person to explore a new CRT display in some unstructured trial-and-error manner could be a very inefficient way of constructing new schema.

It also tends to be more efficient to select users who already have a particular set of desirable schema. The schema-related requirements of an activity are usually clearly outlined on a statement of minimum qualifications in the selection criteria. These requirements could include a knowledge of keyboard operation, a general knowledge of the activity to be performed, and perhaps other specific capabilities. A period of formal training will then follow with the express intent of building, in some systematic way, the existing schema related to performance with the new CRT display. Once the basic schema (which will include experience with the CRT displays in a variety of different contexts) have been formed, the person is then considered minimally proficient and is able to proceed with performing the activity. We have used the schema framework for discussing the development of a perceptual skill.

Illusions and Errors

Although people are usually quite good in judging shape, size, and distance, we may be fooled by illusions. Illusions are the result of errors in the perceptual process. Illusions may be produced when we make assumptions about "how things usually are" (i.e., we rely too heavily on inappropriate schema) which causes us to distort or misunderstand the information we actually receive. Many automobile, airplane and other accidents are no doubt contributed to by illusions.

An interesting illusion, closely related to the real-world activity of proofreading, is the one shown in Figure 7-8. Look at the figure and as quickly as possible count the number of "f's."

FINISHED FILES ARE THE RE-
SULT OF YEARS OF SCIENTIF-
IC STUDY COMBINED WITH THE
EXPERIENCE OF MANY YEARS.

Figure 7-8. How Many f's Are in This Sentence?

People usually respond by saying there are three "f's." In actuality there are six "f's." The difference probably occurs because the perceptual process, in an effort to make the statement meaningful, is unable to ignore the sound of the words. It overlooks those "f's" that sound like "v's," the ones in the "of's."

There are certain types of perceptual errors that commonly occur in processing information that may be related to illusions. This type of an error is particularly easy to identify in computer systems where people are dealing with letters and numbers. If, for example, a person looks at a "5" and writes down an "S," or looks at a "2" and writes down a "Z," or hears an "h" and keys an "a," we would say that he or she has been fooled into misreading or mishearing the correct character.

A designer's primary objectives in making decisions that will ensure the adequate operation of the perceptual process should be to ensure that perceptions are accurate, and to provide means for building perceptual skills. The latter objective is obviously related to providing training and other experiences to assist a person to build the necessary schema. Keep in mind that accurate perceptions are most likely when a person encounters data, information, or conditions that are familiar and consistent with past experiences.

Incidently, the answers to the messages in Figure 7-7 are (left-to-right, top-to-bottom): sandbox, man overboard, I understand, reading between the lines, long underwear, crossroads, downtown, tricycle, bi-level, three degrees below zero, highchair, neon lights, circles under your eyes, paradise, touchdown, mind over matter, backward glance, he's beside himself, G.I. overseas, low profile, six feet underground, icebox, just between you and me, life after death, and five pounds overweight.

INTELLECTUAL PROCESSING

Three Mile Island

At 4:00 a.m. on March 28, 1979, a serious accident occurred at the Three Mile Island (TMI) nuclear power plant near Middletown, Pennsylvania. Mechanical malfunctions caused the initial problems and the inadequate response of plant operators made things much worse. In the minutes, hours, and days that followed, events escalated into the first major crisis ever experienced by the United States nuclear power industry (Kemeny, 1979).

The control room at TMI, even under normal conditions, gives the impression that much is going on. The plant's paging system continuously spouts messages, and panel upon panel of red, green, amber, and white lights, and alarms sound or flash warnings many times each hour.

Besides the nuclear reactor itself, probably the most important element in the TMI system is water. Water in the form of steam runs the turbine to produce electricity and water also keeps the fuel rods from becoming overheated. On the morning of March 28, the water pumps stopped working and the flow of water to the steam generators halted. At this point three emergency feedwater pumps automatically started.

There were two operators in the control room when the first alarm sounded followed by a cascade of alarms that numbered 100 within minutes. One of the operators later reported, "I would have liked to have thrown away the alarm panel. It wasn't giving us any useful information." Fourteen seconds into the accident, one of the operators noted the emergency water pumps were running. He did *not* notice two lights that told him a valve was closed on each of the two emergency water lines and that no water could reach the steam generators. One light was covered by a yellow maintenance tag. No one knows why the second light was missed.

At this point a relief valve should have closed to deal with ever increasing steam pressure. It remained open. However, a light on the control room panel indicated the valve had closed. With the relief valve stuck open, the pressure and temperature of the reactor coolant erroneously dropped. About a minute and 45 seconds into the incident, the steam generators boiled dry because their emergency water lines were blocked.

Five minutes and 30 seconds into the accident, steam bubbles began forming in the reactor coolant system, displacing the coolant water in the reactor itself. The operators still thought that there was plenty of water in the system. With more water leaving the system than being added, the core was on its way to being uncovered.

Eight minutes into the accident, the operators finally discovered that no emergency feedwater was reaching the steam generators. One operator scanned the lights on the control panel that indicate whether the emergency water valves are open or closed. He checked a pair of emergency water valves, which are always supposed to be open except during a specific test of the emergency water pumps. The two were closed. He opened them allowing water to rush into the steam generators. The major effect of the loss of emergency water for the first eight minutes was that it *confused and misled the operators as they sought to solve their primary problem.*

Later investigation would show that throughout the first two hours of the accident, the operators failed to recognize the significance of several things that could have warned them that they had an open relief valve and a loss-of-coolant accident. For example, one set of emergency instructions states that a pipe temperature reading of 200° F indicates the relief valve is open. But the operators later testified that the pipe temperature normally registered about this high because the valves leaked slightly. Recorded data during the accident showed that the pipe temperature actually reached 285°.

At 4:11 a.m., an alarm signaled high water in the containment building's sump, a clear indication of a leak or break in the system. The water had come from the open relief valve. At 4:15 a.m., a disc on the drain tank burst as pressure in the tank rose, sending more radioactive water onto the floor and into the sump. At 4:20 a.m., instruments measuring radioactivity inside the core showed a count higher than normal, another indication, unrecognized by the operators, that cooling water was being forced away from the fuel rods. During this time, the temperature and pressure inside the containment building rose rapidly from the heat and steam escaping from the open relief valve.

Shortly after 5:00 a.m., TMI's four reactor coolant pumps began vibrating severely. This was another unrecognized indication that the reactor's water was boiling into steam. The operators feared the violent shaking might damage the water pumps, so they shut down two of the pumps. Twenty-seven minutes later, they turned off the two remaining pumps, stopping the forced flow of cooling water through the core.

At about 6:00 a.m. someone observed that the block valve, a backup valve that could be closed if the relief valve stuck open, had *not* been closed. The block valve was shut at 6:22 a.m. The loss of coolant was finally stopped. *It had taken two hours and 22 minutes to solve the problem.* Even so, the crisis continued for several more days, causing considerable mental stress for people living close by the facility. The direct financial cost of the accident was estimated to be in the range of $2 billion.

Making Design Changes

The fact that TMI operators failed to realize that they were faced with a loss-of-coolant accident indicates a deficiency in their ability to identify the symptoms of such an accident, and consequently in their inability to quickly solve the problem.

The best solution, however, may be to change the *design of the system,* not to improve on their training. It should be fairly clear that expecting someone to efficiently solve a problem such as this under such unfavorable conditions reflects a lack of sensitivity on the part of the system designers. Operators in the control room later described the accident as a combination of events they had never experienced, either in operating the plant or in their training simulations. The performance required of these people was problem-solving performance. They were dealing with *new* information, one of the most difficult of all intellectual activities.

The operators would have performed much better had the designers developed the situation to be one of *decision making* rather than *problem solving.* Expecting people to effectively "problem solve" in the TMI situation is like expecting them to "problem solve" as their airplane dives toward Earth to crash. If at all possible, that is the time for deciding among a limited set of alternatives with all possible help from a computer. Had the activity been designed for decision making, the operators would have had the easier task of choosing among a finite set of preselected, alternative solutions. The way this system was designed, the operators had to use a series of unreliable cues to find one correct solution. No amount of training, better written instructions, or more carefully developed interfaces (particularly displays) could have overcome this basic design flaw.

INTELLECTUAL SKILL

People are continually making decisions, solving problems, and reasoning in one form or another. This includes daily decisions of what to wear as well as momentous decisions as what college to attend. Good decision making, problem solving, and reasoning come with experience, and constitute what we will refer to as the "intellectual skill." Business requires highly skilled decision makers, and research organizations require outstanding problem solvers. Systems should be designed to encourage the development and use of the intellectual skills related to the activities to be performed. Two of the most important intellectual processes are problem solving and decision making. We will first present some information related to problem solving and creative thinking, and then the processes involved in decision making. These two major intellectual processes are closely related to having successful systems. The third process, reasoning, is more difficult to relate to user performance and will not be dealt with in this discussion.

PROBLEM SOLVING AND CREATIVE THINKING

Problem solving and creative thinking are among the highest and most complex forms of human mental processing. Because problem solving often includes creative thinking, we will consider the two processes together as problem solving.

Problem solving is defined as the combining of existing ideas to form a new combination of ideas. Problem solving requires, then, a considerable accumulation of knowledge as a basis for recombining ideas, and forming solutions.

A *problem* is a situation for which the human does not have a ready response (Davis, 1973). Being confronted with a problem makes an individual uncomfortable and, therefore, frequently provides some of the motivation to find a solution. In other situations, users may attempt to solve problems because the problems provide a challenge, an opportunity to carry out a fantasy or resolve curiosity (Malone, 1980). In addition, a person with more experience may have a ready response for a situation that confuses someone with less experience.

We must first become aware of a problem and then proceed to solve it. Some have observed that problem solving usually occurs in more than just these two steps. Wallas (1926) lists the probably best-known set of steps:

1. *Preparation* - clarifying and defining the problem, along with gathering pertinent information.
2. *Incubation* - a period of unconscious mental activity assumed to take place while the individual is doing something else.
3. *Inspiration* - the "aha!" experience which occurs suddenly.
4. *Verification* - the checking of the solution.

Kingsley and Garry (1957) suggested another way of looking at the steps of problem solving:

1. A difficulty is felt
2. The problem is clarified and defined
3. A search for clues is made
4. Further suggestions appear and are tried out
5. A suggested solution is accepted (or the thinker gives up)
6. The solution is tested.

People solving problems, usually, will overlap these steps. When people begin to think of solutions, they frequently backtrack to earlier stages, perhaps finding it necessary to redefine or further clarify the "given," or they may even skip some steps entirely.

Solving problems in the real world frequently follows this sequence (Shulman, Loupe, and Piper, 1968):

1. *Problem sensing* - a person initially detects, to his or her discomfort, that some kind of problem exists.
2. *Problem formulating* - the person subjectively defines a particular problem, and develops his or her own anticipated form of solution.
3. *Searching* - the person questions, gathers information, and, occasionally, backtracks.
4. *Problem resolving* - the person becomes satisfied that he or she has solved the problem, thus removing the discomfort.

Skills associated with problem solving can be developed and improved by learning and practicing methods and techniques for combining ideas.

Barriers to Problem Solving

Two main barriers to effective problem solving are habit and pressure to conform. Habit and conformity are also referred to as rigidity, fixation, mental set, predisposition, resistance to change, and fear of the unknown. Perhaps the most difficult challenge for a designer is to develop ways for people to overcome the rigidity which causes them to use the incorrect, but habitual solution to a problem. One of the best illustrations of the existence of this rigidity or fixation with traditional solutions is the water jar problem (Luchins, 1942).

Shown in Table 7-1 are a number of problems that a subject can solve by using available empty jars to obtain a prescribed volume of water. For example, consider the first problem shown in the table. The subject has a 32-quart jar, a 4-quart jar, and all the water that he or she needs. The problem is, "How can a person measure exactly 20 quarts of water?" How would you solve the problem?

Table 7-1. Water Jar Problems

Problem No.	*Jars Regarded as Given* A	B	C	*Required Amount (quarts)*
1	32	4		20
2	100	20	3	74
3	14	163	23	99
4	18	43	10	5
5	9	42	6	21
6	20	59	4	31
7	23	49	3	20
8	15	39	3	18
9	28	76	3	25
10	18	48	4	22
11	14	36	8	6

The solution for this first problem requires a person to fill the 32-quart jar, then fill the 4-quart jar three times from the large jar: 20 quarts will then remain in the large jar. The second problem, to measure exactly 74 quarts, can be solved in an equally simple way. The 100 quart jar is filled and from it, the 20-quart jar is filled once and the 3-quart jar twice. This will leave 74 quarts in the large jar. All of the remaining problems can be solved using this same general method.

After working some of the problems, most people develop a fixed way of solving them and continue to solve all of the problems in the same way. Only a few people recognize that for a couple of the problems a better solution is available. For example, consider the seventh problem on the list. Again, the problem can be solved by the method just described: the 49-quart jar is filled and from it the 23-quart jar is filled once and the 3-quart twice. However, by studying the problem for a moment it is apparent that there is a simpler solution. One can fill the 23-quart jar and then fill the 3-quart jar from it. This also gives 20 quarts. The existence of this same type of fixation or rigidity hampers much problem solving.

The common expressions made by people who have adopted a set of rigid, long-standing habits and who resist the introduction of new ways to solve problems include: "It has been done the same way for 20 years so it must be good," "We have never used that approach before," "We are not ready for it yet," "Somebody would have suggested it before if it were any good," "I just know it won't work," "You'll never sell that to management," or "You don't understand the problem."

Without concluding that all problem solving is the same, we can recognize some commonalities because of several important and identifiable dimensions of problem solving. The designer should focus on these commonalities when attempting to design a system that will enable an acceptable level of creative problem-solving performance.

Dimensions of Problem Solving

Davis (1973) has suggested three problem-solving dimensions. The first of these attempts to answer the question, "Is the problem really a problem?" Recall the definition of a problem as a situation to which a person does not have a ready response. Simple arithmetic "problems" and simple questions such as "Who flew the first airplane?" are not problems at all to most adults. Remember, some situations that represent true problems when a person first begins working in a system may not be problems after he or she gains experience.

One of the most difficult and most critical features of problem solving lies in defining problems. The designer should develop a system that makes problems as simple as possible. This sometimes requires that a large, single problem be broken into subproblems—each a subject for problem-solving. For example, thinking of ways to improve the design of an automobile may be too complex a problem. Broken into subproblems, users could focus on improving seating, the driver's station, ways of entering and exiting, and engine performance. Whenever a problem is simplified, however, it must be defined broadly enough to allow totally new approaches to appear. For example, if the problem were stated as, "How can we build a better bus?", we'd limit our view to one approach, that of transporting people on a bus. More broadly stated, the real problem is, "How can we best move people from one place to another?", which opens new ways of dealing with the problem.

The second dimension addresses the question, "Does the activity require some type of systematic, organized approach to problem solving, or are problems solved through trial and error searching?" From a designer's point-of-view, the problem-solving approach makes a difference. Trial and error solutions are usually unplanned and random. Problems that are best solved using trial and error methods should be accompanied by design decisions to facilitate that kind of performance. On the other hand, if the problem-solving behavior is to be more systematic and organized, then the activities should be designed to assist *that* approach. Human performance problems may arise, for example, when a designer assumes that a solution will be systematically arrived at, but because of inadequate training the user relies totally on trial and error. Or in situations where a designer has given little thought to the best means of solving certain types of problems, which may force many users into trial and error behavior. A designer should know which general approach is best, and design the system to facilitate that approach.

Consider the anagram problem shown in Figure 7-9 (Smith, 1943). Can you solve it by rearranging all the letters to form one word?

LOVE TO RUIN

Figure 7-9. Anagram Problem

The solution to this problem can be brought about in a variety of ways depending on the design of the system. In the typical case, a person usually rearranges the problem letters mentally in some systematic way to find the solution word. However, if a designer elects to provide a user with pencil and paper, it may encourage the anagram-solving activity to become trial and error. Frequently, a problem that can be solved systematically will be solved using trial and error if a designer supplies the appropriate material and the time to do so. A basic design decision must be made to determine whether the best human performance can be obtained systematically or through trial and error problem solving. In one case, the systematic solution may come more quickly but with more errors. However, if a trial-and-error solution of the problem is attempted, the errors may be reduced but the time to arrive at a solution may be much greater. A design should provide the necessary materials to accommodate the type of problem solving selected. By the way, the solution to Figure 7-9 is "Revolution."

Problem Solving and Short-Term Memory

Larkin, McDermott, Simon and Simon (1980) observed that limited short-term memory capacity constitutes one of the most severe constraints on problem solving. Their suggestion is to always consider providing paper and pencil or some other means to extend the problem solver's short-term memory. Of course, the "cost" associated with writing down information is that it will probably take more time to solve the problem. Even so, writing notes is usually much faster than taking the necessary time to memorize (i.e., put in long-term memory) all relevant information. A designer should decide whether the short-term memory capacity of a user is adequate for the kinds of problems to be solved, or if a means for assisting memory (e.g., paper and pencil) should be provided. Consider for example, the time required to solve the problem in Figure 7-10 with and without writing. The problem solving strategy tends to be different when paper and pencil are used.

What do these two phrases have in common?

Name no one man

Madam, I'm Adam

Figure 7-10. Palindrome Problem (Adapted from Smith, 1943).

Problem-Solving Strategies

Systematic problem solving usually requires much training and a considerable amount of experience. In some cases, a system designer may be willing to accept trial and error problem-solving performance initially, but as the user becomes more experienced, the designer may accept only systematic problem-solving performance. In this case, the design of the system should reflect the original trial-and-error performance and then make provision for a smooth transition to systematic problem-solving performance as the necessary skill is developed. Conditions, materials, and facilities that enhance trial and error performance will not necessarily enhance systematic problem-solving performance.

Even when solving relatively simple problems, different systematic means or strategies may be used to arrive at solutions (Gilmartin, Newell and Simon, 1975). For example, in the Tower of Hanoi Puzzle, the solver must move a pyramid of disks from one to another of three pegs under the conditions that only the top disk on any peg may be moved and that a larger disk may never be placed on a smaller one (as shown in Figure 7-11). At least four different strategies can be used to solve this problem, each of which makes different demands on the human and thus would require different design decisions to assist the user. It can be solved by a *recursive* strategy (i.e., to move k-1 disk from B to C, move k - 1 from B to A, then from C to A, then k - 2 from B to C, etc.); a *perceptual* strategy (i.e., if k is the largest disk not on the target peg, and is the largest disk that blocks the movement of k then the goal is to move it out of the way); a *pattern-following* strategy (i.e., move the disks in the order: 1 2 1 3 1 2 1 4 1 2 1 3 1 2 1 ...); and, finally, a *rote* strategy where the exact sequence of moves is memorized. It is interesting that the recursive strategy requires several unrealized goals to be held in memory, while the others do not, and that the perceptual strategy requires the solver to notice the "largest blocking disk." The others also have characteristics that are peculiar to the problem-solving strategy selected.

Number of Solutions

The third question is, "Does the activity require one correct solution or many?" A distinction exists between a simple problem-solving task that requires only a

Figure 7-11. Tower of Hanoi Puzzle

single correct solution, and more complex activities that require many original ideas or solutions. A person working in this system may be required to come up with a single solution in some circumstances, and multiple solutions in others. A designer must accommodate both situations. For example, a user (telephone engineer) attempting to determine the best layout of telephone equipment and facilities (e.g., lines, cables, poles) for a new subdivision will think of a number of possibilities, but will ultimately select and implement only the solution that best meets the telephone requirements of the new subdivision. The system designer should provide all that is necessary to increase the probability that the user (the telephone engineer) will arrive at the best solution. Designers must determine ahead of time the best means for solving different types of problems and not leave this decision to users—the system should facilitate user decisions. People can be trained to apply different problem-solving strategies. But this is only possible if the best strategy is determined in advance and made part of the design of the system.

Problem-Solving Techniques

Brainstorming

One or more of the following techniques can improve users problem-solving skills. *Brainstorming* is probably the best known form of "forced" creative problem solving. The designer can provide for brainstorming as part of system performance. Brainstorming can be done by an individual writing out a large number of possible solutions or a small group of people contributing several ideas for a solution.

If brainstorming is to be used, the following rules should be kept in mind (Davis 1973).

First, criticism is ruled out, and adverse judgments of ideas must be withheld until later. The goal of a user is to produce a large number of ideas or possible solutions. Designers can provide systems that encourage individuals to brainstorm by requiring that they generate a long list of ideas as possible solutions to a system-related problem. Second, originality is desired. In fact, the wilder the possible solution, the better. Reducing or eliminating ideas is easier than coming up with good ones in the first place. Creative, even wild, ideas can become imaginative problem solutions. Of course, many of the really impractical solutions will not be used, but a small percentage could justify the exploration. Third, the greater the number of ideas, the better. Ideas seem to become progressively more original when more and more are listed.

Finally, a designer should provide for a situation where users can combine and improve proposed solutions. In addition to improving ideas of their own, users in small groups can suggest how ideas of others can be turned into even better ideas, or how two or more ideas can be joined into still a better solution. If designers provide systems that encourage originating and combining previously unrelated ideas, the chances of having creative problem solutions seem to improve.

Davis (1973) offered other conditions that may improve problem solving with the brainstorming technique. Users may want to gather in groups of 10 to 12 members to increase the variety of ideas. Helpful, also, is that members of such groups have mixed backgrounds, training, and experience and are advised of the problem at least 48 hours in advance of a brainstorming session. These group sessions should last only 30 to 45 minutes. So, in a system that requires problem solving either as the sole output or as one of the tasks to be performed, the designer should evaluate the requirements needed to solve problems. He or she may decide that brainstorming could be useful as part of the system design. This may include training, practice, performance aids, and other materials to help make the brainstorming as profitable as possible. Although good evidence suggests that brainstorming is effective, some studies show that brainstorming may not work in all situations. Designers must use caution in deciding whether or not the brainstorming technique is appropriate for the kinds of problems to be solved. This may mean conducting some performance tests to determine the effectiveness of the techniques in the new application.

Attribute Listing

Attribute listing, another idea-finding technique, also yields novel idea combinations or problem solutions. Crawford (1954) has indicated that each time an improvement is made in a product or system, it is done by changing an attribute. Original invention occurs by improving the attributes (parts, qualities, characteristics) of an object or by transferring attributes from one situation to a new situation. For example, a pencil's attributes are size, color, and shape. Each of these attributes can be altered to develop new kinds of pencils. A pencil can be skinny or fat, one color or striped, circular or hexagonal, etc. By changing one or more attributes, the item itself changes. This also applies to systems. A system may be improved by changing one or more of its attributes. Designers should develop systems that encourage users to critically consider attributes of situations when solving problems.

Checklists

Whenever we read through the Yellow Pages to locate, for example, automotive repairs, we are using an *idea checklist*. Checklist strategy amounts to examining some kind of list that suggests solutions suitable for a given problem. Once an auto mechanic is located, the problem is solved. Designer-provided checklists can provide possible solutions directly or indirectly. Paging through the Yellow Pages, a catalog, or a thesaurus is a direct means of finding a problem solution. We can use checklists indirectly to stimulate production of new ideas beyond those provided in the list itself. Osborn (1963) devised the following list to help inspire solutions:

> *Put to other uses?* New ways to use as is? Other uses if modified?
>
> *Adapt?* What else is like this? What other idea does this suggest? Does past offer parallel? What could I copy? Whom could I emulate?
>
> *Modify?* New twist? Change meaning, color, motion, sound, odor, form, shape? Other changes?
>
> *Magnify?* What to add? More time? Greater frequency? Stronger? Higher? Longer? Thicker? Extra value? Plus ingredient? Duplicate? Multiply? Exaggerate?
>
> *Minify?* What to subtract? Smaller? Condensed? Miniature? Lower? Shorter? Lighter? Omit? Streamline? Split up? Understate?
>
> *Substitute?* Who else instead? What else instead? Other ingredient? Other material? Other process? Other power? Other place? Other approach? Other tone of voice?
>
> *Rearrange?* Interchange components? Other pattern? Other layout? Other sequence? Transpose cause and effect? Change pace? Change schedule?
>
> *Reverse?* Transpose positive and negative? How about opposites? Turn it backward? Turn it upside down? Reverse roles? Change shoes? Turn tables? Turn other cheek?
>
> *Combine?* How about a blend, an alloy, an assortment, an ensemble? Combine units? Combine purposes? Combine appeals? Combine ideas? Courtesy Schribner

The main objective of Osborn's checklist is to suggest improvements for objects or processes; however, the questions may evoke ideas for other kinds of problems. Perhaps a designer should provide such a checklist for problems in the new system.

Table 7-2 shows another idea checklist presented by Davis and Houtman (1968).

CHANGE COLOR?	NEW SIZE?	CHANGE SHAPE?	NEW MATERIAL?	ADD OR SUBTRACT SOMETHING?	REARRANGE THINGS?	NEW DESIGN?
Blue	Longer	Round	Plastic	Make Stronger	Switch	From Other Countries?
Green	Shorter	Square	Glass	Make Faster	Change Pattern	Oriental Design
Yellow	Wider	Triangle	Fiberglass	Exaggerate Something	Combine Parts	Swedish Design
Orange	Fatter	Oval	Formica	Duplicate Something	Other Order of Operation	French Design
Red	Thinner	Rectangle	Paper	Remove Something	Split Up	Eskimo Design
Purple	Thicker	5-Sided	Wood	Divide	Turn Backward	Russian Design
White	Higher	6-Sided	Aluminum	Make Lighter	Upside Down	American Design
Black	Lower	8-Sided	Nylon	Abbreviate	Inside Out	Indian Design
Olive Green	Larger	10-Sided	Cloth	Add New Do-Dad	Combine Purposes	Egyptian Design
Grey	Smaller	Lopsided	Gunny Sack (Burlap)	Add New Smell	Other Switcheroo?	Spanish Design
Brown	Jumbo	Sharp Corners	Cardboard	New Sound		
Tan	Miniature	Round Corners	Steel	New Lights		From Other Time?
Silver	Other Size?	Egg-Shaped	Leather	New Flavor		Old West
Gold		Doughnut-Shaped	Copper	New Beep Beep		Roaring Twenties
Copper		"U" Shaped	Rubber	New Jingle		Past Century
Brass		Other Shapes?	Other Material?	Subtract The Thing That Doesn't Do Anything		Next Century
Plaid			Combination of These Materials?			Middle Ages
Striped						Cave Man
Polka-Dotted						Pioneer
Flowers						
Speckles						From Other Styles?
Paisley						Hippie
Pop Art						Beatnik
						. . .
Other Colors?						Ivy League
Color Combination?						Secret Agent
Other Patterns?						Elves & Fairies
						Clown
						Football Uniform

Table 7-2. Idea Checklist (Davis and Houtman, 1968)

Although a checklist may not always be appropriate for system users, printed or computer-generated checklist performance aids often can facilitate problem solving in many systems. These techniques should shorten the time taken to arrive at solutions. Incidentally, the answer to Figure 7-10 is that each phrase can be spelled backwards and have the same meaning.

Analogical Thinking

Analogical thinking using similarities between objects or ideas, can be a powerful technique in attempting to solve certain types of problems. We will discuss two techniques—synectics and bionics. Both these methods use analogies to produce creative problem solutions.

Synectics is a method that uses at least three types of analogies: direct analogy, personal analogy, and fantasy analogy (Gordon, 1961). To find a direct analogy one would ask how animals, birds, insects, flowers, or trees have solved similar problems. For example, Gordon (1961) discusses the problem of designing a new roof that would turn white in summer to reflect heat, and black in winter to absorb heat. The problem solver used the flounder as a direct analogy. This sea-bottom fish adjusts its light and dark coloration to match its surroundings. The color changing devices are chromatophores, tiny sacks containing black pigment which are connected to the spinal cord. The chromatophores contract to push the black pigment toward the surface, darkening the fish's exterior. Conversely, relaxing the chromatophores retracts the pigment and lightens the fish's coloration. The result of applying the flounder analogy to the roof problem was an all black roofing material impregnated with little white plastic balls that would expand when the roof became hot, changing the surface to a heat-reflecting white.

An example of personal analogy is envisioning oneself as one virus in a crowded culture dish. Such imagining may lead to unusual solutions. Using the fantasy analogy, a user searches for an ideal but farfetched solution. The user might ask, for example, how a particular problem could solve itself. Davis (1973) illustrates this type of analogy by suggesting that asking how to make a refrigerator defrost itself, how to make tires repair their own leaks, and how to make an oven clean itself may have led to the invention of self-defrosting refrigerators, self-sealing tires, and self-cleaning ovens.

Using these analogies can be complicated and usually requires more training than that required when using checklists.

Bionics

Bionics has been defined as the use of biological prototypes for the design of systems for use by humans. In this case, the structure, function, and mechanism of plants and animals are studied to gain ideas for system design. As a problem-solving method it resembles direct analogy. This approach is most useful when specialists in the functioning of plants and animals are used as expert consultants. Davis (1973) reports how bionics has been used to solve some difficult engineering problems. For example, studying the frog's eye led to both a new type of telephone filter and a radar scope, both of which suppress background noise while amplifying the active signal. The infrared sensors of the rattlesnake respond to a heat change of 0.001^{o}F, which may be the temperature difference between a sun-heated rock and a rabbit. The rattlesnake's infrared sensor led to the Sidewinder missile, which trails the hot exhaust of a target aircraft, exploding in its tailpipe. Thus, analogies drawn from plants and animals have assisted problem solvers in combining existing ideas into new ideas and in some cases reaching creative solutions.

In some cases, computers have been designed based on the way people process information; in addition, the study of computer "behavior" has shed light on how people process information (cf. Newell and Simon, 1972).

Designing for Effective Problem Solving

We should design systems that encourage efficient problem solving. All of the problem-solving techniques discussed can be used in systems to help ensure effective and imaginative problem solving. A designer's responsibility is to be aware of these and other techniques and to incorporate them as an integral part of the system. This means that problem solvers should have the necessary instructions, performance aids, training materials, or computer outputs to aid acceptable problem-solving performance. A problem that helps to illustrate the ideas just discussed is shown in Figure 7-12. See how long it takes to reorganize your own ideas and arrive at the solution.

> How quickly can you find out what is so unusual about this paragraph? It looks so ordinary that you would think that nothing was wrong with it at all and, in fact, nothing is. But, it is unusual. Why? If you study it and think about it you may find out, but I am not going to assist you in any way. You must do it without coaching. No doubt, if you work at it for long, it will dawn on you. Who knows? Go to work and try your skill. Par is about half an hour.

Figure 7-12. "Not So Easy" Problem

What could a designer have done to help you arrive at the solution sooner?

When considering problem solving, designers need to have some idea of the extent of knowledge users have concerning a particular problem situation. Another and equally important issue is how efficiently other relevant information can be organized and accessed so that it can be brought to bear easily on specific problems. This may require, for example, a separate computer data base or an extensive library of relevant documents. The answer to the problem in Figure 7-12, by the way, is that there are no "e's" in the paragraph.

Computers and Problem Solving

Computers can be of considerable value in assisting users to solve problems. Ramsey and Atwood (1979) have outlined several steps in the problem-solving process and related these to computer-based aids that designers could provide.

Problem Recognition

As indicated earlier, one of the first steps in problem solving is to recognize that a problem exists. People are frequently slow to react to, or at least to recognize problems. This is especially true in situations in which a person must monitor the current state of the world and detect or react to critical changes. A primary need is for an aid that alerts the problem solver to "relevant" changes. Current status displays, historical displays, and aids for dealing with degraded data can be useful.

Problem Definition

After a problem is recognized, the problem solver must determine how to formulate, or represent, the problem. In most cases, there are several alternative formulations for a given problem. Frequently, success of problem solving strongly depends on selecting an appropriate formulation. Designers should provide aids that provide a change in problem representation (e.g., graphical displays). Developing alternative representations requires a thorough understanding of the specific problem and the problem-solving processes that are most appropriate. Allowing the problem solver to decompose a problem into subtasks and recombine these subtasks in various ways can be useful in problems with relatively independent tasks.

Goal Definition

In some cases, the goal to be achieved is predefined. In other cases (e.g., planning) the problem solver must select an appropriate goal. A selected goal must be not only appropriate, but also attainable (cf. Malone, 1980). In fact, one of the primary difficulties in many situations is that a selected goal may not be attainable. It may be useful to aid the problem solver in generating several alternative goal structures and encourage the delay of selecting a specific goal until later in the problem-solving process.

Strategy Selection

Strategy selection is concerned with determining the general approach that will be used in problem solving. In some cases, a certain strategy is dictated by the problem representation that is selected. In general, strategy selection is based on previous experience with a given class of related problems. The majority of strategy-selection aids are concerned with specific problem domains. In domains in which problems can be decomposed into fairly independent subproblems, aids that allow the user to select strategies for these subproblems independently before combining them into an overall strategy can be very useful. Additional research is needed on the nature of specific problem-solving tasks and the strategy selection heuristics used by expert problem solvers. This would enable the construction of techniques to aid the less experienced user in this phase of problem solving.

Alternative Generation

The problem solver is frequently expected to generate all alternative actions that may be appropriate. This differs from a "decision maker" who has the alternatives identified in advance by a designer. If an activity is not well-defined, the problem solver may not be able to generate appropriate alternative actions. Even when alternatives are identified, if there is a large number of alternatives the problem solver may not be able to retain all alternatives in memory for later evaluation. Aids that store a large number of user-generated alternatives can be easily developed and also can be effective. The principal need is for aids to suggest alternatives that the user would not normally consider, or for well-defined tasks, to present algorithmically determined alternatives. Such aids have been developed for training applications and for cases in which the computer has been programmed to generate optimal solutions without explicit user interaction. For ill-defined task environments, aids that suggest hypotheses to be tested may aid in alternative generation.

Alternative Evaluation

When it comes time to evaluate alternatives, problem solvers actually become decision makers. Both problem solvers and decision makers are generally good at evaluating alternatives in a manner consistent with their perception of the problem and the goal to be achieved. However, if the alternatives have far-reaching consequences or if they must be evaluated with respect to a large number of factors, the problem solver's memory and processing limitations may be exceeded. In well-defined situations, aids have been developed that allow the user to simulate the consequences of various alternatives. Aids that capture the user's evaluation heuristics and then filter information to be consistent with these heuristics and sometimes even present alternatives considered to be optimal are especially useful when a large number of evaluation heuristics must be applied. This type of aid is effective, but may require a great deal of effort to implement.

Alternative Selection and Execution

The last step in both problem solving and decision making is to select and implement the solution or decision. Designers can provide ways for automatically executing user-specified actions. Such aids should permit users to name the desired action without explicitly carrying out the steps involved in the execution.

DECISION MAKING

The reader should be cautioned that the contrasting of problem solving and decision making as two separate processes is not always possible. Obviously, in some situations decision making may be a special case of problem solving, and in other situations problem solving may involve multiple decisions. The problem solving/decision making distinction is used here to help identify and emphasize many of the different types of design decisions associated with intellectual processing. For our purposes, *problem solving* usually involves the discovery of a correct solution in a situation that is new to an individual. *Decision making* involves the weighing of known alternative responses in terms of their desirability and then selecting one of the alternatives.

Uncertainty

Decisions are usually made along a continuum that ranges from absolute uncertainty at one end to absolute certainty at the other. Uncertainty, here, refers to the uncertainty of the consequences of the decision. Purchasing stock in the stock market is an example of decision making with uncertainty. The future state of the world is not known when the decision is made; thus, the consequences of the decision are not known, and there is high uncertainty.

Examples of making decisions with less uncertainty concerning the consequences include deciding whether to walk up two flights of stairs or to take an elevator. The consequences of the decision are more or less known, and the issue is focused more on which is the best decision. Designers must deal with the full range of decision uncertainties. Where uncertainty is low, a designer should make provisions for a correct decision based on available information. Frequently, using decision tables, logic trees, or contingency tables accomplishes this. Appropriate uses of these techniques are discussed in Chapter 19. A simplified decision table is shown in Table 7-3.

Table 7-3. Decision Table

If this condition exists	Take this action
A	1
B	2
C	3

In general, a decision is easier to make if the decision maker knows or has a fairly good idea what the outcome will be. Suppose we have to decide whether or not to carry an umbrella when leaving home in the morning. There are two possible actions: take an umbrella or leave it home. Likewise, there are two states of the world that are of most interest: either it will rain or it will not rain. If the person knows for certain that it will rain, then he or she will carry an umbrella. If the person knows for certain that it will not rain, then he or she will leave the umbrella at home. In both of these situations, an individual makes a decision under conditions of little uncertainty.

However, the world is not known for certainty. On some mornings we may not be at all certain whether or not it will rain that day. Even after listening to a weather report on the morning news, looking out the window at the sky, and considering what happened during the night, there may still be some uncertainty as to whether it will rain. After collecting all this information, an individual can establish a personal probability for rain during the day. The person may reason that the probability of rain is 75 percent.

In general, uncertainty adds complexity to decision making. If a designer determines that a decision situation contains uncertainty, she or he should provide instructions, performance aids, additional training, and any help to reduce the uncertainty as much as possible in a form that aids decision making. The designer may require the use of a computer to help decide among alternatives.

Decision Making Skill

Individuals tend to differ greatly in their decision-making capabilities. One major reason seems to be that some people have developed their decision-making skills to a greater degree than others. To ensure an acceptable level of decision making in a system, the designer should rely on both selection and training—select those with the most developed decision-making skill, and then provide skill-building practice for them.

Since people can be trained to be better decision makers, designers should make proper training materials available to users to help ensure adequate training and practice. If good materials are provided, user decisions will be made more quickly, more accurately and with less training, and the decision-making performance itself will be satisfying.

Computer-Aided Decision Making

A designer should provide all of the information and tools necessary to make good decisions. In computer-based systems, the computer should be used to provide decision alternatives so that a decision maker can make use of as much information as is available.

People can integrate only a limited amount of information to make a meaningful and reasonable decision. Systems that require people to integrate large masses of data before making a decision, will have degraded performance.

Experienced users can make good estimates of the likelihood of different events occurring if given sufficient information to do so. The *computer* can then take these likelihood estimates and, using Bayes' formula (Winkler, 1972), or some other appropriate means, combine them. This enables the final decision maker to decide among alternatives based on the probability of each alternative under different sets of circumstances, and relieves the decision maker of a considerable amount of assimilation and storage of data.

In some systems, decisions are made sequentially, with later decisions being contingent upon the consequences of earlier decisions. Generally in these cases, the user is more concerned with the outcome of a series of sequential decisions rather than the outcome of any single decision. People are fairly good at evaluating a series of events occurring as a result of decisions and accumulating this knowledge in order to make later decisions. However, a designer can assist users by providing a computer for storing and adequately presenting the results of earlier decisions in a way that facilitates future decisions.

Risk

When people make "risky" decisions they usually want to know what is at stake and what are the probabilities for certain outcomes. The designer can help an individual evaluate both stakes and the probabilities of outcomes by providing as much information about these two items as possible. Generally, people can assess what is at stake quite accurately, but do not assess outcome probabilities well, even when considerable information is available.

In decision making one assumes that a person "weighs" the stakes against the probability of occurrence. If the stakes are low (that is, there is little to be gained or a small payoff) and the probability of an undesirable outcome is high, the individual will select an alternative course of action, if one is available. For example, a person may be driving a car with limited visibility due to a heavy rain. Passing a car on a narrow road without a clear view ahead means that the possibility exists that the driver may be killed. The gain, in this case, would be a few minutes saved. Most people would elect not to pass, but some people would make the decision to pass the car. Each decision is very personal.

In order to adequately weigh the stakes against probable outcomes, one must have a true appreciation of the actual risk involved in deciding on one course of action over another. It appears that people are not equally equipped to make this evaluation. In fact, the evaluation of risk may, in itself, be a skill that is developed with experience. A child may not perceive a situation as risky, while an adult may recognize the risk.

Other Considerations

Designers should keep in mind that early decisions tend to restrict the range of choices for future decisions, and, consequently, many users are tightly bound up due to decisions they have made in the past.

Finally, there are at least four decision-making characteristics that designers should recognize when developing a system:

- Users usually wait longer to decide than needed (over-accumulate information)
- Users tend not to use all available information
- Users tend to be hesitant in revising original opinions, even if new information warrants revision
- Users usually consider too few alternatives

White Pages Example

An example of decision making in systems is provided by the White Pages system for telephone listings. Here, telephone company people are charged with the responsibility of reviewing each new account. One consideration of new accounts is whether or not the name submitted by a customer is the actual name of the business. The owner of Smith's Cycle Shop might try to list his business with the telephone company as the AAAAA Bicycle Shop in order to have the listing appear in the front of the White Pages and in the appropriate category in the Yellow Pages. The telephone company has a long-standing rule that the listing in the directory should be the same as the true name of the business. Therefore, someone at the telephone company reviews the names of new customers and makes a decision as to whether the name given is the *true name* of the business.

When a person must look at incoming information (in this case the names of customers) and decide if that information is valid, what can a system designer do to help ensure that this decision is made quickly and accurately and at the same time require the least amount of training time? Traditionally, system designers have given users little assistance in performing these activities. One user may work out clever and effective ways of gathering information and making appropriate decisions, while others may never be able to do so. Even the good decision maker will be replaced someday, which means the next individual may not have the same ideas and may be unable to benefit from the knowledge of previous workers.

In the telephone company example, a designer could assist the decision maker by programming into the computer a listing of all legitimate businesses in the general geographic area. The decision maker then needs only to enter the name requested by a customer into the computer, and the computer will match it against names that are known as legitimate. If there is no match, the user can follow up in other ways.

FOR MORE INFORMATION

Coombs, C. H., *Mathematical Psychology: An Elementary Introduction,* Englewood Cliffs, New Jersey: Prentice-Hall, Inc., 1970.

Davis, G. A., *Psychology of Problem Solving,* New York: Basic Books, 1973.

Lindsay, P. H. and Norman, D. A., *Human Information Processing: An Introduction to Psychology,* New York: Academic Press, 1977.

Neisser, U., *Cognition and Reality: Principles and Implications of Cognitive Psychology,* San Francisco: W. H. Freeman and Company, 1976.

Winkler, R. L., *An Introduction to Bayesian Inference and Decision,* New York: Holt, Rinehart and Winston, Inc., 1972.

8

MEMORY

INTRODUCTION

Human performance is frequently degraded because people forget. One is reminded of a conversation between the King and Queen in Lewis Carroll's *Through the Looking-Glass:*

> The King was saying, "I assure you, my dear, I turned cold to the very ends of my whiskers!"
>
> To which the Queen replied, "You haven't got any whiskers."
>
> "The horror of that moment," the King went on, "I shall never, *never* forget!"
>
> "You will, though," the Queen said, "if you don't make a memorandum of it."

Designers should be aware of at least three types of human memory: sensory memory, short-term memory, and long-term memory.

SENSORY MEMORY

Sensory memory can be demonstrated in a number of different ways. Move your finger rapidly back and forth in front of your eyes; you should see it in more than one place at a given time. Or note the trail that a Fourth-of-July sparkler or lighted cigarette leaves when waved in the dark. These and other more controlled demonstrations show that an image persists (is briefly stored) after the stimulus is no longer present. This persistence of information makes it available for further processing even after the stimulus has moved or terminated. In the case of vision, the persisting information seems to be stored in the sense itself rather than in the brain (Sakitt, 1975). This also may be the case for the other senses. We will refer to this persistence of a stimulus that probably takes place in the sense itself as *sensory memory* (Sperling, 1960; Averbach and Coriell, 1961).

Sensory memories are known to exist for vision, audition, and (possibly) touch. They are characterized by being very brief, and at least in the case of vision, as being a literal representation (i.e., a more or less photographic image) of the stimulus. Items in sensory memory quickly fade or are "erased" by new inputs. In the case of vision an image usually persists for about a quarter of a second or slightly longer (Averbach and Coriell, 1961; Dick, 1969, 1970; Haber and Hershenson, 1973; Dick and Loader, 1974). In his original studies, Sperling (1960) reported a duration of about one second. Others have observed visual sensory memory durations in excess of two seconds where the stimulus-background contrast was great (Averbach and Sperling, 1961; Mackworth, 1963). This means that a visual stimulus is available for cognitive processing for at least a quarter of second up to about two seconds after the stimulus is removed.

The auditory sensory memory appears to last for at least a quarter of a second (Massaro, 1972), and may last as long as one to five seconds (Glucksberg and Cowen, 1970; Kubovy and Howard, 1976). In addition to vision and audition, there is some evidence for a sensory memory for touch that lasts for about 0.8 second (Bliss, Hewitt, Crane, Mansfield, and Townsend, 1966). Similar memories may exist for other senses as well.

It is still not altogether clear what role sensory memory plays in cognition. Glass, Holyoak and Santa (1979) suggest that since auditory information such as speech is spread over time, it is necessary to preserve brief segments so they can be processed as units. For example, in English a person can signal a question simply by raising the intonation pattern, as in "You are tired of *studying?*" Here it is necessary to preserve segments long enough to tell that the intonation is in fact rising. The auditory sensory memory makes this possible. The function of the visual and touch sensory memories is even less clear. Jonides (1979) has proposed that visual sensory memory gives people time to detect, switch attention to, and process events that occur off the center of the field of vision. Massaro (1975) believes that the visual sensory memory assists in performing activities such as reading.

There do not seem to be many design related issues associated with sensory memory. Designers should note, however, that the duration of sensory memory may be somewhat under their control. At least for vision, the duration of sensory memory can be lengthened by optimizing the stimulus-background contrast. Difficulties with visual sensory memory would likely show up as an increase in errors. These errors would be characterized either by a lack of pattern (Miller and Nicely, 1955) or an excess of visual confusions (Keele and Chase, 1967). If a typist or key operator were typing codes, say seven character codes such as DLTRVSA, we would expect that there would be more errors in the right-most positions and fewer in the left positions (Bryden, 1960; Bryden, Dick and Mewhort, 1968; Mewhort and Cornett, 1972; Heron, 1957), and more errors in the center positions than in the end positions (Merikle, Coltheart and Lowe, 1971; Merikle and Coltheart, 1972; Coltheart, 1972).

The first prediction concerning errors comes from studies where investigators have been unable to find visual sensory memory-related "auditory confusions." Auditory confusions seem to be more characteristic of short-term memory experiments (cf. Conrad, 1964). At the same time investigators have found evidence for visual errors and errors with no pattern. The second prediction comes from research where researches have established that subjects tend to "readout" data from visual sensory memory to short-term memory in either a left to right or an "end-first" manner. Under these conditions, the last items out of the sensory memory show the greatest influence of decay—i.e., the most errors (Dick, 1967). Lowe (1975), for example, reported that the processing of end letters on displays terminated shortly after display offset, while processing of the center letters continued for at least 500 msec.

SHORT-TERM MEMORY

A second memory store, one that has even greater human performance implications, is *short-term memory*. People use short-term memory to hold information temporarily, usually for a few seconds. Hundreds, or even thousands, of tasks performed each day require this type of remembering. Many of these tasks require memory to hold information only a few seconds. Conceivably, most of the information dealt with daily is "throw-away" information, actually meant to "pass in one ear and out the other." We would soon become over burdened if every sight and sound—road signs, telephone rings, radio messages—were somehow stacked in our memory. A temporary memory store—what we are calling a short-term memory—is, therefore, convenient.

Information stored in short-term memory appears to come from both external and internal sources. External information comes into the short-term memory through the senses and the perceptual process. Internal sources include the results of reasoning or the outcome of a problem-solving task.

Encoding

The exact visual, auditory, or kinesthetic image or message is not directly stored in short-term memory. Rather, the information stored must first be encoded. Information is converted into a form that is consistent with human physiology and that aids further processing of the information. The precise physiological form of the code is not important, but understanding that the information has been encoded and, therefore, is in a form much different from the original stimulus image is important. Evidence suggests that some of the visual information received by an individual is encoded in short-term memory in *auditory* form (Sperling, 1960; Conrad, 1964). These studies have shown that some errors made in visual tasks are errors that would more likely occur in an auditory task when sounds, not shapes are confused. For example, an A substituted for a K or a B substituted for a C. The issue of encoding, then, has at least two dimensions that interest us: The information is encoded into a form that can be conveniently stored in human memory, and, probably, some visual information is transferred into auditory form (possibly to assist rehearsal). The important point is that the brain deals with information in some encoded form much different than the original stimulus.

Capacity and Duration

Probably the two best-known characteristics of short-term memory are capacity and duration. The short-term memory store is small and can hold about six or seven units of information. A unit is any organization of information that has previously become familiar—such as familiar words, or a familiar configuration of pieces on a chessboard. For example, if someone looked up a seven-character telephone number and stored this information in short-term memory for a few seconds while dialing, he or she would usually store and correctly recall the seven characters. If, however, the person had looked up a twelve-character telephone number and tried to remember it while dialing, his or her performance would almost surely have errors. In the latter case, the person expected the memory store to hold more than its capacity.

It is curious that in the seven-unit capacity, each unit may be made up of one, two, five, ten, or more items. A familiar telephone number may use up only one unit of space, whereas an unfamiliar telephone number may use up all six or seven units of space. Shepard and Sheenam (1965) conducted a study in which subjects reproduced from memory eight-digit numbers of two different kinds. In the "prefix" kind, the last four digits of each number were selected at random while the first four digits were selected from two familiar four-digit prefix sequences. In the second kind, this order was reversed so that one of these two familiar sequences *followed* the four random digits as a "suffix." They found decreases of 20 percent in response time and 50 percent in errors when the last four digits were familiar rather than the first four. This finding has implications for codes that have some parts that are more familiar than others. To improve human performance in activities that use short-term memory, one must take into account the capacity limit of short-term memory and ensure that it is never exceeded. In fact it is probably best to stay well under the known capacity.

The relationship between the number of items to be remembered and the length of time they will persist in short-term memory is shown in Figure 8-1. In one case, a single unit of memory was used. In a second case, three units of memory were used. In both cases, people performing a difficult irrelevant task during the recall interval minimized rehearsal. Obviously the single unit was retained more accurately and longer than the three units. If the designer needs to have information remembered, he or she should probably keep the messages as brief as possible and ensure little interference. People forget longer messages sooner. Therefore when working with short-term memory, *the shorter the code the better.*

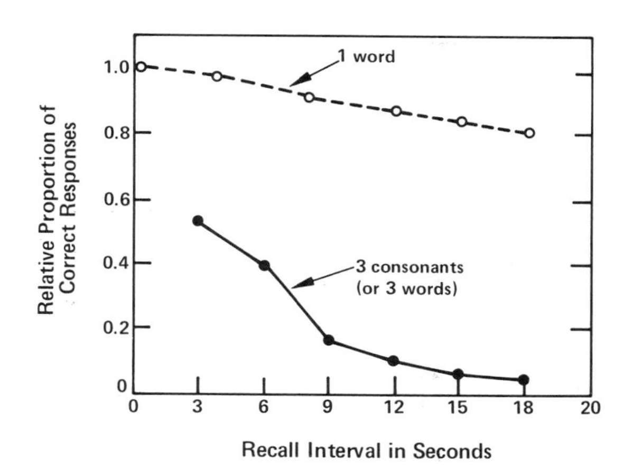

Figure 8-1. Short-term Memory Studies

One study involving memory capacity in chess found that about 90 percent of the time a grand master can reconstruct a position with approximately 25 pieces after having seen it for only five or ten seconds, while an ordinary player could remember the locations of only 6 pieces after the same exposure (Chase and Simon, 1973). That short-term memory has a capacity of about six units explains this phenomenon. When the grand master looks at a chess position in a game, he or she recognizes it as being made up of familiar patterns of pieces (perhaps, 3 to 5 pieces each). Since the grand master can easily hold this much information (six units) in short-term memory, he or she can remember and reconstruct the position of all the pieces on the board. The ordinary player, however, not having the same repertory of familiar units stored in long-term memory, encodes the position as a configuration of 25 pieces, but he or she can hold only about 6 of these in short-term memory. The chess master does as poorly as an ordinary player when the pieces are placed in random positions.

Simon and Gilmartin (1973) simulated this situation using a computer that had a limited short-term memory capacity. At the outset, the computer could store information for only about six pieces in a short-term memory that simulated the ordinary player's performance. The program then was exposed to a large number of configurations of pieces that commonly occur on the chessboard during games, and the computer gradually stored these patterns in long-term memory. After this "training," the computer could recognize these patterns when it encountered them in a new chess position, it could encode them in short-term memory as single symbols, and it could "point" to the associated information in long-term memory. With about 1300 familiar patterns stored in long-term memory, the program approximated the performance of a Class A player, but not a master. Simon and Gilmartin estimated that familiarity with about 50,000 patterns would be required to match the performance of a chess grand master.

Rehearsing

Rehearsal retains information in short-term memory. When a designer must build longer codes, he or she should divide them into groups of three or four to help rehearsal. Designers should also keep in mind that during rehearsal of a code or other information, other intellectual activities cannot take place. If an individual is rehearsing in an attempt to remember long enough to key something, or to speak to someone, or to write down an item number or other code, other activities should not interfere with this rehearsal process. Rehearsal tends to cause material to be stored in the practiced form. The most likely mistakes are those related to the rehearsed sensory features (Conrad, 1964; Conrad and Rush, 1965; Locke and Locke, 1971). By rehearsing we ignore new inputs and may even determine new patterns and possible rhythms to help maintain the items already in short-term memory. Consider the code 427947247. This may be recalled by establishing a pattern of three characters at a time, keeping the last digit of each three-character set the same: 427 947 247.

Patterns

Such creative rehearsal, a skill that can improve with practice, becomes very important in situations where material must be remembered for a short time. Designers also can facilitate remembering this kind of information by building patterns into codes and then teaching users how to work with the patterns.

The code NTH EDO GSA WTH ECA TRU can be extremely difficult for most people to remember, even for a very short period. In fact, it appears to require 18 units of memory when only about 7 units are actually available. Some people may try to remember each of the 6 three-character units and may be able to do so successfully. Other people may stare at the 18 characters and look for ways to make the code more familiar, so that fewer memory units will be required. This attempt to make use of known rules for "expanding" the short-term memory is much like a problem-solving situation. This particular code can be organized to the point where only one or two units of memory are required. By changing only one letter, it actually contains six words: "The dog saw the cat run." Once this pattern is recognized, the load on the short-term memory is drastically reduced and this particular code can be recalled easily with few errors.

Serial Position

Two other conditions related to short-term memory should be considered. The first has to do with serial position errors. This means that if a seven-character code is to be remembered for a short time and then written down, dialed or keyed into a CRT, errors will tend to occur in certain character positions more frequently than in others. For example, in a seven-character code, most errors tend to occur in the fifth position and the fewest usually occur in the first position. Figure 8-2 shows a set of serial position curves.

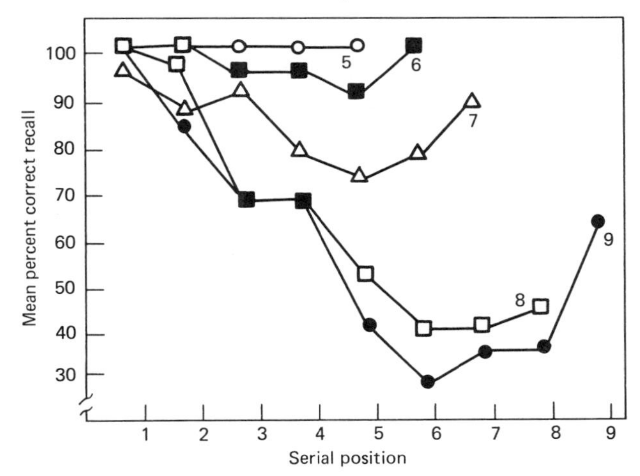

Figure 8-2. Effect of Serial Position on Short-term Memory for Consonants (Adapted from Jahnke, 1963)

It is interesting that whether or not an item in the seventh character position will be recalled depends on whether the original stimulus was auditory or visual (Conrad and Hull, 1968). With auditory stimuli, the last character tends to be recalled about as well as the first character, but with a visual stimulus, the last character tends to be mistaken (see Figure 8-3).

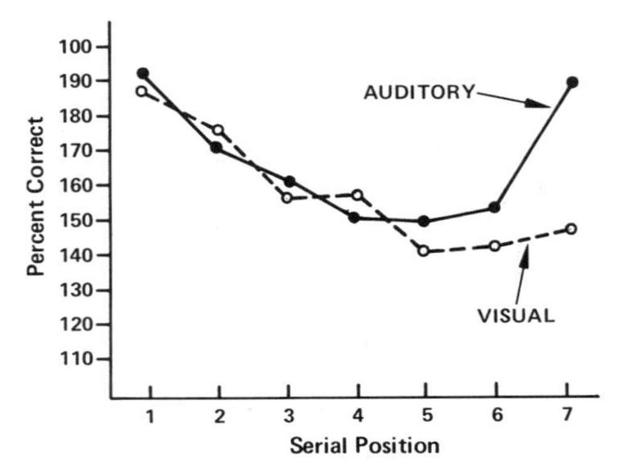

Figure 8-3. Illustration of the Serial Position Curve (adapted from "Input Modality and the Serial Position Curve in Short-term memory," by R. Conrad and A.J. Hull, in *Psychonomic Science*, 1968, *10*.)

von Restorff Effect

The second condition that shows up in many systems is called the von Restorff effect (von Restorff, 1933). The von Restorff effect relates again to the serial position of characters. Figure 8-4 shows two curves: the auditory curve just discussed, and the von Restorff curve. The latter curve shows a different probability of recalling the characters in a code, depending on the makeup of the code itself. Codes such as 3714B92 tend to elicit the probability of errors shown in Figure 8-4. That is, the "B" tends to be recalled almost as well as the first and last characters probably because the "B" is different from the other characters.

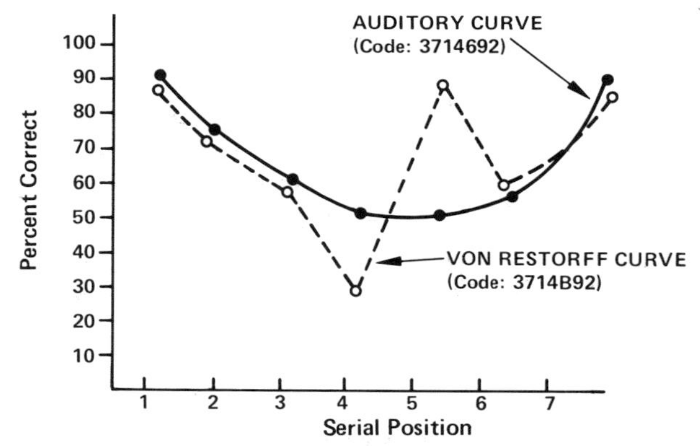

Figure 8-4. Illustration of the von Restorff Effect

Possibly more important to a system designer, however, is that the character just *prior* to the different character is usually recalled *less accurately*. The designer needs to evaluate trade-offs here and to be aware that a "different" character actually could elicit more errors in the character just before it.

Conclusions

Obviously, designers should not require users to retain even a small amount of information for over 20, or even 10 seconds if rehearsal is not possible. And they should not require users to retain easily rehearsed information beyond 30 seconds. Note in Figure 8-1 how the proportion of correct responses declines after even the first few seconds.

Primarily, we lose material from short-term memory by replacing it with new information. We think that new information pushes out old information because of the restricted number of units that short-term memory can hold at any one time. One of the best ways for a designer to ensure that information in short-term memory persists is to provide for no interference from new information.

Designers should particularly keep in mind problems associated with short-term memory such as limited capacity (seven or fewer units), relatively short duration (usually less than 20 seconds), and the requirement of continued rehearsal without interference to maintain information in short-term memory.

LONG-TERM MEMORY

We consider long-term memory a more or less permanent memory store. The concepts of learning and memory are closely related—the results of learning must be remembered for experience to accumulate. Long-term memory is essentially unlimited; at least it seems capable of storing all memories that come about in the lifetime of an individual. Unlike items stored in short-term memory, items stored in long-term memory appear to last forever.

Items tend to make it into long-term memory if the item can easily be "hooked-up" or linked with something that is already there. Thus, long-term memory relies heavily on organization to build and maintain content. Although we cannot be sure information is ever actually lost from long-term memory, some stored memories may become less accessible, or less easy to retrieve.

When most people refer to learning, remembering, and forgetting they are really thinking of long-term memory. Designers should be interested in knowing what already has been learned and stored in the memories of users, and what types of information should be added to memory (what must be learned by users) for users to perform adequately. Ensuring that information is properly stored and easily accessible is important also. We usually say a person has forgotten information when it would be more accurate to say he or she *has lost access to it*.

A discussion of memory is also a discussion of forgetting since the terms are complementary: The amount forgotten equals amount learned minus amount retained. Designers deal constantly with users who forget. If people did not forget, a designer would need to present information to system users only once to ensure that appropriate knowledge was gained for performing the activity.

Two of the most persuasive lines of evidence underlying the belief that people do not lose information stored in long-term memory are related to brain stimulation and hypnosis. The first of these relates to some experiments done in the 1940s in which electrodes were used to penetrate a small portion of the inner brain. The patients were awake during this procedure. When these small areas were stimulated, some patients reported having vivid recollections of situations, feelings, and odors relating to experiences they once had had but supposedly had forgotten (Penfield, 1958; Penfield, 1969; Penfield and Perot, 1963). However, very few of 1132 (3.5 percent) experienced this phenomenon. Twenty-four claimed to have had only an auditory experience, where they "heard" a voice, music, or some other meaningful sound. Nineteen patients claimed to have had a visual experience, where they "saw" a person, group of people, a scene, or a recognizable object. Only 12 reported a *combined* auditory and visual experience. Furthermore, detailed examination of these few patients' reports suggests that they could have been reconstructions or inferences rather than memories of actual events (Loftus and Loftus, 1980).

The second form of evidence involves the use of hypnosis. Frequently, hypnotized people can recall vividly experiences that occurred when they were much younger. People seem to have difficulty recollecting these experiences without hypnosis. Barber (1965), however, pointed out that no evidence exists to support the view that recall under hypnosis is superior to recall under ordinary waking conditions. Maybe people report more *willingly* rather than more ably when under hypnosis. Although memories are not always accessible, the evidence is unclear whether these memories are lost.

Why then do we seem to forget material from long-term memory? To provide even the briefest answer, one must understand that at least three operations take place related to remembering and forgetting: encoding, storage, and retrieval.

Encoding

Consider the following example. We have a filing system organized, in part, according to the categories shown in Figure 8-5. We file new articles pertaining to these categories in the appropriate section. A new article on handprinting errors, for example, would be placed in a file cabinet under the "handprinting" category. From time to time an article appears that does not fit neatly into one of the existing categories. We must decide where to file the new report. For example, an article on whether or not to use boxes on forms to reduce handprinting errors could be filed under either "handprinting" or under "form design." This categorization process must be consistent to work. Problems may arise later if the new article is placed under "form design" and another similar article is placed under "handprinting." In human memory, we refer to this process of deciding how to classify information as *encoding*. It requires an individual to perceive the information and determine one or more essential characteristics of the information. (Of course, the information being perceived has already changed form, that is, it has been encoded. It now must be encoded again according to its storage address.)

HUMAN ERROR
History of Human Error
General
Theory
Detection (Manual and Computer)
Correction
Costs
Reporting
Error Rates

HUMAN INFORMATION PROCESSING
Processing Capacity
Short-Term Memory
Perception
Memory
Speed-Accuracy Tradeoff
Bayesian Processing/Decision
Problem Solving

DATA ENTRY
Handprinting/Writing
Keyboard/Keying
Optical Recognition
CRT (Displays)
Forms Design
Character Design
Code Design

Figure 8-5. Filing System Categories

Occasionally, what actually is stored is not the same as what was perceived originally. Maybe only the essence of what was sensed will be encoded. This is a basic difference between the filing system in our example and human memory. What is in our hypothetical filing system is the exact article that was originally received and not an abstract or other approximation.

Possibly the encoding process becomes more efficient with experience—some people learn how to learn better (or accumulate many more associations from which they can retrieve). This means that being exposed to different types of training may enable people to encode more effectively.

Storage

The second necessary operation in long-term memory is storage. In our example, we assume that a new article is stored in the filing cabinet and not placed in the back of a desk drawer or accidently thrown out. Furthermore, we assume that the print on the paper will not disappear with time so that the information once put into the file cabinet stays in the same form with the same amount of clarity as when it was filed. We assume that the same can be said for human long-term memory, although we have little information on the form and clarity at filing time (after encoding).

Retrieval

The third long-term memory operation is called retrieval. Retrieval is the opposite of storing. If we wished to find the reports on the use of boxes with handprinting, we would have to decide exactly where to start looking in the file system. Is it filed under "handprinting" or under "form design"? Or, has there been a new category established that covers in greater detail this particular topic of interest? There are many potential locations for this article, and it may not be clear exactly where the article has been stored. The person who originally filed the article may be able to retrieve it quickly, but only if not too many articles have been filed in the meantime. Someone unfamiliar with the system might have to search each of the categories until he or she finds the article. People usually are not aware of searching in different locations when trying to remember something stored in long-term memory, but it seems that this is precisely what we do. In some cases, people try to remember the storage location by recalling how the information was originally filed, including the circumstances surrounding the original filing. In many cases, if the original means of storing can be recalled, the appropriate storage area can be identified quickly.

Forgetting

Forgetting may be due to a failure of any of these three operations (and possibly others) in long-term memory. Original encoding may be incorrect (information may be stored under another category), information may be in some way degraded during storage, or information may be difficult to retrieve because the search process takes place in the wrong part of the file.

For many years it was thought that forgetting was primarily a result of disuse or decay. This became known as the "leaky-bucket" description of forgetting. This concept of forgetting suggested that learning was the result of practice or use, and that when information was not used, forgetting occurred.

Therefore, *disuse* was considered as the main cause of forgetting. Disuse was related to time and, consequently, the longer the time interval the more information was supposed to have leaked out or decayed. Forgetting in long-term memory, according to this view, was due to a failure of storage, with neither encoding nor retrieval playing a major role.

In fact, forgetting from long-term memory usually does not occur in this way. It is possible to demonstrate that forgetting is most affected by what a person does between learning and performance. For example, people forget less if they sleep after learning than if they are awake and engaged in some activity (Jenkins and Dallenbach, 1924; Ekstrand, 1972). Forgetting varies according to the nature of the activity that takes place after information is stored and before it is used. The leaky-bucket or disuse idea of forgetting as a complete explanation is not acceptable. A designer should keep in mind that although disuse does not account for all forgetting, it may account for some forgetting.

Others have suggested that forgetting takes place because there has not been ample opportunity for new information to be consolidated with past learning (Muller and Pilzecker, 1900; Hebb, 1949). The idea is that normal activity produced by learning tends to continue (i.e., perseverate) after the end of the presentation of new material. Again, this concept (lack of consolidation) may account for some forgetting, but is inadequate to account for all forgetting.

Proactive Interference

Probably the most important concept pertaining to forgetting has to do with interference. There are two kinds of interference of interest to designers: proactive interference and retroactive interference. Proactive interference suggests that material learned *prior* to the learning of new material may interfere with the use of the new material in a performance situation. In studies of proactive interference, a group of people learn to perform task A, then learn task B, and then actually perform on task B. Another group of people will *not* learn A first, but will learn B, and then perform on B. The performance of the two groups is then compared. The people who learned A and then B do not perform as well as those who had learned only B. Certain kinds of material (A) seems to interfere with future learning and performance of similar material (B).

By learning A first and B second, people tend to forget items closely related to what was learned in A. Occasionally, people regress to using information learned in A rather than information learned in B. To use a simple example, people learn to turn on a house light by flipping a switch up (situation A), and many people have learned this over a period of years. Some new light switches require the depression of a button (situation B), and these same people then learn to use the button. People may press the button the majority of the time, but occasionally (particularly in times of boredom or stress) may attempt to flip the button up like the switch in situation A. This suggests a regression to earlier learning. This, then, is the idea of proactive interference.

Designers must be alert when determining the "best way" to train people to accomplish an activity. People will always bring with them a set of previously learned ways of doing things. Designers must find out what kinds of responses have already been learned, and then incorporate as much as possible the same kinds of responses in the new system. This will decrease the effects of proactive interference. Problems associated with proactive interference present some of the greatest difficulties when trying to design for an acceptable level of human performance.

Retroactive Interference

Retroactive interference, on the other hand, occurs when one learns to perform task A, then learns task B, and is then expected to perform task A. This is usually the easier of the two to understand because the interfering condition is introduced after the subject has learned to perform the original task. In some systems, designers take great pains to train people to perform a certain task and then, before these people have an opportunity to develop skills in performing that task, train the people on a second task. At the end of this dual training session, the trainees are expected to perform primarily the task for which they were originally trained. Frequently they do not do well. The interference from learning the second task may affect the long-term memory of an individual, and one of the predictable outcomes is an increase in errors due to forgetting.

Measuring What Is Remembered

Before we discuss problems related to the retrieval of information, the designer should be aware that there are two commonly used ways of measuring remembered information: *recall* and *recognition.* In a typical recall situation, a person is presented with a list of instructions on how to perform on a certain type of terminal. After these instructions have been presented, the individual is asked to recall as many of the instructions as possible. This procedure places considerable emphasis on the actual retention of the instructions originally presented.

Another way of measuring retention is recognition. With recognition, an individual must select from among two or more alternatives, such as on a multiple-choice test. Recognition is usually considered the most sensitive of the direct measures of retention, since it often demonstrates retention even when a person is unable to recall material. Recalling information is much more difficult than recognizing information. A designer should make full use of this fact and reduce the number of times people need to *recall* information or instructions, and increase the number of opportunities for them to *recognize* one or more available alternatives.

Inappropriate Retrieval Strategies

What is stored determines what retrieval cues are effective in providing access to what is stored (cf. Tulving and Thomson, 1973). When we forget something, it does not necessarily mean that the memory is lost; it may be merely inaccessible because the context in which we are trying to remember it does not permit retrieval strategies consistent with the strategies originally employed at the time the information was learned. Thus, in many situations, forgetting may be understood as being due to a lack of appropriate retrieval cues at the time of attempted retention. There is evidence to suggest that people can remember better if they are expected to do so in the same context as that in which learning originally took place.

One frequent problem when trying to remember is the apparent "blocking" of the correct response in favor of no response or an incorrect one. There are many who feel that probably the main cause of forgetting is not a failure of storage, but competition between alternative responses when an item in storage is trying to be remembered. This type of remembering difficulty is a very common experience to most people. They may find themselves trying to remember a name or a fact while "hanging on" to some obviously incorrect response. In situations such as this, a stimulus may cause a person to seek a particular memory but take the wrong route and end up at an incorrect storage place. The erroneous information is fed over and over again into consciousness, and an individual continually rejects it. The difficulty comes in trying to reroute or rechannel toward a different place in the memory store. Again, a designer can reduce the number of retrieval problems by emphasizing recognition rather than recall of information.

Intentional Forgetting

Some information may be intentionally forgotten, particularly those things that are painful to remember. Freud (1901) suggested that some memories are intentionally (though unconsciously) blocked, and this blocking inhibits recall. He called this "repression." There are also numerous conditions generally related to disturbances of the brain that can lead to forgetting. Forgetting that is due to tumor, head injury, disease, or old age is frequently referred to as "amnesia." Generally, designers are not concerned with the effects of intentional forgetting or of amnesia on system success.

Conclusions

A consideration of each explanation for forgetting should help designers reduce the amount of forgetting that occurs by people working in a system. We have considered the "leaky bucket" theories that suggest that in some way and for some reason information gradually disappears from memory. More likely, however, some form of the material placed in long-term memory is retained throughout a lifetime. Interference, whether by something that was learned before or something that was learned after the pertinent material, interests designers considerably. Both complete and partial inability to retrieve can be dealt with by making design decisions that emphasize recognition over recall. Also, to enhance retrieval of information a designer may have people work and, consequently, retrieve information in the same or similar setting in which information was acquired originally.

SENSORY-RELATED CHARACTERISTICS OF MEMORY

For memory that begins with visual stimuli, the evidence suggests a division into three different memory stores, sensory, short-term and long-term, each with a specific set of unique characteristics. Similar distinctions are not quite so clear for stimuli received through other sensors. The characteristics of human memory are closely related to the type of material (e.g., visual, auditory, touch, smell) being stored.

Auditory stimuli seem to follow much the same pattern as visual stimuli. There is evidence of a sensory memory that is slightly shorter than visual sensory memory and that lasts for about 100 ms. (Baddeley, 1976). Estimates for this memory also exist that range from 50 ms. (cf. Loeb and Holding, 1975) to 250 ms. (Massaro, 1970). In addition, the short-term memory associated with the auditory sensor seems to be shorter than for the visual sensor. Auditory short-term memory appears to last for only about 5 seconds (cf. Pollack, 1959; Wickelgren, 1969; Glucksberg and Cowan, 1970). Forgetting in short-term memory seems to occur by displacement of existing items by new auditory items rather than decay.

Auditory long-term memory clearly exists. Without it we could not recognize voices on the telephone, songs heard on the radio, or identify the animal associated with a "bark" or "meow." Two very different fields have contributed to most of what is understood about auditory long-term memory. The military has been training people for years to make auditory discriminations when listening for sonar signals that designate a submarine (cf. Corcoran, Carpenter, Webster, and Woodhead, 1968). The other application is in the field of music, where researchers have studied, among other things, the ability of people to learn and remember specific combinations and sequences of tones (cf. Deutsch, 1973).

There are two other senses that have received some study in terms of memory: touch and smell. The touch sense seems to have a sensory memory that lasts about 800 ms. (Bliss, Hewitt, Crane, Mansfield, and Townsend, 1966). Evidence for a touch short-term memory indicates that forgetting occurs fairly rapidly over about 45 seconds (Gilson and Baddeley, 1969). Together, these studies suggest that the sensory and short-term memories for touch are both longer than those generally associated with vision or hearing.

Again, the existence of a tactile long-term memory is widely accepted because of familiar experiences that we all have with it. Most people are able to recognize in the dark or with their eyes closed the feel of such things as velvet, metal, wood, a favorite vase, or the headlight switch in an automobile. One interesting idea that is emerging from research in this area is that the right hemisphere of the brain seems to be much more important than the left hemisphere in terms of tactile memory (Milner and Taylor, 1972). For example, one study has shown that blind subjects appear to be more efficient at reading Braille with the left hand (and hence the right hemisphere) than with the right hand.

It is commonly believed that learned smells are never forgotten. Baddeley (1976) has noted that smells also appear to be extremely good retrieval cues for apparently forgotten events. In Victor Hugo's words, "Nothing awakens a reminiscence like an odor." What little research is available suggests that there are no sensory or short-term memories associated with this sense. In addition, there is no evidence of forgetting, no suggestion of a difference between short-term and long-term memory, and there appears to be nothing equivalent to rehearsal (Engen and Ross, 1973).

MEMORY SKILL

Mnemonics

Remembering names, dates, events, formulas, etc., can be improved with proper training and practice. This suggests that there is a remembering skill. People have been interested in improving this skill since ancient times. As long ago as 500 B.C., the Greek poet Simonides was teaching how to develop a "trained memory." Today, this memory skill is developed most dramatically by the use of *mnemonics*.

According to Baddeley (1976), mnemonics represent cognitive schemes for helping to ensure the retention of material which would otherwise be forgotten. They range in complexity from a string tied around a finger to complex visual imagery schemes. In essence, mnemonics are cognitive performance aids. Most mnemonics require a person to either *reduce* or *elaborate* on information being received (Baddeley and Patterson, 1971).

Reductions

Reductions frequently take the form of acronyms. When trying to remember the names of the Great Lakes a person may think of the word "homes." The word "homes" will help in remembering Huron, Ontario, Michigan, Erie, and Superior. Or when trying to remember the colors of the spectrum and the order in which they appear, a person may used the acronym, "ROYGBIV" (red, orange, yellow, green, blue, indigo, violet).

The tying of a piece of string around a finger to help remember to make a phone call is a way of reducing the memory load. Another technique that is used to help people remember the number of days in different months is shown in Figure 8-6 (Lorayne, 1976). Close both hands into fists and place the fists side by side, back of hands up as shown. Starting on the knuckle of your left little finger, moving from left to right and including the valleys or spaces between the knuckles, recite the twelve months. Thumbs are not counted. The months that fall on the knuckles are months with thirty-one days. The valleys represent the months with thirty days or less.

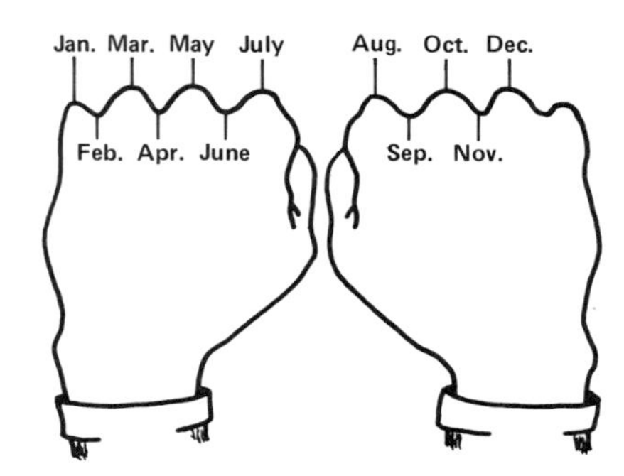

Figure 8-6. A Mnemonic for Remembering the Number of Days in a Month

Another commonly used memory aid is the use of rhymes like "I before E except after C," which helps people remember how to spell such words as "believe" and "receive." The alternative to remembering that rhyme is to memorize all words that contain "ie" or "ei". Thus, reducing the memory load from several hundred words to six words helps to have an acceptable level of performance without extensive training and memorization.

Many codes that are developed and used in systems, including abbreviations, reflect this same idea of having reduced the amount of information to be remembered. Performance aids are frequently used as memory aids, in which case they contain, in some abbreviated form, a portion of a considerably greater amount of material that a person has learned in training.

In the case of reduction, there is a danger of reducing the information so much that it is no longer possible to reconstruct the original. Even so, people seem to have a remarkable ability to make use of relatively simple memory aids to assist performance. We use this ability frequently in remembering to perform certain activities. It is not a foolproof method, as evidenced by the many times we still forget. But it is interesting that we more often than not do remember what the memory aid means, and then perform accordingly.

Elaboration

Elaboration on the other hand, involves *adding* information in order to make the material easier to remember. This can be done in at least two ways: (a) by taking greater advantage of meaning that already exists in remembered words, phrases, or concepts, and (b) by using visual imagery. In other words, there are two basic forms of elaboration: the first depends on *verbal cues,* the second on *visual imagery*. In many cases the two are combined.

Verbal Cues

The use of existing words, phrases or concepts is based on a basic rule of long-term memory. In order to remember any new piece of information it must be associated with something the person already knows. Association, as it pertains to memory, simply means the tying together or connecting of two or more things. A person beginning to study music may be told by the teacher to learn the lines of the treble clef, E, G, B, D and F. The teacher helps the student remember these letters by thinking of the sentence, "Every good boy does fine." The teacher is doing nothing more than helping the student to remember something new and meaningless by associating it with something the student already knows.

Baddeley (1976) gives us another example of making use of words, phrases, or concepts to remember. He provides a rhyme, similar to the one below to help in remembering pi to the first 20 decimal places.

> Pie.
> I wish I could recollect pi.
> Eureka cried the great inventor
> Christmas Pudding Christmas Pie
> Is the problem's very center.

The rule here is very simple: one simply counts the number of letters in each word and comes up with the answer 3.14159265358979323846.

When considering pi, it is interesting to note that pi has been evaluated many times to see if there is a pattern to the numbers. There is none. One of the most important testimonials to the extent in which the memory skill can be developed is that there are a few people that have used elaboration mnemonics to memorize and accurately recall pi to over 10,000 decimal places.

In some cases, elaboration requires a person to remember substantially more material than is necessary. This is well illustrated in a "peg system of memory" described by Lorayne (1976). Based on a technique published as early as 1849 by an English schoolmaster named Brayshaw (Hunter, 1957), the technique involves associating each of the digits 1 through 0 with one or more consonants. Lorayne proposes that the digits and consonants be matched as follows:

Table 8-1. Lorayne's Technique for Improving Memory

Digit-Consonant	*Hints for Remembering the Digit-consonant Matching*
1=T (or D)	A t has one downstroke.
2=N	A small n has two downstrokes.
3=M	A small m has three downstrokes.
4=R	The word "four" ends with the R sound.
5=L	The Roman numeral for 50 is L.
6=J (or SH, CH, soft G)	A capital J is almost a mirror image of a 6 (ᑯ)
7=K (or hard C, hard G	A capital K can be formed with two 7's (Ⱪ).
8=F (or V)	A handwritten small f and a 8 have two loops, one above the other (ꞗ).
9=P (or B)	The letter P and the number 9 are almost exact mirror images (ℯ).
0=S (or Z, soft C)	The first sound in the word "zero" is Z.

The consonants are then incorporated in a key word. All of this is necessary because it is difficult for most people to remember numbers. But by knowing the sounds of the consonants associated with each number it is possible to combine the sounds in words in order to remember them. If a person wants to remember the number 39, all that he or she must do is come up with a word that uses both M and P, such as "Mop." The M is for 3 and the P is for 9. Knowing the ten digit-consonant combinations gives a way to make any number meaningful; that is, it gives a way to associate the number with the set of familiar verbal materials already in memory.

A person can convert numbers to letters as needed or else have a specific word associated with one and two digit numbers. Lorayne suggests that the best performance comes from memorizing the following "peg words," each of which is constructed using the following digit-consonant combinations:

1.	tie	26.	notch	51.	lot	76.	cage
2.	Noah	27.	neck	52.	lion	77.	coke
3.	Ma	28.	knife	53.	limb	78.	cave
4.	rye	29.	knob	54.	lure	79.	cob
5.	law	30.	mouse	55.	lily	80.	fez
6.	shoe	31.	mat	56.	leech	81.	fat
7.	cow	32.	moon	57.	log	82.	phone
8.	ivy	33.	mummy	58.	lava	83.	foam
9.	bee	34.	mower	59.	lip	84.	fur
10.	toes	35.	mule	60.	cheese	85.	file
11.	tot	36.	match	61.	sheet	86.	fish
12.	tin	37.	mug	62.	chain	87.	fog
13.	tomb	38.	movie	63.	jam	88.	fife
14.	tire	39.	mop	64.	cherry	89.	fob (or fib)
15.	towel	40.	rose	65.	jail	90.	bus
16.	dish	41.	rod	66.	choo choo	91.	bat
17.	tack	42.	rain	67.	chalk	92.	bone
18.	dove	43.	ram	68.	chef	93.	bum
19.	tub	44.	rower	69.	ship	94.	bear
20.	nose	45.	roll	70.	case	95.	bell
21.	net	46.	roach	71.	cot	96.	beach
22.	nun	47.	rock	72.	coin	97.	book
23.	name	48.	roof	73.	comb	98.	puff
24.	Nero	49.	rope	74.	car	99.	pipe
25.	nail	50.	lace	75.	coal		

The list above can be used in several ways. One of the most common would be to remember a long list of items with associated numbers. Assume there are 31 steps to be done in order to complete a preventive maintenance routine. Assume also that the maintenance needs to be done in a place where having a written performance aid is not possible. The user must memorize the maintenance steps and their order.

The user memorizes the peg words during training, along with their associated maintenance activities. For example, if three of the mid-steps are:

1.
2.
.
.
.
14. grease rear joints
15. replace filter
16. reset dial
.
.
.
30.
31.

And if the three peg words are 14=tire, 15=towel and 16=dish, the maintenance steps could be remembered as "a tire covered with slippery grease", "a towel with sand being sifted through it" and "a dish with a digital readout in the middle." The associations between tire and grease, towel and filter, and dish and dial are established.

The user would work through the peg words until the maintenance was completed. If someone later were to ask what step number 15 was, the user would immediately envision the towel with the "filtering sand" and reply, "replace the filter." This is one way that has been proposed to use the verbal form of elaboration to improve performance.

There are other techniques for establishing peg words. One that is frequently used requires that the peg word rhymes with the number it represents, as for example, in the following list:

Peg Words That Rhyme

One is a bun
Two is a shoe
Three is a tree
Four is a door
Five is a hive
Six is sticks
Seven is heaven
Eight is a gate
Nine is wine
Ten is a hen

This technique requires, first, the memorization of the ten words rhyming with numbers one through ten. Suppose the first item to be remembered is telephone. Then one might imagine a bun with a telephone in it. Bun is always used for the first item, since it rhymes with one. If the second item to be remembered is dog, one might imagine a dog standing in some jogging shoes ("two is a shoe"). When trying to remember, one thinks of the number, then the associated rhyming word, and finally the associated image. Using this technique, and if given enough time to construct images, people are able to remember better (Bugelski, Kidd and Segman, 1968).

Ericsson, Chase and Faloon (1980) report on a person who was able to increase his memory span from 7 to 79 digits, but only after more than 230 hours of practice (one hour a day, three to five days a week for about one and one-half years). He was read random digits at the rate of one per second; he then recalled the sequence. If the sequence was reported correctly, the next sequence was increased by one digit; otherwise it was decreased by one digit. This individual categorized each number into groups of three or four, and then applied a mnemonic that was usually (90 percent of the time) based on running times and ages. For example, the number 3492 was remembered as "3 minutes and 49 point 2 seconds, near world-record mile time;" and 893 was "89 point 3, very old man."

Baddeley (1976) points out that one important advantage to using a verbal mnemonic lies in the fact that it is multidimensional. He uses as an example his semantic concept of a cat, which involves not only the word and its various verbal associates (dog, mouse, cream, etc.) but also what cats look like, sound like, the texture of their fur, what it feels like to pick a cat up or to be scratched by one. Baddeley suggests that this multidimensional coding has two major advantages: It allows greater discriminative capacity, and it allows redundant coding.

The story is told that many years ago, before people kept carefully written records, witnesses for major events such as births and weddings were young men and women. The younger, the better, so that the memory of the event would last for more years. The witnesses were often beaten to make the day even more memorable. This represents another albeit crude way of providing or building into the system multidimensional cues for memories that must later be recalled.

Information that is coded along several dimensions is less likely to be forgotten than information coded unidimensionally. The greater the number of dimensions, the greater the probability that at least one cue will survive to provide an available retrieval route (cf. Bower, 1967). Consider the last time you tried to remember the name of a person that seemed to be "on the tip of your tongue." You may have tried alternative means of retrieval such as thinking of the color of hair, height, facial expressions, or voice. The name was probably remembered only after reviewing several dimensions of the person.

It is also possible that high-imagery words are easy to learn at least partly because they have both a verbal and a visual content. The greater the number and range of encoded features, the greater the probability that one of these will be accessed, hence allowing the item to be retrieved.

Imagery Cues

Another form of elaboration is to use *visual imagery,* where a mental picture is created and "viewed." In this case, the images are created without the benefit of direct sensory data. It is a cognitive experience with the new image constructed from an item or items already in memory. Probably the main value of using imagery in remembering is to have an object that a person can "see" and that can be cognitively manipulated to be linked to other objects or cues. This process helps in the storage and retrieval of the information being manipulated. Imagery requires the memorizer to remember a great deal of additional information about the location representing the item, the elaborate image signifying it, and the links between location, image, and information.

To illustrate the extent that imagery can be developed, consider some of the phenomenal memory feats described by Luria (1968) for a person referred to simply as "S". Luria began his study of S in the mid-1920s. S was working at the time as a newspaper reporter. His editor noted that he never took notes but was able to remember a full day's worth of assignments. The editor tried to embarrass him one day by having S write down the assignments after they were given. S responded by writing down *exactly* what the editor had said. The editor encouraged S to visit Luria to give the psychologist an opportunity to study this remarkable memory.

S's capacity for memorizing material appeared to be unlimited either by the number of items or by the time span over which the items were remembered. For example, S could remember perfectly material he had memorized as long as fifteen years earlier. Not only was S able to remember the test material provided by Luria, he was also able to describe what the psychologist was wearing that day, where he was sitting, and so on.

One of the problems given to S was to remember a list of 50 numbers like that shown below:

6680542616
8479354237
3803470283
6095050176
2732573504

S required about 2.5 to 3 minutes to study this list. He could then exactly reproduce the table in about 40 seconds. It would take him about 30 seconds to reproduce only the second column in reverse order. The same set could be recalled in order several months later or up to 15 years later.

The key to S's memory feats was his use of imagery. As he was memorizing he might imagine, for example, a familiar street in his hometown. While walking along the street, he would place images of objects to be recalled in various places along the way. Usually, the objects—such as names, numbers, or events—were placed on the front porch, a gate, in the window, on the lawn, or another conspicuous place. Each object was placed at only one house, and in the proper order. To recall items, S merely had to "walk" along the street and observe what had been placed at each house. This type of visual imagery served to improve his memory considerably.

The few errors that S made are very interesting. Most of the errors were omissions. The errors usually could be attributed to him doing such things as placing a white object on a white background, such as a bottle of milk in front of a white house, or placing a dark object in a darkened passageway. The error then occurred when he did not "see" the item during recall. Eventually S learned to avoid such errors by being more careful where he placed items.

S's skillful use of imagery to improve his memory enabled him to remember considerably more material than he could before the skill was developed. However, the vivid imagery that he learned to use so well tended to interfere sometimes with his ability to understand the meaning of a message. His use of imagery often led him to think about irrelevant details of images rather than the message they conveyed.

Bower (1970, 1972) has identified the following principles for use with imagery:

1. Both the cues and items must be visualized.
2. The images must be made to interact so as to form a single integrated image. When subjects are told to imagine two items on opposite walls of a room, no facilitation occurs.
3. The cues either must be easy for the subject to generate himself or must be provided by the experimenter during recall.
4. Semantic similarity among the encoding cues impairs performance.
5. Contrary to popular belief, there is no evidence that bizarre image combinations are more effective than obvious relationships.

Imagery mnemonics seems to provide a retrieval plan which directs a person to the appropriate storage location. Having a good familiar set of locations provides a learner with a clearly labeled set of files in a well-organized file cabinet. Visual imagery then provides a powerful link between the cue and the item.

Keele (1973) has noted that imagery alone is not sufficient for improving recall. Imagery must be accompanied by organization. Atwood (1969) conducted a study where subjects were led into his laboratory, noting various places (such as doorways and desks) that could later be used for memorization. Subjects in one group were then asked to mentally put images of various objects to be recalled in the different places in the laboratory—a knife on the desk, perhaps. Presumably, those subjects would be able to recall the information stored by mentally retracing their steps through the laboratory and recalling the objects in the different places.

Another group of subjects was also told to construct images, but the places in which they mentally put the objects were not organized in any particular way—for example, they imagined a knife sticking in the ground. The group that put the images in places in the laboratory were able to remember much better. Atwood also reported that if people were asked to imagine printed, abstract words (such as "metaphysics") in different locations instead of imagining objects, recall was very poor. Good organization of poor images is not sufficient for good recall.

Conclusions

Designers can take steps to improve the memory of users in at least three ways. First, by ensuring that people are not expected to perform functions that put an unreasonable demand on human memory. As memory limitations are better understood they should be reflected in this early design activity. Second, designers should develop *interfaces* that support and encourage an acceptable level of remembering. Shakespeare's Globe Theater, for example, was called a "memory theater" because there were special cues in the architecture that served as pegs to help actors remember their lines. Designers can do similar things when developing a system. Cues to assist human memory should be everywhere. They should be built-in as part of displays, controls, workplaces, human/computer dialogs, codes, computer command languages, and so on.

Finally, the memory of users can be improved through the design of good *facilitators*. That is, providing training, instructions and performance aids that make full use of the mnemonic concept. The designer can assist users in creating and organizing meaningful images to help in remembering certain items over a long period of time. In other words, a design should help users develop their memory skills. (Some users may not even be aware that this is possible.) And after a user constructs and practices appropriate images, the designer should ensure that performance aids will complement and support the initial imagery training.

FOR MORE INFORMATION

Baddeley, A. D., *The Psychology of Memory,* New York: Basic Books, Inc., 1976.

9

MOTIVATION

INTRODUCTION

Motivation is any influence that gives rise to performance. Assuming equal ability, skill, and knowledge, high motivation usually results in high levels of performance, and low motivation results in low levels of performance. Given two people with equal ability, skills, and knowledge, but substantially different motivation levels, the person with the higher level of motivation will perform best.

The strength of motivational influences can range from virtually none to very high. At virtually none, a user could have the skills and knowledge necessary to perform an activity but not do anything. He or she might not show up to work at all (go fishing or stay in bed), or come to work but not feel like performing. At a very high level of motivation a user may arrive early, work hard all day (take no breaks and only ten minutes for lunch), and then take work home to do in the evening. People in this latter group are sometimes referred to as "workaholics." Most users are motivated to perform at a level somewhere between these two extremes.

Traditionally, designers have not taken much of an interest in user motivation. This aspect of performance was left to management of the new system to develop and implement. Recently, however, we have discovered that the nature of the work in a system can affect the motivational level of at least some users. The better the work design, the higher the motivation level of many users.

INTERNAL VS. EXTERNAL MOTIVATION

The reason that the design of work affects some users more than others appears to be related to whether internal or external motivational influences most affect a user. Internal motivational influences are closely related to performing the work itself. External motivational influences are related to conditions outside the actual performance of the work. Probably the best-known external influence in money.

With many useful classification schemes, the distinction between internal and external motivation can sometimes become fuzzy. In general, an activity is labeled internally motivating if its completion brings no obvious external reward. Conversely, an activity is labeled externally motivating if it leads to external rewards like food, money, or social reinforcement. Generally, another person or machine dispenses an external reward in a way that is not separate from the activity itself (cf. Condry, 1977). Some problems with this distinction are immediately apparent. For example, there may be potential future rewards for an activity that are not obvious to an observer—learning a skill may lead to external rewards in future situations where that skill is valuable.

Instead of defining internal motivation as the absence of rewards, we can also define it in a positive way as a need for competence and self-determination (White, 1959; deCharms, 1968; Deci, 1975), or as a search for an optimal amount of psychological incongruity (Hunt, 1965). Still another way of defining internal versus external motivation is to let the user make the distinction, that is, to allow users to determine whether their actions and consequences are largely under their own control or are primarily determined by external forces (Rotter, 1966; deCharms, 1968). In spite of these ambiguities, the distinction between internal and external motivation is still useful.

A study reported by Tannehill (1974) illustrates the difference between internal and external influences. The experiment involved a college psychology professor and college sophomores. The professor took two groups of 12 students to the woods. A low hill separated the groups so that, although aware of each other, neither group could see what the other was doing.

To the first group the professor assigned the task of chopping wood. He provided them with axes and agreed to pay them $2.00 per hour to chop, split, and stack logs. He assigned the second the same task at the same rate of pay, except that they could *only* use the *blunt* edge of the ax blade. They were *not* to use the sharp blade of the ax under any circumstances.

The first group, using the sharp edge, enjoyed the work and friendly competition sprang up. They could see the results of their efforts and began to take pride in the growing stacks of chopped and split logs. The second group, required to thump the logs with the blunt edge of the ax, began to mutter about what the professor could do with this experiment. When they threatened to quit, the professor offered them more money to stay.

More money worked at first. However, soon even the increase in money was not enough to hold the second group to the task. The professor then continued to raise the hourly rate until he was paying $12.00 per hour when the last student ended the experiment, threw down his ax in disgust and stalked off.

This example clearly portrays both internal influences (those related to the performance of work) and external influences (money). A combination of both influences motivates users, but the designer has more control over internal influences.

Even though designers cannot be held totally responsible for user motivation, they should try to identify work, interfaces, and aids that motivate favorably. For instance, designers should want to find out why people work at all, what influences them to work hard, and what influences them to care about and be proud of their performance. The following sections contain some discussions of motivation.

DIFFERENCES IN USERS

According to *Work in America* (1974), the work population is changing. New people entering the work force (many of whom work in newly designed systems) are different from those of even ten or twenty years ago. They tend to be younger, better educated, and less conforming than the previous generation—and many more are women. What might have motivated workers ten or twenty years ago may not do so now. In general, workers expect their jobs to provide more satisfaction and personal development. Consider the following example:

L. S., age 58, entered the work force 29 years ago; he speaks English (a second language) only moderately well, and has an eighth-grade education. His son, T., is 21, has a college degree, and is actively interested in politics. The same company employs both men. L. S. works a molding machine and T. manages the stockroom.

T. heard from plant gossip that his father's work was secretly being damaged so that the company could claim a tax loss that it badly needed. T. confronted the plant supervisor, and was angered to have the rumor confirmed.

He exposed the situation to his father, who was considerably less concerned than his son. L. S. felt that as long as he was paid for his work he had no reason to care what happened to the items he produced. T. disagreed

Workers entering today's work force often have values and needs different from those of previous generations.

and soon left for a job with another company. L. S. continued to work at his molding machine and to collect his pay every Friday afternoon. Different reasons motivated the two men to perform.

The father seemed to be at least partially motivated by the external influence of pay. He did not care so much about what happened to his work once he finished it. He simply did what was required and collected his pay at the end of the week. On the other hand, the internal influence of the work itself motivated his son—the feeling of pride and satisfaction from a job well done. T. was motivated by a different purpose and toward a different set of objectives than his father.

INSTINCTS, DRIVES, AND MOTIVES

The Greeks, almost 2,500 years ago, believed that people did things for two reasons—to seek pleasure or avoid pain. Unfortunately, some managers still feel that *all* user motivation in systems can be explained in this way—users try to avoid the pain of work, while they attempt to gain maximum pleasure through the highest possible earnings and benefits.

Instincts

The development of the concept of instincts was another attempt to explain motivation. Instincts were thought to be an unconscious, basic part of our human nature. People were felt to have little choice over whether or not they would try to satisfy their instincts. Instincts were considered natural forces that drove people to behave in certain ways. However, one of the outstanding characteristics of the whole idea of instincts was the wide diversity among psychologists in trying to identify a basic set of instincts universal to all people. Sigmund Freud, for example, attempted to explain people's performance in terms of three essential instincts: sex, self-preservation, and death. By 1924 psychologists had identified nearly 6,000 separate human activities that supposedly arose from instincts. Shortly thereafter the term became unpopular.

Drives

The idea of instincts was replaced with the concepts of "drives" and "motives." The drive concept infers that certain impulses within the individual induce very specific behaviors. This concept holds that an attempt to reduce a physiological drive or an aroused bodily condition supposedly motivates people to behave in certain ways. Adherents of this concept believe a person becomes motivated whenever an imbalance exists. For example, when a person's body needs water, a thirst drive is created and a person's behavior is directed at reducing the thirst by turning on a faucet to draw a glass of water. Drinking restores the balance and thirst diminishes.

A worker's basic physiological need for rest or sleep should not be overlooked.

Many drives have been studied, including hunger, thirst, sleep, and temperature regulation. For our purposes it is sufficient to understand that in some situations motivation is like filling in a gap, a gap created by some specific physiological need. System designers should be aware of the fact that people may be motivated in some situations by basic physiological drives. The design implications here are simple: Design systems so that people's basic physiological needs can be routinely satisfied. For example, users should not be deprived of sleep. Lack of sleep and the upset of certain other body rhythms, such as those that accompany the phenomenon of "jet lag," may produce degraded performance. People who are thirsty, hungry, or need to use a rest room may have difficulty concentrating on the performance of a system activity.

Motives

Incentives that are not physiologically based but are learned are called "motives." Motives are seen as a function of experience, usually some sort of social experience. Consider an athlete who practices and trains for several hours daily. Surely he or she is not motivated to do so because of a physiological imbalance or drive. Similarly, a student who puts in extra effort to get to the top of the class is not attempting to fill a basic physiological need. In both cases, it seems some special desire on the part of the individual to excel motivates the behavior. People frequently do things because they feel good doing them, because these activities represent a positive value for them, or because they simply have learned to enjoy performing the tasks.

Integration of Need Theories

Psychologist Abraham Maslow was perhaps the first psychologist to really pull together what was known about drives and motives as motivators. Maslow (1954, 1970) suggested that people have five major *needs* in varying degrees and intensities. See Figure 9-1.

Figure 9-1. Maslow's Hierarchy of Motivational Needs

Maslow contended that when the lower order needs—such as the physiological needs and the need for safety—are satisfied, they lose their potency as motivators and new, higher order needs emerge. The importance of higher order needs does not decrease, Maslow says, even when they are satisfied. In fact, for self-actualization, increased satisfaction leads to increased need. Although some specific ideas of Maslow's have received little empirical support, the designer can benefit from his conceptualization of human motives. In particular, a designer should attempt to create a climate in which users can develop their potential. Such a proper climate usually includes provision for greater worker responsibility, autonomy, and variety. In Maslow's terms, these provisions would provide for higher order need acquisition and satisfaction. The lower order needs must be accommodated before needs higher up the scale will even be considered. Designers should see to a person's basic physiological and safety needs by providing proper work schedules, sufficient heating and air-conditioning, opportunities for recreation, adequate safeguards against dangerous equipment and so on. A designer should not expect users to seek to satisfy higher level needs—feel good about themselves and their work—until their physiological and safety requirements are met.

MOTIVATING SYSTEM USERS

History of Work Motivation

Even though there has been considerable research done on motivation, until relatively recently there was little systematic thought given to motivating system users. People seldom asked why they or others desired to perform their jobs at a certain level.

Historically, fear has been a potent motivator in work situations. For centuries fear controlled people. Supervisors in some work situations still use fear. Another major motivation was allegiance. Consider the prodigious building feats of medieval times and the craftsmanship that went into the construction of cathedrals and castles. Allegiance to the church, king, or country motivated people to accomplish remarkable feats of building and craftsmanship. And the same allegiance motivated them to defend their structures, land, and beliefs.

The history of the Industrial Revolution in the United States presents a record of a lack of respect for the worker. It is, like all history, a mixed story. But one central theme was the exploitation of many workers, not as a human resource, but as an element in a mechanical work system. In those days concern for workers was not that important. Machinery was simple; almost anyone, including children, could be put to work. Units of production were small, and production techniques were easy to understand. People viewed worker motivation as unimportant in a time when laborers were abundant. There were exceptions. Some of today's approaches to motivating workers had their beginnings in the policies of a few enlightened employers of 100 years ago.

The growth of technology, combined with the increasing competitiveness of the marketplace, produced the need for greater efficiency and led to mass production. Attention focused more directly on the individual worker who came to be seen as the weak link in the productivity chain. People were neither as efficient nor as reliable as employers wished them to be. Questions about the workers' motivation levels became critical.

Frederick Taylor in the early 1900s proposed several ways to increase the worker's efficiency. One of his ideas was to give the worker a single best method for doing a job and then to use money as an incentive. Taylor erroneously assumed that workers were uniformly motivated by a desire for money and that other motives were unimportant.

In the early 1930s a Harvard sociologist, Elton Mayo, set out to uncover workers' needs. Mayo (1933, 1945) argued that it was necessary to consider the "whole person" on the job. He shifted attention from only the relatively limited human-machine interactions toward trying to gain a more thorough understanding of interpersonal and group relationships at work. The famous Hawthorne Studies (conducted by Mayo and his colleagues) found that workers participating in the study, at least partially because of having "control" of their workday and by being treated as important and unique, exhibited increased motivation. Mayo interpreted these results in terms of the human's need to belong to, and to be regarded as a significant member of a group. Subsequent analysis revealed that other factors were also involved (cf. Gellerman, 1963).

All the major approaches for improving user motivation since about 1940 have had little, if anything, to do with the design of work in a system. Consequently, they were beyond the control of a designer. These approaches included the following:

1. Reduced hours and longer vacations
2. Increased wages
3. Better benefit packages
4. More profit sharing
5. Better training of supervisors in:
    a. Human relations skills
    b. Sensitivity toward others
    c. The art of leadership
6. Employee counseling services
7. Organizational planning.

In the late 1950s Hertzberg (1957) conducted a review of the literature on job satisfaction studies. Based on the review, he and his co-workers interviewed a large number of employees, asking them to talk about periods at work when they felt exceptionally good or exceptionally bad.

For the workers they interviewed they concluded that feelings of strong job *satisfaction* come principally from performing the work itself. The actual work motivators seemed to be:

- The actual achievements of the employee.
- The recognition received for the achievement.
- Increased responsibility because of performance.
- Opportunity to grow in knowledge and capability at the task.
- Chance for advancement.

On the other hand, feelings of *dissatisfaction* are more likely to be attached to the context or surroundings of the jobs, including:

- Company policies and administration.
- Supervision, whether technical or interpersonal.
- Working conditions.
- Salaries, wages, benefits, etc.

Hertzberg found that conditions related most closely to user satisfaction are not simply the opposite of those related to dissatisfaction. Herzberg's findings have direct implications for system designers. Among other ideas, he suggested that motivating influences can and should be built into a system. Providing a computer operator with *feedback* of status information during excessive delays, or allowing a welder to leave his or her "mark" on a finished product as a symbol of recognition and responsibility are possibilities for improving motivation. Satisfying work is often seen as work that has value in itself—work that appears *worthwhile* to the user. Satisfying work also seems to be work that is complete in itself so that each user is responsible for a *recognizable product*.

Robert Ford (1969) in a series of studies conducted at the American Telephone and Telegraph Company, incorporated Herzberg's ideas in the actual redesign of work. In an effort to make some jobs more satisfying, Ford created work modules aimed at increasing task meaningfulness and feedback, as well as employee responsibility, achievement, recognition, and autonomy. Billing clerks, for example, who worked as a team, were responsible for sending out charges on toll bills at staggered dates through the month. The team was in charge of five subsections of a large geographical area. Productivity was low, due dates were missed, and costs were high. The team approach was discarded. Each worker was given responsibility and total control of getting out bills for a given subsection on time. User performance seemed to improve when each worker was given the freedom to organize his or her own bills.

Motivation and Basic Design Decisions

Function Allocation

As early in the design process as *allocating functions* to people, the designer must consider user motivation. People generally cannot reliably detect very low levels of visual, auditory, and tactual stimuli. Even so, designers commonly make the mistake of routinely allocating such difficult functions to people regardless of the potential motivational effects. Consider how difficult it would be to come to a job every day and listen for eight or more hours for barely detectable infrequent sounds or stare at a CRT for hours searching for tiny blips. Allowing a few signals to go undetected (by allocating the function to a machine) might be better than having to deal with degraded performance, costly turnover or absenteeism that often accompany poor user motivation.

Human Performance Requirements

Once we identify the major functions that a human will perform, we must further consider each function in terms of its *human performance requirements.* Setting out human performance requirements requires an understanding of user motivation, since these requirements include statements about errors, manual processing time, training time, and job satisfaction. If these requirements are not realistic, motivation is affected. For example, to require the total training for a helicopter pilot to be completed in just two days is absurd. Such a requirement, if taken seriously, could have a tremendously negative impact on the motivation of trainees exposed to a training program with little chance for success. A similar effect on motivation could be predicted if the allowable time to complete one's work was far too short, or the required accuracy was unrealistically high for the conditions. Designers should take time to study similar systems, and to use such historical data when developing realistic human performance requirements.

Designing Work Modules

The next major step in the design of systems is the *analysis and synthesis of tasks* included in the system and their *integration into work modules.* This process includes determining the system structure, identifying tasks, preparing a task flowchart, and identifying work modules. These activities constitute a significant part of the basic design process that can lead to acceptable levels of human performance. User motivation considerations must also occur in this stage of system design. Designers should build into the design of work modules those characteristics that will best ensure motivation in users. It is here that designers can have the greatest impact on motivation.

Motivation and Work Characteristics

Many users will respond favorably to work that has the following characteristics (Porter, Lawler and Hackman, 1975):

1. The user feels responsible for a *meaningful portion of the work.*
2. The performer considers outcomes of the performance *worthwhile.*
3. Sufficient *feedback* is provided for the user to adequately evaluate his or her performance.

Meaningful Portion of Work

Turner and Lawrence (1965) have suggested the term "autonomy" to describe the degree to which users *feel personally responsible for their work.* Work designed to have high autonomy will provide users with the feeling that they "own" the outcomes of their performance. A clearly identifiable product will result—a product that the user can initial and in which the user can take pride when all goes well, or feel discouraged when it does not work out well. For example, a key operator can take pride in a totally accurate set of input, a telephone operator can take pride in an exceptionally well-handled call, or a factory worker can take pride in a well-constructed part. In all these cases, the designer provided users with an identifiable "something" that results from good performance. On the other hand, work with low autonomy causes users to feel that the same results will be accomplished (good or bad, more or less work) no matter who performs the task. Low automony work generates no feeling of ownership, and no identification with a product or contribution. The implication for designers is clear. The performance required by each work module should lead to the development of an identifiable product. The term "product" is used here to indicate any identifiable outcome of performance, but the outcome should be clearly associated with the performer. The product could be a typed letter, completion of a maintenance routine or computer program, or the building of an entire generator.

McGregor (1960) argued that the more a person becomes involved in his or her work, the more meaningful and satisfying the job becomes. McGregor believed that employees are quite capable of making significant decisions in their work and that such autonomous decision making is in the best interest of a company. He felt that management's role was not to manipulate employees to accept orders from above, but to assist workers in developing their abilities, skills, and interests so that they maximally contribute to organizational effectiveness. McGregor's approach suggests some interesting implications for motivating workers through the design of systems. For example, what motivates one person may have little effect on another. Designers should design systems to be as flexible as possible to account for these individual differences. Probably the optimal condition in most systems occurs when the overall structure exists, but the worker has great flexibility to act as an individual within the structure.

Weiner (1973) proposed another approach to understanding autonomy in motivation. Weiner found that people usually attribute success and failure to one or more of the following sources: ability, effort, task difficulty, or luck. Ability and effort are called internal factors because they are part of the person. Task difficulty and luck, on the other hand, are referred to as external factors. He suggests that people are more motivated to continue work when they can attribute success on a job to ability and effort rather than to task difficulty and/or luck. Consider the frustration of a young worker assembling electronic components, who has just put together his first module and cannot figure out how he did it. Surely his motivation to continue is far less than a more experienced co-worker who has just established a new piece-rate record for the plant. In the first case success was attributed to luck, while in the second, it was attributed to ability and effort.

Worthwhile Work

A designer should also provide work that appears *worthwhile* to a user. If a person must perform activities that appear to have little value to anyone, it is unlikely that the performance will be taken very seriously by anyone, including the performer. Porter, Lawler and Hackman (1975) proposed two ways a designer can develop work that will be experienced as meaningful by users.

The first concerns providing the user with a *complete* or "*whole*" *piece of work*. The work should provide a clear beginning and ending, with the user having *total responsibility* for what takes place in between. And most importantly, a product's change due to the user's performance should be *highly visible* to both the user and others.

In a telephone company, for example, one person often has total responsibility for producing a telephone directory for smaller towns. The person begins with a collection of names of telephone subscribers, and using a computer, works with a printer and others to produce an error-free directory on schedule. The user has total responsibility from start to finish for production of the directory. An older, and alternative approach, that did not take into account this "wholeness" principle, had different people responsible for various pieces and parts of the total process. In the latter case, several people produced the directory, and no one person had the full feeling of satisfaction for an accurate directory produced on schedule.

When others consider the performance worthwhile, they also consider the performer worthwhile. When this occurs, along with open feedback channels that function correctly, the user feels worthwhile and motivated to continue the high-level performance. This assumes of course that a product is available to be evaluated. To simply observe the behavior of an individual does not provide much of an insight into a person's performance. Recall how difficult it was for Barton Hogg, who had the students digging in the sand for lead, to judge their performance by observing their behavior. To judge how good a performance is, a product must be produced. A designer should provide for the production of a worthwhile product.

A second way to help ensure that work appears worthwhile to a user is to provide for *variety in performance*. Performing a variety of different activities gives a person the chance to develop and exercise different skills. Many people remain interested in the work performed as long as skill building continues. Once the skills are fairly well developed, the work may not be as interesting. By providing a variety of different experiences, a designer can considerably extend the time for skill-building.

As skills develop, performance becomes more and more automatic. When this occurs, a user has more time for thinking, reading, or socializing while performing the work. In one computer-based system, the users' performance became so skilled (automated) that they could read magazine articles while effectively taking reports from customers over the telephone. Obviously, the work developed for these users contained a limited variety of different experiences.

When skills have been developed and the work is routinely, almost automatically, performed, many users seem to gain little satisfaction from the work itself (internal influences). These users may begin to receive more satisfaction from other aspects (external influences) in the work situation, influences external to the actual work performance. In some work situations thinking about water-skiing after work, reading a novel, watching television, playing a radio, doing crossword puzzles, and visiting with co-workers are all possible while performing the assigned work.

Socializing, in particular, is very common in many systems. In one word-processing unit, the users were all highly skilled at keying and would spend a good deal of time talking with one another and using the telephone. Neither their typing performance (accuracy) or productivity (key strokes per day) were questioned. Even so, it was very interesting to watch what happened when a tape containing dictation needed to be typed. The dictation was always passed to the newest person in the word processing center. As one typist observed, "it takes too much thinking to type from a dictated tape, you get left out of things that are going on around you." Typing from the dictated tape is potentially more challenging which requires the building and exercising of new skills. But, in this situation, once a person has given up on internal motivating influences in favor of external ones, the external influences seem to be preferred. In fact, people may view "meaningful work" as an interruption and even an irritant once relevance shifts to external influences.

When considering both "wholeness" and "variety" the designer usually has only partial control over the outcome. To help overcome this problem, designers should develop relatively small work modules—ones that contain few tasks. This enables the management of a new system to combine the smaller work modules in ways that best satisfy the requirements of wholeness and variety. Management usually has one critical additional piece of information that most designers never have, they know *exactly* what kind of people were hired to perform in a system.

Designers, then, should develop meaningful work modules, each with a definite product, and provide the modules to management who match the precise work required with the specific people hired to do it. Management may find "whole" jobs do not motivate some users. In this case, management may more profitably have several people performing smaller work segments. On the other hand, some users may respond very well to the "wholeness" idea and in these cases, a combined set of modules can give these people work that is intrinsically satisfying. The same logic can be followed for giving users work with or without variety. From a designer's point of view, the most reasonable approach is to provide management with the flexibility to go in either direction.

Feedback

A designer also has a major responsibility to ensure that users receive adequate *feedback* from their performance. If a key operator inputs large quantities of material into a computer each day and never receives any idea on its acceptability—was the volume great enough and accurate enough—then the key operator would have little chance to feel worthwhile or to know if the work is important. Clearly, a product exists and somebody most likely considers it important, but if the user does not know this, then it is all meaningless and the potential for motivation is nil.

Feedback can come from the work itself, but only if the designer has taken time to develop a way for this to happen. For example, a computer could be programmed to inform a key operator when certain errors occur, or even to give an evaluation of the volume and accuracy at the end of each day, compared against other days or other operators. The designer has total control over this level of feedback. It should be provided for each work module. This means that at a minimum, each work module should result in a worthwhile product and provide feedback to users about their performance.

Skinner (1953) contends that improved performance depends on feedback because any improvement results from how we were rewarded for doing the same or similar things in the past. Performing in an acceptable way depends on our past experience and the "contingencies of reinforcement" (reward or punishment) we have learned. In designing a computer system, for example, verbal messages praising a naive user at various stages of interaction can have a reinforcing effect. The user will likely be motivated to continue interfacing with computers because he or she found it rewarding to do so in the past.

Manipulating the contingences of reinforcement is a powerful way to motivate people to achieve an acceptable level of human performance, particularly while learning a new system. Appropriate rewards built into the ongoing operation of a system can help ensure an acceptable level of motivation over a long period of time. Remember, the total work experience should be positively reinforcing to a user. A poor situation cannot be patched-up by an occasional, and perhaps not even work-related, pat on the back.

System feedback often makes delays seem more tolerable to system users.

A system could be developed to provide feedback on user performance to management, with the hope that users will receive the necessary feedback from management. Unfortunately, management does not always make proper use of this type of feedback. Management may forget to tell a user, not report on a regular basis, figure that no benefit can be gained from telling users about their performance, or use the feedback information to punish certain users. Thus, a designer may provide the opportunity for management feedback to take place, and later find that it is never used or not used properly.

To help promote motivation, designers should provide feedback on user performance to both the user and management. If only one can be provided, the best way to ensure that feedback takes place is to build it into the work itself.

Responsiveness of Users

Remember, not all users will consider the existence of internally motivating influences as desirable. The design recommendations just discussed, that provide for a clearly defined product, worthwhile work and the necessary feedback, apply primarily to users who are still sensitive to internally motivating influences.

Condry (1977) and Lepper and Greene (1979) explored conditions where external influences appeared to destroy the effect of internally motivating influences. Lepper, Greene and Nisbett (1973), for example, found that when school children who liked to use marking pens received a promised reward for doing so, they later used the marking pens *less* than another group that received no reward.

We have reason to believe that as users gain experience performing work that does not contain internally motivating influences, over a period of time the users come to rely heavily on externally motivating influences—such as money, time off, opportunity to talk to others. A study by Kornhauser (1965) of automobile assembly-line workers suggests that when work is designed without internally motivating influences, users will eventually devalue internal influences in favor of external influences (cf. Porter, Lawler and Hackman, 1975).

Consider this in regard to the example reported earlier of the students chopping versus thumping wood. Assume that no other jobs were available and the "thumpers" had to perform the activity on a regular basis. These students would soon give up expecting any motivating influences from the work itself and come to depend totally on externally motivating influence (money, other benefits, socializing). Furthermore, even if given the opportunity to work with the "choppers" who were performing work that *did* contain motivating influences, the "thumpers" would continue to rely on *externally* motivating influences and ignore the internal influences that were part of the work. They had become dependent on external influences and the internal influences were no longer effective as motivators. Unfortunately, we have gone through several years (even decades) where people were expected to perform work that was totally devoid of internally motivating influences.

This being the case, a designer may frequently inherit people from other systems where designers had ignored the idea of building in internally motivating influences. It may take quite some time before performance improvements related to internally motivating influences will result for large numbers of workers. Nevertheless, a designer has a definite responsibility to develop work activities that take into account what is now known about motivating potential users.

Work Environment and Motivation

The physical and social working environment also should concern the designer, since the environment may affect the extent to which a user finds his or her job dissatisfying. Anyone who has experienced working under "sweat shop" conditions can attest to the diminution of the motive to perform in such an environment. Designers should be reminded of the importance of having clean, well-ventilated, and properly illuminated workplaces. Anything less may have deleterious effects on human performance in the system.

Motivation also can be considered through the concept of *arousal*. People seem to require working conditions that are "optimally arousing." Motivation, emotion, and stress all seem to be related to arousal. Shortly after the turn of the century, Yerkes and Dodson described the relationship between performance and arousal. Poulton (1972) and Welford (1976) present more recent discussions. Figure 9-2 illustrates what has come to be known as the

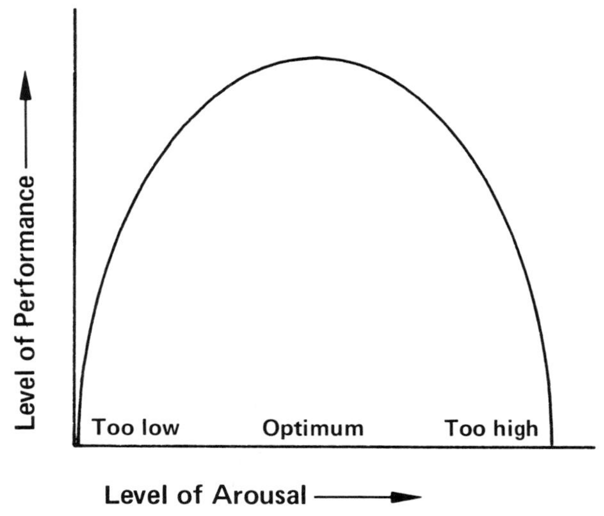

Figure 9-2. Relationship Between Arousal and Performance According to the Inverted-U Model

inverted-U model of performance.

Basically, the inverted-U model suggests that at low levels of arousal, task performance is also low; a condition we experience as *boredom.* As arousal increases, we reach a point where adaptation occurs. We call this arousal level optimal because it results in the best task performance. Increasing arousal past this optimal level results in *stress,* and stress may result in performance degradation.

Air traffic controllers provide an illustration of the arousal concept. A typical working day for M., a controller, begins at 2 p.m. when, customarily little traffic appears on her radar display. This situation continues for the next 45 minutes. Boredom sets in and M. tries to provide stimulation by daydreaming. Things begin to pick up at about 3:30 when a number of interstate flights begin their afternoon runs. Suddenly, M. receives a distress call from a DC-10. Just 22 miles away from the control tower, it is experiencing engine trouble and is requesting emergency clearance to land. At the same time four other aircraft appear on the radar screen in the same vicinity as the crippled aircraft. M.'s arousal level rises rapidly and exceeds the limits of adaptation—she is undergoing stress. M. tells herself to remain calm; she ignores several unnecessary messages and tries to reduce her high arousal level. For the next 20 minutes M. just barely manages to avoid an air catastrophe. Things calm down considerably afterwards, and at 9 p.m., after seven hours of continuous work, she leaves for home.

All people, at some point, usually experience boredom, stress, or fatigue in their work. System designers should assume the responsibility of minimizing these effects on performance. Sometimes, degraded user performance due to boredom, stress, and/or fatigue can be reduced or eliminated by allocating tedious, boring tasks to machines or by providing periodic rest breaks. The important point to remember is that people seem to work best at an optimal level of arousal. Designers should assist users in achieving such a level.

Excessive arousal due to worker overload can lead to performance deterioration.

MOTIVATION, SATISFACTION, AND PERFORMANCE

For years psychologists and others have been interested in the relationship between motivation, satisfaction, and performance. Research generally shows that whatever relationships do exist often vary from individual to individual and from situation to situation. According to some researchers (e.g., Brayfield and Crochett, 1955; Vroom, 1964) little evidence exists to suggest that satisfied users consistently exhibit higher levels of motivation. One can easily envision a user who is quite satisfied with some aspect of his or her job but who is not motivated to perform for any number of reasons—for example, he or she likes the social aspects of the job but does not care about the actual work.

One reason for this dichotomy is the difference between job satisfaction and performance satisfaction. A person's job includes much more than a person performing an activity in a particular context. A job is frequently considered in terms of (a) the time and difficulty to get to it and back home, (b) the number and types of other workers, (c) opportunities to participate in nonperformance-related activities, (d) length of breaks and cleanliness of the break area, (e) quality of food in the cafeteria, etc. A designer cannot hope to positively affect all aspects of a job, but he or she can strive to elicit satisfaction from the much narrower consideration of the actual work performed.

It *has* been demonstrated that increased motivation does lead to improved job performance (cf. Hertzberg et al., 1959). Users motivated by the work generally perform well. However, designers should not confuse user satisfaction and motivation. Design decisions that lead to increased motivation will not necessarily lead to increased satisfaction. Motivated workers in a system tend to be effective performers (assuming, of course, that they have developed the needed skills), while in many cases the most that can be said about satisfied workers is that they are not dissatisfied. Satisfied workers may or may not be motivated toward a high level of performance. It seems to depend on whether or not their satisfaction results from motivating influences that come from the work itself or if the satisfaction results from motivating influences that are external to the work. People may be satisfied because their jobs give them the opportunity to be around and visit with other people. The ideal situation from a designer's point of view is depicted in Figure 9-3. In this case the user is motivated by internal influences of the work, performs well, and feels satisfied with his or her performance.

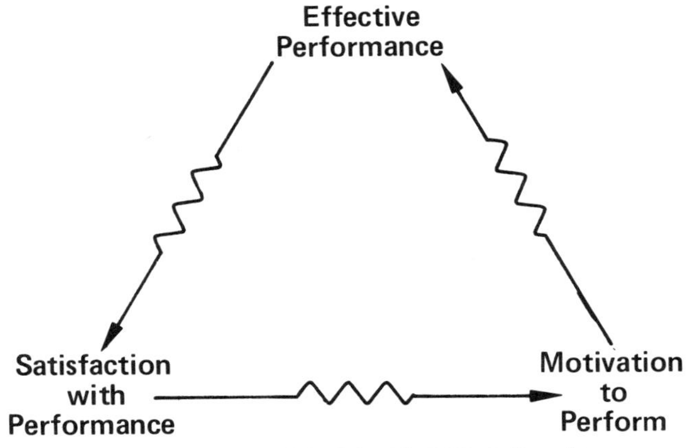

Figure 9-3. A Model of the Ideal Work Situation

According to Skinner's reinforcement approach to motivation discussed earlier, the likelihood that a user will be motivated to perform well depends upon the extent to which the activity was found to be rewarding or satisfying to perform in the past. Designers can develop work situations in which individuals gain satisfaction from the work itself. If this satisfaction from work performance is rewarding to the user, the user probably will be motivated to continue performing at about the same level.

FOR MORE INFORMATION

Hackman, J. R. and Oldham, G. R., *Work Redesign,* Reading, Mass.: Addison-Wesley, 1980.

Steers, R. M. and Porter L. W., *Motivation and Work Behavior,* New York: McGraw-Hill, 1979.

10

DESIGNING FOR PEOPLE

SYSTEMS

Almost everyone is familiar with the term "system." People speak of a school system, a transportation system, or an economic system. Shoppers have systems for buying and gamblers have systems for beating the odds. For our purposes, however, we will define a *system* as an entity that exists to carry out some purpose. It consists of people, machines and other components that interact to produce a result that these same components could not produce independently. Inherent in this definition is the notion that a system does not just happen. Rather, the concept of a system implies that we recognize a purpose; we carefully analyze the purpose; we understand what is required to achieve the purpose; we design the system's parts to accomplish the requirements; and we fashion a well-coordinated system that effectively meets our purpose.

If the term "system" sounds mysterious, think of it as it's applied to something as familiar (although complex) as an airline system. Two out of many of an airline's objectives are to make a profit and to transport people and cargo safely and comfortably. An airline achieves these objectives by flying and maintaining airplanes and other equipment, selling tickets, advertising, feeding passengers, maintaining communication systems, providing meteorological data, conforming to government regulations, bargaining with employees, etc.

What makes the airline a system is hundreds of individual activities designed to fit together in a coordinated fashion to achieve the airline's objectives. No matter how efficiently an individual activity is carried out, an airline's business depends on a coordinated *system*.

There are many ways to design a system. From the beginning all of the approaches should include a careful consideration of human performance. The human, by far the most complex of all system components, requires considerable attention to achieve acceptable performance.

TOTAL SYSTEM DESIGN

Total System Design (TSD) is a developmental approach that is based on a series of clearly defined development stages. Frequently in the past, people made system design decisions without giving ample attention to human performance. TSD implies that all major system components (the people, hardware and software in a computer-based system, for example) are given adequate consideration in the developmental process. An integral part of the approach involves giving serious attention to human performance from the beginning of the design process. The result is a system as carefully balanced as an expensive stereo system, where all components uniquely contribute to achieving the system's objective.

Most good system development efforts are based on at least two main principles. First, clearly defining the problem, and second, planning and carrying out a strategy to attain the desired solution. Defining a problem helps establish the need for a new system. The emphasis then shifts to determining and carrying out a useful development plan or strategy. Nearly everyone is familiar with the necessity for good planning, and many have benefited from their own plans or the planning of others. Building a backyard patio, going on a camping trip, transferring to a different city all require considerable planning and much attention to detail. As complicated as such activities can be, the development of a system can be considerably more complex. Consequently, it is helpful to have special, structured ways of planning and executing the design process so that it can be carried out as efficiently as possible.

DESIGNING IN STAGES

Partitioning the system development process into a series of meaningfully related groups of activities called *stages* makes the process more manageable. Each stage contains a set of activities that usually require completion before moving on to the next stage of activities. There are numerous activities to be completed in each stage.

For example, there have been numerous computer-based systems developed at Bell Laboratories. The major human performance-related activities performed during the development of these systems were compiled by a technology and training advisory group (Fleischman, Levin, Manley, Peavler, Yavelberg, Rebbin, Schwartz, Soth and Vassallo, 1979). All of these activities are well documented, and many of them have detailed methodologies established for carrying them out. The fact that these particular activities have been effectively used in the design of many systems provides a measure of their validity and their potential usefulness to other designers. Several of the most important human performance related activities are listed below according to the system design stage in which they are usually performed. We will discuss these activities, and others that designers may find valuable in more detail in this chapter and in the following chapters.

- While system objectives and specifications are being developed:
  - — Determine user needs
  - — Determine user characteristics
  - — Determine organizational characteristics
  - — Determine the work flow
  - — Determine human performance measurement procedures and parameters.
- During the system definition stage:
  - — Define functional requirements
  - — Define operational (performance) requirements.
- During the basic design stage:
  - — Allocate functions
  - — Design work procedures
  - — Design performance feedback mechanisms.
- During the interface design stage:
  - — Design interfaces
  - — Design work areas.
- During the facilitator design stage:
  - — Develop staffing requirements
  - — Design and develop instructions

— Design and develop performance aids

— Design and develop training.

- During the evaluation stage:

  — Develop testing specifications

  — Conduct test sessions

  — Perform system evaluations.

These stages are briefly described below, along with some suggested activities related to human performance to be carried out during each stage. This discussion will focus on activities that have a strong impact on human performance once the system is operational.

SYSTEM DEVELOPMENT STAGES

Stage One: Determine Objectives and Performance Specifications

System objectives and performance specifications should be clearly identified before the design of a system begins. System *objectives* usually are stated in general terms to avoid their revision in cases where performance requirements may be altered in response to technological, budget, or other constraints. Examples of system objectives are: prepare a White Pages telephone directory, develop a new automobile, or create a space vehicle that will go to and return from Mars. System performance *specifications* should state what the system must do to meet its objective. Generally, these requirements are derived from a careful study of user needs typically involving interviews, questionnaires, site visits, and work studies.

The system performance specifications usually include a set of requirements and a set of constraints. *Requirements* help to identify the purpose of the system as a whole, and the operational characteristics or performance requirements which detail specific objectives. System *constraints,* on the other hand, are limits within which accomplishment must take place.

System specifications help clarify the system's objectives, and are related directly to the operational performance of the system. Building on the objectives already presented, examples of performance specifications include: the White Pages directory must be produced in three months; the automobile must get 100 miles to the gallon and comfortably carry eight people; the space vehicle must carry a crew of three and a payload of 10,000 pounds. From a human performance point of view, during this stage the designer is primarily concerned with:

1. Identifying the intended users of a system (both those who will do the work, and those who will receive outputs from the new system).

2. Studying how the work is presently being done (interview and observe present users and prepare flowcharts detailing their work).
3. Understanding the activity-related needs of users to which the system will be addressed (identify opportunities for user satisfaction in the current work, and existing standards for accuracy and timeliness).
4. Ensuring that the system objectives reflect these user needs (fulfillment of important user needs can be a system objective itself).

What is done in this stage and how well it is done will affect not only the next design stage, but will impact in various ways on the entire development process. In this stage, designers must make decisions that could have considerable impact on human performance, and others that will have little or no impact on human performance. We will emphasize only those activities that could have a substantial impact on human performance.

A system may be developed to achieve a variety of objectives such as to do things better, faster, or at less cost. Achieving these objectives requires that system users be provided with information that is necessary, as well as accurate, timely, and in useful form. These *needs* of users should be clearly spelled out in this stage. It will be more costly in terms of time and money to do so later. Designers should be particularly interested in identifying the specific requirements users need to adequately complete work activities. Designers have a major responsibility for ensuring that the requirements of users are clearly and comprehensively defined. In the next stage (Definition), these requirements will be further refined and considered in relation to other factors. But by far the designer's most important consideration in this stage is to identify, to understand, and to accurately document the *needs* of users.

Simply knowing what work has to be done (activities) and how to do it (methods) is not enough. The designer also needs to know what system inputs will be available to carry out the activity and what the products (outputs) of all activities will be.

Potential users include all people who may be involved with the operation of the new system. It is important to identify the specific needs of different types of existing and potential users. If these needs are well understood, the new system can be designed to support the users as fully as possible. One way to start this process is to consult the person or persons who first identified the need for a new system. Additional information on user needs may be collected through observation, interviews, and questionnaires.

The development of system objectives is an iterative process which extends throughout this stage. Since this activity occurs long before the assignment of functions to people or machines, the designer must be content with ensuring that all important user needs have been identified and are reflected in the stated system objectives.

If the system is developed to meet a set of needs that users in the existing system do not regard as critical, the users could reject the new system as inappropriate. Users can "reject" a system in a wide variety of ways ranging from acts of sabotage to the quiet use of an alternative (usually manual) system. Other forms of rejection are to work slowly, purposely make excessive errors, never quite "catch on" to what is required, and to constantly complain of how dissatisfying it is to work in the system. These responses are all related directly to human performance.

Determine Alternatives

We should point out that many designers tend to figure out how to develop a system and *then* proceed to identify objectives and performance specifications. This approach creates two problems. The first, and most obvious is that system objectives are defined to fit a preconceived design. It is best to identify all objectives and performance specifications and then work in the design characteristics.

The second problem is that a premature design discourages the development of design alternatives. It is usually best to develop several alternative designs before selecting one to use. Crossley (1980) reports on a study where four experienced designers were asked to independently design a simple bell-crank lever. Not surprisingly, the four people produced significantly different designs. What interests us even more, is that none of the designers produced what turned out to be the best design. This and other studies suggest that the first design solution to come to mind is rarely the best.

Generally, a designer must consider several alternative systems to identify the one with the highest potential for the best human performance. Consider, for example, that one design alternative may require only clerical personnel, a second may require only management personnel, and a third may require both clerical and management people, plus some technical specialists. A designer must consider the availability of the different groups of people, the available training resources, the general types of work to be done, etc. before making an informed decision on the alternative that will most likely lead to the best human performance.

Determine System Performance Specifications

Once system objectives are identified, system performance specifications can be determined. The purpose of this activity is to develop a quantitative description of:

1. What is to be done
2. How well it is to be done
3. How to measure what is done and how well it is done.

This activity is iterative with the preceding activity, the development of system objectives. As the objectives are increasingly refined, the performance specifications should be refined also.

The kinds of system performance criteria related to human performance that we are considering tend to be quite general, such as cost and time savings, decreases in accidents, reductions in required people, improved customer service, and better employee work opportunities.

The designer should be especially careful when evaluating the potential impact of the new system on human performance. Keep in mind that the functions have not yet been allocated, making it difficult to make any precise estimates. Even so, the designer must be careful not to establish system performance specifications that will hinder future decisions concerning achieving acceptable human performance.

When considering system performance specifications, the designer should always take into account the context in which the system will operate. From a human performance point of view, this includes, for example, the nature of the labor market, existing labor agreements, skills in the existing work force, the existing organizational structures, etc. Characteristics of a system's context can greatly affect human performance. A designer would not want to develop a system, for example, that required large numbers of skilled people if the system is to be located in a sparsely populated rural area.

A designer might decide that to the greatest extent possible the same kinds of people employed in an existing system will be employed in the new system. To evaluate this decision, the designer would have to have a good understanding of the existing user population, as well as a good idea as to what such a decision implies in respect to interface and facilitator design and development.

Another possible decision may be that the new system will reduce the number of people needed by 50 percent. Such a decision suggests that machines will perform many more functions than in the past. The designer should be aware of the major human performance implications of this decision. Some (perhaps many) of the existing people will lose their jobs once the system is ready. This could cause considerable hostility toward the new system.

A Bank Example—System Objectives and Specifications

As an example of the types of activities that can take place during development of system objectives and performance specifications, consider a situation where a bank is having an extremely difficult time because of large numbers of customer overdrafts. In most cases these overdrafts are not due to fraud. In fact, the bank conducted a study and found that the vast majority of people who had overdrafts simply did not know how much money was left in their bank account when the checks were written. The bank's management decides that it needs a way of helping people keep better track of their bank balances. They decided to develop a new system that would reduce the number of overdrafts—a system that would help customers know how much money they had in their checking accounts at any given time. After much discussion, the bank's management agreed on the following objective for their new system.

System Objective: Provide a means for encouraging (assisting) customers to keep a more accurate accounting of their bank balance.

This objective is general enough to allow designers considerable freedom in finding a good solution, but specific enough to convey to the designers exactly what the system ought to do. The designers also are provided with the following set of system performance specifications that include a list of requirements and constraints.

System Performance Specifications

- Requirements:
    1. Should be convenient for most customers to use.
    2. Most customers should be able to accurately use the new approach.
    3. Most customers should be interested in using the new approach.
    4. The system should be easy to use by most customers.
- Constraints:
    1. Customers' understanding of what is required by a bank concerning their account varies widely.
    2. Customers have a wide range of math skills.
    3. Customers have a wide range of attitudes toward maintaining an accurate balance (from having no interest to having personal accountants).

The system designers took the objective and performance specification (and all other information they could collect), and came up with several alternative ways of designing a new system. Some of the alternatives are listed below.

1. Customers fined $100 for each overdraft.
2. Customers rewarded by a substantial reduction in their monthly service charge for having no overdrafts.
3. Customers encouraged to buy and use a hand-held calculator for updating their balance as checks are written. The encouragement is done through extensive advertising.

4. A special calculator (designed, programmed and issued by the bank) is given to each customer.

5. Customer telephones the bank and updates all transactions once a day and receives balance at that time (the telephone call is required or the bank will not honor any checks written that day).

6. Bank mails daily statement to customer that shows all transactions and the balance.

7. Bank sends balance and all transactions to the customer's home computer on a daily basis (the bank buys a large number of home computers and makes them available as rentals or for sale at reduced prices).

8. Bank provides extra money that is then borrowed by the customer to cover overdrafts (this allows overdrafts to continue, but with a slight penalty to the customer).

9. Customer is required to telephone for clearance before each check is cashed. Each check must contain the "clearance number" to be honored. Balance is given when call is made.

10. Customers are allowed to cash checks only at places of business that have a direct hookup to the bank, so that the balance is adjusted as the check (more like a credit card) is cashed.

Stage Two: Definition of the System

Determine and Describe Inputs/Outputs and Functions

Once system objectives are set, and system performance specifications are determined, the factual basis of the system can be established. At this point the basic layout of the system usually has been decided. This means (we hope) that the best of several alternatives has been selected. System-level inputs and outputs now should be defined, and the work activities (functions) required to convert inputs to outputs specified. A "function" is simply a statement of work that must be accomplished for the system to meet its objectives. Designers usually identify these functions, and prepare system-level flowcharts that show each of the functions roughly in the order they will be performed. This flowchart represents a first cut at determining the system configuration.

Detailed narrative descriptions can be prepared for each of the functions and for the system configuration itself. For human performance reasons, descriptions should be as specific as possible concerning available inputs, required outputs, and the functions.

Avoid Allocating Functions

A designer should be aware of the tendency in this stage to go beyond a simple description of the work to be performed and to make decisions about *who* will perform each of the functions (people or machines). Some designers prepare functions that describe not only *what* will be done, but also *how*. It is not uncommon to find that when functions are assigned to people or machines during this early stage, wholesale assignments are made to machines leaving the leftovers for people to do. *These assignment decisions are premature,* and could have a considerable negative effect on human performance.

Functions can be more systematically analyzed, with allocations more appropriately made to people or machines, later in the developmental process. The best approach is for a designer to stick with establishing the factual basis of the new system (i.e., ensure that all inputs, and outputs are identified, and all required functions are known), and not begin thinking about what machines and people will do.

The Bank Example—Definition

In the previously mentioned bank example, the alternative selected was to *design, program and issue a special calculator*. The calculator helps change a user's account balance as deposits are made and checks are written. (To simplify the example, problems relating to a monthly reconciliation of the checking account will not be addressed.) Recall that the three major human performance-related activities during this stage include identifying system inputs, outputs and functions. The major functions are:

1. Determining the account balance
2. Adding money to the account (making a deposit)
3. Withdrawing money from the account (writing a check)
4. Changing the balance in the account
5. Storing the balance until needed.

Each of these functions has an input and an output. For example, the input for the function "changing the balance in the account" relates to money added or withdrawn. The output of performing the function is a changed balance in the calculator. We do not know whether the user or the calculator performs the functions. At this point, it is more important just to ensure that the list of inputs, outputs and functions is *exhaustive*. We should emphasize *what* must be done to satisfy the system objectives, taking into account the system requirements and constraints.

Collecting People-Related Information

Another main activity of a designer in the Definition stage is to identify where information about people is needed and to gather it. Two kinds of information the designer will wish to collect for use in later phases of development include:

- Availability of human resources (numbers and types)
- Basic characteristics of people in the potential user population, including information on sensing, cognitive, and responding capabilities.

For example, suppose the policy has been adopted that the new system will be designed so that it can be operated by existing employees. In such a case the designer will want to make sure that good information is available about the existing work force. The designer will want to ensure that information about these human capabilities and limitations is available for use in the next stage where designers will be deciding which functions will be done by people and which by machine.

The Design Process

From a human performance point of view, we consider the actual design of a system to take place in at least four distinctly different stages: *basic design, interface design, facilitator design,* and *testing.* Beginning the actual design of a system greatly expands the human performance responsibilities of the designer. We will deal with each of the major design stages in more detail in later chapters; however, some of the most important activities are summarized on the following pages.

Stage Three: Basic Design

Early Decisions

Once system objectives and performance specifications are set and the factual basis of the system is defined (functions and associated inputs/outputs), the actual design process can begin. Much of the information developed in the Definition stage can serve as the basis for what we need to accomplish in this third stage of system design. Incomplete or inaccurate material from the Definition stage—particularly information related to potential users and system functions—will be reflected in the quality of all future design decisions.

The system develops rapidly during the actual design process. As the system takes shape, unforeseen changes and modifications may become necessary. A designer should be constantly alert that human performance decisions are compatible with hardware and software decisions. Remember, the Total System Design Strategy means all major components are given adequate consideration. Some designers have a strong tendency to ignore important human performance considerations at this point in favor of ensuring adequate design of the software and hardware. When this occurs, users are left to fend for themselves because designers gave little or no thought to human performance considerations.

Since human resources must be available to perform the functions that will be allocated to people, these resources must be accurately assessed. Earlier data collection during the Definition stage provided general information on potential users. Now, much more specific information about abilities, skills, knowledge and motivation must be identified. Informed predictions should be made concerning the impact of labor market conditions and employee turnover on the pool of human resources over the expected life of the system. The identified human resources should then be used in the best possible way to help ensure an acceptable level of human performance.

Function Allocation

With this information in hand, designers develop criteria for conducting a *function allocation*—dividing functional responsibilities between people, hardware and software. Some functions are relatively easy to allocate. Their allocation will be clear from stated system performance specifications—for example, to preserve current personnel—or because there is clearly only one reasonable choice for the allocation. Other functions will be more difficult to allocate. The designer may need to perform a more detailed analysis of potential human functions to (1) determine whether such functions *can* be successfully handled by the expected personnel resources, and (2) to make predictions as to whether people *will* perform the activities over a long period of time. Remember, some people may have the ability, skills and knowledge, but not the motivation to perform adequately.

Human Performance Requirements

After the designer identifies human functions, he or she should determine a set of *human performance requirements* for the functions. These requirements relate specifically to human performance (unlike system requirements) and address such performance characteristics as accuracy, speed, time necessary to develop unique skills, and user satisfaction. A designer must know these requirements to make informed decisions concerning work modules, interfaces and facilitators.

Task Analysis

When the designer knows the human functions and human performance requirements, he or she is ready to perform the *task analysis*. Task analysis breaks down and evaluates a human function in terms of the abilities, skills, knowledge and attitudes required for performance of the function. The analysis consists of identifying a function's inputs and outputs, and identifying the lower level activities necessary to change the inputs into outputs. Next the inputs and outputs of each lower level activity are identified. The process is repeated until the activity is broken down to a level best understood and performed by a specific user. We refer to this level as the "task" level, and use it as the basis for evaluating interfaces and developing facilitators.

Designing Work Modules

Tasks are combined into *work modules.* A work module is a grouping of one or more related tasks for assignment to a particular type of user. Frequently more than one person performs the work of any given work module. A work module is generally smaller than a job; in fact, each job consists of 1,2,3, or more work modules.

Designers perform many other activities during this time that are not as directly related to achieving an acceptable level of human performance. Completion of some of these activities may necessitate modification of some of the work modules the designer has developed. For example, designers should undertake studies of how system elements interact during representative routine and nonroutine modes of operation. Particularly important is the effect of contingencies—infrequent events—that may arise. The designer should carefully evaluate the potential impact of contingencies on human performance. Keep in mind the major human performance problems at the Three Mile Island nuclear facility that were due, at least in part, to designers not having thoroughly considered contingencies.

The Bank Example—Basic Design

Returning to our bank example, designers allocated the functions to either a *user* or *calculator* as follows:

1. Determine bank balance
    - Take out calculator (user)
    - Report balance (calculator)
    - Read or listen to display (user).
2. Deposit money and make a record
    - Fill out deposit form (user)
    - Give money to cashier (user)
    - Take out calculator (user)
    - Turn on (calculator)

- Use number keys to enter amount and then press DEPOSIT key (user)
- Turn off (calculator).

3a. Write a check and make a record (alternative 1)

- Take out checkbook (user)
- Write current date (user)
- Write recipient's name (user)
- Write amount in numbers (user)
- Write amount in script (user)
- Sign (user)
- Give check to recipient (user)
- Replace checkbook (user)
- Take out calculator (user)
- Turn on (calculator)
- Use number keys to enter amount and then press CHECK (user)
- Turn off (calculator).

3b. Write a check and make a record (alternative 2)

- Take out combination checkbook and calculator (user)
- Turn on (calculator)
- Insert check in calculator (user)
- Print current date (calculator)
- Use number keys to enter amount and then press CHECK key (user)
- Write recipient's name (user)
- Sign (user)
- Give check to recipient (user)
- Replace combination checkbook/calculator (user).

4. Change and store balance (calculator)

Notice that some of the first of the original five functions were broken into lower level functions to have activities assigned to either the customer (user) or the calculator. Two of the original functions were combined into a single function (change and store balance) and allocated to the calculator.

A close examination of the allocations indicates that the designers elected to assign numerous functions ordinarily performed by the customer to the calculator. The primary reason was to reduce the skill level required of the user. In addition, it was thought that the numerous allocations to the calculator would help speed up performance of the functions as well as improve accuracy. For example, the calculator would automatically turn itself on when the first number was pressed, and turn itself off when either the DEPOSIT or CHECK function keys were pressed. That the balance always appeared in the display also saved time. It was hoped that errors would be reduced by having the calculator automatically do all of the calculations. User satisfaction would be increased by providing the alternatives of either reading or listening to the display, and by developing a checkbook/calculator combination especially suited for conveniently writing and recording checks. Generally, a good practice is to provide alternative allocations as the designers of this system did with the "write checks and make a record" function. The two alternatives can then be evaluated against other human performance considerations and the best one selected.

Also in our bank example, the human performance requirements were outlined as follows:

1. Customers would receive *no training* on how to use the new calculators. The only "how to use" information would be provided in a short set of written instructions.
2. Customers should be able to read their balance in 2 seconds or less, enter information for a deposit in 15 seconds or less, and write a check in 30 seconds or less.
3. Customers should be able to read their balance accurately 999 times out of every 1000, and accurately enter all information for making a deposit or writing a check 99 times out of every 100.
4. A high percentage of use by customers, and the results of periodic questionnaires concerning the new system, should indicate customer satisfaction.

The functions assigned to customers were further evaluated in terms of the human performance requirements, and other important criteria. The functions were broken down into the actual tasks to be performed by customers. These tasks were further analyzed and compared with the skills known to exist in the user population. The tasks were organized into logical modules of work. These systematically derived work modules will help customers quickly learn about the new system, and efficiently use the calculator.

Stage Four: Interface Design

Overview

After making the basic decisions concerning the nature and structure of work in the new system, the designer switches focus to interface characteristics. Here, he or she identifies and deals with the most important human interfaces. These include the layout of workspaces, controls, displays, human/computer dialogs, forms, etc. The aim should be to identify areas where system performance will be improved by well-designed interfaces. Among other things, this means that interface points are compatible with human capabilities and limitations.

The Bank Example—Interface Design

In our bank example, probably the three most important interfaces are: the visual and auditory calculator displays; the size, shape, and location of the keys; and the type of customer/calculator dialog selected. Each of these considerations requires considerable attention to develop interfaces that contribute to meeting the human performance requirements discussed earlier.

The display and keyboard interfaces were developed by first using human factors data to arrive at a good basic design, and then trying various configurations on typical users to find the best possible layout. A similar method was followed for determining the best customer/calculator dialog. Three major alternatives were considered. First, a telephone-type keyboard with both letters and numbers on the keys, that customers would use for entering amounts and spelling DEPOSIT and CHECK. Second, the customers would use the same type keyboard but only key a "d" for deposit and a "c" for check (or some other prescribed abbreviation). The first alternative would provide for English instructions to the calculator (easily remembered), but would require quickly and accurately depressing several keys. The second alternative would reduce the number of keys to be pressed, but would require the customer to remember an abbreviation. A third alternative—eventually selected—was to have ten keys for entering numbers, and two special keys (function keys), one labeled DEPOSIT and the other labeled CHECK.

Stage Five: Facilitator Design

Overview

In the Basic Design stage, functions were allocated between people, software, and hardware. In addition, the designer identified human performance requirements, conducted a task analysis, and initiated the development of work modules. In the Interface Design stage the designer carefully considered all important interfaces and made appropriate decisions. Finally, in the Facilitator Design stage, the designer is primarily interested in planning for a set of materials that will encourage acceptable human performance. Some of the activities include:

1. Determining the best "mix" of facilitator materials
2. Developing a preliminary set of selection requirements
3. Identifying where instructions, performance aids, and training will be needed and the best use for each.

In this stage, then, the designer focuses upon a rigorous analysis and several critical design decisions concerning the facilitators associated with work people will perform in the system.

The Bank Example—Facilitator Design

In the bank example, the designers knew from the beginning that customers could not be selected (screened), and would receive no formal training on how to use the system. This meant that human performance had to be facilitated by a clear and concise set of written instructions. The designers first came up with what they determined was a complete and meaningful set of instructions and then had potential users read them and do what they said. The instructions were modified several times until the majority of potential users could read and perform adequately. To help ensure that the instructions would always be available, they were etched on the back of the calculator.

It should be noted that the thoughtful allocation of functions between a customer and calculator, the good organization provided by the task analysis and work module identification, and the clever use of keys with numbers and two function keys, all contributed to having a brief but meaningful set of instructions.

Stage Six: Testing

As the developmental process progresses, design specifications are converted into reality—coding specifications become a computer program, blueprints become mockup and/or actual equipment, and hardware and peripherals are purchased. In addition, the selected controls and displays are purchased, as well as other interface-related items. Actual command languages, human/computer dialog structures, and computer messages are developed and are tried. Facilitators in the form of detailed selection criteria, instructions, performance aids and training materials are prepared.

An integral part of the development of products is the extensive use of testing. As interface and facilitator materials become available, they require systematic evaluation and testing. Programs are debugged, combined, and "string" tested; equipment is bench tested; CRTs are evaluated; and, along with command languages and computer messages, the human interface is tested. Instructions, performance aids, and training are tested, first separately and then together. As larger portions of the system (i.e., subsystems) are completed, they are tested. Once all components are ready, the entire system is tested as a unit. The development of products and testing cannot be easily separated. Development usually leads to testing, and the test results frequently lead to further development.

A final caution: For a new system to reap the full benefit of efforts to ensure acceptable human performance, human performance considerations must begin in the earliest stages of development. In most cases, it will not be possible to overcome faulty function allocations by providing superior training; or to overcome an insufficient labor force by having a good interface design. Early attention to human performance is the solution.

FOR MORE INFORMATION

Kirk, F. G., *Total System Development for Information Systems,* New York: John Wiley and Sons, 1973.

11

BASIC DESIGN

INTRODUCTION

In the previous chapter, we discussed the overall system development process. Several activities were outlined that should be performed as early as possible in the development of a new system to help ensure an acceptable level of human performance. Once the actual design of the system begins, at least four major processes should take place during the *Basic Design* stage. These were briefly discussed before, and will be dealt with in more detail now.

The first major process is *function analysis* and *allocation*. This involves analyzing and describing all functions and finally allocating their performance to either a machine (possibly a computer) or a human. In the second process, the functions assigned to people are evaluated further to determine *human performance requirements* for the system.

During *task analysis*, the third major process, the functions are broken down to determine how best to configure the work to ensure an acceptable level of human performance. An integral part of this process is to synthesize or *combine tasks into work modules* that are designed and developed further to ensure an acceptable level of human performance. A final process performed in some systems involves combining work modules in order to *form jobs*—job design. However, job design is a responsibility that may be best performed by the management of a new system.

FUNCTION ANALYSIS AND ALLOCATION

Once the need for a new system has been demonstrated, it is necessary to identify the major functions that the system will perform. *A function is a statement of work,* and if all functions identified are successfully completed then the system should meet its objectives. Functions usually reflect the most general, yet differentiable, means whereby the system requirements are to be met. There was a time when people performed all functions in a system. Changes came about when animals and machines began to be used to supplement human muscle power. It was then possible to divide work between people, animals, or machines.

Functional Thinking

One of the most important advantages of the stage approach to system development is that it forces designers to concentrate on the most important decisions as the developmental process proceeds. Thinking in terms of "functions" enables designers to separate work from the means used to achieve it.

Singleton (1974) suggested two advantages to functional thinking in this stage of the design process. First, it produces a common language for all specialists. For example, power generation may prompt a chemist to think of a chemical reaction, such as the burning of gasoline, as a source of power; an electrical engineer to think of an electric motor that gets its energy from a wall socket; a mechanical engineer to think of the wind and windmills; and a solid state physicist to look to solar cells. Each specialist has her or his own idea of what power generation entails, but each clearly understands the abstract concept.

Second, Singleton sees functional thinking as the means to an integrated design. He views many current designs as primarily done in one physical discipline, with other disciplines tagged on as an afterthought or as a quick fix of an unforeseen problem. An example might be the catalytic converter included in many new automobiles. This is a quick, chemical solution to the old problem of clean, efficient personal transportation. An alternative approach would be to attack the problem from a more fundamental position, which may lead to better solutions such as fuel injection, the Wankel engine, or electric cars. If a functional approach had been taken from the beginning and if all the necessary specialists were thinking in functional terms and working toward an objective together, a more integrated design may have resulted.

John (1980) suggested another advantage to functional thinking. She observed that many theories of creativity and innovation regard the definition of the problem in the broadest terms possible as the first step in achieving true novelty in a solution. Thus, if a system is expressed in terms of its functions, its problems are expressed in their most abstract form and the best chance for innovation exists from the beginning.

Determining System Requirements

Frequently, the original set of the functions must be analyzed and broken down to where a set of system level requirements can be clearly established. Successful completion of functions is measured by evaluating whether or not the established requirements for each function are met. Keep in mind that in this phase we are primarily interested in *system level* requirements. *Human performance* requirements cannot be established until functions have been allocated to people or equipment.

Function Analysis

Before an allocation of functions can be made, each function must be well defined and reduced to a level at which the allocation process is meaningful. Thus, one of the first steps in dealing with functions is to identify, analyze and describe each of the functions to be performed in a system.

Function Allocation Strategies

Singleton (1974) provided an interesting discussion of recent attempts at function allocation. In the early 1950s there were numerous attempts at cold, logical comparisons of people and machines. The problem was seen as a simple one of determining the relative performance of people and machines in a variety of different activities. This idea for allocating functions resulted in several lists comparing human capabilities with machine capabilities. The original developers of these lists envisioned a designer considering a system's functions one at a time. Each function would be carefully analyzed and then compared with established human and machine performance criteria and allocated accordingly. This *comparison allocation* approach has proven to be of limited value.

Another still popular early allocation strategy is *leftover allocation.* With this approach as many functions as possible are allocated to a machine (computer) and the functions leftover are done by people. The (erroneous) assumption is that people can do and are willing to do any type of work assigned to them. Thus, they are allocated the "leftovers".

Later, because of escalating costs, the primary criterion for allocating functions became economic. In its simplest form, the question was "For each function, is it less expensive to select, train, and pay a person to do the task, or design (or acquire) and maintain a machine to do the same task?" This *economic allocation* approach is still used in many systems.

The designer must consider the *overall cost* of a system. This includes the cost of *design* and *operation.* Money saved by shortening the design process may be lost many times over the life of a system due to expenses caused by inappropriate allocation of functions. For example, to keep costs down, the time for the design process of one system was cut from 24 to 20 months. Certain allocations, that should have been made to computers, were made to people. Over the 20-year life of the system, these inappropriate allocations cost many times the amount saved by shortening the design process.

A more recent concept is the *humanized task* approach. Singleton (1974) reasons that with greater use of the function allocation process in non-military design problems, jobs began to be viewed in terms of a life's work rather than for a period of military service. The important considerations are much different. In life-long jobs, time spent in training becomes an advantage, and job satisfaction is vital.

The humanized task concept essentially means that the ultimate concern is to design a job that *justifies* using a person rather than a job which merely can be done by a human. With this approach, functions are allocated and the resulting tasks designed to make full use of human skills and to compensate for human limitations. The nature of the work selected for people should lend itself to internal motivational influences.

More recently, some function allocation problems have been effectively dealt with using a *flexible allocation* approach. With flexible allocation of functions, users can vary the degree of their participation in an activity. In other words, *users* allocate functions in the system based upon their values, needs, and interests. This concept is particularly relevant in appropriately designed computer-based systems where a person can allocate his or her own functions by making the necessary adjustments to the software. This makes possible wide differences in human involvement. At one extreme, a job could be designed for performance by automatic, repetitious routine; at the other extreme, the same set of activities could be made to require continuous attention and considerable decision-making.

The flexible allocation approach suggests that it is possible to allow a user to choose the tasks and the task difficulties either temporarily or permanently. In some cases a machine can be designed to take over those things that a user finds difficult, as well as those things the user simply does not feel like doing at any specific time. A good example in common use is the autopilot in a commercial aircraft. Once the aircraft is airborne and headed in the right direction the pilot can switch to an autopilot that carries out routine inflight functions. This is probably the most elegant solution to the function allocation problem: to design a system so that certain functions can be *allocated by users.*

The Duration of Function Allocation

The function allocation process actually takes place throughout the design of a system. Usually, designers make the most substantial allocations early in the development of a system, and make minor changes later to finely tune the system. The allocation process is iterative. A designer originally allocates each function either to a human, machine or human/machine interaction. That allocation is then used as a working model until more is learned about system characteristics, when some changes to the original allocation should be made.

The allocation process is very forgiving, particularly when dealing with software or computer-based systems, which usually can be changed more readily than solely hardware-based systems. A designer should review the allocations critically several different times until the system is operational. It is particularly important to carry out a detailed evaluation during system testing to determine whether or not an early function allocation is degrading human performance and hampering the overall operation of the system.

There was a time when all early design decisions were considered "cast in concrete." This substantially reduced the amount of freedom associated with later decisions. In many hardware-based systems this is still the case. A particular function may be allocated to a machine. A lengthy design process ends with test of the prototype of the machine. It is usually too late to make major changes to the allocation. As a consequence, many machines and equipment have been designed, developed and marketed where people were expected to make the best of the original poor design decision. As noted earlier, the situation is much different in computer-based systems where it is easier to make changes to software and even to certain hardware components.

Function Allocation in System D

Let us step through a function allocation process to provide an example of what should take place during this phase of system development. Figure 11-1 represents a hypothetical system with numerous functions delineated. If each of these functions is performed, the system will meet its objective. The functions in this hypothetical system are listed as F_1, F_2, and so forth. At this point, we have simply a listing or set of functions, each representing a statement of work done, and there is no reason to believe that any one function will be done by a person or by a machine. We will refer to this hypothetical system as System D.

F1	F2	F3	F4	F5	F6
F7	F8	F9	F10	F11	F12
F13	F14	F15	F16	F17	F18
F19	F20	F21	F22	F23	F24
F25	F26	F27	F28	F29	F30

Figure 11-1. A Set of Functions in a Hypothetical System (System D)

System designers may deal with functions at a macro (high) or micro (low) level, as long as the functions are at a level where they can be readily assigned to people or machines for completion.

Levels of Allocation

Often, by the time a designer can systematically allocate some functions, others already have been allocated either directly or indirectly. Figure 11-2 shows a set of functions that have been allocated for our hypothetical System D. Allocations may take place in at least four different ways. The first deals with certain allocations that a designer's management makes. For example, management may decide that a certain set of activities must be computerized or, another set must be mechanized using special machines and other apparatus that would need development. Remaining functions would be more systematically allocated. Interestingly enough, many of these original decisions may not have been analyzed for cost or other benefits. The cry to "computerize everything" is a good example. Designers frequently find themselves in a situation where certain functions have been allocated long before the designers become involved.

Allocated to machine by management	M	M	M	M	M	M
Allocated to human or machine by requirements	H	M	M	M	M	M
Allocated to human or machine by a systematic allocation procedure	M	M	M	H	H	M
	M	H	M	M	H	M
Unable to allocate						

Figure 11-2. A set of Allocated Functions, "M" indicates that the function was allocated to a machine, "H" means it was allocated to a human.

A second form of allocation is determined by the requirements levied against the system. If, for example, a certain function must be reliably performed in less than 100 milliseconds then it certainly will not be assigned to a human. The requirement establishes that it will be done by a computer or other type of machine. Thus, another group of functions may be allocated indirectly by setting system level requirements that are so stringent that a designer no longer has an option.

A third situation is the traditional function allocation that designers are supposed to perform. Ideally, *all* functions should be systematically allocated by designers. In many systems, however, functions are allocated by the three sources just discussed: management decision, system requirements, and, finally, designers.

Those functions in Figure 11-2 that the designer was unable to allocate require further analysis, perhaps even further breakdown into functions at a lower level, so that allocation can be completed. In some cases, once these functions are further broken down it will become obvious which are human functions and which should be assigned to a machine. However, certain functions can best be accomplished only by a close (inseparable) human/machine interaction.

Probably the best example of a human/machine function is where there is an intricate give-and-take (dialogue) between a human and a computer in attempting to satisfy some system objective. Identifying this function either as an exclusively human function or exclusively computer function would not lead to an acceptable design solution. It must be dealt with as a true interaction. The designer who is responsible for these functions must have a good understanding of both machine (computer) functions and human performance.

At the end of the systematic function allocation process each function will be assigned to a human, machine, or a human/machine interaction. It is a good idea for designers to document the major reasons for each function allocation. This documentation becomes invaluable, particularly if new designers assume responsibility for the system from the original designers. Frequently, after the system has been under development for even a few months, it is difficult to tell why a function was originally assigned to machines as opposed to a person. If allocation changes seem warranted as development proceeds it is helpful to go back to the original allocation to determine its rationale.

Considering Alternative Allocations

An example of alternative ways to accomplish a particular function in a system is shown in Figure 11-3. The work in this function concerns determining who will correct errors that the computer detected.

	Alternative 1 (clerk)	Alternative 2 (clerk/computer interaction)	Alternative 3 (computer)
CLERICAL ACTIVITIES	• Reads error notification • Analyzes notification • Decides who will correct • Determines error made • Fills out form • Photo Copies form • Sends materials to error-maker	• Analyzes error • Decides who will correct	
COMPUTER ACTIVITIES	• Prepares and prints notification	• Displays error message to clerk • Prepares notification • Stores copy in memory • Determines error-maker • Routes to error-maker	• Prepares notification • Stores copy in memory • Determines error-maker • Routes to error-maker

Figure 11-3. Alternative Ways to Accomplish a Function for Reconciling Errors

In the first alternative, the approach is primarily manual, a clerk receives the error message from the computer, reads it, analyzes it, decides who can best make the correction, fills out a form with comments concerning the errors, makes a photocopy and then either corrects the error or mails the materials to the appropriate person for correction.

The second alternative splits the tasks between the person and the computer: The person is assigned the analytical tasks associated with the error, and decides who can best correct it. After displaying the error message to the clerk, the computer prepares a notification, stores a copy, informs the clerk who the error-maker was, and automatically routes the error to the error-maker or anyone else, all at the direction of the person, in this case the clerk, who does the analysis of the error. Here, the human makes the decisions and the computer performs the clerical work.

The third alternative has the computer handle the reconciliation exclusively. The computer automatically prepares a notification, stores a copy, determines the error-maker by a code on the original input form, and automatically routes each error to the error-maker. In the latter case, the error-maker always corrects the error.

Alternative 1 saves a great deal of programming, whereas Alternative 3 requires additional programming as well as a considerable number of advance decisions. Alternative 1 relies heavily on people making good decisions and doing much clerical work. Alternative 3 takes the human out of the loop and has the computer perform both the analytical and clerical tasks. Alternative 3 tends to be less flexible in that all errors are dealt with in the same way. Alternative 2 probably offers the highest degree of flexibility and efficiency because people do the analytical and decision-making tasks, whereas the computer automatically cares for clerical tasks after people make decisions. Even so, any one of the alternatives is an acceptable way to accomplish the function.

Designers may prefer Alternative 1 in a system where fewer than 10 errors occur in a day and where people that are *regularly* assigned to other tasks can easily handle these errors. Alternative 2 is probably best in situations where numerous errors occur daily—from those errors that are easily and quickly corrected to those that require considerable correction activity, usually by the error-maker. Alternative 3 will also probably work best in a system where many errors occur daily—the kinds of errors that are to be corrected by the error-maker. As mentioned earlier, probably the ideal allocation decision is to build a system where a person, on any given day, can elect to perform all of the activities shown in Alternative 1, or elect to do only the analysis and the decision making as in Alternative 2, or have the computer deal with all detected errors. In fact, this decision could be made at different times during the day, for several days at a time, weeks at a time or for each individual for as long as they are in the job.

Considering Alternative Configurations

In most applications one of these three alternatives would be selected early in the design process and would never change as the design matured. In the past the chances of making a change from one alternative to the other was usually small unless system testing demonstrated a significant degradation in system performance directly attributable to a faulty allocation decision. Designers should make every attempt to break away from this traditional approach in favor of one with more flexibility to meet the needs of individual users.

Alternative 1 could be assigned to one possible configuration along with other similar decisions—an emphasis on using people to accomplish these types of tasks. Likewise, Alternative 2 could become part of another configuration and Alternative 3 part of still another configuration. All of these decisions are made to help ensure an acceptable level of human performance, given the system objectives and requirements. The designer can then use *other criteria* to select the best configuration.

Criteria that can be used for evaluating alternative configurations include cost (developmental and/or operational), meaningful maintainability (over 5, 10, or 20 years), portability (i.e., usefulness in different divisions or companies), producibility, and safety. The cost consideration is obvious. As we mentioned earlier, in many systems it is used as the primary method of allocating functions. This means that the configuration selected usually costs the least in money and resources to design. Unfortunately, the operational costs are not always traded-off against the development costs. In many cases, decisions are made to reduce development costs with little understanding or appreciation for the ultimate costs associated with operating or maintaining the system. This is particularly true when trying to determine whether a function should be programmed or done by people. We often decide to program what can be done easily and quickly and leave the remaining functions for people. This may substantially shorten the developmental time but could have disastrous effects on system operation and maintenance costs.

Telephone company studies of several computer-based systems have shown that it costs at least four times as much for people in a system as it does for the computer processing costs. In many well-designed systems it costs about the same to develop computer programs as it does to develop user material. When considering costs associated with the allocation of functions, a designer should carefully consider both developmental and operational costs.

Selecting from among alternatives is an extremely valuable approach because it enables a designer to make the original allocation decisions based solely on achieving an acceptable level of human performance. Two, three or more alternative configurations could result from these decisions. Then, using a separate set of criteria the best of the alternatives could be selected.

Designers seldom take time to systematically develop and compare alternative ways of meeting system objectives. There never seems enough time to adequately consider alternatives and for this reason a single "best way" is usually chosen early in the developmental process and from this point on, decisions are made to facilitate these original choices. Human performance in many systems could greatly improve if designers were forced to consider alternative allocation configurations from the very beginning and the alternatives evaluated by advocacy, each design team (out of possibly two or three) arguing their strengths and citing the weaknesses of the other.

Other Considerations

After the functions are allocated and a configuration is selected, further design and development considerations should focus on achieving acceptable performance levels for both the human and machine functions. In the case of functions requiring close human/machine interaction, special considerations are required.

As the design development proceeds and uncertainty about each function lessens, the allocation of some functions may change. As was mentioned earlier, a good designer's prerogative, duty, and responsibility are to make changes to the allocation where these changes will enhance human (or system) performance.

Frequently the time needed for carefully and systematically analyzing functions is not available. In many cases the requirement to meet a schedule simply does not allow sufficient time for all required development processes to be done adequately. When this occurs the function allocation process may be greatly abbreviated. But keep in mind, systematically or not, functions are always allocated.

Another reason the function allocation process is generally not performed as well as it should be is that it is difficult in the early stages of design to make good decisions regarding function allocation. There is frequently a lack of adequate information on which to base many allocations. The little information available does not always apply to the many new situations designers encounter. For example, some human versus machine comparisons continually change. If a system required either a human or a computer to recognize a specific pattern, such as handprinting, people would probably be assigned the task. Machines do not recognize handprinting as well as people. However, if the new system takes five years to develop, during that time a computer could be developed that recognizes handprinting as well as people. Thus, any comparison between what humans do well and what machines do well always needs careful evaluation in terms of continuing, rapid changes in technology.

If done well, a function allocation will lead to a smooth integration of human and machine components. Ideally, we would always allocate activities to people that they do well and enjoy doing. We would try to avoid allocating activities to people that would be boring or confusing to them and that may lead to a general feeling of dissatisfaction. It is curious that we have the power to create enjoyable jobs for people, yet frequently we end up with systems where half the jobs are menial and tedious, and could be performed by chimpanzees or pigeons, while many more are so difficult they require a Ph.D. in physics. In these cases one wonders how carefully designers considered the needs of potential users.

Ideally, functions would be allocated to the human and/or the machine based on which will perform the function best. However, if the function contains work that we would generally not feel good about people performing, it should be allocated to a machine. When the best potential performer is not clear, the designer should allocate the function to the human only if performance of the activity has a high probability of bringing about user satisfaction. Because the best way to perform functions varies with the objectives and requirements of a particular system, it is difficult to establish a firm, clear set of guidelines for making these decisions. Even so, allocation decisions should certainly take into account the designer's understanding of the strengths, limitations, and satisfaction tendencies of potential users.

It is a serious responsibility to create the work that others will do. Allocating functions is one of the most important activities designers ever perform.

HUMAN PERFORMANCE REQUIREMENTS

We can best determine human performance requirements only after we have a good set of allocated functions. Before that point we have difficulty identifying and dealing effectively with human performance requirements because we do not know what functions people will be expected to perform.

Once we have established a human performance requirement, we should always develop a way of measuring it. If the requirement relates to accuracy, then there must be a meaningful way to measure errors. If the requirement concerns manual processing time, there must be a meaningful way to measure it—time per customer contact, number of items produced per hour, and average keystrokes per day. Usually the problem is deciding what measure (from among many possibilities) gives the best indication of the performance. With a little imagination, any human performance can be meaningfully measured (cf. Gilbert, 1978, p. 29).

As a minimum, human performance requirements should include statements concerning errors, manual processing time, training time necessary to ensure the minimum skills, and job satisfaction. If we do not clearly state the requirements from the beginning we cannot expect human performance considerations to be taken seriously. In fact, in the absence of human performance requirements, other system requirements will take precedence—those related to software or hardware development. When this happens the entire development process will focus on meeting these other requirements. And when the system is operational, people will be left to perform as best they can without adequate provision for ensuring an acceptable level of human performance.

For example, a new computer-based system was being designed to provide telephone installers with the information needed to make a residence telephone installation. The work order used in the old manual system contained many different items of information. If one or more of the information items on the work order were incorrect, the installer had to stop work, call back to the office and wait for the correct information. The time lost in correcting the error meant that other customers could not be serviced. If all information on the work orders was always correct then an installer would be able to make each and every installation without ever checking back with the office. The actual payoff to the telephone company of having accurate information on each service order was the servicing of more customers per day. Obviously, the more customers serviced per day by each installer, the fewer installers, trucks, and other personnel required.

The system was evaluated, and it was determined that the probability of having an obstruction-free installation was about .81 (81 times out of 100). This accuracy level was not acceptable to potential users of the new system. They wanted the installer to make at least 90-95 percent of the installations without having to stop and call back for accurate information—the information had to be made more accurate. It was found that the accuracy level for the items contained on the work order would have to average about 99 percent to have an obstruction-free installation 90 percent of the time, and 99.5 percent to have an obstruction-free installation 95 percent of the time. As the new system was being designed, considerable effort was made to ensure these high accuracy levels.

In another system the average time to correct errors was fifteen minutes per error. This was an unacceptable amount of time for efficient system operations. This resulted in a human performance requirement in a new system of ten minutes to correct an error. Reducing by five minutes the average time to reconcile errors also reduced the costs of operating the system during a one-year period from $144,000 to $96,000.

Other human performance requirements could be associated with training time (e.g., total time to train clerical personnel to perform the basic activities should not exceed three weeks). Another human performance requirement could relate to job satisfaction (e.g., after performing an activity for six months, employees should respond in a positive way to their work, as measured by a questionnaire).

Having identified the manual functions and associated human performance requirements we are ready to continue the system development process.

TASK ANALYSIS

Task analysis is intended to match the work to be done with the kinds of people who will do it. The process has four main parts. The first part of the process, determining a *system structure,* gives the designer an overall view or objective for the analysis. The second part, *identifying tasks,* breaks down the original functions into smaller pieces and ultimately arrives at a level called tasks. The third consists of describing each task and organizing all tasks into a *flowchart* that will accommodate the variety of different transactions a new system must accomplish. The fourth part, identifying *work modules*, synthesizes the tasks previously identified into manageable modules of work. Having systematically derived work modules assists in the design of interfaces, and the preparation of facilitator materials, such as instructions, performance aids, and training.

In the early 1900s, Frederick Taylor (1911) proposed the general structure of this approach for developing work. He called this new way of dealing with work "scientific management." The design of jobs was central to his notion of scientific management, as illustrated in the following statement:

> Perhaps the most prominent single element is the task idea. The work of every workman is fully planned out by the management at least one day in advance, and each man receives in most cases complete written instructions, describing in detail the task which he is to accomplish.... This task specifies not only what is to be done but how it is to be done and the exact time allowed for doing it. These tasks are carefully planned, so that both good and careful work are called for in their performance. (p.59)

Basically, this approach suggested that the work should be carefully analyzed, and then assigned to appropriate workers with specific instructions on how the work could be accomplished most efficiently. Once the work is designed and people selected to perform the work, the new workers are to be adequately trained to ensure that they perform the work exactly as planned.

System Structure

Determining the system structure can enable a designer to develop jobs that contain simple or complex tasks. Designers can break down tasks to a level where they can all be performed by "off-the-street" new hires, or a level where they can be performed by people who have acquired many high-level skills. The designer has considerable flexibility in determining which functions will be broken down into low-level tasks and which will be developed as high level tasks.

Higher level tasks require little training if a person has acquired the necessary skills elsewhere. On the other hand, if a designer elects to design all high-level tasks and at the same time does not have users who have gained skills in other systems, then a tremendous amount of training and experience will be required for the people to perform the more complex activities.

Three possible system structures are shown in Figure 11-4. The diamond-shaped structure is one where most tasks will be moderately difficult with a few low level, and a few high level. The inverted pyramid structure on the other hand, has a large number of complex tasks. In the latter case, the designer plans to transfer into the new system people working elsewhere. The designer assumes that their other work experiences have enabled them to develop skills to perform these high-level tasks with a minimum of training.

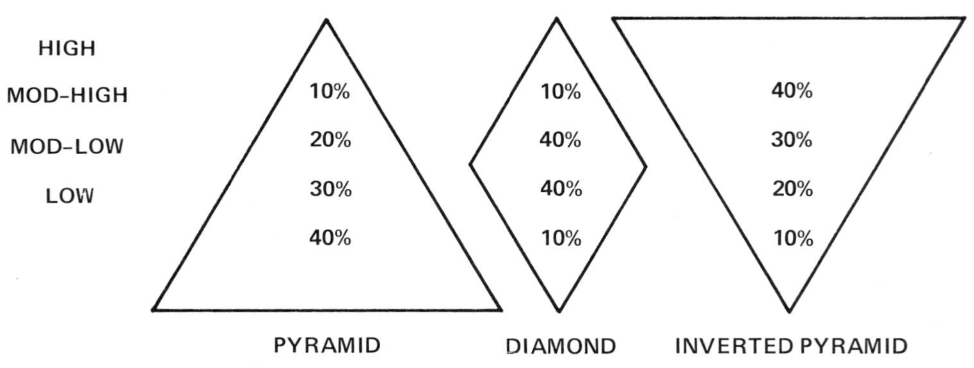

Figure 11-4. Three Possible Structures for a System.

It's easy to see that if a designer assumes an inverted pyramid as the final goal of task analysis and then finds that in the real world most of the people coming into the system are new hires, there will be considerable difficulty with human performance. The converse is also true. In this latter case, the difficulty will stem more from having people perform work that is too simple, and therefore boring or tedious. This concept of different structures forces designers to *think in advance* about where they would like to end up and to not just let the system evolve. The relative proportion of difficult, moderate, and easy tasks should be planned in advance.

The pyramid structure also enables a person to come into a system and work at lower-level tasks until he or she can develop the necessary skills. A person can then move to new, more complex jobs without having to transfer out of the system to find something more challenging.

Identifying Tasks

Identifying tasks has essentially nine primary considerations associated with it. These include determining existing knowledge and skills, deriving skill level categories, identifying outputs and inputs, deriving lower level activities, ensuring that the activities are mutually exclusive and exhaustive, and matching activity complexities with previously determined skill levels. The process may be repeated any number of times until each activity is assigned a single skill level. Secondary considerations in the identification of tasks include meeting the system structure objective just discussed, as well as meeting a full-advantage objective. A designer attempts to develop tasks that will ultimately take full advantage of the user work force. This is difficult to quantify, but during the analysis process, most designers gain a feel for what is meant by taking full advantage of the skills available in their user population and this should be reflected as the tasks are being identified.

Task Complexity Levels

For a designer to determine existing user knowledge and skills, he or she must become very familiar with the potential work force. In some cases this requires considerable data collection. Originally, a designer can use a rough description of existing user skills to help derive skill level categories which he or she then can use to break down tasks. Figure 11-5, for example, shows the estimates made by one designer for the user population at the beginning of a system design. In this case she assumed the skills were loosely related to the time that the person had spent with the company. The designer then assigned four different task complexity levels. The most complex tasks related to people who had extensive skills developed through more than three years experience in many different areas of the company. The majority of these experiences were in some way related to developing skills associated with the tasks to be performed in the new system.

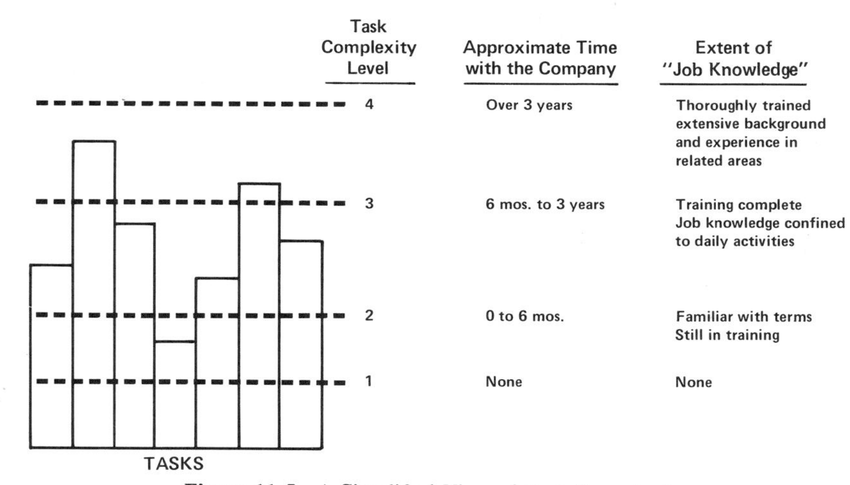

Figure 11-5. A Simplified View of Task Complexity

At the other end of the continuum, the designer assumed that the lowest task complexity could be assigned to those people who had few, if any, job relevant skills. These people would be new hires with no company experience and the tasks associated with having few, if any, relevant skills would be given a task complexity level of 1. Consider the tasks in Figure 11-5. Of these seven tasks one had the task complexity level of about 4, one was close to the task complexity of level 1, three had the task complexity level of 3, and two were at a task complexity of level 2.

Converting Functions to Tasks

Once some rough task complexity levels are established, it is possible to start identifying tasks. The task complexity level is critical to this process because it is used as a "stopping rule." There are many ways to carry out a task analysis, the value of the approach described here is that it generally can be done very quickly. The process of breaking down functions into subfunctions, and then breaking down each subfunction into lower level subfunctions makes it difficult to know when to stop the breaking down process.

Tasks are important and will be a key feature of all the materials to be prepared at a later time. Obviously, if the designer were to take the lowest level entry and describe all that was involved for each part of the body to even do something as simple as filling-out a form, it would take several volumes of material just to provide instructions, performance aids, and training.

A designer must develop a heuristic to determine when to stop breaking down a function. The stopping point is usually where a work description can be read and understood adequately by a user with a minimum of training. If the level is too high, users will not be able to effectively deal with it, and if the level is too low, the user may feel it is too simple, perhaps even demeaning.

The actual process of breaking down functions into subfunctions and, ultimately, to a task level is sometimes referred to as an input/output analysis. An example of inputs and outputs is shown in Figure 11-6. In this case, the output is 6 and the input is the scores of 5, 6, 7, 8, 4, and 6. The function to be performed is to determine the average. The difficult part comes in identifying all the subfunctions required to convert the input to the output. In this example, three things must be done. First, the scores must be added; second, the number of scores must be counted; and third, the sum of the scores must be divided by the total number of scores. The three subfunctions are each lower level activities; only when they are completed will the function itself or the higher level activity have been accomplished. This is the basic process for conducting a task breakdown.

There are two rules for the subfunctions. First, they must be exhaustive. This means that everything necessary to accomplish the main function must be included in the subfunctions. The function will be accomplished only if each of the subfunctions is performed. Second, each of the subfunctions should be independent of the others.

As another example of the relationships of functions and subfunctions, consider the following. Suppose the original function is to *prepare and serve a*

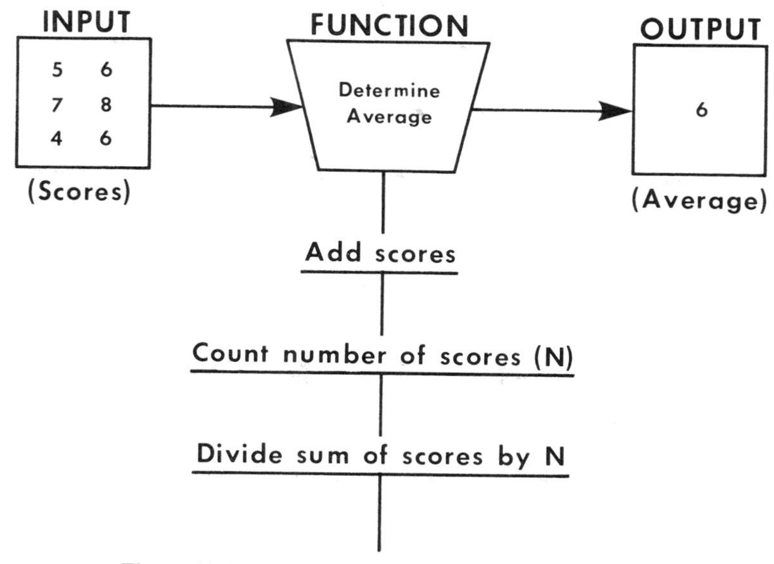

Figure 11-6. An Example of an Input/Output Analysis

salad. In this case, the designer derived three subfunctions necessary to accomplish this effort (shown in Figure 11-7).

Figure 11-7. An Example of Breaking Down a Function into Subfunctions

First, note that in our example the three subfunctions are all at about the same level of detail, i.e., the same hierarchical plane. Notice also that these activities are mutually exclusive—that is, they are separate and distinct from one another with no overlapping activities. In addition, they are exhaustive; on this level they represent all the activities required to achieve the goal.

Assume that the designer focuses on the "prepare salad" function first. In order to break down the function further, or determine more detailed activities (or subfunctions), he or she must first make a design decision. What kind of salad is wanted: lettuce, jello, Waldorf? This information should be available in the system objectives. If the salad is to be a Waldorf, for example, the designer can move to a more detailed level of functioning, as shown in Figure 11-8, as well as into a new region of definition.

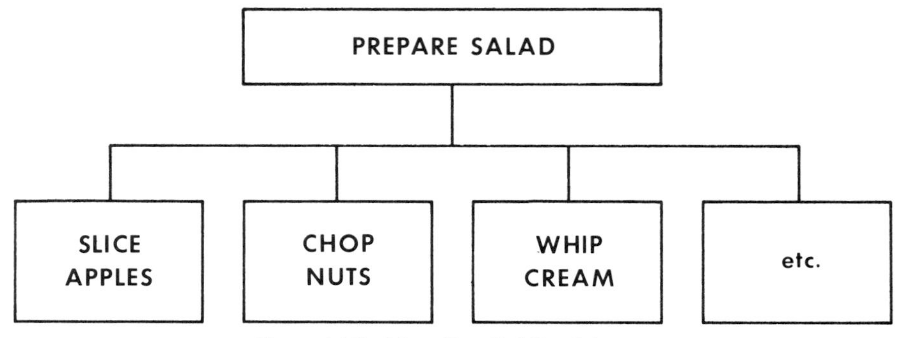

Figure 11-8. More Detailed Breakdown

Figure 11-7 and 11-8 are examples of moving to different levels of functional detail. Those functions at a more detailed level are an explanation of the whole from which they are derived. For example, "prepare salad" is related to "slice apples," "chop nuts," "whip cream," activities inherent in preparing Waldorf salads. Notice that these functions are not exhaustive, as implied by the box labeled "Etc." All the subfunctions are mutually exclusive and at about the same level of detail.

The activity of breaking down functions into subfunctions continues until we have identified all tasks. *A task is always defined in terms of the potential user.* For highly skilled users, the task level would be much higher than for inexperienced users. Thus, we continue to break down a function into subfunctions until we reach a level where we can match each task against the desired skill levels, a more precise set of user characteristics, or possibly even a particular user. When an activity can be comfortably assigned a skill level, such as the complexity levels 1 through 4 discussed previously, then, by definition, it is at the task level.

Describing Tasks

Once the task or task levels have been defined, they should each be described and a flow chart prepared that shows the interdependencies of all tasks as they relate to various uses of the new system. This activity generally identifies duplication of tasks and may result in adding, dropping, or combining of tasks as a flow chart is prepared. This approach to task analysis tends to be very forgiving, and duplications and omissions can be taken care of with little loss in efficiency. Once all tasks are accounted for in a task level flowchart (see Figure 11-9), the process of synthesis into work modules can begin.

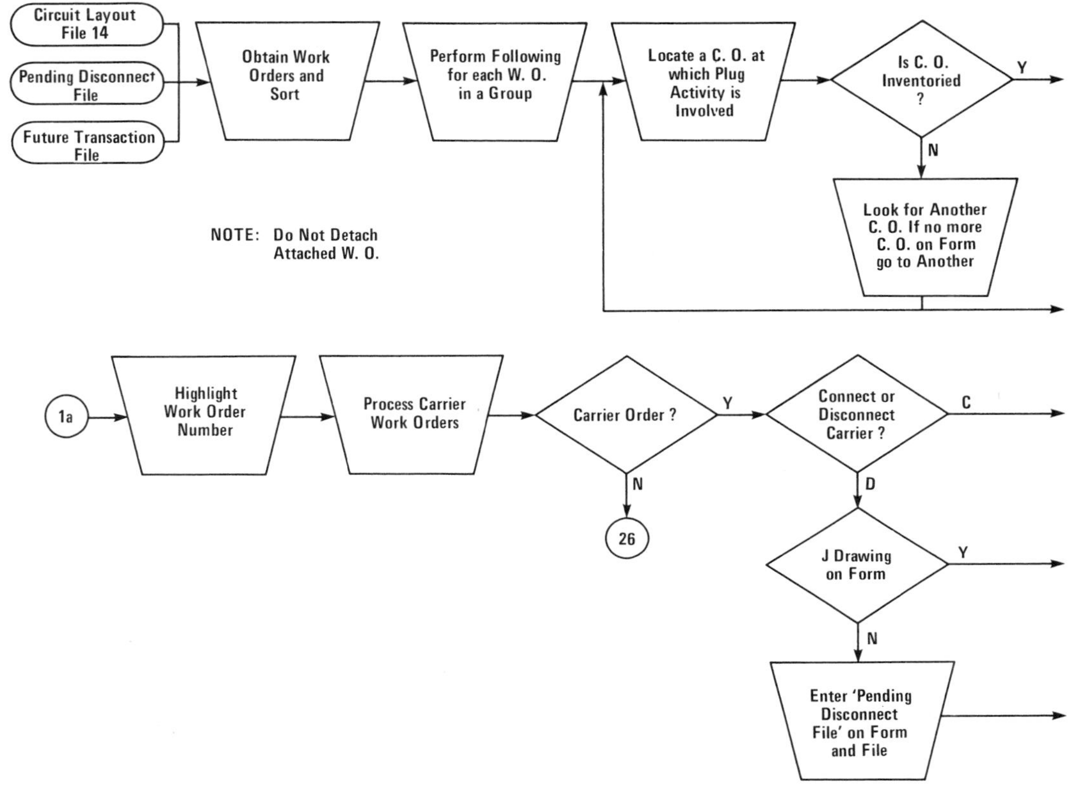

Figure 11-9. An Example of Task Analysis Flow Chart

Work Module Design

Once the functions have been properly allocated, the human performance requirements established, and tasks identified, adequately described, and flowcharted, the next logical step toward ensuring the desired level of human performance is to identify *work modules.*

A human work module is a set of tasks that a user accomplishes as a part or all of his or her job (Soth, 1976). It is a basic unit of work. Usually, one or more work modules are combined to form a job.

Work modules can be made up of tasks derived from several *different* functions (see Figure 11-10). In this figure, work module A consists of 5 tasks from function 1. However, work module C consists of two tasks from function 1, two tasks from function 2 and one task from function 3.

Recall that a function is nothing more than a statement of required work. A work module, however, is a description of work to be performed by a specific type of person. The main objective of the designer is to effectively match the work with potential users.

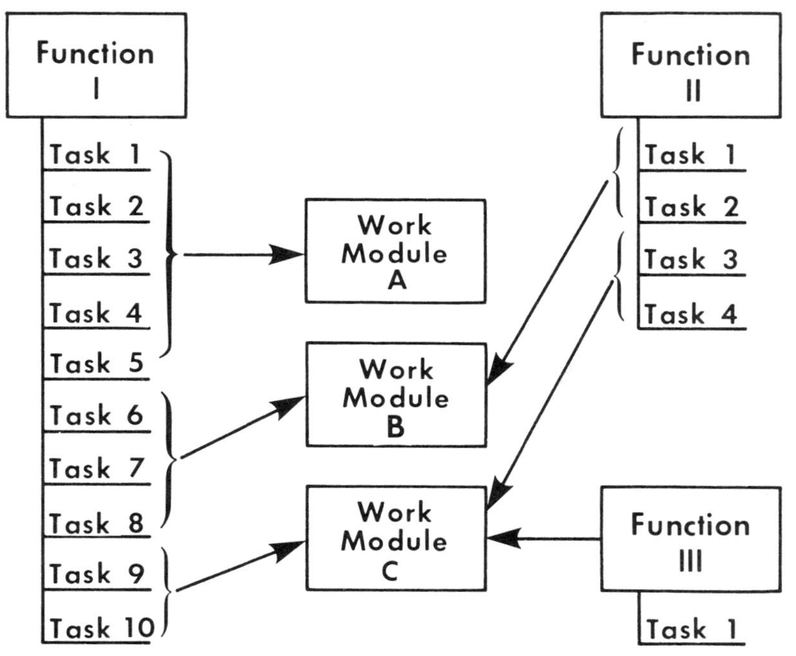

Figure 11-10. The Relationship of Functions to Work Modules

Consider the following when grouping tasks together into a work module:

- Data relationships
- Skill level for a task
- Task relationships and sequence
- Time dependencies
- Human/machine interface considerations

Tasks that have the same or closely related data items used together are candidates for the same work module. For instance, task 1 might require the use of the billing name of the customer account, and task 4 require the use of the billed amount. These tasks use related data and should be considered for inclusion in the same module, providing the other four considerations are met.

Tasks of the same or almost the same skill level are also candidates for combination. If tasks 1, 2, and 4 require a high skill level and task 3 requires a moderate skill level, these can probably be combined, since a high skill level would be required for the module. However, it is usually not practical or economical to combine high skill level tasks with low skill level tasks.

One reason for this is that it makes it easier to prepare instructions, performance aids and training materials. Another reason is that users are given an opportunity to perform activities at about the same difficulty level. For example, a work module in a system with severely degraded human performance was analyzed. The work module was found to contain a total of 38 tasks. Even worse, the tasks ranged from being very simple to extremely complex, as shown in Figure 11-11. The manual processing time was excessive and the training time was considerably greater than expected. In fact, when some students finally learned how to perform certain tasks in the work module they had already forgotten how to perform others. The tasks were analyzed again by designers. As a result, some tasks were combined and three new work modules were developed from the original, poorly designed work module. These three new work modules are illustrated in Figure 11-12.

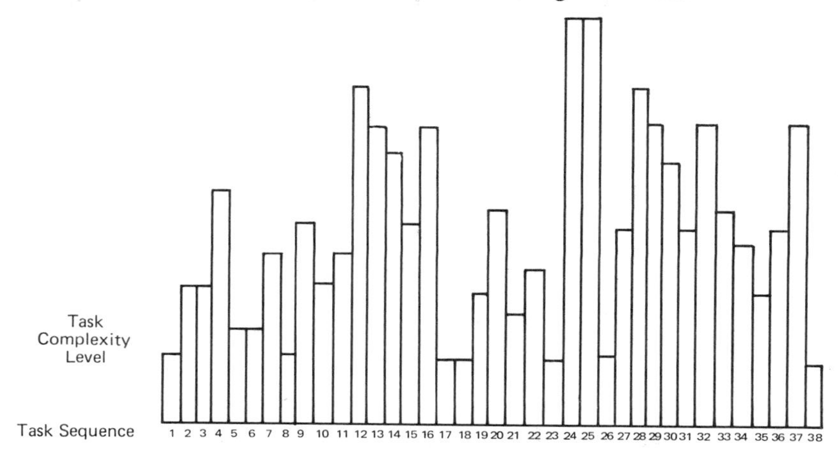

Job: Prepare Manual Service Order Input

Figure 11-11. Example of an Exceptionally Large Work Module in a System with Degraded Human Performance

It was not possible to have the tasks in the redesigned work module at exactly the same level, but the differences were minor when compared with the original work module. There are fewer tasks in each of the three new modules, which greatly assisted training for each of the modules. As people gained experience on the first work module, some received the opportunity to perform on the second module also. As the new system matured, some of the people who had begun on work module 1 and progressed to work module 2 eventually progressed to perform the work contained in work module 3.

Tasks that are related, whether by data or nature of the work, or that have a sequence dependency should be considered for inclusion in the same work module. This indicates that the normal flow of the work must be anticipated

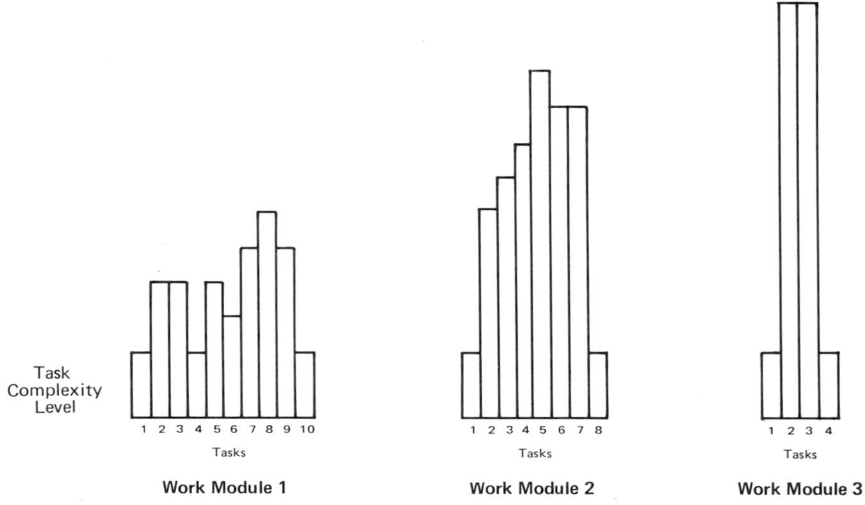

Figure 11-12. An Example of Three New Work Modules Made From One Large Work Module

and kept in mind when forming human work modules. (Keep in mind that work modules that are combined usually do not include tasks for transmitting information from one to another.)

These criteria can be used to construct a cohesive and logical set of related tasks for assignment to one person—work module. This is a convenient way to design work since these basic units can be assembled and built into a variety of jobs either by designers or by the users who have responsibility for operating the system.

The number of tasks in most work modules ranges from four to nine. Probably the ideal size is about seven tasks. If the volume of input to a work module is great, more than one person can perform the same human work module. In fact, under conditions of heavy volume one work module may become the job for several people.

FACILITATING HUMAN PERFORMANCE

Once good work modules are developed, many types of materials can be developed to help ensure an acceptable level of human performance. All of these materials are based on the task analysis results and are usually prepared to support the work module instructions. We will discuss some of these materials to give the reader some appreciation for the kinds of products that can ultimately result from a well done task analysis. Each will be discussed in greater detail later in the book.

Statement of Minimum Qualifications

Once the designer identifies tasks and determines work modules, he or she must identify the specific skills and knowledge required for each work module. This amounts to stating his or her assumptions of the precise qualifications of the person to perform the work module. As the work modules are designed, the designer should have a set of user characteristics in mind—the design should clearly envision the person performing the work. When this information is written down, it is called a *statement of minimum qualifications* or SMQ.

If the SMQs for all work modules were combined into a single document it would provide an accurate description of the types of people needed to perform in the new system. The number of each type of person then can be estimated from human performance test results to be discussed later. By having a good idea of the number and type of people required to perform in a new system, personnel information can be made available to users long before the system is ever operational.

The SMQ is a detailed description of minimum acceptable qualifications in terms of skills and knowledge required to efficiently and effectively perform the work outlined by the work module. It provides a way of screening people in order to have the exact kinds of people in the new system that the designer had in mind when he or she completed the task analysis and designed the work module.

Instructions, Performance Aids and Training

Once the designer has in mind the potential user and has designed the work modules, then he or she can proceed with developing instructions, performance aids, training. Not all systems will require all three of the facilitators. In some systems, only performance aids are needed; in other systems both instructions and training may be necessary. In fact, in some systems the collection of instructions contained in work module descriptions provides a good (and sometimes the only) overall description of manual functions to be performed in the system. The work module instructions can be used as the basis for training development and, in some cases, may be a major component in a training package. The work module instructions may also be used as the basis for designing performance aids and other reference materials.

Performance aids are used frequently to help reduce the memory requirements in systems. Having a list of frequently-called telephone numbers on your desk is an example of a performance aid. The alternative would be to memorize the numbers or to look them up in a book whenever they are needed. In both of these situations human performance would be degraded by requiring longer manual processing times.

Most systems require some training to reduce the time necessary for start-up. Training materials are based on the work module instructions and enable a person to quickly and efficiently go from their entering skill and knowledge level to the point of performing adequately. This may be accomplished by a simple reading and rereading of the instructions. But an individual is likely to attain proficiency faster if provided with set of training materials based on a well-developed set of training objectives using appropriate media. In some cases this may require a computer.

BEGINNING EARLY

It is vitally important that human performance considerations begin at least by the time functions are being allocated (even earlier, if possible). Having a good set of human performance requirements is also critical. A systematic and detailed analysis of the tasks to be performed is also very important, no matter what procedure is used for this particular analysis. Finally, identifying small manageable modules of work that can be used to construct jobs is the culminating analytical activity in this developmental approach. If these design considerations are well done, then a strong foundation is established for developing interfaces and facilitators, including human/computer dialog structures, instructions, performance aids, training materials, and forms.

FOR MORE INFORMATION

Meister, D., *Human Factors: Theory and Practice,* New York: Wiley-Interscience, 1971.

Meister, D., *Behavioral Foundations of System Development,* London: John Wiley and Sons, 1976.

Singleton, W. T., *Man-Machine Systems,* Middlesex, England: Penguin Books, Inc., 1974.

12

DISPLAYS, CONTROLS, AND WORKPLACE DESIGN

INTRODUCTION

An advertisement in a major magazine uses human performance considerations as a principal reason why people should buy the BMW automobile manufactured in Germany. Part of the advertisement reads:

> All seats are orthopedically molded; all individual seats are infinitely adjustable. Controls are within easy reach and all instruments are instantly visible in an innovative three-zone control panel that curves out toward the driver in the manner of an airplane cockpit.
>
> So thorough is the integration of human and machine that the driver literally functions as one of the car's working parts—the human part that completes the mechanical circuit.

As this advertisement correctly states, three of the most important human performance considerations concern displays (instruments), controls, and the workplace layout.

Displays allow a person to *monitor* the status of a system and controls allow a person to *change* the status of a system. A driver monitors the speed of an automobile by observing the speedometer (display) and determines the direction the vehicle will travel by moving the steering wheel (control). Well-designed displays and controls are usually critical to achieving acceptable human performance. The relative positioning of displays and controls, as well as the design characteristics of such items as desks, chairs, and files must be carefully combined into the best possible workplace layout.

USER EXPECTATIONS

Probably the designer's most important consideration when designing displays, controls, and workplace layout is to make decisions that are consistent with what the user expects. With controls, for example, some direction-of-motion expectancies seem almost natural--pushing a throttle forward to increase forward speed or turning a wheel clockwise to turn right. Certain other direction-of-motion expectations have become traditional such as turning a faucet clockwise to shut off water. Designers should avoid relationships between controls and displays as well as between controls and vehicle motion that yield an unexpected direction-of-motion relationship. When the user expectations are not obvious, a designer should conduct a study to determine what the expectations are.

VISUAL DISPLAYS

This section will cover some major human factors considerations in the design of visual displays. The requirements for displays are developed by combining characteristics of human information processing with information gathered during task analysis. In fact, determining the need for a display, as well as the types of information each display should present, is an integral part of the task analysis process. The most common visual displays are dials, indicator lights, digital displays, computer printouts, paper forms, and cathode ray tube (CRT) displays. Guidelines for computer printouts and CRT displays will be discussed in more detail in Chapter 15.

Task-Related Considerations

To design good displays, a designer should know as much about the tasks to be performed as possible. Designers should consider carefully the way each item on the display will be used, its importance, when it will be used, and specific characteristics of users, such as visual acuity, or skill level. Different uses of the same display during normal operation versus an emergency situation must also be considered. It is usually a good idea to obtain a sample group of potential users to test prototypes or mock-ups of the displays. The tests should be conducted in a context that reflects as close as possible the real-world uses of the displays.

Selecting the correct type of display for a given application is one of the most important decisions a designer can make. First, the designer decides what sense to use. If, for example, the eyes are already heavily occupied (as when driving an automobile in heavy traffic), then it is probably best to select a display that uses the ears, fingers, or nose. If, however, the eyes are generally "free" and the task requirements tend to favor visual information, then visual displays may be advantageous. The designer should not make this decision unless he or she has a good overall idea of the required activity.

Favor visual presentation of information if:

1. The auditory sense of the user is overburdened.
2. The message is complex.
3. The message is long.
4. The message deals with a specific location on a panel.
5. The message does not require immediate action or will be referred to later.
6. The user is allowed to remain in one location.
7. The receiving location is so noisy that some auditory messages may be missed.

Other Major Considerations

Three of the most important variables we should consider when designing or selecting visual displays are:

1. Type of display (based on what the user needs to know)
2. Information format (should it be an exact replica or coded?)
3. Physical characteristics (character design and size, background brightness, density or clutter).

Electronic Displays

Advances in the field of electro-optics has led to the expanded use of electronic displays for the presentation of a wide variety of information. Cathode ray tubes (CRT), light emitting diodes (LED), liquid crystal displays (LCD), gas discharge panels, and other types of light emitting displays are being used for information displays varying from single character displays to those with 40 or more lines of data to maps and pictures.

Probably the most popular is the CRT. Research and development has provided a broad base of application for the CRT. The primary advantages of using CRTs are high writing speed, high resolution, simple addressing, full color capabilities, full range of gray scales, storage, large range of screen size, and high luminous efficiency (Considine, 1976). Disadvantages of CRT use include bulkiness of the equipment, curvature of the screen, high voltage required, relatively delicate equipment (vacuum tubes), and limitations of maximum screen size. The development of CRT flat panels may help alleviate some of the problems associated with bulk, screen curvature, and high voltage requirements.

Display Technologies

Obviously, numerous display technologies are available to designers. Eight of these technologies will be briefly described, and then a procedure presented to assist designers in selecting the most appropriate technology. The designer can make similar types of tradeoffs for display technologies not discussed here.

Cathode Ray Tube (CRT)

As indicated above, the (CRT) is the workhorse of displays, and has been adapted or modified to meet many requirements. It is fairly reliable with an acceptably long life, and is produced quite inexpensively. Among knowledgeable system designers, however, the CRT is chosen for numerous applications because of its tremendous flexibility. CRTs are available in a variety of sizes and shapes, provide gray scale and color, can have reasonably good resolution, can provide a storage capability, and can be addressed with both raster and stroke patterns (Sherr, 1979).

Flat-Panel CRT

The conventional CRT has great flexibility in information display. However, a major disadvantage in some applications is its depth—as the displayed image size is increased, so is the length of the tube. Flat-panel CRTs have been developed that are as thin as 5 cm (Goede, 1973).

Light-Emitting Diode (LED)

The light-emitting diode (LED) has been used successfully in calculators, wristwatches, instrumentation readouts, and discrete miniature lamps. Its popularity is based upon the combination of good luminance, low cost, low power, high reliability, and good compatibility with integrated circuit technology. Larger one- and two-dimensional arrays of LEDs have been developed for message readout and graphics displays.

Electroluminescence (EL) Panels

The electroluminescence, or more properly field-excited electroluminescence display is potentially compatible with requirements for alphanumeric readout and graphics. EL panels can produce a reasonable gray scale and dynamic range; has a wide acceptance angle for viewing; can be fabricated in sizes ranging from a few centimeters to greater than a meter; and are potentially capable of very high element density (Kazan, 1976).

Plasma Displays

The plasma display panel offers a good alternative to the CRT in some applications, possibly for large screen picture-on-the-wall television. Plasma displays have been used for alphanumeric readouts in single rows, multiple rows, and larger matrix panels for graphics and TV applications. Alphanumeric indicators are available in both dot matrix and seven-segment forms (Weston, 1975).

Liquid Crystal Displays (LCD)

The liquid crystal display (LCD) technology is one of the most popular and most developed of the flat-panel display types. Rather than emitting light energy, the LCD controls or modifies the passage of externally generated light (Goodman, 1975).

For many applications, the LCD is a good choice for a single or multiple character alphnumeric readout. The characters can be made any size, the contrast is typically adequate, costs are very low, and voltage/power requirements are compatible with battery sources. Graphic and video displays are also possible.

Electrochromic Displays (ECD)

The electrochromic display (ECD) is another type of light modulating display device. Like the LCD it has no light generating or emitting properties; rather, it modulates the ambient (or reflected) illuminance. As compared to the liquid crystal technology, however, the ECD device is generally transparent and absorbs only a selected portion of the visible spectrum upon application of an electric field. ECDs have found limited acceptance in alphanumeric display applications, mostly in battery-driven situations such as wristwatches and calculators.

Electrophoretic Displays (EPID)

The electrophoretic induced display (EPID) is also a light modulating, rather than a light emitting, display. It results from the process of electrophoresis, the movement of charged particles suspended in a liquid by the application of an electric field. The pigmented particles are selected to be a different color or optical density than the suspending liquid, so that the migration to the front surface of the display cell permits the user to "see" the particles, whereas migration to the rear surface of the display causes the user to see only the suspending liquid (Ota, Ohnishi, and Yoshiyama, 1973). EPIDS can be used for on-off displays, seven-segment alphanumerics, and matrix displays. EPID displays tend to be esthetically pleasing, due in part to the color combinations available to the designer.

Selecting a Display Technology

This section suggests a procedure by which a designer can select one or more CRT or flat panel display technologies for a specific application (Snyder, 1980). The procedure also can be used to evaluate specific display designs against system and user requirements. The procedure requires knowledge of the information to be displayed, the environment in which the display will be used, the layout of the user's work station, and any voltage/power constraints.

Define Display Functional Requirements

The first step in any display selection/evaluation process is the specification of the functional requirements of a display. In this step, it is then necessary to answer the following questions:

1. What symbology must be displayed?
2. Is dynamic presentation required? If so, at what data rates?
3. Is the displayed information alphanumeric, vectorgraphic, or pictorial, or some combination of these?
4. How much of this information must be presented simultaneously, and in what format?
5. What are the workspace layout constraints?
6. What environmental/power constraints exist?
7. What is the nature of the ambient illuminance?

These functional requirements should then be used to generate more specific design or performance requirements.

Establishing Design Requirements

The functional requirements indicate what information is to be displayed, where, when, and how often. The *design* requirements, on the other hand, specify the exact design variables to which the hardware (and software) must conform to assure an acceptable level of user performance.

Some of the most important human performance related considerations that should be specified in the design requirements are:

- application characteristics
- display size
- color capability
- luminance capability
- resolution

Tables 12-1 and 12-2 compares each of the eight display technologies with these considerations. Once the design requirements have been established for a particular application, the actual selection process can begin.

Table 12-1. Comparison of Applications Characteristics of Eight Display Technologies

Display Type	Single Alphanumeric	Matrix (graphic)	Matrix (TV)	Large-screen, direct viewing	Large-screen, projection
Cathode Ray Tube (CRT)	possible, but not practical	yes	yes	yes, but low luminance	yes, with light valve
Flat-Panel CRT	yes	yes	yes	no	no
Light-emitting diode (LED)	yes	available, but expen-sive	no	no	no
Electrolumines-cent (EL)	yes	yes	mono-chrome only	no	no
Plasma	yes	yes	yes	possible	no
Liquid Crystal (LCD)	yes	yes	yes	no	yes, as light valve
Electrochromic (ECD)	yes	no	no	no	no
Electrophoretic (EPID)	yes	possible	no	possible	no

(adapted from Snyder, 1980)

Table 12-2. Qualitative Comparison of Eight Display Technologies by Design Variables

Technology	Size	Power/ Voltage	Color Capability	Luminance Capability	Resolution
Cathode-Ray Tube (CRT)	Miniature to large projection	high	yes	low to high	high
Flat CRT	small	medium	yes	low to medium	medium
Light-emitting diode (LED)	small	low	limited	low to very high	high
Electrolumines-cent (EL)	small to large	medium to high	limited	low to high	high
Plasma	small to medium	high	possible	medium	medium
Liquid Crystal (LCD)	small to medium	low	limited	n/a	medium
Electrochromic (ECD)	small to medium	low	discrete	n/a	unknown
Electrophoretic (EPID)	small to medium	low to high	discrete	n/a	medium

(adapted from Snyder, 1980)

Technology Selection

The design requirements should be compared directly with the capabilities and limitations of the alternate technologies to eliminate those technologies incapable of meeting specific design requirements. Following the steps shown on the flowchart in Figure 12-1 should result in logical elimination of totally unacceptable technologies or devices. At each of the decision points, design variables are used to eliminate candidate technologies/devices based upon design requirements.

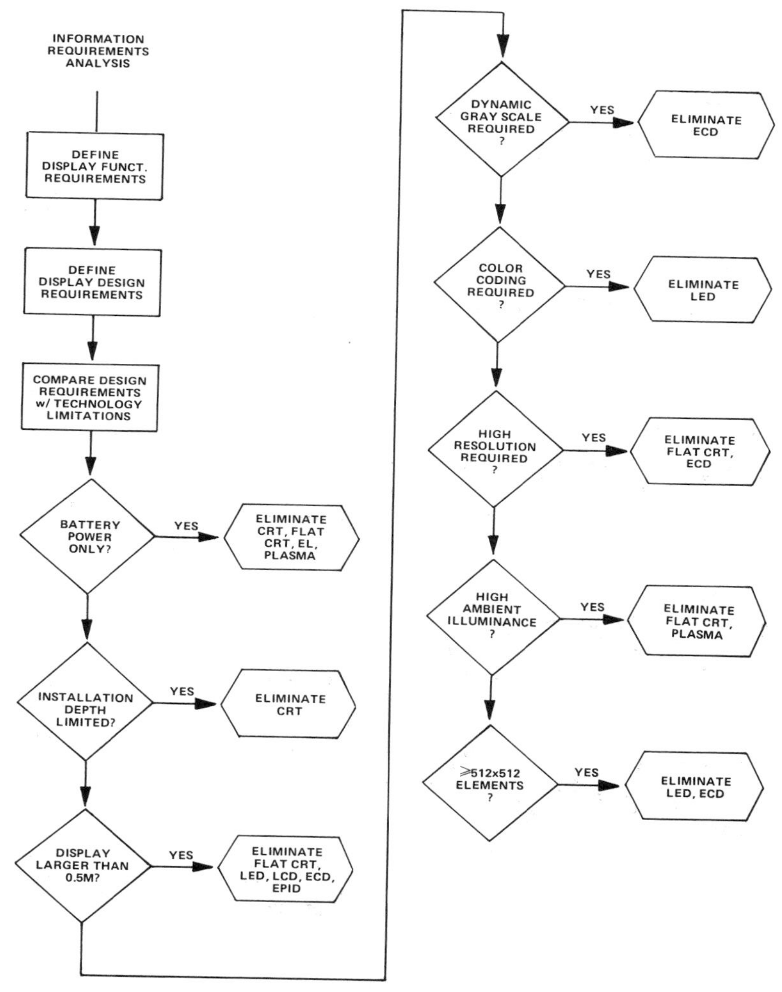

Figure 12-1. Flowchart Illustrating Display Technology Selection (adapted from Snyder, 1980)

Display Types

Once the designer decides to use visual displays, he or she selects the best type of display for the activity.

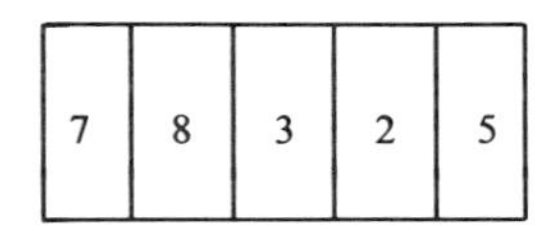

Figure 12-2. Digital Display

Digital Displays

Displays frequently present quantitative information. For reading precise, static or slowly changing numerical values, or making exact numerical settings, a digital display, such as that shown in Figure 12-2, is usually preferred. Digital displays require little space and the amount of accuracy available is usually only limited by the number of digits displayed. However there are some disadvantages:

- Determining the rate of change is difficult
- Reading rapidly changing display values is difficult
- Interpolating is difficult when two numbers are partially visible in a window
- Gauging distance to a boundary (control limit, danger zone) is difficult

Fixed-Scale/Moving-Pointer Displays

Where interpolation between numbers is required, or rate or trend information is important, a fixed scale with a moving pointer is usually preferred. The fixed-scale/moving-pointer display is particularly good for making check readings, and it usually has a simple and direct relation between pointer motion and motion of a setting knob. This type of display comes in various forms as shown in Figure 12.3.

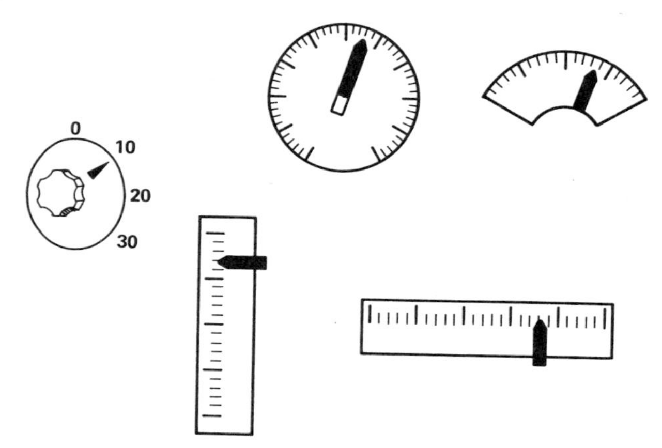

Figure 12-3. Examples of Fixed-Scale/Moving-Pointer Display

A moving pointer against a fixed scale is also good for qualitative check reading. If several dials are to be scanned rapidly, orient pointers so that the "normal" position of the pointer is at 9 o'clock (Figure 12-4). Remember to orient dial scales so that the increase in range will be read from left to right or from bottom to top.

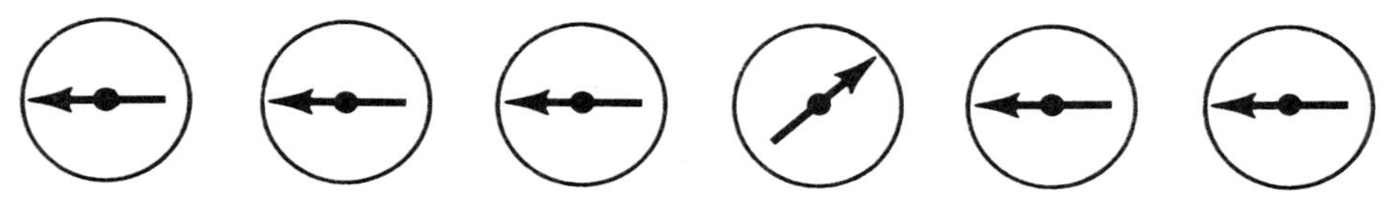

Figure 12-4. Preferred Layout for Checkreading Several Displays

Two disadvantages of fixed-scale/moving-pointer displays are that they frequently require more panel space than some other types of displays, and the scale length is fixed and thus limited.

Displays and Redundancy

For information such as start-stop, on-off, up-down, forward-back, safe-overload, use displays that present large differences in the categories of interest. These differences can be emphasized by using *redundant coding* methods, such as position, color, or labels. Figure 12-5 shows a familiar display using redundant coding. Redundancy with the "stop" light, for example, includes color (red), position (top) and the word "stop."

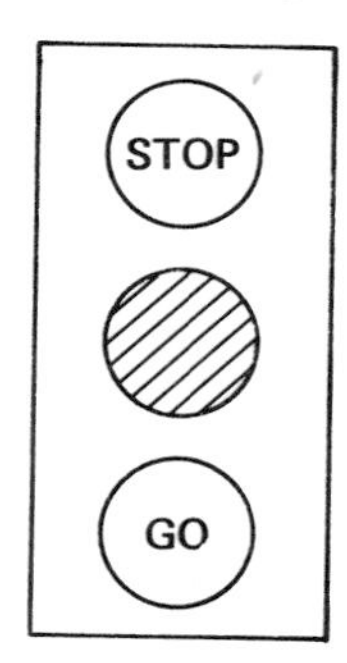

Figure 12-5. A Traffic Light Showing Redundancy When Designing Visual Displays

Information Format

The information format of a display is determined by how the information needs of a user best can be presented. Does the information have to be an exact replica of the source data, or can it be coded?

Color Coding

Frequently, displays require a certain amount of coding. Color coding is widely used, with colors representing various categories of information.

Colored lights are generally used in situations that change and where the environment is not fixed, such as changes in the condition or status of various pieces of equipment. Lights can also flash on and off as an attention-getting device. Brightness and rate of flash are only moderately useful when coding information, so it is best to think of lights as having only two major properties that can be varied: color and "on/off" status.

Some standard practices dictate the use of certain colors to code information. Generally, red represents a warning. Amber indicates caution and green indicates that a system is operating normally, or is in a ready or available status. Therefore, these colors should *not* be used to indicate conditions other than those that they typically represent.

The introduction of color CRTs will lead to more frequent presentation of words in color. Designers of color displays should consider the interaction of a label's color with the color associations of words in the label. GRASS, for example, is often associated with green (Warren, 1974). Scheibe, Shaver and Carrier (1967) found that a mismatch between the color in which a word is presented and the color with which it is commonly associated tends to slow down identification of its color. This sort of mismatch also slows down reading and may impede the performance of concurrent activities. Mismatches may also contribute to misidentification of a word. To ensure rapid and accurate user response to alarm and status displays, it is important to avoid such color mismatches. A correct match of color and label might improve both the speed and accuracy of identification and contribute to the smooth flow of work. If so, color of presentation and association should be matched whenever possible.

Warren (1980) conducted a study to determine the degree of association between color and words in alarm and status labels. With few exceptions the words had one or more colors strongly associated with them. These strong color associations should be considered when displays are developed, both to help user performance and to avoid user identification errors. Table 12-3 recommends the color to use when presenting certain words. For example, CRITICAL, ALARM, and POWER should always appear in red: ON, RUN and ACTIVE in green; and STANDBY, MINOR, and AUXILIARY in yellow.

Table 12-3. Recommended Colors for Alarm and Status Words

Word	*Color*
active	green
alarm	red
clear	white
critical	red
disable	red
emergency	red
enable	green
failure	red
major	red
minor	yellow
normal	green
off	black
on	green
on-line	green
power	red
run	green
standby	yellow
stop	red

(adapted from Warren, 1980)

To illustrate the extent of today's use of inconsistent color coding schemes, a study of color-coding of displays in nuclear power plant control rooms found that red denoted "on" or "flow" and green denoted "off" or "no-flow" (Parsons, Eckert and Seminara, 1978). In addition, the investigators found that this use of red-green coding had been mingled with military and other color-coding schemes to the point where almost every designer used their own preferred scheme. To avoid this kind of confusion, designers should rely on standard color coding schemes.

Keep in mind when using color coding in displays that some colors are not discriminable by some users. Some users will have defective color vision.

Redundant Coding

Designers should never use color as the only means of coding. Color should always be used with other forms of coding (refer again to the example in Figure 12-5). To be effective, color coding requires an adequate amount of light, preferably white light to show the true hues. If the visual displays are not lights themselves (e.g., signal lights), be sure that adequate lighting will always be available when the displays are used. For example, if people need to read maps or charts in an aircraft cockpit, automobile or truck cab under subdued lighting, or even red lighting (to help preserve night vision), the color coding will not convey information well.

Limit Number of Displays

Designers should use warning lights sparingly to avoid the "Christmas tree" effect. Too many lights on a panel make it difficult to determine what any one of the lights means. During the 1979 Three Mile Island nuclear power plant accident, over 100 alarms, most of them visual, went off with no way of suppressing the unimportant ones and identifying the important ones (Kemeny, 1979). Therefore, it is usually desirable to use lights only in situations where a preferable way to give the user information is impossible.

If both information and warning lights are used on a panel, the warning lights should be a different color or three to five times brighter than the information lights. In addition, warning lights should be well within the user's range of vision. We can learn a lesson from the Three Mile Island accident. Some of the key visual displays that would have helped early diagnosis of the problem were actually located on the back of the control panel (Kemeny, 1979). Important warning displays should also be located within the primary visual area—the area where the user is looking most often.

The effects of having too much coded information are pronounced when work load is high and exposure time is short. It is important not to overburden a user with cluttered displays. When considerable data is presented in a visual display, the display should be formatted so that a user can easily locate and read relevant information.

Size Coding

Size coding is another way to display various information categories. A small square, for example, might be used to represent one data category and a large square might represent another. Dot coding is useful in representing multiples of the same category. One dot on a map might represent a population of 3000; two dots might represent 6000; and so forth.

Geometric Shape Coding

Geometric shapes can also be used to code various categories of data. Circles and triangles provide good discrimination, good road maps make use of these symbols. Pictorial representations provide effective category distinctions. For example, the number of telephones installed in a given period could be shown by a telephone silhouette and a number associated with it. Pictorial representation can be used effectively in CRT displays for certain applications.

Numbers, Letters and Words as Codes

Much of the information on visual displays is in the form of words, numbers, or some combination of letters and numbers. A major problem in visual display design is keeping the number of words to a minimum, to help reduce the display size (display space is usually at a premium). Codes, including abbreviations reduce display size. However, if codes for items must be used, these codes should be clear and meaningful to users. The less meaningful the code, the more training required to learn that code, and the greater the number of errors that can be expected when it is used.

Selecting a Coding Technique

There are several factors involved in choosing the appropriate coding technique in display design. The first is the kind of information to be displayed. A lot of information usually requires an alphanumeric code. On the other hand, the status (on/off) or the presence or absence of trouble usually requires a light.

A second factor is the amount of information to be displayed. Some codes are better than others in yielding a maximum number of items that can be reliably represented. By assessing both the kind and amount of information to be displayed, a designer can make an initial estimate of the coding possibilities and eliminate those that appear unacceptable.

A third factor, the space requirement for the code, may be critical. To use an extreme example, a designer could indicate danger or a malfunction by a red warning light, or a message in large letters spread across a CRT screen. Obviously, the red warning light will occupy much less space, and would be the signal to use if space is at a premium.

Another factor to consider is ease and accuracy of understanding. The coded information should be understood promptly and accurately. This means simplifying display coding. The greater the amount of data on a display, the more difficult and time-consuming it becomes to search out relevant codes. When a great deal of data is displayed, the user may fail to detect some of it, or may misinterpret irrelevant data as relevant. Errors of this type are likely to increase as the amount of data increases, available search time decreases, and the user's work load increases.

A final consideration in the selection of coding techniques for displays is the interaction among displays at any given time. Designers should consider the possibility of other displays causing distraction. Warning lights should quickly alert users to something wrong—this is a useful distraction. However, a status light representing "power on" should not interfere with the reading of other displays.

The designer must consider code compatibility and code discriminability. Code compatibility is good correspondence between the data to be coded and how it is coded. For example, exact quantitative data, such as the number of telephones installed, should be displayed by numbers. On the other hand, a qualitative item, such as the functioning mode of a system, should be represented by a color coding technique that has appropriate qualitative connotations—red for danger.

Code discriminability permits the observer to distinguish one coded value from another. This requires recognition of the word, character, or symbol used for coding. The design, familiarity of the character or symbol, and the number of distinct categories used affect recognition.

Physical Characteristics

The majority of non-CRT displays have similar physical characteristics that are related to human performance. To best understand these characteristics a designer should be familiar with the following terms (see Figure 12-5):

1. Scale Range—The numerical difference between the highest and lowest value on a scale.
2. Numbered-Interval Value—The numerical difference between adjacent numbers on a scale
3. Graduation-Interval Value—The numerical difference represented by adjacent graduation marks.

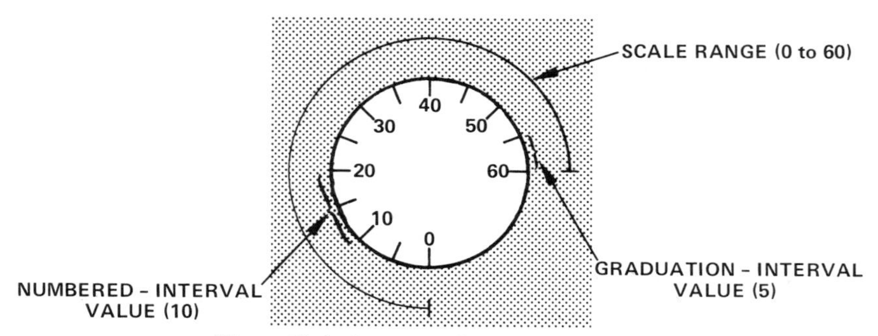

Figure 12-6. Important Scale Characteristics

Scale Selection

Before selecting a scale for a non-CRT indicator, a designer should decide on the appropriate scale range and should estimate the reading precision required. Figure 12-7 gives examples of different levels of scale precision.

If possible, all displays should indicate values in an immediately usable form so that users need not perform a mental conversion. Transformed scale values can be found in jet aircraft engine tachometers that have been calibrated in percent r.p.m. rather than actual r.p.m. For the pilot, this has several advantages. Maximum r.p.m. differs for different engine models and types. Transforming the scale values into percent r.p.m. relieves the pilot of the necessity of remembering operating r.p.m. values for different engines. In addition, the range from 0 to 100 percent is more easily interpreted than a range of true values, such as 0 to 8000 r.p.m., and the smaller numbers on the dial make a more readable scale. In Figure 12-8 the two tachometers illustrate these advantages. The dial on the left can be read more easily and precisely

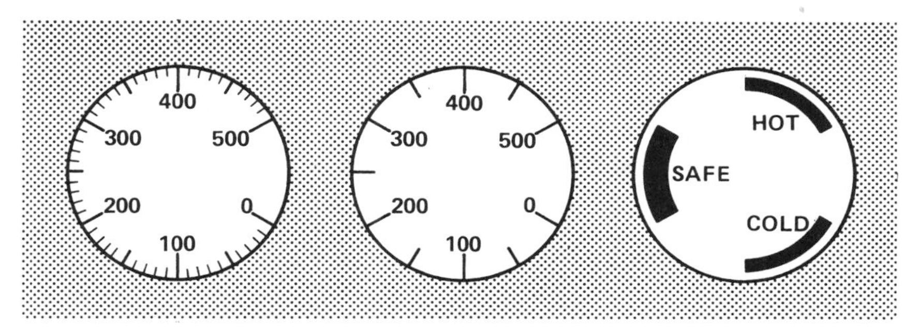

Figure 12-7. Examples of Scale Precision (adapted from Van Cott and Kinkade, 1972)

than the one on the right.

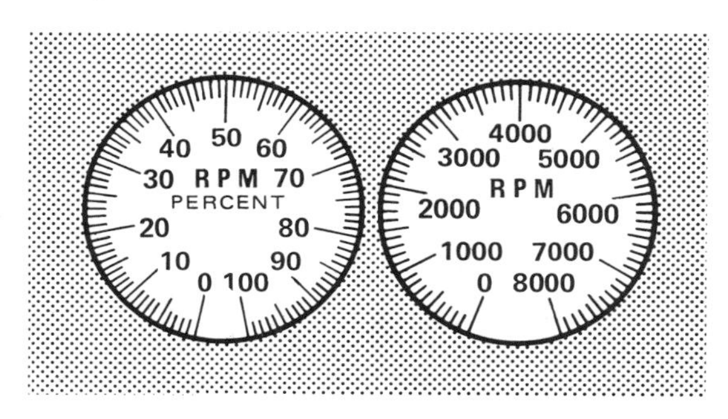

Figure 12-8. Sample Tachometer Dials

Scale Design

There must be enough separation between scale indexes to make reading easy. In addition, cues should be provided for determining differences between major and minor graduation marks. More specific recommendations for scale dimensions depend on the illumination level at the dial face.

For normal illumination: The following recommendations apply to indicators that are reasonably well illuminated. Assuming high contrast between the graduation marks and dial face, adequate illumination levels on the dial face, and reading distances of 13 to 28 in., the following recommendations should be observed (see Figure 12-9):

1. The minimum width of a major graduation mark should be 0.0125 in.
2. Although graduation marks may be spaced as close as 0.035 in., the distance should not be less than twice the stroke width for white marks on black dial faces or less than one stroke width for black marks on white dial faces.

3. The minimum distance between major graduation marks should be 0.5 in.

4. The height of major, intermediate, and minor graduation marks should not be less than 0.22, 0.16, and 0.09 in., respectively.

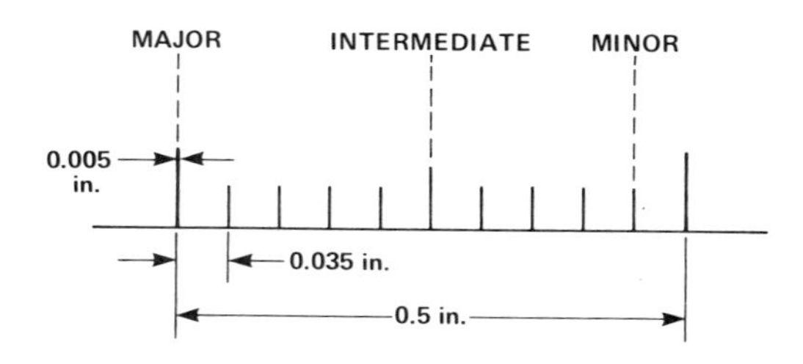

Recommended minimum scale dimensions
for high illumination (28 in. viewing distance).

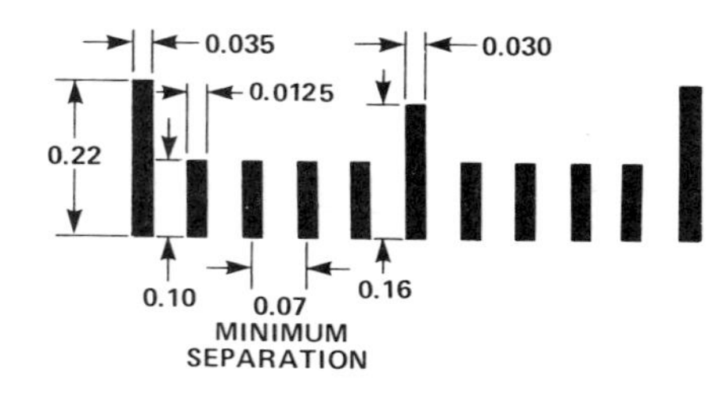

Recommended minimum scale dimensions
for low illumination (28 in. viewing distance).

Figure 12-9. Recommended Scale Dimensions (adapted from Van Cott and Kinkade, 1972)

For low illumination: When indicator scales must be read in lower than normal illumination, the scale design must result in maximum readability. Under these conditions, the additional aid of varied stroke widths of major and minor graduation marks becomes important.

The recommended minimum dimensions shown in Figure 12-9 apply to scale design for low illumination levels and should be followed whenever possible. These dimensions should not be considered as fixed values in the sense that other factors, such as scale size, number of graduation marks, and importance of indication, are not given equal consideration, but should be considered as models for relative dimensions. For instance, we assume a reading distance of 28 in. for the dimensions in Figure 12-9; for other reading distances, assume a proportional increase or decrease in the recommended scale dimensions (Table 12-4).

Interval Values

Some combinations of graduation-interval values and scale-numbering systems are more satisfactory than others. The following recommendations will assist the designer in selecting the most readable scale (see Figure 12-10):

1. The graduation-interval values should be one, two, five, or decimal multiples thereof. Graduation-interval values of two are less desirable than values of one or five.

2. There should be no more than nine graduation marks between numbered graduation intervals.

3. Normally, scales numbered by intervals of 1, 10, 100, etc. and subdivided by ten graduation intervals are superior to other acceptable scales.

4. Ordinarily, scales should be designed so that interpolation between graduation marks is not necessary; but when space is limited, it is better to require interpolated readings than to clutter the dial with crowded graduation marks.

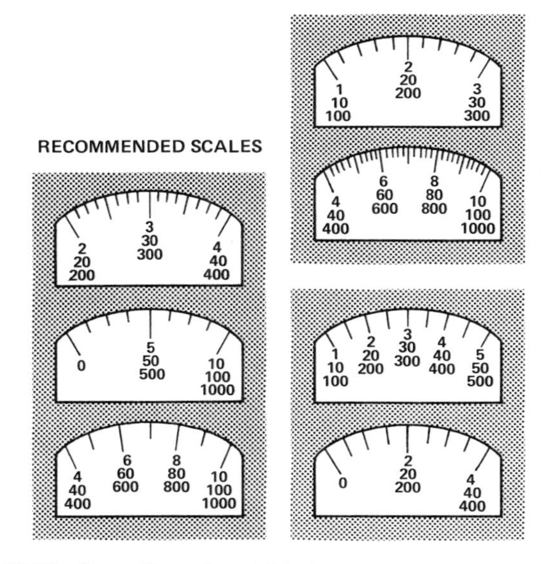

Figure 12-10. Some Examples of Displays with Acceptable Graduation Interval Values and Scale-Numbering Systems (adapted from Van Cott and Kinkade, 1972)

With this information in mind, the designer can select the most suitable scale from Figure 12-10.

Scale Interpolation

Quantitative scales should be designed for reading to the nearest graduation mark. For instance, a scale with a range of 50 read to the nearest unit should be numbered by tens with a graduation mark for each unit as shown in Figure 12-11(A). If less space was available for this scale, it would appear as in Figure 12-11(B). But the graduation marks on this scale may be too crowded for accurate and rapid reading, particularly under low illumination. Such situations call for a scale that may require interpolation as, for example, in Figure 12-11(C). This scale has a graduation-mark spacing that is more acceptable for low illumination. Also, this scale requires only a simple interpolation of one unit between graduation marks. Limited space might require interpolation in fifths or even tenths of a unit, but such interpolation could increase reading errors.

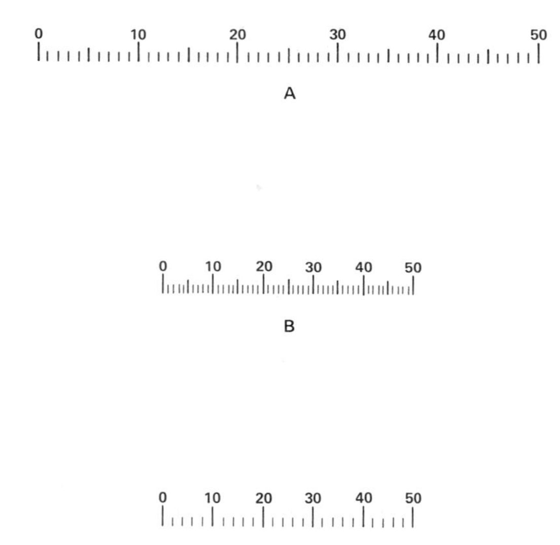

Figure 12-11. Sample Quantitative Scales

Number and Letter Size and Style

Designers should ensure that numbers and letters on indicator dials, panels, and consoles are as clear as possible, taking into account space restrictions and range of illumination. Recommendations in the following paragraphs apply to general flood-lighted and integrally lighted indicators as well as those viewed under ordinary ambient lighting.

Table 12-4. Scale Numbering Recommendations

Nature of Markings	Height (in.)*	
	Low luminance	High luminance
Critical markings, position variable (numerals on counters and settable or moving scales)	0.20-0.30	0.12-0.20
Critical markings, position fixed (numerals on fixed scales, control and switch markings, emergency instructions)	0.15-0.30	0.10-0.20
Noncritical markings (identification labels, routine instructions, any markings required only for familiarization)	0.05-0.20	0.05-0.20

* For 28-in. viewing distance. For other viewing distances, increase or decrease values proportionately.

(adapted from Van Cott and Kinkade, 1972)

A simple character style is best. The width of all numbers should be 3/5 of the height (see Table 12-4 for recommended heights), except for the "4," which should be one stroke width wider than the others, and the "1," which should be one stroke width wide. In addition, the stroke width should be from 1/6 to 1/8 to the numeral height.

The width of upper-case letters should be 3/5 of the height except for the "I," which should be one stroke width, and the "m" and "w," which should be about 1/5 wider than the other letters. Again, the stroke width should be from 1/6 to 1/8 of the letter height.

Scale Layout

Numbers should increase in a clockwise direction on circular and curved scales, from bottom to top on vertical straight scales, and from left to right on horizontal straight scales. (See A in Figure 12-12.) Except on multi-revolution indicators, such as clocks, there should be a scale break between the two ends of a circular scale. When the scale has a break in it, the zero or starting value should be located at the bottom of the scale (B), except when pointer alignment is desired for check reading. In this case, the zero or starting value should be positioned so that the desired value is located at the nine o'clock position (C). The zero or starting value on multi-revolution indicators should be at the top of the scale (D).

In general, on circular scales it is better to place numbers inside of the graduation marks to avoid constricting the scale. However, if ample space

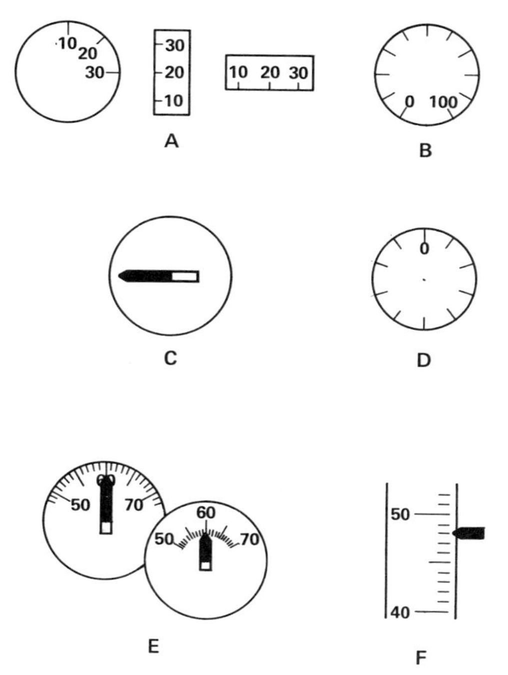

Figure 12-12. Examples of Scale Layouts (adapted from Van Cott and Kinkade, 1972)

exists, place the numbers outside the marks so they are not covered by the pointer (E). On vertical and horizontal straight scales, place the numbers on the side of the graduation marks opposite the pointer, and align the graduation marks on the side nearest the pointer and "step" them on the side nearest the numbers (F). The pointer should be to the right of vertical scales and underneath horizontal scales.

Zone Marking

Zone markings indicate various operating conditions on many indicators such as operating range—upper, lower — or danger limits, caution, etc. These zone markings may be color coded.

Figure 12-13 shows recommended colors associated with various operating conditions. If colored light, particularly red, illuminates the indicator, do not use color coding.

Pointer Design

Simplicity is the basic principle in pointer design. Some specific recommendations for pointer design are shown in Figure 12-14.

1. Pointers should extend to, but not overlap, the minor scale markings (A).

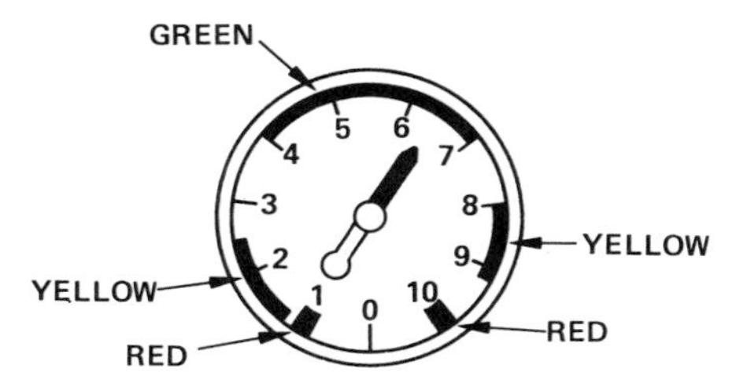

Figure 12-13. Example of Zone Markings (adapted from Van Cott and Kinkade, 1972)

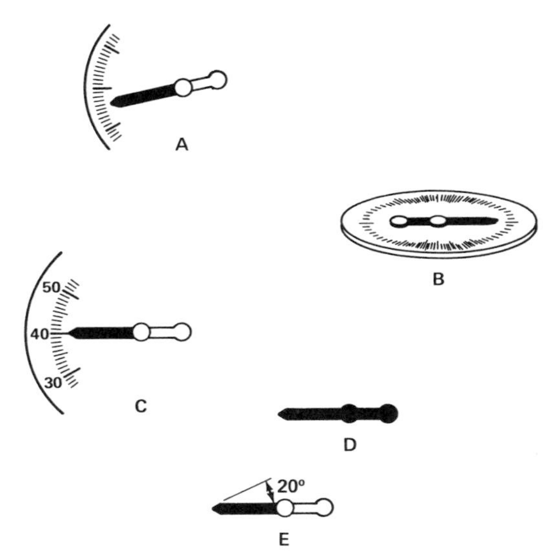

Figure 12-14. Recommendations for Pointer Design (adapted from Van Cott and Kinkade, 1972)

2. The pointer should be as close to the dial face as possible to minimize parallax (B).

3. For most applications, the section of the pointer from the center of rotation to the tip should be the same color as the dial markings. The remaining portion of the pointer should be the same color as the dial face (C).

4. For indicators designed for horizontal pointer alignment, the tail end of the pointer should extend beyond the center of rotation by an amount equal to about one half of the head of the pointer (D).

5. Recommended pointer tip angle should be 20° as illustrated in (E).

AUDITORY DISPLAYS

In addition to visual displays, systems frequently use many auditory displays. Auditory messages are suitable when:

1. The message is relatively short.
2. Response time to the message is important (auditory messages have good alerting characteristics).
3. The message need not be referred to later.
4. The vision of a user is already overburdened.
5. The receiving location, or environment is not suitable for the reception of visual communication (e.g., too light or too dark).
6. The user's job requires considerable movement (visual messages require that the receiver must be looking where the message will be displayed).

A common use of auditory displays is for alarms and warnings. When designing an auditory message to indicate an equipment malfunction, a designer should consider the following alternatives:

1. If the user will ever be a considerable distance from the equipment in the performance of other tasks, the signal should be loud, but low frequency.
2. If the user goes into another room or behind partitions, the signal should be low frequency.
3. If there is substantial background noise, the signal should be of a readily distinguishable frequency.
4. If an auditory signal must attract attention, and the above design features are not adequate, a designer might consider modulating the signal to make it even more noticeable.
5. The alarm should cease *only* after the user responds appropriately.

CONTROLS

People use controls such as steering wheels, knobs, levers, pushbuttons and toggle switches to interface with systems. Controls usually enable a user to make a change in the system and often are used with displays. A designer can determine the best type of control only after he or she knows exactly what is required of the user.

Most controls serve four kinds of functions:

1. *Activation*—an "on" or "off" condition or some other binary action.
2. *Discrete setting*—a control set to a position representing any of three or more discrete system responses. Gear level positions such as "park," "neutral," "reverse," and "drive" fall into this category.
3. *Quantitative settings*—individual settings of a control device that vary along some continuous quantitative dimension. The volume adjustment on a radio is usually a quantitative setting.
4. *Continuous control*—constant control of equipment (e.g., steering an automobile, maintaining a constant water pressure).

Selecting Controls

Since the type of control used will vary with task requirements, designers must know how to select controls for particular situations. This selection can be done best only after a thorough task analysis.

When selecting an appropriate control, consider four major areas:

1. Function of the control
   a. What is its purpose?
   b. What does it affect?
   c. What must be controlled?
   d. How important is the control?
   e. Is it critical to the successful operation of the system?
   f. Does it provide only a minor adjustment?
2. Task requirements
   a. What degree of precision is required?
   b. How fast must the setting be made, or the control activated?
   c. Is it an emergency control—one that must be reached immediately after a particular condition is noticed?
3. User information requirements
   a. What must be done to help the user locate the control?
   b. Is it a single control on a small panel, or is it grouped with many other controls on a large panel?
   c. What are the user's needs for determining the control setting or the effects of changed settings?

d. How quickly must the user be able to determine its existing setting or enter a new setting?

4. Work layout

   a. Where should the control be located?

   b. How much space is available?

   c. How important is the control?

   d. How does its positioning affect operator efficiency?

Designers should distribute controls so that no one limb is overburdened. When accuracy and speed of control positioning are important and when it is not necessary to apply moderate-to-large forces, use hand controls instead of foot controls. And for precision and speed, use one-hand controls instead of those operated with both hands. Radio tuning for example, is a fine adjustment task best suited to finger operator/rotary knobs. Even though it is the best to operate steering wheels with two hands, two-handed controls most often work best when large forces are required.

Consider foot controls when:

1. The application of moderate-to-large forces (greater than about 20 to 30 lb.), whether intermittent or continuous, is necessary.
2. The hands are overburdened with controlling other tasks.

Assign controls requiring large or continuous forward applications of force to the feet. Although a considerable number and variety of controls can be assigned to the hands, each foot should not have more than two controls assigned to it, and these should require only fore-and-aft or ankle flexion movement.

Identifying Controls

Controls should be easily identified. This can be done, in part, by standardizing their locations. All critical and emergency controls should be both visually and tactilely distinctive (shape coding). However, identification information should not hinder the manipulation of the control nor increase the likelihood of accidental activation.

Orienting Controls

Designers should make sure they orient controls so that their motion is compatible with the movement of the associated display element, equipment component, or vehicle. Finally combine functionally related controls to reduce reaching movements, to aid in sequential or simultaneous operations, or to economize in panel space.

Coding Controls

After the designer selects the appropriate control, he or she must decide how a user will know what function it controls. Making controls easy to identify decreases the use of a wrong control, and reduces the time required to find the correct control. Proper control coding not only improves user performance, but also reduces training time. There are a variety of ways to do this.

The five most common methods for coding controls are labeling, color, shape, size and location. Several methods of coding should be combined to achieve maximum differentiation and identification. An emergency control labeled EMERGENCY that is larger than surrounding controls, and is red, would be more readily identified than if only one of these coding methods were applied.

The choice of coding method depends on the following:

1. The total demands on the user when the control must be identified.
2. The extent and methods of coding already in use.
3. The illumination of the user's workplace.
4. The speed and accuracy with which controls must be identified.
5. The space available for the location of controls.
6. The number of controls to be coded.

Labeling

The simplest way to indicate a control function is to label it. Well-designed labels aid initial learning and subsequent performance by allowing the user to immediately identify the control being used to carry out a particular function.

Indicate the following three things when designing a label for a control:

1. Function controlled
2. Direction of control movement
3. Result of control movement.

Essentially, such information tells the user that if he or she manipulates a control in a certain way, it will produce a specific effect.

Observe the following general recommendations for labeling:

1. Locate labels systematically in relation to controls (usually above the controls).
2. Design labels to tell what is being controlled—for example, gear position, brightness level.

3. Make labels brief, but use only common abbreviations.
4. Only employ unusual technical terms when absolutely necessary and only when they are familiar to all operators.
5. Do not use abstract symbols (squares, stars, etc.) when they require special training. Use common symbols in a conventional manner—a red cross, a poison symbol, an arrow, international pictograph sets.
6. Use a letter and number style that is standard and easily readable under all conditions.
7. Locate labels so the operator can observe them while adjusting controls.

Labeling allows rapid visual identification of the control. But it could require a large amount of panel space to avoid a cluttered appearance. When a user operates a control or positions other devices on the panel, he or she may obscure a particular control. Nevertheless, the advantages of labeling usually far outweigh the disadvantages.

As a rule, label any control that appears on a panel. A written description in the form of a label should tell the function and describe the setting of the control (if appropriate). Follow consistent labeling practices. Labels should be clear, unambiguous, and present only essential information.

Avoid abbreviations unless space limitations make them necessary and then use only standardized abbreviations. It is not always possible to determine standard abbreviations for certain functions or certain settings of a control, but at least the abbreviation should be standardized within the system so that all the users of the system can attach the same meaning to the same abbreviation.

Labels should be on or very near the items they identify, and they should be placed horizontally to facilitate reading. When space is limited, place labels vertically only if they are not critical to safety or performance. Location of labels should be consistent—do not place a label above one display, but below another. Label every console/rack, panel, functional group, control/display, and control position with labels graduated in size, increasing approximately 25% from smallest to largest in the following order: (a) control position, (b) control/display, (c) functional groups, (d) panel, (e) equipment console or rack. For a normal 28-in. viewing distance, the characters of the smallest label should be approximately 1/8-inch.

Place nonfunctional labels—nameplate, manufacturer, or part number—inconspicuously so they will not be confused with operating labels. Avoid placing labels on curved surfaces. If necessary, make sure the significant portion is directly visible.

Identify groups of related controls by enclosing them within a borderline and labeling the groups by common function. If there is a group of controls for tuning some device, these controls should be grouped together and enclosed by a thin line. At the top of the enclosing line, the label TUNING should also appear.

The designer must also consider letter height, width, spacing and style, and contrast between letters and background. If labeling alone is not enough to distinguish one control from another, varying the color, shape or size of the control may prove useful as an additional coding technique. Vary the color of the panel surface to indicate that the controls located in that portion of the panel have a particular characteristic or that they control a particular portion of the system.

Color Coding

Color coding is most effective when a specific meaning can be attached to the color—for example, red for emergency controls. As pointed out earlier, the use of color coding depends on ambient or internal illumination. Keep in mind that color should not normally be used as the *only* or even the primary method for coding controls.

The color coding scheme should be standardized for the equipment or family of related equipment. System-wide standardization is recommended so users moving from one job to another can readily learn to do the new job. A function represented by a blue control on one job should also be represented by a blue control on another job in the same system. The same color code should be applied for all functionally related codes. Certain colors have associations that should be reinforced. Red, for example, should be reserved for emergency controls. Green is frequently used to indicate safety equipment of various sorts. Green may also be used for important or frequently used controls if consistent with other system controls. The connotation of green as a *go* condition or a *safe* condition must be kept in mind when applying green as a color code for controls.

Shape, Size and Location

In addition to labeling and color coding, controls can be coded by means of shape, size, and location. Using different shapes and sizes is especially useful when controls must be identified without the use of vision. Using different shapes provides for tactile identification of controls, as well as aids in identifying them visually. When feasible, select shapes that suggest the purpose of the control. In addition, it is important to use shapes that are easily distinguished from one another when touched.

Controls can be coded by size, but if the operator must rely on touch alone, usable sizes are quite limited. However, the ability to discriminate size by touch is relatively independent of shape discrimination; hence, a limited number of different sizes can be used with different shapes. Keep in mind that different sizes and shapes are less effective if the user is wearing gloves. When the user cannot feel all the controls before selecting the proper one, only two or at most three different sizes of controls should be used (small, medium, and large).

Designing or Selecting Controls

Follow these general principles when designing or selecting controls.

1. Critical and frequently used controls should be located within easy reach of a user.
2. The force, speed, accuracy, and range of body movement required to operate a control should never exceed the capability limits of the *least capable* user. In fact, these performance requirements should be considerably less than the abilities of the least capable operator.
3. The total number of controls should be kept to a minimum.
4. Control movements should be as simple, easy, and as natural as possible (i.e., they should conform to user expectations whenever possible).
5. Control movements should also be as short as possible and exhibit the necessary "feel" to allow the operator to make accurate settings.
6. When a control needs to be powered (provided with a mechanical advantage—perhaps hydraulically or electrically), artificial resistance may be required.
7. Control actions should result in a positive indication to the user.
8. Control surfaces should be designed to prevent the finger, hand, or foot, from slipping.
9. Controls should be designed and located to prevent or reduce the probability of accidental operation. When accidental operation would result in a critical situation such as personal injury, these controls should be provided with guards.

Common Controls

The following design recommendations apply to commonly used controls, such as pushbuttons, toggle switches, rotary selector switches and knobs. Some general comparisons are shown in Table 12-5.

Pushbuttons

Pushbuttons are of three major types: latching (push-on, lock-on), momentary (push-on, release-off), and alternate action (push-on, push-off). In most applications pushbuttons are operated by the fingers. Finger pushbuttons can be coded by labeling, color, size, shape, or location, while foot pushbuttons are usually coded by location only. Finger pushbuttons can be operated quickly and simultaneously with other pushbuttons in an array and are identified easily by their position within an array or by their associated display signal. One of the disadvantages of many pushbuttons is that the control setting is not easily identified, visually or tactilely.

Table 12-5. Comparison of Common Types of Controls

	Type of Control			
Characteristic	*Push button*	*Toggle switch*	*Rotary selector switch*	*Knob*
Time required to make control setting	Very quick	Very quick	Medium to quick	-
Recommended number of control positions (settings)	2	2 to 3	3 to 24	-
Likelihood of accidental activation	Medium	Medium	Low	Medium
Effectiveness of coding	Fair	Fair	Good	Good
Effectiveness of visually identifying control position	Poor	Good	Fair	Fair
Effectiveness of check-reading to determine control position when part of a group of like controls	Poor	Good	Good	Good

(adapted from Van Cott and Kinkade, 1972)

Some design recommendations are discussed below.

1. Resistance should start low, build up rapidly, then drop suddenly to indicate that the control has been activated.

2. The surface of the pushbutton should have a high degree of frictional resistance to prevent slipping. For finger-operated buttons a concave surface is even more practical.

3. To indicate that the control has been activated, provide positive feedback to the user in either a visual, auditory or kinesthetic form.
4. A series of pushbuttons should be mounted in a horizontal (rather than vertical) array.

Toggle Switches

Toggle switches are best used as two-position controls; however, three-position toggle switches are in common use. Toggle switches require only a small amount of space for their location and operation, and can be operated quickly and simultaneously with other toggle switches in a row. They are identified easily by their proximity to the associated display or by their location within an array.

Toggle switches should have a vertical orientation: up for "on," "go," or "increase;" down for "off," "stop," or "decrease." Install toggle switches for horizontal orientation *only* if necessary to be consistent with the orientation of a control function, equipment location or display, or to prevent accidental activation. Provide an audible click to indicate that the control has been activated in areas where the ambient noise level is low enough for a click to be heard. A series of toggle switches should be mounted in a horizontal (rather than vertical) array for speed and ease of operation. The toggle switches should be arranged for easy position checking.

Rotary Selector Switches

Rotary selector switches usually have from 3 to 24 control positions. Speed and accuracy of setting controls and checking their position is sometimes difficult if there are more than 24 settings. Keep in mind that rotary switches require considerably more space than the size of the switch because the size of the hand that turns the switch must be taken into account. (See Figure 12-15.)

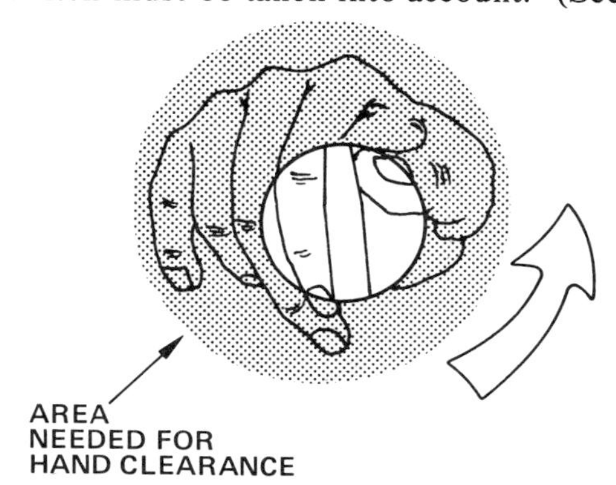

Figure 12-15. Clearance Requirements for a Rotary Selector Switch (adapted from Van Cott and Kinkade, 1972)

Some design recommendations for selector switches are given below:

1. For most applications, selector switches should have fixed scales and moving pointers. The entire scale should be visible.
2. Detents should be provided at each control position (setting).
3. For speed and ease of operation, there should be no less than 15° between settings.
4. When few control settings are required, the settings should be about 30° from each other. The arrangement should be as shown in Figure 12-16.

Figure 12-16. Selector Switch Control Settings

5. For most applications the moving pointer should be a bar-type knob with a tapered tip.
6. Setting values should increase with a clockwise rotation.
7. Minimize parallax by designing the end of the pointer to come close to the scale index.
8. Index numbers should not be obscured when the hand is on the control.

Control Knobs

Control knobs are generally used for adjusting displays. Control knobs are used for making small turning movements that do not require large forces. They can be used for either gross or fine positioning over a wide range of adjustments.

Recommendations when using control knobs include:

1. The scale should be visible when the operator's fingers are on the knob.
2. Setting values should increase with clockwise rotation of the knob.
3. If a pointer or index is used on the control, it should be close to the scale index mark to minimize parallax.
4. If the control is not used for multi-turn operation, start and end stops should be provided.

Special Controls

Hands and feet traditionally have been the only parts of the body used for operating controls. Theoretically, at least, any of a large number of body responses—movements of the knees, hips, elbows, shoulders, head, or eyes—could be used to actuate control devices. In fact, knee levers have been used for years as a standard control on certain sewing machines, and head movement is now used to aim guns in some helicopters.

Designers should be alert for situations where a user may not be able to use hands or feet. Some users may lack the use of one or more limbs either because they are multiple amputees or are paralyzed. Designers should take into account the potential for using less conventional control methods. A nod of the head, for example, could be used to type a character, start a motor, summon an elevator, open a refrigerator door, or turn on a television set.

ARRANGEMENT OF CONTROLS AND DISPLAYS

Even if all displays and controls are properly designed, unless they are organized in a logical manner, a user is likely to read the wrong displays, or lose valuable time hunting for the proper control. Thoughtful grouping of controls and displays is essential to having acceptable human performance.

When using a large number of controls and displays, their grouping should aid in determining:

1. Which control affects which display.
2. Which control affects which equipment component.
3. Which equipment component is described by each display.

Controls and displays are most commonly grouped by function and by sequence of use. In functional grouping, controls and displays related to one function are put together in one panel area. In addition, *sets* of controls and displays within functional groups also can be put together.

Grouping of controls and displays by sequence of use helps to provide efficient user movements and reduces the requirement of retracing or "skipping" around the panel. This type of grouping should provide for movements from left to right and from top to bottom.

Population Stereotypes

Ensure that the movement of a control leads to the most expected movement of a display or vehicle. The recommended direction-of-movement relationships are usually intended to ensure natural relationships. Natural relationships refer to control-movement habit patterns that are consistent from person to person without special training or instructions; they are responses that individuals make most often, and are sometimes referred to as *population stereotypes*.

The following general direction of movement rules are applicable:

1. The direction of movement of a control should be considered in relation to: (a) The location and orientation of the user relative to the control; (b) The position of the display relative to the control and the nature and direction of the display's response; (c) The change resulting from the control movement, either in terms of motion of moving components (landing gear, automobile wheels, etc.) or in terms of some dimensional quantity (volume of a radio receiver or brightness of a CRT screen).
2. The preferred direction of movement for most hand controls is horizontal, rather than vertical, and is fore-and-aft, rather than lateral.
3. All equipment that the same person uses should have the same control-display motion relationship.
4. Control-movement relationships are particularly important when they result in vehicle movement. A movement of a control to the right should result in a movement to the right, a right turn, or right bank of the vehicle, etc.

Recommended relationships between control movement and system or component response are shown in Table 12-6. If there is any doubt as to which direction-of-movement relationships are called for in a particular situation, do a study to determine the best way.

Table 12-6. Recommended Control Movements

Function	*Control Action*
On	Up, right, forward, pull
Off	Down, left, push
Right	Clockwise, right
Left	Counterclockwise, left
Up	Up, rearward
Down	Down, forward
Retract	Rearward, pull, counterclockwise, up
Extend	Forward, push, clockwise, down
Increase	Right, up, forward
Decrease	Left, down, rearward.

(adapted from Van Cott and Kinkade, 1972)

Locating Controls

Frequently used controls should be in an easy to use location. Users should not have to hold their hands, arms or legs in awkward positions for long periods. Controls requiring motion excursion should be located within comfortable arm reach; the user should not have to move the body to initiate or complete the control motion (as is usually the case when turning on headlights in automobiles).

When the application of considerable force is required, controls should be located so that maximum use of combined sets of muscle groups, and support from a seat back or other structure are possible. If quick response is important, locate controls at the user's fingertips.

Adequate spacing between controls or adjacent structures will help prevent inadvertent activation and ensure that controls can be operated without hampering human performance. For spacing, consider:

1. The body member being used
2. Control size and amount of movement (displacement)
3. Requirements for "blind" reaching (i.e., being unable to see the control)
4. Effects on system performance of inadvertently using the wrong controls
5. Clothing that might hinder control manipulation (e.g., pressure suit, gloves, boots).

Locating Displays

Visual display elements range from a natural display of objects as seen from a vehicle window, to specific hardware displays such as dials, indicator lights, signs, CRT displays, and status and map boards. Rules for arrangement of visual displays are generally quite obvious. They should be placed:

1. In front of the user
2. As nearly perpendicular to the line of sight as possible to reduce parallax
3. At a distance for adequate resolution of visual detail
4. Out of any direct light that may cause reflection and/or glare.
5. When a control is associated with a specific display, it should be located below or to one side of that display so the operator will not obscure the display.

Location of auditory displays is less critical than with visual displays. However, the direction from which a sound comes may be used as one of the cues for differentiating one signal from another.

Summary

To summarize, controls and displays should be arranged so that a user can see all displays from a normal working position, without excessive shifting of the head or the body. In addition, they should be arranged to elicit efficient patterns of movement. Controls and displays should be adequately identified so a user can find an individual control or display without error. Control and display arrangement should provide expected direction-of-movement relationships. Finally, controls should be spaced far enough apart and away from adjacent structures, to permit adequate grasp and manipulation through an entire motion range. If gloves are to be worn, larger controls and control spacings will be required.

WORKPLACE DESIGN

Good workplace design will improve human performance. The two most important questions that should be asked by designers are "Who will be the specific user of the workplace?" and "How will the workplace be used?"

To design an adequate workplace, a designer should first consider the physical dimensions of the people who will use the system. In selecting the appropriate measurements, determine if the users will be sitting, standing, or moving about. Design workplaces that will accommodate at least 95 percent of the potential users (see Chapter 5). Design for smaller users when considering reach, strength, or any other tasks involving human limitations. Consider the larger users when designing doors, passageways, or anything where the large size of an individual may hinder performance. Large operator dimensions should define clearance requirements; those of smaller operators should define reach requirements.

The three main user-related considerations that influence workplace configuration are:

1. Reach and clearance envelope
2. User position with respect to display area (in an office, factory, laboratory, etc.) and/or field of view (in a vehicle)
3. The position of the human body (standing, seated) and means of support (chair, stool).

Reach and Clearance

Clearance at various levels is important for access to and from a workplace, for ease in grasping and operating controls, for ease in adjusting the body properly to the visual-control task, and for isolation of the operator from physical discomfort or injury. These factors may be influenced by the manner in which the operator is restrained for protection (e.g., seat belts) or by special clothing required for the job. In establishing clearance requirements, whether related to access, control manipulation, body position, or operator isolation, the designer must recognize the specific needs of the user.

User Position and Vision

Since designers must choose among competing priorities in workplace layout, Woodson (1972) has proposed that the main consideration be given to ensuring that a user can see what needs to be seen. To look out of a windshield, or to view a CRT, eye position relative to the task should establish the basic layout reference point. These important areas of vision are referred to as the *primary visual areas*. Secondary and tertiary visual areas also should be considered. For example, when driving an automobile, the area outside of the window is defined as the primary visual area; whereas, a secondary visual area might consist of the instruments or lights on the control panel. The primary visual area normally establishes the principal orientation of a workplace layout with other visual functions integrated into it.

The primary visual areas of control/display panels center around the user's normal line of sight—approximately 10° down from the user's horizontal line of sight. Primary displays, including emergency or hazard indicators, critical monitoring displays and displays where color identification is critical, should not require excessive movement of the operator's head or eyes from normal line of sight. All warning displays (those indicating a present or potential system failure or personnel/equipment hazard) should be within 30° of normal line of sight or 45° for a sit-stand workplace. Secondary displays may require movement from the normal line of sight, but not head movement. Finally, auxiliary displays, including infrequently used displays such as console power indicators and maintenance displays, may require operator head and eye movement from normal line of sight.

The primary visual area should be free from obstruction. In automobiles, certain window structures create "blank spots" causing serious loss of visual information. Windshield support posts should either be placed where they will not obstruct vision, or redesigned so that the area blocked is small enough to allow operators to see around the post.

Position of Body

A seated position for users is advantageous when the following requirements exist: a high degree of body stability and equilibrium, long work periods, the use of both feet to accomplish the task, and large force on foot controls. A standing position for users is preferred in the following situations: a person must be mobile to reach and monitor different controls and displays; precise manual control actions are not required; foot control actions, other than simple on/off actions are not required. Generally, having users stand for long periods to perform any type of work activity is not recommended.

A combination of sit-stand user positions is sometimes recommended. For example, the user may require the stability provided by being seated for precise control actions and the mobility provided by standing for monitoring a large control panel. In some cases, it is advisable to provide a high chair or stool so that a user who must sit and stand alternately can do so with the least amount of effort. Such a high chair or stool should enable the user to maintain his or her seated height at approximately the same level as when standing.

A workplace consists of numerous elements that should be designed individually and arranged together in a way that encourages efficiency and acceptable human performance.

Workplace Layout and User Movements

Tichauer (1969) proposed several brief rules for workplace layout that take into consideration the movements made by users. These rules have been adapted and are presented below:

1. The user should be able to perform his or her task equally well, seated or standing. He or she should be able to change from sitting to standing and vice versa at will, without altering the normal vertical distance between elbow and principal plane of work. For this reason, the working table should be high enough to permit work in comfort while seated at the same table.

2. The best height for a working table is approximately 38 inches for women and 41 inches for men. The height of the seating surface of a stool or chair to fit these tables should be 30 inches for women and 34 inches for men. Such chairs should be provided with foot rests if these are not incorporated in the structure of the work tables.

3. If chair and table are well proportioned, then the level of the user's elbow should be about two inches above the work surface.

4. The seating accommodation should be fitted with a comfortable back-rest to permit correct posture. The dimensions of the back-rest should be such that no stresses are placed on any part of the user's back.

5. The seating surface of a chair should be flat, 12 inches wide and 10 inches deep. So-called "shapes to suit body contour" carved into hard seating surfaces can lead to discomfort and fatigue. Properly upholstered, padded seats can aid comfort.

6. No work should be performed on a strip of 3 inches from the border of the desk that is closest to the user and the majority of work should be performed in the area encompassed by a $\pm 30^o$ angle from the line projected through the user's mid-sagittal plane.

7. All necessary movements should be performed without excessive flexion of the trunk.
8. The user should have ample space to move his or her elbows at will.
9. All points of pick-up and delivery for any object at the work station should be situated within easy reach.
10. The excessive use of short and primitive movements, especially in tedious and monotonous repetition, can lead to a lack of precision in performance. If long series of repetitious movements are required adequate rest pauses should be provided.
11. Receptacles or at least fixed points should be provided as repositories for tools that are not in actual use (e.g., pen or pencils, screwdrivers, soldering irons, etc.).
12. Prepositioning should be provided for tools and components where possible, in order to eliminate fumbling, hesitation, and transfer grasps.
13. Unused tools should never be placed between the user and a place where work is to be performed. Tool repositories should always be more distant and more lateral from the user than the main working area, but preferably not located in such a manner as to require the user to perform angular, vertically curved or close reaches across materials or components, while performing the pick-up of tool or implement.
14. Movements should be curved with the curves following the kinematics of the human skeleton. Straight motions tend to be slow and lack precision.
15. Sudden changes in velocity and/or direction of a movement should be avoided for they take up too much time and cause fatigue.
16. Wherever possible both hands should move during an operation. The movements should be simultaneous, symmetrical and opposite in direction.
17. At times, leg movements can be employed with advantage to relieve the hands of excessive work. This is frequently good practice in situations where the hands will do assembly, positioning or holding work while the feet produce action of machinery (e.g., the sewing machine, or the punch press).
18. Foot motion slows down hand motions. It is preferable that the hands do not move while the foot or the leg moves. Consider the difficulties typists have with foot switches on recorders.

19. Where possible, both feet should not move simultaneously.

20. When foot or leg motion is required, then the seating surface of the chair should be sufficiently short (about 7 inches) to permit freedom of movement.

21. In situations involving the use of pedals, there should be sufficient space provided for knees and feet to move freely in all directions.

22. One of the main causes of hesitation is the fear of accidents for self or others due to unguarded and dangerous parts of equipment. A user trying to avoid a danger spot could add considerably to the time taken to perform a particular operation (e.g., using an unguarded soldering iron).

23. When working at a table, desk or workbench, the highest point to which a user's hand should be raised should not exceed 8 inches above elbow level. This point should be approximately 14 inches away from the border of the desk nearest to the user and this maximum operation height range extends for a further 6 inches to a point approximately 20 inches away from the near border of the desk. In other words, the maximum height of 8 inches above elbow level may be reached in the distal part of the tactile operations area and throughout the tool pick-up area. When operating nearer or further away from this range the hands should be operating at a substantially lower level.

24. When handling loads of medium weight while standing, and if these loads have handles, all movements should be ballistic. The areas of ballistic swing should be compatible with human kinesiology. It is just as difficult and time-consuming to handle a load that is too close to the body as one that is too far away.

25. When lifting loads that have no handles, then the center of gravity of the load should be as close as possible to a line situated in the central sagittal plane of the body, passing through the midpoint of the heels and the most posterior part of the neck.

26. Unnecessary motions of the torso, especially bending down, are time consuming and lead to rapid fatigue.

27. All color schemes at the workplace should result in pastel shades for background colors and hard shades giving good appreciation of equipment contours, articles, levers and tools.

Workplace Environment

A main objective of workplace design should be to create user acceptance. Users tend to be motivated if the workplace is well organized, convenient, and attractive. If a user has difficulty getting into position, or seeing certain displays, or reaching and monitoring controls because of poor arrangement, his motivation to work may be reduced.

The designer should strive to eliminate or minimize the effects of an undesirable environment on human performance, including excess noise, light, and heat. For example, proper orientation of a display panel may reduce the effects of glare from an ambient light source. Independent seat suspension, seat padding, and contoured seating may reduce stress from road shock, as well as the fatigue that comes from spending several hours in relatively confined space.

FOR MORE INFORMATION

Chapanis, A., *Ethnic Variables in Human Factors Engineering,* Baltimore: The Johns Hopkins University Press, 1975.

McCormick, E. J., *Human Factors in Engineering and Design,* Fourth Edition, New York: McGraw-Hill, 1976.

Van Cott, H. P. and Kinkade, R. G., *Human Engineering Guide to Equipment Design,* Washington, D.C.: U.S. Government Printing Office, 1972.

Woodson, W. E. and Conover, D. W., *Human Engineering Guide for Equipment Designers,* Second Edition, Berkeley: University of California Press, 1964.

13

SPEECH COMMUNICATION

A MAJOR DISASTER

A KLM Royal Airlines jumbo jet roared down the fog-shrouded runway at Tenerife in the Canary Islands. Suddenly, a taxiing Pan American World Airways jumbo jet loomed in the fog. The KLM captain yanked his control column back in a desperate attempt to take off and fly over the Pan Am plane. But it was too late. The collision claimed 583 lives.

It all began when the Pan Am jet left Los Angeles for Las Palmas, in the Canary Islands, with 373 passengers on board. About the same time the KLM jet left Schiphol Airport near Amsterdam with 234 passengers. While both planes were enroute to Las Palmas, a bomb exploded in that terminal and several planes, including the Pan Am and KLM jets, were diverted to nearby Tenerife. The KLM and Pan Am planes arrived early in the afternoon of May 27, 1977.

The KLM pilot was under considerable pressure. He had been on duty for more than nine hours. If he stayed at Tenerife much longer, the 747 would have to remain overnight in the Canary Islands. Both KLM flight schedules and passengers' plans would be disrupted.

While the KLM and Pan Am planes were waiting, the weather at Tenerife deteriorated. Fog rolled in decreasing visibility. When it finally came time to leave, the KLM plane could not use any taxiways to get to the end of the runway since all were jammed with other planes diverted from Las Palmas. To reach the end of the airport's only runway, the KLM plane had to taxi down *that* runway.

The Pan Am plane was cleared to taxi down the same runway behind the KLM jet. The Tenerife controller ordered the Pan Am plane to taxi part way down the runway to a turnoff and then to exit to a clear taxiway. But the fog became so thick the Pan Am crew missed their assigned turnoff and continued taxiing down the runway behind the KLM jet.

Meanwhile, the KLM craft reached the end of the runway and turned around. The captain still needed two separate clearances for takeoff: the first from air traffic control (ATC) and the second from the tower. Without waiting for the clearances, the captain began to advance the throttles. The following conversation was heard from the cockpit:

KLM copilot: "Wait a minute, we don't have any ATC clearance."
KLM captain, retarding the throttles: "I know that, go ahead, ask."

The KLM copilot requested and received clearance from ATC. He then read ATC clearance back to the tower, but even before he finished, the pilot advanced the throttles again. The copilot told the tower hurriedly, "We are now at takeoff." (That commonly means "takeoff position.")

Tower: "Okay, stand by for takeoff and I will call."

But only the word "okay" was clear. The Pan Am crew, radioed the tower to announce their position on the runway. That transmission caused a squeal that partly blocked the rest of the tower's order to the KLM crew to stand by. Saying, "We go!" the KLM captain gunned the 747 down the runway. The next two transmissions were heard in the KLM cockpit.

Tower: "Roger Papa Alpha (Pan Am)...report the runway clear."
Pan Am: "Okay, we'll report when we're clear."
KLM engineer, in the KLM cockpit: "Is he not clear, that Pan American?"
KLM pilot: "Oh, yes."

In the dense fog, the Pan Am crew could not see the KLM jet roaring down the runway toward them. But the radio transmissions alarmed them.

Pan Am pilot: "Let's get the hell out of here!"
Pan Am copilot: "Yeh, he's anxious, isn't he?"
Pan Am pilot: "There he is—look at him.
Pan Am copilot: "Get off, get off, get off!"

The Pan Am pilot tried to steer his jumbo jet off the runway. But it was useless. The KLM jumbo was going about 130 miles an hour when it struck the Pan Am plane. Most of the 583 who died were killed on impact.

In this situation, adequate communication between the aircraft, tower, and air traffic controller could have saved many lives. The example is unique only because of the large number of lives lost. In system after system there are misunderstandings and missed messages due to degraded speech communication. To avert the possibly disastrous consequences of some breakdowns and guarantee an acceptable level of human performance in all systems that depend on speech, designers should ensure that the important aspects of speech communication are carefully considered.

The speech communication chain is shown in Figure 13-1. A message originating in the brain controls certain muscles in the mouth which modify the air flow passing through it from the lungs. This produces sound waves coming out of the mouth. These sound waves impinge upon the listener's ear, and are converted to neural impulses transmitted to the brain. To complete the communication between two individuals the process is reversed. In general terms, speech communication can be degraded if there are difficulties with the *speaker, transmission* of sound waves or the *listener.*

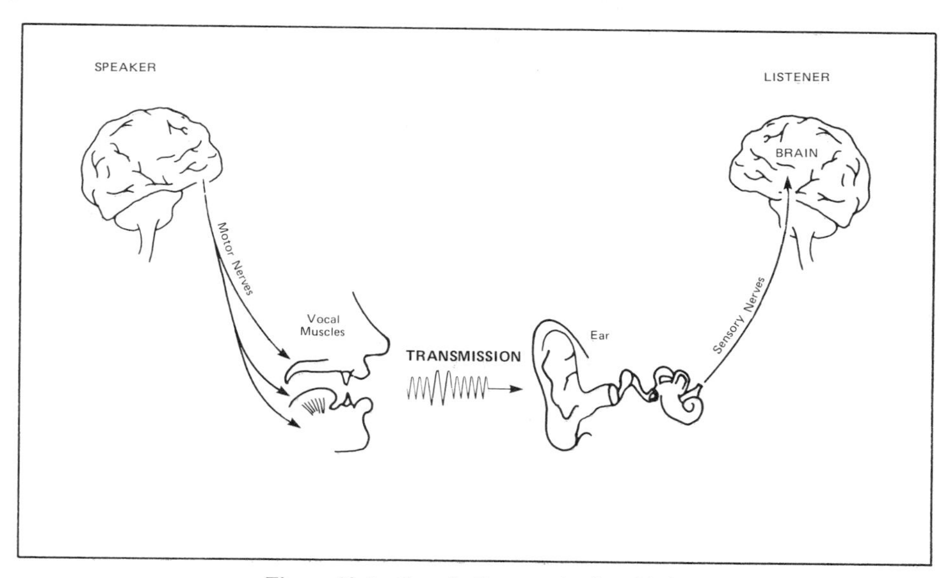

Figure 13-1. Speech Communication Chain

FREQUENCY, INTENSITY AND SIGNAL-TO-NOISE RATIO

A clear separation exists between the physical parameters of sound waves and peoples' perception of them. While the physical sound wave is described in terms of its frequency and its intensity, our perception of it is in terms of its pitch and loudness. It is very important to keep in mind this relationship between the physical characteristics of the sound wave and the perceptual dimensions. Physical properties are inherent in the sound wave and can be measured independently of any human observer. The perceptual dimensions are a function of the sensations evoked in the listener and have to be measured by means of a listener. We will deal here primarily with the physical characteristics of frequency and intensity.

Frequency and Intensity

To understand some of the basic human performance considerations related to speech communications in systems, we need to review at least three basic definitions originally presented in Chapter 4 on Sensing. Recall that the two essential elements of sound related to speech communication are *frequency* and *intensity*. We define frequency as the number of cycles of pressure change occurring in one second. If the sound pressure changes from positive to negative and back to positive five hundred and fifty times a second, the frequency is 550 cycles per second, or 550 Hertz (Hz).

Intensity is defined in terms of how large the sound pressure changes are. The human ear can hear sounds so weak they are almost impossible to distinguish from the random movements of air molecules, and sounds so strong they have an intensity 10^{15} times as great. The range of sound intensities is so large that a special number scale has been developed to measure it.

Sound intensities are measured in decibels (dB). The decibel has two special features—ratio measurement and logarithmic measurement. Table 13-1 shows some sample decibel computations and illustrates how logarithms compress some very large numbers into smaller ones. In each case the number of decibels is equal to $20 \log_{10} \frac{P_1}{P_2}$, where P_1 and P_2 equal the sound pressure of two different sounds.

Table 13-1. Some Sample Computations Using the Decibel Scale

Sound pressure of sound 1 (P_1)	Sound pressure of sound 2 (P_2)	$\frac{P_1}{P_2}$	$\log_{10}\frac{P_1}{P_2}$	$20 \log_{10}\frac{P_1}{P_2}$	Number of decibels (dB)
10	1	10	1	20	20
100	1	100	2	40	40
1,000	1	1,000	3	60	60
10,000	1	10,000	4	80	80

(adapted from Chapanis, 1965a, *Courtesy Human Factors*)

To provide the decibel scale with an absolute meaning, a reference sound pressure level is used as a common anchor point. That reference pressure is 0.0002 dynes per square centimeter or 20 micro Newtons per square meter. This reference level was chosen at what researchers originally thought was the average threshold for perceiving a 1000-Hz tone.

A great many frequencies and intensities make up most of the sounds we hear. If these frequencies and intensities are related in some systematic way to each other, the resulting sound is potentially "meaningful." On the other hand, if the frequencies and intensities are unrelated, we generally perceive the sound as noise. It is convenient to describe a sound or noise using a chart that shows the combinations comprised of all frequencies and intensities. Figure 13-2 shows such a chart of noise in a telephone company business office.

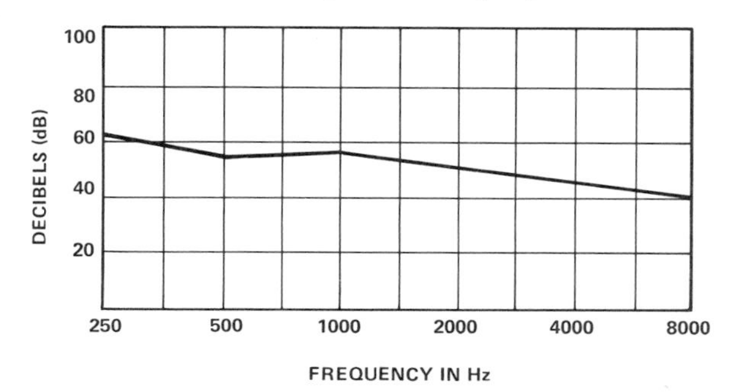

Figure 13-2. Spectrum of Noise in a Telephone Company Business Office

When we consider human performance problems related to speech communication, we must be aware of the differing sound level intensities at the various frequencies, for both background noise and speech. Figure 13-3 show the average spectral density for continuous speech. Note that most of the speech sounds are between about 200 and 4000 Hz, and 30 to 60 dB.

Signal-to-Noise Ratio

A third definition that is useful in understanding speech communication is the *signal-to-noise ratio*. The signal-to-noise ratio, abbreviated as S/N ratio, is a measure of the relative sound intensity of the signal, speech, to the background noise. For example, if the average intensity of speech is 75 dB and the average intensity of the noise in which the speech is spoken is 70 dB, we have an S/N ratio of +5 dB. If the average intensity of the speech is 60 dB and it is heard in an environment with an overall noise level of 60 dB, the S/N ratio is 0 dB. Finally, if the average intensity of the speech is 50 dB and it is heard in an environment with noise at 80 dB, the S/N ratio is -30 dB. To compute the S/N ratio, subtract the number of decibels in the noise from the number of decibels in the speech signal.

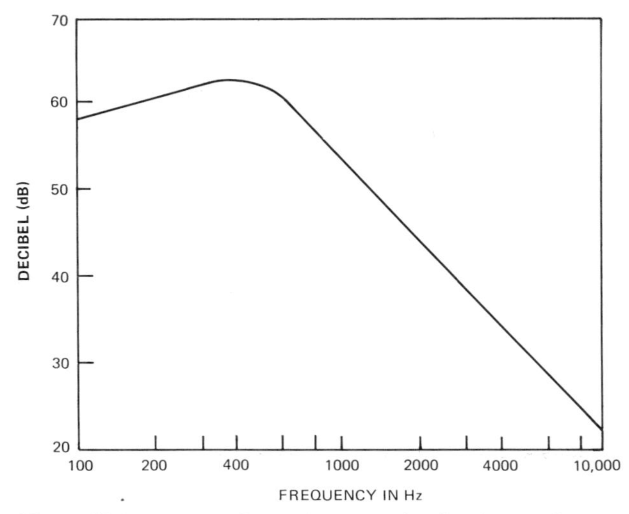

Figure 13-3. Average Spectral Density for Continuous Speech (Silbiger, 1973)

INTELLIGIBILITY TESTING

The smallest unit of speech is called a *phoneme*. Phonemes may be combined to make up *syllables*, syllables to make up *words*, and words in turn may be combined to form *sentences*. Usually, each syllable contains a consonant, a vowel and a consonant (CVC). The ability to recognize correctly the speech sounds when spoken is of primary importance in speech communication, and is termed *intelligibility*. The process of measuring intelligibility is also often called articulation testing. Articulation tests may be either syllable (CVC), word, or sentence tests. These tests have consistently shown that sentence intelligibility is always higher than word intelligibility, and intelligibility for meaningful words is usually higher than intelligibility for meaningless syllables.

To ensure that the speech material used in an intelligibility test contains the relevant sounds of the English language, special words and sentence lists have been constructed. These are called phonetically balanced (PB) words or sentences. An example of a phonetically balanced word list is shown in Table 13-2. Other lists are available, along with a more detailed discussion of articulation testing in Kryter (1972).

Table 13-2. Example of a Phonetically Balanced Word List

1. smile	14. box	27. are	40. fuss
2. strife	15. deed	28. cleanse	41. folk
3. pest	16. feast	29. clove	42. bar
4. end	17. hunt	30. crash	43. dike
5. heap	18. grove	31. hive	44. such
6. toe	19. bad	32. bask	45. wheat
7. hid	20. mange	33. plush	46. nook
8. creed	21. rub	34. rag	47. pan
9. rat	22. slip	35. ford	48. death
10. no	23. use(yews)	36. rise	49. pants
11. there	24. is	37. dish	50. cane
12. than	25. not	38. fraud	
13. fern	26. pile	39. ride	

(adapted from Egan, J. P., Articulation testing methods, *Laryngoscope, 58:* 955-991, 1948.)

Intelligibility tests are usually conducted using a group of highly trained listeners and having them listen to a set of CVC syllables, PB words, or PB sentences. They then write down what they hear. This is scored into a percent correct.

French and Steinberg (1947) developed a calculation method for intelligibility that is called the Articulation Index (AI). This method has been standardized by the American National Standards Institute. To calculate the AI, one has to know the speech spectrum and the noise spectrum of the transmission system involved. Details of this calculation will not be covered here, but can be found in Kryter (1972). Designers should be aware, however, that for high intelligibility, a considerable fraction of total speech bandwidth must be delivered to the listener and the signal-to-noise ratio at the ear must be reasonably high. If the speech peaks are 30 dB or more above the noise throughout the frequency band from 200 to 6100 Hz, the listener will make essentially no errors (AI=1.00). If the speech peaks are less than 30 dB above the noise in any part of the speech band, *some mistakes* will be made (AI=0.5). If the speech peaks are never above the noise, the listener will rarely be able to understand anything (AI=0). The relationship between AI and the intelligibility of various types of speech test materials is shown in Figure 13-4.

Intelligibility was very important in the early design of telephone transmission systems, but is no longer a useful criterion since telephone transmission is now highly intelligible. Measurement of intelligibility is still important, however, in connection with some crosstalk problems. Because of the close proximity of wires which carry electrical speech signals, it is possible that some of the speech from one set of wires will be transferred to another set of wires at a much reduced level. Speech from the first set of wires may be faintly audible to the person on the second set of wires. This is called *crosstalk*. In order to assess the levels at which this crosstalk could be allowed, two sets of

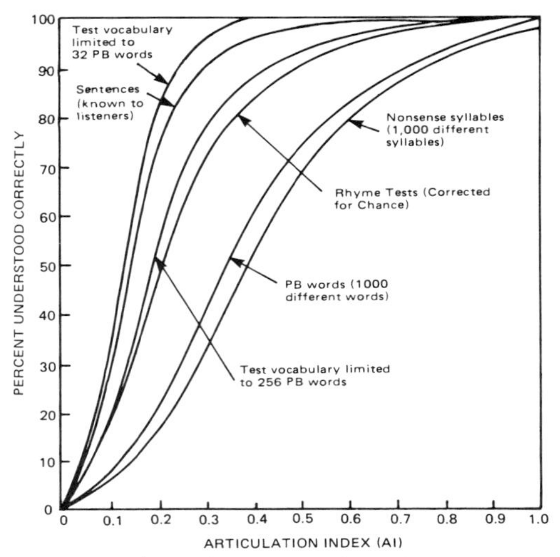

Figure 13-4. Relationship Between the Articulation Index and the Intelligibility of Various Types of Speech-Test Materials (adapted from *Human Behavior: A Systems Approach*, by N.W. Heimstra and V.S. Ellingstad. Copyright © 1972 by Wadsworth Publishing Company, Inc. Reprinted by permission of the publisher, Brooks/Cole Publishing Company, Monterey, California.)

measurements are usually taken. The first measurement is crosstalk detectability, which is whether the listener can detect it. The second one is recognition, whether the listener can understand what is being said, which is a function of its intelligibility.

DESIGNING SPEECH TRANSMISSION SYSTEMS

Reducing Noise

There are some important considerations that apply in situations where users are expected to perform using speech. The first, best and most effective way to protect speech from noise interference is to control and isolate noise at its source. Keep excessive noise out of the communication channel altogether.

Improving Hearing

Where noise is a problem, the effects of noise on speech communication can be reduced somewhat by using ear-protective devices, better quality microphones, automatic gain controls, high-pass filters, peak limiters, loudspeakers and headsets. Chapanis (1965) provided a brief list of recommendations for designing systems requiring speech communication.

1. When noise levels are high (about 80 dB or more) or when speech levels are high (about 85 dB or more), ear plugs will usually improve speech intelligibility in face-to-face communication.

2. The most effective microphones are those that (a) have high sensitivity to acoustic speech signals, (b) transduce acoustic signals faithfully, and (c) reject other acoustic signals and noises at the speaker's location (so-called noise-canceling microphones).

3. Loudspeakers rather than headsets should be used when (a) environmental noise levels are low, (b) listeners must move around or would otherwise be hampered by wires and cables, or (c) many listeners must hear the message.

4. Headsets rather than loudspeakers should be used when (a) environmental noise levels are high, (b) different listeners must hear different messages, (c) reverberation in the room is a serious problem, (d) the listener must wear special equipment (such as an oxygen mask), or (e) the power output is too low to operate a loudspeaker.

EVALUATING SPEECH TRANSMISSION SYSTEMS

Loudness Loss

The design of telephone transmission systems is to a large extent based on considerations of *loudness loss,* i.e., the difference in the loudness of speech between the sending talker and the receiving listener. This is based on the fact that the loudness of the received speech has a great influence on the listener's judgment of transmission quality.

In order to better understand "loudness loss" measurements, it is important to distinguish between *speech level* and *speech loudness.* Speech level is an objective measure obtained by making an electrical measurement across the wires on which speech is carried. The currently preferred speech level measure is the *equivalent peak level (epl).* Speech loudness is a subjective measure based on reports by a person on his or her perception of an acoustic stimulus.

In telephone system design, loudness of the received speech has been determined by a subjective method called "loudness balance." In this method, a reference speech signal is compared with the signal under test. The subject then adjusts the test signal until it is equally loud to the reference signal. When the loudness of two transmission systems is being compared this task is not too difficult, since both signals can be produced in the same receiver on the same ear and the ability of subjects to equate the two is relatively good. However, if the receiver itself is under test, a procedure called "binaural loudness balance" is usually employed. This is a much more difficult task for people, since they have to compare the signal in one ear with a comparison signal in the other ear.

An objective method for determining the loudness of a speech transmission system is a laboratory system called the Electro-Acoustic Rating System (EARS), which was devised to make objective measurements of partial and overall telephone connections in a manner which reflects subjective loudness loss. The design of the current United States telephone system in terms of its loudness performance is based on measurements made with the EARS system.

Speech Quality

The loudness of the speech signal, while important, is only one factor in the design of speech transmission systems. The overall *quality* of the transmission needs to be considered. This is usually measured by evaluating user preferences.

Preference is defined as the proportion of a listening group, expressed in percent, that prefers a speech test signal to the speech reference signal as a source of information. Speech quality is thus defined in terms of preference. A telephone transmission system which is preferred over another system is the one that has higher quality. Determination of speech quality therefore depends on the measurement of preference.

There are two principal methods for determining preference. One is the *comparison* method, and the other is the *category judgment* method. In the comparison method two speech sounds are compared with each other. The subject indicates which of the two is preferred, according to some criterion such as: "Which would you prefer to listen to?" or, "Which would you prefer to use?"

The category judgment method employs a scale, usually with verbal labels, such as *excellent, good, fair, poor,* or *unsatisfactory*. The subject rates the speech that was heard by checking one of these adjectives.

Most transmission performance objectives are stated in terms of this five-category rating scale. The system is then described as having a certain percentage of judgments of "good or better," meaning "good" and "excellent." The percentage "poor or worse," meaning "poor and unsatisfactory," is also often reported.

Listening Tests

Many types of speech transmission systems may be evaluated by what are commonly called subjective listening tests. In this type of test a large group of people can simultaneously listen to speech samples while their judgments are recorded.

Listening tests are generally used to evaluate those types of impairments which do not depend upon the listener interacting with the system. Examples of this are effects of analog noise, digital noise, speech level received, crosstalk and listener echo. Tests can be of the category judgment type, comparison type or threshold type. The comparison and category judgment methods were previously discussed. Threshold type tests may be used to determine whether the impairment is perceptible at all.

Some types of impairments can only be determined by the interaction of the listener with the system. That is, the listener has to talk to experience the impairment. Impairments such as these are talker echo, delay, and voice-switched gain. Although the subjects are usually given a task which will generate a conversation, the person conducting the study has little control over what is actually said or over the speech level. If the impairment depends upon interaction, it is important that the subjects be given a task which will cause conversation between them.

In addition to the laboratory methods for obtaining speech quality information, field data also may be obtained by customer interviews. A customer may be contacted immediately after completing a call in which a particular type of impairment was present. The customer may then be asked whether any difficulty of talking or hearing over the connection was experienced, or be asked to describe the call in terms of the usual category judgment scale of excellent, good, fair, poor, or unsatisfactory. Call back interviews of this type may be used as a check on the laboratory experiments. Because of the large expense of such field interviews, they are usually used when simulation in the laboratory environment is difficult, or during field trials of new devices or systems.

SPEECH INTERFERENCE LEVELS (SIL)

In face-to-face conversation, when a talker is speaking to someone in a quiet office and the listener is about one meter away, the level of conversational speech is usually about 65 dB. Given a good estimate of the expected background noise level in a new system, the designer can evaluate the estimated speech signal (65 dB) against the estimated noise level using the S/N ratio to help determine if there will be difficulties in speech communication.

As the noise level increases, most users can also increase their voice level (within certain limits). To take this into account, some designers prefer to use an evaluation method called the "speech interference level." This is a relatively simple method for predicting the intelligibility of face-to-face speech communication, and one that is usually better than making estimates based only on the S/N ratio. The speech interference level method was devised for use in contexts where noise is relatively continuous, and the intensity and frequencies remain fairly constant—ventilation noise in offices, noise in engine rooms, aircraft noise, and noise around milling machines, for example. This method yields the maximum noise level that will permit correct reception of about 98 percent of speech.

To find the speech interference level of a given noise:

1. Measure the sound pressure level of background noise in frequency bands of 600 to 1200, 1200 to 2400, and 2400 to 4800 Hz.
2. Determine the arithmetic average of the decibel levels in the three octave bands. This average value is the speech interference level.
3. Consult Table 13-3 to find the maximum distance between talker and listener at which about 98 percent of speech will be heard correctly.

Table 13-3. Speech Interference Levels that Permit Reliable Speech Communication (Beranek, 1949) for About 98 Percent of Test Sentences

Distance between Talker and Listener (ft)	*Speech-interference level (dB)*			
	Normal	*Raised*	*Very loud*	*Shouting*
0.5	71	77	83	89
1.0	65	71	77	83
2.0	59	65	71	77
3.0	55	61	67	73
4.0	53	59	65	71
5.0	51	57	63	69
6.0	49	55	61	67
12.0	43	49	55	61

Courtesy John Wiley & Sons

Consider, for example, a new system located in a room with several printing machines. The system requires people to accurately communicate while operating their respective machines. The users can raise their voices to speak, but it is not desirable to have several people speaking in very loud voices or shouting. The noise level is fairly fixed due to the construction of the machines. The sound intensity levels for the three frequency bands were determined to be:

Frequency Band *(Hz)*	*dB*
600 - 1200	68
1200 - 2400	62
2400 - 4800	53

The average is 61 dB, which is also the speech interference level used to enter Table 13-3. The designer consulted Table 13-3 and noted that the people would have to be no more than 3 feet apart if they were to accurately communicate using *raised voices,* and if they were six feet apart it would require the use of *very loud voices.* Conversation using a normal voice only would be possible if the people were standing within one or two feet of one another. This information helped the designer determine the placement of the printing machines and people. It also suggested to the designer the necessity of having a special "quiet room" available for communicating verbal instructions concerning each new copying job.

LANGUAGE CONSIDERATIONS

A designer should know the conditions under which speech communication will be used in a system. This includes not only being aware of the noise levels in which speech must be heard, but also the number and types of verbal exchanges that will take place. If the conditions surrounding speech communication are well understood, it may be possible for the designer to modify the existing language or even construct a specialized language that will help improve the intelligibility of the speech.

Depending on the criticality of the messages and the severity of the noise level, designers can take steps to improve speech communication on at least three levels: sentence (phrase), word, or character. In some work situations it may be sufficient to have a listener understand most sentences; on the other end of the continuum, each and every letter may be critical to accurately understanding the message (e.g., a coded part number).

Matching Expectations and Message

Speech communication is influenced by a set of expectations that exist merely by knowing what activity is being performed. A physician returning a telephone call usually means that the listener expects some specific medical advice. The same person receiving a call from a dentist's receptionist is likely to expect a message concerning an upcoming appointment or an overdue bill. After we drive into a gas station, we expect the attendant to ask, "How much do you want?" In any given situation, we call up the "educated expectation," or schema, we think is most appropriate for the particular activity. It is this schema that helps us to understand the message. If the schema is not appropriate, or if the message is totally unexpected or bizarre in its form, it may hinder understanding of the message. Consider trying to understand the message from the physician, when thinking that it is the dentist's receptionist calling. To help prepare the listener to receive the message, a designer should provide as many cues as possible *before* the message begins.

Vocabulary Size

One important way to increase the intelligibility of a message (particularly in the presence of noise) is to limit the size of the vocabulary. For example, if the number of words to be transmitted over a noisy communication channel is kept small, and if the entire list of these words is known to both the listener and the talker, the chances of accurately communicating are greatly increased. The smaller the vocabulary set, no matter whether it is made up of sentences, words or letters, the better the communication. Keep in mind that to take full advantage of limited set size, both talker and listener must know the set thoroughly. The difference between very small (2-word) and moderate size (1000-word) vocabularies can change by 18 dB the signal-to-noise ratio required for a given level of intelligibility (Howes, 1957).

Familiarity and Length of Words

Other things being equal, the more frequently a word occurs in everyday usage, the more readily it is correctly identified when transmitted using speech. Kucera and Francis (1967) provide a frequency listing of over one million present-day American English words.

The length of the word also influences its intelligibility; the longer the word, the more readily it is correctly identified. The listener is able to identify a long word by hearing portions of it, particularly a familiar, highly probable word, whereas, missing one syllable of a short word is more likely to prevent the identification of the entire word. Both familiarity and length factors can change by 10 to 15 dB the signal-to-noise ratio required for a given level of intelligibility (Howes, 1957).

Word Context

Another factor influencing the intelligibility of speech in noise is the context in which the words are heard. For example, a word is harder to understand if it is heard in isolation than if it is heard in a sentence. Listeners can identify words contained in meaningful messages at signal-to-noise ratios that would be unacceptable if the message consisted of an equal number of unrelated words. For example, under noise conditions that permit correct reception of only about 75 percent of phonetically balanced words, over 95 percent of test sentences are correctly understood (Kryter, 1966). A temporary employee from Texas once asked for permission to fly home early from New York to attend a "fire." When asked to provide more information she replied, "Fire, fire, fire—don't you like to go to fires?" Further questioning determined that she was not a pyromaniac who enjoyed watching fires burn, but that she wanted to attend the Texas State *Fair*. Whenever possible, designers should provide for having words always contained in a meaningful context.

Speech Sound Confusions

Situations may arise with a noise level so great that certain speech sounds may be confused. To illustrate how speech sounds can be confused, Miller and Nicely (1955) conducted a study where subjects listened to consonants against a background of random noise. As the noise increased it soon became impossible for people to distinguish the different letters. Kryter (1972) reviewed their results and made the following observations:

1. At S/N = -18 dB, all consonants are confused with one another.
2. At S/N = -12 dB, the consonants such as m, n, d, g, b, v, and z are confused with one another, and the consonants such as t, k, p, f and s are confused with one another, but the consonants in the first group are seldom confused with those in the second group.
3. At S/N = -6 dB, although the m and n are confused with each other, they are clearly distinguishable from the other consonants.
4. At S/N = 0 dB, certain individual consonants are still confused.
5. At S/N = 12 dB, all of the consonants are readily distinguished.

When individual characters are to be communicated verbally in noisy environments, a designer may want to have users use the International Word-Spelling Alphabet shown in Table 13-4.

Table 13-4. International Word-Spelling Alphabet

A-Alpha	N-November
B-Bravo	O-Oscar
C-Charlie	P-Papa
D-Delta	Q-Quebec
E-Echo	R-Romeo
F-Foxtrot	S-Sierra
G-Golf	T-Tango
H-Hotel	U-Uniform
I-India	V-Victor
J-Juliet	W-Whiskey
K-Kilo	X-X-Ray
L-Lima	Y-Yankee
M-Mike	Z-Zulu

Multilanguage Communication

There are at least 3000 different languages in the world. Of these languages 130 are used by more than one million people. If there is potential for a new system to be used by people with different language backgrounds, the designer needs to take this into account when making decisions on speech communication.

FOR MORE INFORMATION

Kryter, K. D., Speech Communication, in H. P. Van Cott and R. G. Kinkade, *Human Engineering Guide to Equipment Design,* Washington, D. C.: U. S. Government Printing Office, 1972.

Kucera, H. and Francis, W. N., *Computational Analysis of Present-Day American English,* Providence, Rhode Island: Brown University Press, 1967.

14

HUMAN/COMPUTER INTERFACE

INTRODUCTION

Probably the most significant invention of the 20th century is the computer. It pervades virtually all aspects of life. It is a rare individual indeed who is not affected in some way by a computer.

Designing for an acceptable level of human performance in computer-based systems continues to be difficult. Not considering human performance in the human/computer interface frequently results in large numbers of errors, requires huge amounts of training time, and causes user frustration and dissatisfaction.

Unfortunately, designers do not yet have a complete set of general design principles that, if followed, would result in the desired level of human performance. When confronted with human/computer interface problems many designers simply do the best they can. They have few meaningful human performance principles to help in accurately converting their ideas into reality, and, consequently, have no set of standards by which to criticize and improve their performance (cf. Davis, 1966; Turoff, 1967). Most available information is in the form of opinions.

COMPUTER USERS

The longer computers are with us the more heterogeneous becomes the group of computer users. Users were once a select group of mathematicians sharing a common interest in computer technology. Today this group includes scientists, engineers, administrators, clerical personnel, educators, and students, all of whom view the computer not as an end in itself, but as a *tool* for getting a job done.

When attempting to optimize human performance, the designer constantly looks toward the ultimate user of the system and makes design decisions to best benefit that user. The designer recognizes that a new system cannot elicit an acceptable level of human performance from all people in the world. Therefore the designer must know in advance the design-related characteristics of the user group. This becomes an extremely difficult problem when designing computer-based systems. Any computer-based system will have users with various characteristics, and these characteristics change as each user gains experience.

Ideally, as people change, the system will also change. One design objective should be to build human/computer interfaces with critical elements that change as the users become more experienced in the human/computer relationship.

Types of Users

Stewart (1974) distinguishes three types of users who may have quite different requirements even when using the same computer-based system. These are managers, clerks, and specialists. Managers solve problems and make decisions while clerks handle data. It is not uncommon for managers to have little interest in using a computer. On the other hand, most clerks are not in a position to make such a decision, and usually must use the computer whether they want to or not.

The specialist operates primarily within a particular discipline or problem area and has at his or her disposal a specialized body of knowledge and a set of techniques applicable to these problems. A specialist tends to view the computer as a tool and spends whatever time is necessary to master the tool so that the job will proceed to a successful conclusion.

If we consider only these three major types of people we have three very different types of users, each one having different needs and requirements for a system. Managers may need relevant, easy-to-use systems; specialists may need specialized systems, possibly with dedicated programmer backup; and clerks may need a system that provides protection from boring, repetitive, computer-created tasks (Stewart, Damodaran, and Eason, 1976).

General Users

In the following discussion we assume a target user who is not a computer professional, but who still must interface in some way with a computer. We designate this group as *general users.* General users range from the manager, specialist or clerk just discussed, to reservationists at an airline counter, to a telephone installer on top of a pole communicating with a computer using a hand-held terminal.

Inexperienced Users

Designing for users who have had little or no computer-related experience can be difficult. Zimmerman (1977) suggested the following principles be applied in designing a system for use by inexperienced people:

1. Initiative should always come from the computer.
2. Each required entry should be brief.
3. Entry procedures should be constructed so that they are consistent with user expectations.
4. No special training should be required.
5. The system should act upon certain entries only after a confirmation by the user (e.g., the user requests that some information be deleted and the computer confirms in some way that this is what the user actually intended to do).
6. Computer messages should be clear and unequivocal.
7. Computer messages should be composed of a simple, generally comprehensible vocabulary.
8. User decision-making should be facilitated by offering a small number of simultaneous options that require only one decision per step.
9. User decision-making should be facilitated by a distinct request from the computer that a decision be made.
10. The user should be able to comfortably control the pace of the human/computer interaction.
11. The user should have the option to easily request both computer and human assistance.

The Computer as a Tool for Users

For all practical purposes users of computer-based systems, particularly inexperienced users, should be required to think of the computer only as a machine that obeys commands that are expressed in strict accordance with a set of rules. It is not necessary for most users to know anything at all about the nature of the machine, just as it is not necessary for the operator of a car to understand the workings of the engine in order to drive it.

What makes the human/computer interaction substantially different from most other types of human/machine interactions is that the human/computer interaction can be an exchange of information that appears very much like a conversation between a person and a computer. The interaction involves a two-way exchange of information in the form of *commands* or *queries to* the computer, and *messages* of a wide variety *from* the computer.

Designers should remember that the computer is simply a tool used for accomplishing some system goal. Users also have other tools available to help meet these goals—pens, pencils, or telephones. The tools themselves are the merely a means to an end. A danger inherent in the use of any tool, including the computer, is that of bending a problem situation to the tools available (Stewart, 1976; Sackman, 1976; Nickerson and Pew, 1971). Such bending can have serious consequences. For example, a problem solver's thinking may be constrained to the point that he or she thinks about a problem in terms of the *computer's* capabilities and limitations rather than in terms of his or her own (human) capabilities and limitations, or in terms of the problem itself.

Users that Change

User's capabilities change rather rapidly once they begin interfacing with a computer. Because people gain new skills as they interact with a computer, a good system should be able to accommodate users at various levels of expertise. This design goal was unheard of with many hardware-oriented systems because of the expense of having two or three different models. But with computer-based systems, it is within the realm of possibility, and a designer can develop a system that accommodates inexperienced, moderately experienced, and experienced users.

A person learning to pound a nail can learn the task by using exactly the same tool he or she will use after mastering the skill. This is not the case with computers. When learning to use the computer, people usually require a tool (particularly software, but sometimes hardware) much different than the one they will use after gaining proficiency. As people change, their computer-related needs change. Figure 14-1 illustrates many ways that users can change (dotted line) as they interact with computers. For example, non-programmers may become programmers, or occasional users may become regular users and regular users may become occasional users.

GENERAL GUIDELINES

Presented below is a set of user-related guidelines for designers of systems with human/computer interfaces. These guidelines are adapted from Nickerson and Pew (1972) and others.

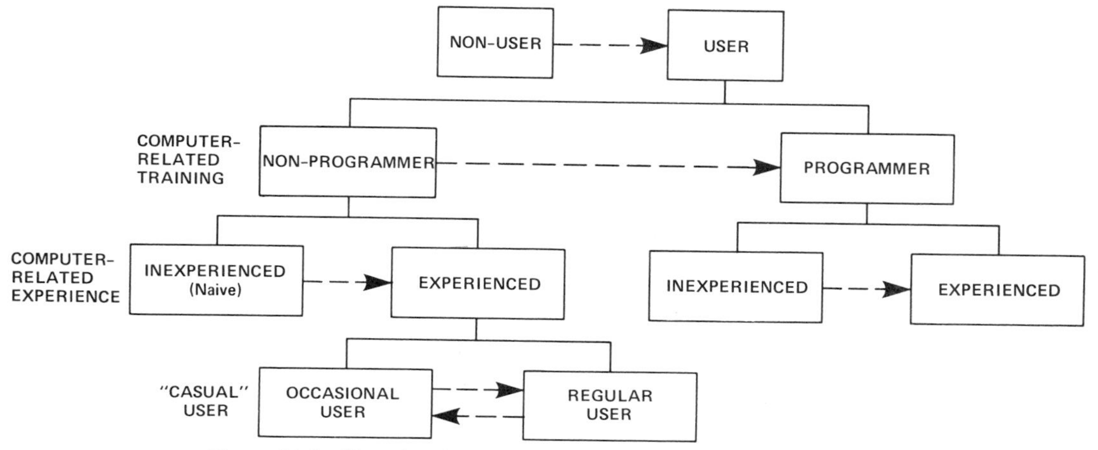

Figure 14-1. Changing Characteristics of Computer Users

Understanding the User, Activity, and Context

A designer should have a good understanding of the people who will be using the computer and the kinds of activities they will be performing. Besides statements of minimum qualifications (SMQs) for users (see Chapter 18), this includes a knowledge of the user's expertise in computer-related specialities; the extent to which the user's job will focus on using the computer actively; the user's level of design-making authority and responsibility; the extent to which the computer will be an option rather than a job requirement; and finally, the user's attitudes toward computer technology in general.

Ensuring User Control

A designer should ensure that the user feels in control of the human/computer interaction. Even though an inexperienced user will have a different set of requirements than an experienced user, both should feel in control—a well-designed system ensures this. Remember to keep in mind how the final system will appear to the user.

Maintaining Consistency

The designer should ensure consistency from display to display, from interaction to interaction, from message to message, and from program to program within the same system. This is particularly helpful for inexperienced users. Experienced users also benefit because it encourages the development of relatively automatic actions, and permits the user to transfer principles learned from the execution of well-practiced commands to the execution of new ones.

Minimizing Processing Time by People

The designer should ensure that people can interact with the computer as quickly as possible.

A designer can save even an inexperienced user time by providing alternative commands that give the option of using either full names or abbreviations (the computer should be able to recognize both). When logging on a system, for example, the user could have the option of stating his or her level of experience and the desire to use names or abbreviations as the commands. In some systems the computer may make such a decision after determining the experience level of a user from data already collected. In addition, a computer could be designed to accept a wide range of names or abbreviations, including misspellings and abbreviations that are "close" but not exact. This alternative probably is preferred because it gives the user the option of using simpler command language as experience builds up, and it also gives the option of using a simple or abbreviated (or even wrong) command in areas that are used frequently (cf. Gilb and Weinberg, 1977).

Keeping the User Informed

The designer should provide adequate feedback to inform the user continually about current system status and the options available. For each user action there should be an appropriate computer reaction. At times there may be a need for an immediate acknowledgment that a request has been received and is being processed, as well as later feedback that the requested action has been accomplished by the computer. When the user's terminal is dormant awaiting completion of an action, the user should be informed periodically that the computer is still working on the problem. Inexperienced users should be provided with a sensible next step at every point in the development of a transaction. Signaling the possible next steps can take the form of a menu of options; illuminating the next function key or set of function keys; calling attention to items on the CRT through color, blinking, contrasting intensity, or inverse video (black on white); displaying current options in a special window dedicated to alerting functions; or by repositioning a cursor to a new location that suggests the next class of actions available.

Improving User Accuracy and Facilitating Error Handling

The designer should make decisions that encourage user accuracy and improve the quality of data already in a computer. A well-designed system will reduce the opportunity for errors, increase the user's ability to detect errors, and provide users with an immediate opportunity to correct any errors that do occur. Many computers use built-in error detection routines that are reasonably effective and can detect as many as half or more of the user's errors. If these errors are detected in real time, an opportunity exists for people to correct the errors and also to take note of what was done to cause the error to occur in the first place. Obviously the main focus of designers should be to prevent errors from ever occurring. However, once an error has occurred, both inexperienced and experienced users should be able to detect and correct it with ease. This may require more than one way for the correction to occur.

Optimizing Training

The designer should take full advantage of computer technology to initially train new users and to provide refresher training for experienced users. The designer should provide users with computer based aids that help recall such things as the language and procedures to be used. A good computer-based system does not require extensive assistance from people.

HUMAN-TO-COMPUTER

Common Input Methods

Keyboards

The familiar standard typewriter keyboard has been around for over a hundred years and probably will continue to be used for many years to come. The arrangement of keys on the keyboard, also known as the "Qwerty" keyboard (for the first six letters on the top row of keys), is very similar to one invented as early as 1872 and used as the first commercial typewriter.

The origin of the Qwerty keyboard layout is not clear. Some say that the arrangement is similar to that of the characters in a printer's case. A story that has become traditional is that the keyboard was originally designed to minimize the possibility of striking conflicting keybars. This was accomplished by separating those keys most often used on consecutive strokes. This means that the keyboard may have been designed to *slow down* keying.

The original keyboard seems to have survived largely because that particular manufacturer continued to hold sales leadership in the new industry. Other keyboards were introduced, but none of the manufacturers were successful enough to have their machines and keyboard arrangements accepted.

From time to time various "improved" keyboards have been proposed but none has gained widespread acceptance. Figure 14-2 shows some examples of keyboard layouts. The Qwerty keyboard arrangement seems to elicit an acceptable level of human performance in most situations. Several generations have learned to use this keyboard with reasonably good speed and accuracy. Having so many people trained on this keyboard makes it even more difficult to have alternatives accepted.

Designers usually do not have the option of designing the basic keyboard in new systems. However, it is conceivable that in the future there will be a requirement to totally optimize the human/computer interface in some very important systems. At that time, it may be possible to change the keyboard to

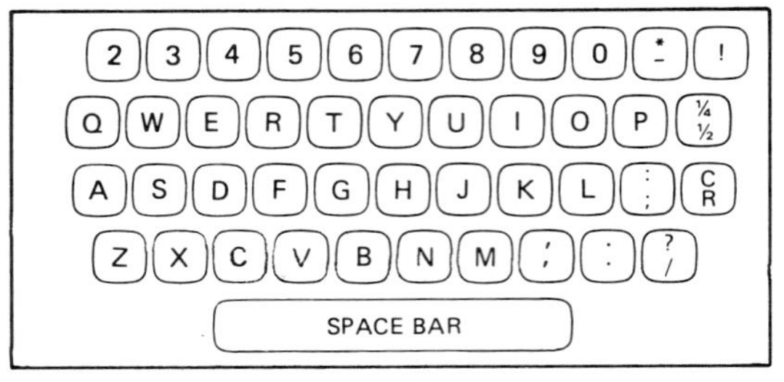

The Standard Qwerty Keyboard Layout

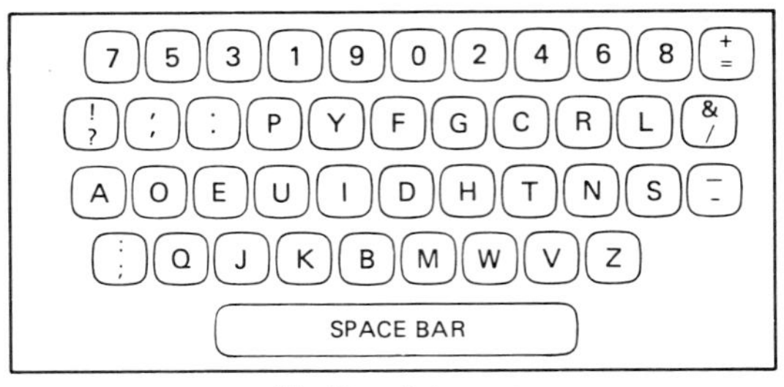

The Dvorak Layout

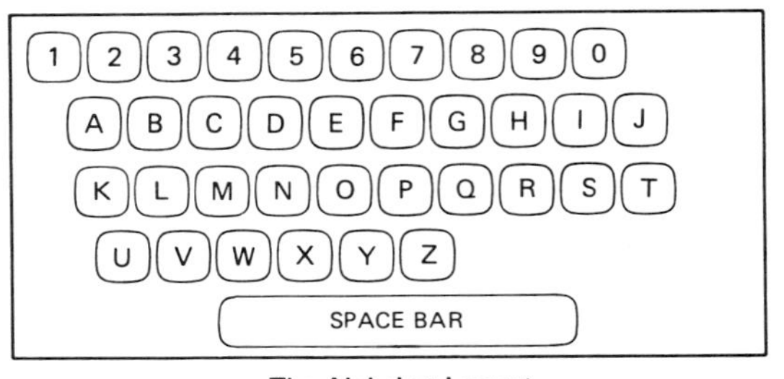

The Alphabet Layout

Figure 14-2. Examples of Keyboard Layouts

match what we now know about the human performance of keying.

Griffith (1949) conducted a study on the Qwerty keyboard and found five major areas where improvements to the keyboard would result in better human performance. He reported that on the Qwerty keyboard 48 percent of the motions to reposition the fingers laterally between consecutive strokes are one-hand motions rather than the easier and faster two-hand motions. Making fast one-handed motions is more difficult than making two-handed motions in the same way that it is more difficult to beat a drum rapidly with one stick than with two. He reported that one-handed motions consume, on an average, 75 percent more time than two-handed motions.

The second problem Griffith identified is that the Qwerty keyboard requires reaching from the home row to another row for 68 percent of the key strokes. He felt that a well-designed keyboard could reduce this reaching, putting at least 70 percent of the strokes on the home row. A third problem he found is that the hardest motions of typewriting are one-handed motions with reaches to one or two rows above the home row. These should be reduced. The fourth problem is that the Qwerty keyboard overloads the left hand with 56 percent of the total strokes. A well-designed keyboard probably should have the left hand carrying somewhat less than half the stroking load. The fifth and final problem is that the Qwerty keyboard elicits many awkward one-handed fingerings and the load on each finger is not equitable. Griffith suggested that on a well-designed keyboard, consecutive strokes on different keys should be done using different fingers and, where possible, different hands.

Probably the most well-known attempt to systematically design a keyboard that takes into account these five concerns is the "simplified," or Dvorak typewriter (see Figure 14-2). This typewriter was designed about 1932 by August Dvorak and William Dealey. Although the Dvorak keyboard represents an attempt to solve many of the problems associated with the Qwerty keyboard, studies done to determine the superiority of this keyboard have produced mixed results. Several studies have suggested that the Dvorak keyboard may improve human performance while other studies, particularly a study sponsored by the General Services Administration of the U.S. Government (Strong, 1956), have failed to show any advantage to the Dvorak keyboard.

Recently, a new question about keyboard layout has arisen. An increasing number of non-typists are required to enter alphabetic data into computers. For these inexperienced people, it seems likely that a keyboard arranged in alphabetical order would be preferable to the standard Qwerty keyboard. It has been shown, however, that an alphabetic layout of the keys is no more advantageous than the standard Qwerty layout (Michaels, 1971). Performance on the two keyboards was essentially equal for unskilled typists. Hirsch (1970) also studied the typing performance of non-typists on the standard Qwerty keyboard and on an alphabetically arranged keyboard. His results suggested that an unskilled person actually could enter correct data faster on the Qwerty keyboard than on the alphabetic keyboard.

We will make one final note on keyboard design. In 1926 a German named Klockenberg published a book dealing with the design and operation of the typewriter. He described how the keyboard layout required the typist to assume postures of the trunk, head, shoulders, arms, and hands that were unnatural, uncomfortable, and fatiguing. Klockenberg suggested a number of improvements to the keyboard layout, some of which are still valid. Probably his most interesting idea was to separate the keyboard sections allotted to the left and right hands to alleviate tension in the typist's shoulders and arms (see Figure 14-3). Kroemer (1972) published a study showing that this layout produced better performance.

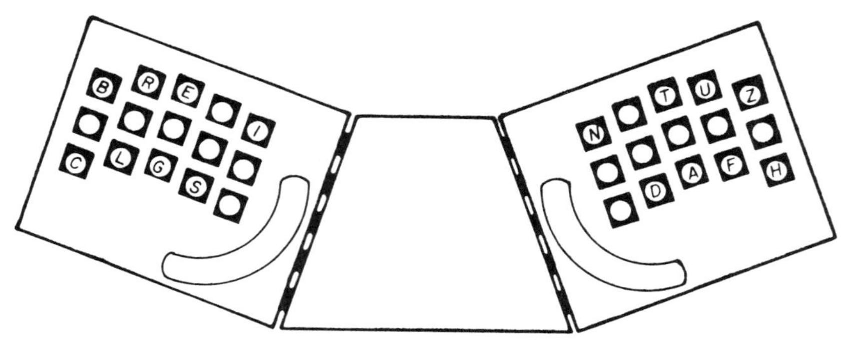

Figure 14-3. Klockenberg Keyboard

The major issues surrounding the layout of numeric keyboards have centered around the differences between the adding machine and the telephone *Touch-Tone* layouts. Studies indicate that the telephone arrangement (with the one, two, and three on the top row) elicits slightly better performance than the adding machine layout (with the seven, eight, and nine on the top row).

As mentioned earlier, keyboards no doubt will be a primary way of communicating with computers for a long time to come. Even though, in most applications, designers do not have the option of designing the basic keyboard, frequently they do have the opportunity to *select* a keyboard from those available. Whether designing or selecting a keyboard, it is important to meet certain design specifications. We discuss some important specifications in the next section. In addition, there are a few good references that provide even more information: Alden, Daniels, and Kanaraick (1970); Seibel (1972); and Klemmer (1971).

Function Keys

Frequently, a designer must decide whether a computer command should be typed out using the entire keyboard or only a single key called a "function key." A good example of using a function key is the ordering process used by some fast food hamburger chains. If you order a hamburger, fries, and a milkshake, the person taking the order depresses a function key for the hamburger, another for the fries, and still another for the milkshake. The prices associated with each of those three items have been programmed into the system, making it unnecessary for the user to enter the actual cost of each item and then press an add button.

Function keys have several advantages. In our example, the operator only has to recognize the hamburger button (possibly identified by a picture of a hamburger); he or she need not know the price of the hamburger. In addition, the numerous alternatives—hamburger, cheeseburger, large fries, small fries—are always visible and readily available. Function keys provide an easy way of learning the full range of entries, make the keying process faster, and, because the number of keystrokes is substantially reduced, minimize keying errors.

Finally, since certain functions that occur most frequently or that tend to occur together may be placed in the same area the keys in this area may be color coded reducing even further a user's time and effort.

It is generally quite easy (and probably will become more so in the future) to alter the programs associated with the different function keys. This can be done, for example, to change the prices associated with each key.

Probably the main disadvantage to a large set of function keys is that a huge keyboard may be required. Also, function keys may be very expensive, since the function keys restrict each keyboard to a special purpose. Without function keys the same terminal can be used for several different applications. If the keyboard is custom designed and loaded with a large number of special function keys, its use generally may be restricted to a single application or system. There is also the danger that over-simplification of a task (by having too many function keys) can result in very boring work.

Sometimes several functions are relegated to a single key. Some of the smaller hand-held calculators have one key that is used to enter the number "9", enter the square root, and to square a number. What this key does depends upon whether another key is depressed before or at the same time. Another example of a single set of keys performing more than one function is on the *Touch-Tone* telephone layout: the same key can be used to signal the number "5" in one situation or the letters "J", "K", or "L" in another situation. Still another example of using a single key for more than one function is the standard typewriter keyboard on which a shift key is depressed to allow a person to make capital letters and certain special symbols with the same keys used for lower case letters. Having two or more functions assigned to one key may substantially increase errors. Hammond, et al. (1980) found that about one-third of the errors in a computer system were the result of mistyping, and *half* of these were due to using the correct key, with the *wrong* shift.

Probably the biggest advantage of assigning two or three functions to a single key is that it saves room.

It is probably appropriate to use function keys when the number of different functions is relatively small (30 or fewer). However, we should point out that one of the main dangers of using function keys, particularly in the design of new systems is that it is difficult to anticipate all contingencies. Therefore, if the design requires major modifications, it may become a matter of redesigning hardware rather than software. If significant hardware changes are required—if, say, ten new function keys must be added to the keyboard—it may be very expensive and also may require a hybrid system that would include the original function keys, plus a set of commands that need to be keyed individually.

Instead of function keys, probably the most frequently used approach for communicating commands to a computer is to type out a keyword (usually a verb) along with an argument (usually a noun), and possibly a set of variables (usually numbers). Computerized text editors and many command languages use this approach. The advantage is that it gives much more flexibility in changing the functions as the system matures. Any changes needed tend to be software changes and not hardware changes.

Of course there are certain disadvantages in using keywords. For example, learning and recalling the full set of commands is much harder than simply recognizing the appropriate function key. In addition, users frequently use only a small subset of all available commands, and so may require a means for reviewing infrequently used commands. Entering more key strokes means that the chance of hitting a wrong key is increased, although making a character error may not be nearly as serious as pressing the wrong function key. Also, depressing several keys is slower than depressing a single function key. Finally, there is a disadvantage to using typed commands when many users are not skilled typists. However, in many systems this is not a significant problem. People who have had minimal exposure to a typewriter and who are expected to enter a relatively small and finite set of commands are usually able to learn to do so in a relatively short period of time.

Other Input Methods

Other methods of input that do not require a keyboard or other type of device (e.g., a light pen) include *handprinting* and *speech*. There are certain advantages to using handprinting or speech instead of a keyboard or other device. Handprinting and speech tend to be more natural and require little (if any) additional learning. However, since so many people know how to type, this is not a strong advantage.

Of course the main problem with using direct human input (handprinting or speech) is that we do not have computers that can recognize and comprehend all the various scribblings or spoken words that people can provide. Thus, systems that are used for recognizing handprinting and/or voices are presently limited in scope. However, this is not to say that in the future there may not be a greater emphasis on the computer recognizing handprinting and voice input. The number of errors expected with handprinting and voice input appears about the same as with keying. The big advantages of using these approaches are the reduced training requirements and the possibility of increased input speed. When entering text, speech input tends to be a little faster than keying, while keying is faster than handprinting.

The effect of voice input on user satisfaction, however, is not very clear. A study that compared reading digits out loud (as though they were being spoken to a computer) versus keying the same digits was reported by Braunstein and Anderson (1959). Four of their five subjects had no prior experience with keypunching. The reading rate was approximately twice as fast as the keying rate. But probably the most interesting finding was that most of the subjects preferred keying to reading the digits, and all of the subjects found reading the digits a tiring task. For the one experienced key operator, voice input did not offer speed or accuracy advantages over the keying input. Welch (1980), in a similar study using actual voice recognition equipment, found that moderately experienced key operators could key digits almost twice as fast as they could read them.

Speech Input

Speech recognition may be the ultimate step in transferring information from the human to the computer (at least until computers learn to recognize human thoughts). Computer speech recognition means that a person can speak commands, text or other information and it will be reliably recognized and interpreted by the computer. Using speech as the form of human-to-computer communication greatly simplifies the requirements placed on the user because he or she uses a well-learned mode of communication.

Two levels of speech recognition can be identified: isolated word recognition and continuous speech recognition. Between these two forms of recognition is a third intermediate form in which a speaker inserts exaggerated interword delays to facilitate the computer's understanding. In some systems, people talk directly to computers. In fact, almost perfect isolated word recognition for vocabularies of 30 to 200 words is well within the state of the art (Flanagan, 1976; Hyde, 1972).

Presently, human-to-computer speech usually involves either relatively small vocabularies (such as digits) spoken by a sizable population of users, or large vocabularies (a couple of hundred words) for a small number of speakers who have allowed the machine to calibrate their voices (Velicko and Zagoruyko, 1970; Tsemel, 1972). Speech recognition systems have been successfully implemented in several practical situations since the early 1970s. Industrial applications of speech data entry include quality control, incoming inspection, receiving accountability, part and serial number verification, warehousing, and sorting. In computer-aided design situations, speech data entry is being used for text annotation in printed circuit board component layout, preparation of program tapes for numerically controlled machine tools, generation of wire lists, civil engineering drawings, structural drawings, and bills of material. Word processing applications that have been adapted to speech input include data retrieval/paging, selection of function modes, and report generation. Results from these applications indicate, for example, that factory workers with minimum training can use voice input quite successfully.

Speech input is particularly advantageous in situations when one or more of the following conditions apply:

- The worker's hands are busy.
- Mobility is required during the data entry process.
- The worker's eyes must remain fixed upon a display, optical instrument, or some object to be tracked.
- The environment does not allow use of a keyboard.

Voice input is suitable for these applications because it requires neither hands nor eyes for its operation. Mobility up to 1000 feet can be provided by a dedicated VHF or UHF-FM radio link, and the only parts of the voice data system that must be exposed to the work environment are a microphone and a display for feedback (Welch, 1980).

Reaching an acceptable level of human performance in speech recognition systems requires that the computer always or nearly always recognize what a person has said. From a human performance point of view, some of the most critical or difficult areas related to optimizing this type of human/computer interface are shown below. The designer should:

1. Make provisions for users to separate words by gaps of silence of 100 milliseconds or more (Welch, 1980). This means that the computer must be able to reliably identify word boundaries. Continuous speech recognition is a more difficult task than single word or even short sentence recognition primarily because of the uncertainties involved in obtaining the correct word segments (Reddy, 1975).
2. Make adjustments for the unique characteristics of different speakers.
3. Take advantage of the redundancy of human speech by taking into account that vowel shapes are influenced by adjacent consonants. Use this information to improve vowel recognition and obtain information about consonant identity (Broad, 1972).
4. Select vocabulary elements to eliminate easily confused commands (Glenn, 1975).
5. Eliminate or control interfering background noise, possibly, by placing people in acoustically shielded environments or eliminating unwanted noises at the microphone itself (Martin, 1976).
6. Provide a reject capability so that inadvertent sounds, such as sneezes, coughs, and throat clearings, do not produce incorrect recognition decisions.
7. Give feedback to the user so that he or she receives positive acknowledgement that the computer has recognized the input. Some caution is necessary since Welch (1980) found that auditory (vocal) feedback for the computer increased errors in some situations.

8. Make provision for user intervention at critical stages, for example, to correct errors (Hill, 1971).

9. Provide voice prompting from the computer in situations where there is an advantage to freeing the user from reading (Welch, 1980). Along these lines, Hammerton (1974) has concluded that "instructions should be heard and data seen."

Speaker Recognition

Speaker recognition (or verification as opposed to *speech* recognition) is another capability now used in many systems. In this case, a brief message *and* the voice that carries the message must be recognized by the computer. The same human performance considerations that were related to speech recognition are also important to speaker recognition. However, speaker recognition systems have an additional unique set of requirements. In most speaker recognition systems, the speaker enters his or her identity and then speaks a short prearranged verification phrase. The computer makes measurements on the offered voice sample and compares the results to stored reference data. The computer must then decide whether or not the claimed identity is valid. A variation of this is a speaker identification situation where no claimed identity is volunteered by the user, and the question directed at the computer is "Who am I?" The computer must then examine its stored reference patterns, make a certain number of comparisons, and indicate an answer.

Evidence suggests that some computer-based systems can identify speakers better than people can. Lummis (1972) found that computers were significantly better than humans in speaker verification tasks. He suggested they always should be preferred for speaker verification where minimizing errors is important. If a computer-based system is used for this kind of activity, then the designer must optimize the human/computer interface by making decisions that ensure that the highest possible percentage of speakers are recognized by the computer.

To reach this objective a designer should require longer speech utterances over shorter ones and, by selecting appropriate filters, try to minimize interference from speech irregularities caused by nasal congestion, laryngeal inflammation, diplophonia (raspy voicedness), and stuttering. Other difficulties include identifying the differences in male and female speech and in closely related members of the same family, particularly identical twins.

Computer-based speaker recognition systems have been used to control access to facilities, as a tool for law enforcement agencies, and in some banking and business transactions. Even so, computer-based speech recognition is still a relatively untapped resource in many systems.

Command Languages

Introduction

A command language is used to transmit instructions from a human to a computer. It is one of the most used and critical elements of a human/computer interface. For systems consisting primarily of experienced users, access to a computer is, in almost all cases, by way of a command language (whether implemented by keyed commands, function keys, or speech). Frequently, the command language is unique to the particular system in which it is used. No other feature is as important in determining the effectiveness of the human/computer interface.

A typical command structure has a one-word command (usually an action verb), often followed by one or more "arguments," which specify various options or alternatives for realizing that particular command. In addition, the command may include a set of parameters or variables (usually numbers representing some quantity). There are at least two distinct methods of formatting commands and arguments: *positional format,* in which specific kinds of information must be assigned in a fixed, or relative, position in the argument string; or *keyword format,* in which arguments may be given as permissible strings of special words indicating the argument type as well as its value. The positional format imposes a greater memory load on a user since the values of the arguments must be remembered as well as the position (Miller and Thomas, 1977).

Command languages are always one-way. People use command languages when communicating *to* a computer. Messages *from* the computer are a different story and must be designed using a different set of guidelines. This requirement of using one language in one direction (command) and another language in the opposite direction (computer messages) means that most of the time users are forced to use two languages in carrying out a human/computer dialog. This modifies somewhat the goal of having the human-to-computer language be a natural language. That there is usually not a balanced two-way communication link has important implications. In this unbalanced condition, the computer is always the "listener" and the user is always the "speaker." If the rules were reversed, and the computer only communicated with people using a programming language, such as Fortran, COBOL, or PL-1, most people would have great difficulty in understanding the computer.

The two most critical problems with most command languages occur when a person is learning to use them and, later, when an individual has to recall and use a command that he or she has not used for a long while. Therefore, a well-designed command language is one that is easily learned and not so easily forgotten. Boies (1974) found that most users tended to use only a small number of the commands available and would apply most frequently the simplest, least powerful form of such commands. Such findings could result from the fact that the design of the commands made it difficult for users to recall them when needed. We should not lose sight of the fact that the primary concern of most users is to accomplish the task at hand. Even the most elegant of computer data bases may be inaccessible due to a faulty command language.

Jones (1978) and others have suggested that the human interface with the computer is best when it resembles human communication as much as possible. The behavior of the computer in *2001: A Space Odyssey* (Clarke, 1968) provides a pattern for human/computer interaction that many designers try to emulate. Consider the following exchange from "2001":

Computer: We have a problem.

Astronaut: What is it?

Computer: I am having difficulty maintaining contact with earth. The trouble is in the AE-35 Unit. My fault prediction center reports that it may fail within 72 hours.

Astronaut: We will take care of it. Let's see the optical alignment.

Computer: Here it is, Dave. It's still OK at the moment. (A screen display is given.)

Astronaut: Do you know where the trouble is?

Computer: It's intermittent and I can't localize it. But it appears to be in the AE-35 Unit.

Astronaut: What procedure do you suggest?

Computer: The best thing would be to replace the unit with a spare so that we can check it over.

A well-designed command language is logical, flexible, and provides a degree of forgiveness—that is, it allows for some human error. Unlike most of the languages in the world today that have evolved over long periods of time, each command language is the conscious creation of one or more designers. Because most command languages tend to be restricted in their application, designers can create languages that do not contain many of the disadvantages of the evolved languages. The designer has the option of providing a very precise vocabulary, but the vocabulary that is to become part of the command language should make allowances for the existing natural language of the system user.

To make the human-to-computer communication process as natural and predictable as possible for the user, the input command should be conveyed in a language in which he or she is most comfortable. The user's native language is the most natural language. In some cases, however, strict natural language statements to the computer are not always the best choice. In many specialized applications, such as nuclear research, a mathematics-based language is probably a more efficient form of communication than, for example, English words. Bobrow (1967) observed that the user's natural language is the most useful form of communication only in situations where little jargon exists, where messages are seldom repeated, or where the idea to be transmitted is defined precisely. When dealing with computers, there will always be both semantic and syntactic limitations to any natural language. Gould et al. (1976) found that computer users readily accepted restricted syntax languages when the languages were adequate to describe the tasks to be performed.

Jones (1978) has pointed out some discrepancies between designers' assumptions and the reality of language. Frequently, designers tend to assume a one-to-one correspondence between newly formed command statements and their meanings, whereas users may see many of the commands as having the same meaning. In addition, designers tend to believe that nothing is assumed by users except what is expressly stated about a command, whereas innumerable unstated assumptions may exist. Finally, designers tend to assume that deductions are made from an absolutely unvarying frame of reference, whereas, in reality, the meaning of a command to a user depends strongly on the immediate context.

Inexperienced users seem to prefer commands that are as complete and unambiguous as possible for the application. But that is not the case for highly experienced users. Whenever possible, designers should provide experienced users with a good set of abbreviations for frequently used commands. Streeter, Ackroff, and Taylor (1980) had a group of subjects generate abbreviations for a number of commonly used commands. They then derived a set of rules based on how these subjects produced their abbreviations. Rule-based abbreviations were compared to the most frequently given abbreviations for each command. Even without knowing the rules, another group of subjects remembered substantially more of the rule-based abbreviations than did subjects who studied the frequently-produced abbreviations. The researchers concluded that human performance could be enhanced if abbreviations conformed to a set of rules such as the one they derived. Their abbreviation rules for commands are shown below:

- For terms consisting of more than one word:
    - — take the first letter of each word as the abbreviation
- For monosyllabic words:
    - — take the initial letter of the word and all subsequent consonants
    - — make double letters single
    - — if more than four letters remain:
        - — retain the fifth letter if it is part of a functional cluster (such as *th, ch, sh, ph, ng)*

otherwise:

— truncate from the right

— delete the fourth letter if it is silent in the word.

- For polysyllabic words:
  - — take the entire first syllable
  - — if second syllable starts with a consonant cluster, add it
  - — if first syllable is a prefix and the second syllable starts with a vowel, add the second syllable
  - — make final double consonants single
  - — truncate to four letters (but always retain entire first syllable).

Command Language Guidelines

The guidelines that follow were gathered from numerous articles and from direct experience with computer-based systems. Obviously, the list is tentative and many of the guidelines require validation. However, even in their present condition they may suggest to designers some new ideas for improving command languages.

All of the guidelines are related in some way to improving the learning of a command language. In addition, some of them should help improve user satisfaction (particularly items 1, 2, and 3), some should help speed the user-computer interaction (particularly items 8, 17, 18, and 33), and the remainder should help reduce errors.

Keep in mind that these guidelines pertain almost exclusively to commands—information that is transferred from a human to a computer. Information is transferred from a computer to a user (by definition) with "messages."

The guidelines are organized below according to several major command language issues.

Command Style and Structure

1. When using commands, the user should have the impression of having control over the computer.
2. Commands should be meaningful enough to encourage a *natural dialog*. For example, the command language should closely correspond to user activities.
3. A command/action combination should always "make sense" to the user. The action taken by a computer should be totally consistent with user expectations, and should never be counter-intuitive. Good

command/action combinations will lead users to overlook that the dialog is taking place with a machine.

4. The syntax for all commands should be consistent.

5. A command should describe exactly the function to be performed. For example, *print* should refer to printing a letter, word, line or record.

6. Provision should be made for the computer to automatically adapt to the increasing skill level of users. Where this is not done, the designer should provide for at least two levels of commands (low and high), and give the users the opportunity to specify either level.

7. A command should have some representative (inherent) meaning. For example, *add, plot,* or *quit* are preferred to *OPT1, 75,* or *DG6* which have no clue to the nature of the option.

8. For unskilled users, the commands should make considerable use of natural language, and avoid codes (including abbreviations) and highly structured formats.

9. A command should not require the user to enter leading zeros, leave blanks (except to separate command words), or to make specific placement of decimal points.

10. When a system requires a keyboard for input, use a character set that does not require use of the shift key.

11. When non-alphanumeric characters are part of a command, do not use two of them together (e.g., <- or >. or $!).

12. A command should be as short as possible (depending on the experience of the user) to help prevent entry errors and to facilitate the detection and correction of errors.

13. Provision should be made for command modifiers (arguments) and/or data items (parameters) to be entered individually or in a string, depending on the preference of the user.

Command Coding

14. As a user becomes more familiar with a system, he or she should have the option of entering briefer forms of commands (codes); and of eventually being able to progress toward using only highly coded, shorthand commands.

15. If command words are coded using abbreviations, common abbreviations should be used. For example, *y* for "yes" and *n* to represent "no."

16. Codes should not conflict with normal conventions. For example, it would be inappropriate to use a *3* to represent "yes" and *8* to represent "no."

17. Coded commands should be standardized for all systems used by the same person. For example, if *y* represents "yes" and *n* represents "no" in one system, then that coding should be consistently maintained for each system.

18. In some situations provision should be made for individual users to specify their own codes for commands.

Confusability

19. Commands should be semantically unambiguous. For example, a user should not be expected to remember a subtle difference between the two commands *print* and *write*.

20. Commands should be perceptually dissimilar:

    - visually dissimilar (for keyed or handprinted input)
    - auditorily dissimilar (for voice input).

Defaults

21. Default values should be provided where appropriate for commands (keywords), modifiers (arguments), and data items (parameters).

Learning, Retention, and Recall

22. Commands should facilitate learning:

    - prior to use—learning a basic set of commands
    - during use—for progressing from making menu responses to using full commands, or moving from using a basic set to an unexpanded set, or from using full-word commands to coded commands.

23. Commands should be provided that require a level of training that is consistent with the required frequency of the activity to be performed. For example, little used commands should be allotted little or no training time, but should be clearly described in a performance aid.

24. Cognitive processing and memory limitations should not be exceeded when:

    - command/action combinations must be memorized
    - commands are looked-up, but computer actions and computer-based or document-based retrieval methods are memorized

- commands and computer actions are looked-up, but computer-based or document-based retrieval methods are memorized.

25. Commands should facilitate recall of commands and computer actions:

    - by using mnemonics for code/command associations

    - by using highly descriptive words for command/action associations.

26. Users should not be expected to learn, remember, and use more commands than are needed to carry out the basic functions of a system. All other commands should be easily accessible in a performance aid.

27. Only the maximum number of commands that can be readily remembered by users of various skill levels (naive, moderate, high) should be provided.

Power

28. A command should contain the appropriate "power"—micro versus macro-level—to expedite the particular procedure to be carried out.

29. The power of a selected command should be determined by the user. For example, some users may prefer a few high-power commands, while others may prefer several low-power commands to accomplish the same activity. Both options should be available.

Backup

30. A command should have as a backup some form of easily accessible computer-based help in case a user:

    - does not understand the syntax or semantics of a command

    - has forgotten available commands and their actions

    - does not recall the correct codes for remembered commands.

31. At any point in a sequence of actions the user should be able to receive specific assistance either automatically or by entering a "help" command. This means that the computer should continuously track and be aware of the user's location in a system.

32. Ideally, the computer should automatically provide help in structuring and entering commands whenever it determines that the user is in difficulty. For example, when the user response time is substantially longer than the average for the session, or when a command is erroneously entered.

33. A user manual or "command book" should be provided when a computer-based backup facility is not available.

Friendliness

34. The computer should be able to distinguish and accept common misspellings of a command, make the necessary corrections, and then carry out the right action. For example, identify *copu* as *copy, eit* as *edit* and *seend* as *send.*

35. The computer should be able to distinguish both upper- and lower-case characters as having the same alpha meaning.

36. The computer should be able to distinguish the inappropriate use of the shift key (when two characters are assigned to the same key), and indicate to the user what the correct entry should be. For example, when a user incorrectly enters a 4 in a situation when the $ is required, the computer should respond with a message: "Did you mean $?"The user responds with "yes" (or "y") instead of reentering the $.

Error Detection

37. Commands should facilitate error detection by using a restricted character and/or word set that is known to both users and the computer. For example, the characters "a" and "h" sound so much alike, both should not be used for voice input. Likewise, l (lower case "L") and 1 (one) for keyed input.

38. The computer should allow only valid commands at each decision point, with the computer keeping track of where the user is and the valid or acceptable commands at each point. When an invalid command is entered, the computer should respond with a list of acceptable commands, with the most probable listed at the top.

Input Procedures

Perhaps the best way to begin a discussion about what is meant by human-to-computer procedures is to use an example. Grignetti and Miller (1970) described a system in which users were expected to use a set of computer-based procedures, with their accompanying commands, to correct typographical errors they had introduced into sentences. The users could correct the errors by using one of three strategies. The KILL command erased a sentence and required the user to retype the corrected version in its entirety. The DELETE/INSERT command required users to count the number of characters up to the error and to enter this number, the number of characters to be deleted (if any), and the characters to be inserted (if any). The REPLACE command was of the form "replace 'old string' with 'new string'" (where 'old string' includes the error plus any preceding characters that may be necessary to specify uniquely the position of the error and where 'new string' is the corrected version of 'old string'). Users were expected to select the error-correction approach that was best for each type of error. To make the correction required knowledge of both the command language and the *procedures* for implementing the commands. A simple understanding of the command names alone is not sufficient for making the corrections. The language itself could be very well designed while the procedures could be extremely awkward.

Consider the example in Figure 14-4. The language is not too bad, but the procedure leaves much to be desired.

Computer: DO YOU WISH TO MONITOR TEMPERATURES?
User: YES

Computer: HOW MANY POINTS DO YOU WISH TO MONITOR?
User: 4

Computer: PLEASE ENTER POINT NO. 1.
User: 123

Computer: PLEASE ENTER POINT NO. 2.
User: 43

Computer: PLEASE ENTER POINT NO. 3.
User: 762

Computer: PLEASE ENTER POINT NO. 4.
User: 521

Computer: TYPE C FOR CONTINUOUS OR S FOR SINGLE SHOT.
User: C

Computer: PLEASE GIVE MONITORING INTERVAL IN MILLISECONDS.
User: 500

Computer: DO YOU WISH TO MONITOR PRESSURES?
User: YES

Computer: HOW MANY POINTS DO YOU WISH TO MONITOR?
User: 5

Computer: PLEASE ENTER POINT NO. 1.
User: 345

Computer: PLEASE ENTER POINT NO. 2.
User: 678

Computer:	PLEASE ENTER POINT NO. 3.
User:	987
Computer:	PLEASE ENTER POINT NO. 4.
User:	428
Computer:	PLEASE ENTER POINT NO. 5.
User:	34
Computer:	TYPE C FOR CONTINUOUS OR S FOR SINGLE SHOT.
User:	V
Computer:	**********ERROR 27 ILLEGAL CHARACTER
Computer:	DO YOU WISH TO MONITOR TEMPERATURES?

Figure 14-4. An Example of an Awkward Procedure (This example first appeared in P.F. Jones, "Four principles of man-computer dialogue," *Computer Aided Design*, Volume 10, Number 3 (May 1978), pp. 197-202. Published by IPC Science and Technology Press Ltd., P.O. Box 63, Guildford, GU2 5BH, Surrey, UK.)

Rubin and Risley (1976) presented three guidelines for procedures design. First, because beginning users have so much to assimilate, the procedures should be constructed in a way that does not overtax the human memory. A good construction method is to incorporate the memory aids within procedures by using a language with commands that are approximately what a naive user would interpret them to mean. Another is to use common-sense default values for arguments. Another way of minimizing the memory load is to provide display menus whenever the user must choose among several alternatives.

Second, designers should provide computer-based resources so users can gradually increase the complexity and sophistication of their interactions with the computer. This means providing the user with the option of eliminating some of the more mundane procedures that were originally provided for the benefit of novices. It also means giving the user the capability of developing increasingly sophisticated interactive procedures that are specially tailored to his or her own needs.

The third recommendation suggests that designers provide inexperienced users with an introduction to the computer's procedures. This could be a document that allows the user to accomplish a complete interaction with the system. Scripts, that is, recorded interactions between the new system and experienced users, can be used to permit the beginner to observe how an experienced user goes about using the various available procedures to solve problems. Once the full range of procedures is introduced, users can use the computer to try them out on sample problems. In this way the user actually would do several problems (the answers to which he or she would know in advance). In fact, in early problems the computer can lead the user to the right solution in short steps, and, as the user gains experience, the computer becomes less and less involved in prompting the user to learn and use the correct procedures.

There are numerous general types of input procedures. These types differ considerably in the demands they impose on the user. A brief presentation of four frequently used procedures follows.

One of the most common input procedures is the use of a *menu*. Just like the one in common use in restaurants, an input menu usually contains a set of alternatives and the user is expected to select one or more of these. The actual selection process may include the moving of a cursor, the typing of a character, or perhaps the use of a light pen or finger on the screen of a CRT. A sequence of menus may be provided to deal with a tree structure of possible choices. With the use of menus, the naive user is still likely to understand what is needed to make accurate selections from the computer.

The use of *masks* is another commonly used input procedure. This is similar to having a form designed on a CRT screen. The mask may require a single entry or numerous entries. Generally, the headers or the field names are determined and fixed by designers, and the user is unable to change them. The data fields may be either finite or variable. The data entered are usually associated with computer editing. For example, a computer may be programmed to accept only alpha characters in certain fields and numeric characters in other fields. Computers then generate an error message when an error is identified.

Another commonly used input procedure, which may be more interactive than the previous two, is a *question and answer* format. With this approach, the computer usually prompts the user with a question. In this case a sequence of fields is presented one at a time and the user fills in the appropriate response and then is presented with the next field. The desired input may be a fixed-length code or perhaps even an extended piece of text that will be entered but does not undergo automatic editing by the computer. Each of these three approaches is computer initiated, and particularly well-adapted to inexperienced users.

Another input procedure is one where the *user* initiates commands rather than simply responding to requests from the computer. Use of this approach usually requires that users be trained and gain experience in the operation of a particular computer. This input procedure could involve the use of a rather detailed command language. A user-initiated command could begin with a verb such as *edit, compute,* or *write.* The verb may be followed by one or more arguments that condition, limit, or bracket the domain of the action or that specify perimeters of the values to be addressed. To use this type of input procedure the user must learn the command language vocabulary and the command/argument structure as well as the appropriate procedures for using them. This approach is particularly good if the activity is repeated so often that it would be tedious for an experienced user to be forced to use a series of menus, masks, or computer prompts to input information.

A well-designed language and input procedure can help avoid ambiguity and eliminate major differences between what a user intends and what a computer is willing to accept. A poorly designed language and/or input procedures could cause considerable confusion and many errors. In the design of languages and input procedures the designer must be very familiar with the experience level of potential users. Both the command language and input procedures should be consistent with the way the user thinks about the operations performed by the computer.

COMPUTER-TO-HUMAN

Common Output Methods

Information stored in the computer is usually presented to people either by a CRT screen or a hard copy paper printout. Other means are also available, but only the most common output methods will be discussed here. Many output units are off-the-shelf, readily available terminals. Again, as in the case of keyboards, the problem that a designer usually faces is not one of designing the physical characteristics of displays, but of selecting from those available the one that has the characteristics most suitable to a particular application.

Among the most important of the many physical characteristics of the output are: height and width of characters; type font; and the contrast between the characters and their background. These will be covered in more detail in Chapter 15. A second type of characteristic is associated with the type of message being communicated. It includes the length of display lines, indenting, punctuation, field definitions, and other similar considerations. Finally, a third consideration is the content and length of messages. A set of design characteristics for all three considerations are outlined later in this chapter.

This section will discuss some of the considerations that are important to ensure that computer-to-human communication is appropriate, clear, and well-formatted. These considerations are aimed at improving *human performance* as opposed to computer or system performance.

Computer Response Time

The definition of computer response time depends on the nature of the system involved. Computer response time most often refers to the time that the user must wait for a computer response following a command (i.e., the time until a message is *completed* on a screen or paper printout). Many of the considerations in a real-time mode are similar to those in batch mode. The major difference is the time interval between the human input and the computer output. Obviously, if the time interval is shortened from twenty-four hours to one or two seconds, the human and computer are able to exchange more messages in a given period of time. This greater opportunity to exchange messages tends to improve human performance in some systems (cf. Chapanis, 1975).

Of the many significant human/computer interface considerations, no area seems so obvious and straightforward, and yet remains so unclear, as response time. One of the reasons that acceptable response times are so difficult to determine is that people's expectations may differ from situation to situation. Consider, for example, that a particular human action may involve different cognitive processes depending on a set of objectives that the user has internally specified. As a result the particular human action, while it may always result in the same specific computer process, does not always connote the same user purpose. As a consequence, the response time of a computer system for the same specific computer actions may be entirely acceptable to a user with one purpose in mind, but be marginally acceptable or entirely unacceptable when he or she has another purpose in mind.

Too long or erratic response times may have an impact on the user's attitudes toward the system, the user's work habits at a terminal, the type of work the computer is to perform, and even the circumstances in which a user will use a computer. User attitudes can affect not only frequency of use of the computer, but also the types of uses to which the computer will be put.

In all but the simplest applications, the question is not merely one of an *acceptable response time,* but one of a *set* of acceptable response times. Users seem willing to wait varying amounts of time for different types of requests. The amount of time a user is willing to wait appears to be a function of the perceived complexity of the request and the time when the request is made—people will wait longer for "hard" requests or those made at the end of a series of commands, or "closure point," than during the interaction. It is important, therefore, that some recognizable relationship exist between a particular task (or particular set of tasks) and a specific set of computer response times.

Several people have defined the minimum psychological needs that make a human-to-human conversation effective. Of these, *continuity*—a give and take of information without long pauses—appears to be one of the most essential. Conversational type human/computer exchanges should be designed to be more or less continuous. Long gaps between human or computer responses may give the appearance of terminating the exchange. To help ensure continuity, a computer response of some kind is necessary within one or two seconds.

Some educated guesses have placed the longest acceptable response time at from two to four seconds (Miller, 1968); a few milliseconds (Carbonell, 1969); and one second responses to simple inquiries and one to two seconds for information that may be readily learned or looked up by a person (Zimmerman, 1977). Williams (1973) reported that a two-second delay in response time was almost universally acceptable to his test subjects.

Miller (1968) categorized a large number of human/computer interface tasks into seventeen types. In each case, he attempts to establish an appropriate response time on the basis of the most probable cognitive processes involved and on the assumption that the user is operating in an interactive (or dialog) mode with the computer. The focus, in this case, is on those human capabilities that make the term "interactive" meaningful. The most critical assumption in Miller's scheme is that certain actions by the user stand for certain intentions of the user. The second most critical assumption is that displayed information must be meaningful to the user. Even if data is displayed in response times acceptable to the continuation of the human/computer interaction, the data must have relevance to the user. These concepts form the basis for a reasonable definition of response time—the elapsed time between a user request and a meaningful reply.

A set of maximum response times is shown in Table 14-1. These "armchair" estimates are adapted from Engel and Granda (1975) and Miller (1968). Maximum acceptable response times obviously have to be balanced against efficiency of the computer system configuration, the criticality of the "interactive mode," and the "meaningfulness" of the displayed information to the user.

Carbonell, Elkind and Nickerson, (1968) and others have pointed out that it may not be the magnitude, but the *variability* of delays, that is the most distressing factor to users. Other researchers have suggested that in some cases longer computer response time actually might be helpful (for example, in detecting keying errors). Williams et al. (1977) found that inexperienced users using a system for a simple data correction task were able to use time delays effectively for the analysis and planning of error correction strategies. There have been other instances of users reporting dislike of too fast response times (cf. Shackel, 1979).

Keep in mind that providing a timely response to users may mean having the computer present a status message. Users sitting at the terminal and waiting for a response may wonder if the computer is still working, if the terminal is still connected to the computer, if the computer lost the input, or if the computer is in a "never ending" loop. A short message—STILL PROCESSING—that appears every ten seconds indicating that the computer is still working on the problem provides the user with some assurance. As we indicated earlier, the first such message should appear within one or two seconds after the command is sent, should be taken off the screen after five seconds, and then should be noticeably updated in about ten seconds. This cycle should be repeated until the expected response occurs. Another solution to this problem was proposed by Spence (1976) in the form of a "count-down clock" in the corner of the display showing the rate of computer progress and an estimate of how long the user still had to wait.

Table 14-1. Recommendations for Some Maximum Acceptable Computer Response Times

Response Time Definition	*Maximum Acceptable Response Time (secs)*
Key depression until position response; for example, "click"	0.1
Key depression until appearance of character	0.2
End of request until first line is visible	1.0
End of request until text begins to scroll	0.5
From selection of field until verification	0.2
From selection of command until response	2.0
From light pen entry to response	1.0
From input of graphic manipulation to response	2.0
From drawing with light pen to response	0.1
From input to new menu list	0.5
From inquiry input to response	
Simple	2.0
Complex	4.0
From command until display of a message	2.0
From "execute" command until execution begins	10.0

(adapted from Engel and Granda, 1975, and Miller, 1968)

Many problems arise when users attempt to work in an interactive mode with systems that operate primarily with many response times in excess of ten to fifteen seconds. This does not necessarily imply that a system is poorly designed. It may mean however, that the user must adopt different, less efficient approaches to the tasks. Grossberg, Wiesen, and Yntema, (1976) found that to compensate for time delays, experienced users change their strategies in using a computer if the system allows. For example, they learn to perform activities in a different sequence, i.e., one that takes advantage of commands with traditionally short response times, and requires less frequent long responses from the system. In systems where protracted response delays are inevitable, "write ahead" capabilities that allow the user to continue her or his work without awaiting a response should be provided. This approach of changing strategies to accommodate time delays is consistent with the idea that people who are using the computer in the batch mode with 12 to 24 hour turnaround time use different strategies than those who have a one or two second turnaround. As shown by Chapanis (1975), one of the main differences is that as the response time shortens, the number of messages that are exchanged *increases*.

In some situations rearranging tasks may be extremely difficult because inquiries to a system are done serially that is, inquiry 3 is dependent on the answer to inquiry 2, which in turn is dependent on the answer to inquiry 1. In these situations time intervals between input and response output tend to be "dead time" for the user. This situation could cause at least two human performance problems. First, as this time interval increases beyond ten to fifteen seconds, continuity of thought becomes increasingly more difficult to maintain. Second, human performance may be slowed to the point where system performance is affected. Boies and Gould (1971) found that each second of computer response degradation leads to a similar degradation added to the user's time for the next request. Doherty (1979) suggests that this phenomenon (*Boies phenomenon*) exists because users seem to have a sequence of actions in their short-term memory. Increases in computer response time disrupt this storage, requiring time consuming reloading of short-term memory. Doherty observed further that:

> If we were to tolerate a system response time as long as two seconds in our laboratory then Boies' phenomenon shows that it would cost us a minimum of 36 million seconds per month of lost human time. That is 10,000 man-hours, or 60 people lost full time for the month. It is our objective to keep 90 percent of our interactions to a response time of one-half second or less because we believe subsecond response time is an important human requirement. (p. 88)

Messages

Computer-to-human messages can convey a wide range of information. For example, a message can be a prompt or a request for more information, it can be a diagnostic message generated by an error condition, or it can be a system status message. Whatever the type, the message should be concise and clearly understandable to the user. Cryptic codes, acronyms, and abbreviations are not meaningful to most inexperienced users. The person who is learning to use or who infrequently uses a computer tends to want and need considerable help from the computer. On the other hand, the expert or frequent user usually desires nothing from the computer that is not essential to meeting the goal the user has in mind. The experienced user tends to want the message from the computer to be as short and terse as possible and may even be frustrated when forced to watch and wait as the computer outputs messages containing unwanted and unneeded information. However, even in the latter case, detailed messages should be available to be displayed when required by an experienced user.

Rule of Two

One approach that has been relatively successful is to design a set of messages using a "rule of two." This means that the computer is able to operate in either of two modes. In the first mode, the computer has a set of messages that are directed to inexperienced users. In the second mode, the computer has another set of messages that are more useful to experienced users. Users that fall somewhere between these two extreme positions can use messages from either set. The challenge for designers is to satisfy inexperienced and experienced users, as well as all those with needs, skills, and preferences in-between. Nickerson and Pew (1972) suggested several features that can be incorporated in a system to help with this problem. These include user interrupt capability (to cutoff too lengthy messages), and brief computer-to-human messages backed up with a more complete explanations if requested by the user.

Error Messages

Miller and Thomas (1977) discuss recovery philosophies relating to four types of recovery situations. "Recovery" here refers to the reinstatement of some past state of the computer system, usually after some kind of error or malfunction. The four types of recovery situations are:

1. user correction of data input
2. user abolishment of prior commands
3. abnormal exit from within some program or programming language processor
4. system crash.

The first two recovery situations are relatively common and well-known to most designers. They require either the user or the computer to detect the undesired situation and initiate corrective actions. When a problem is detected by the computer, an error message is issued. Error messages are probably the most common of all messages. When errors are detected, the error message should clearly describe what the error is, give the probable cause, and describe an appropriate corrective action. Thus, every error message should state what the error is, why it occurred, and what can be done to correct it. As people become more experienced with computer systems, they usually prefer that long error messages be shortened.

The latter two situations originate with the action of a system itself. In these two situations messages should also provide users with three pieces of information: what happened, why, and how to recover. For example, when there is an abnormal exit from within some program, the user should be given an error message that indicates what rule was violated and where in the program this occurred. Users also should be provided with information indicating what they can do to correct the situation, that is, to continue processing. Many programs today simply give the user a message such as "job aborted."

As suggested earlier, experienced users should have a facility for suppressing or cutting off long messages. The length of the interaction should vary according to the interest level of the user. Most people have had the experience of being with someone who goes on and on in discussions of things they have little interest in. It would be an advantage, particularly with the computer, to have the ability to cut off the computer when enough information has been presented.

Feedback

One of the main considerations with computer messages is that they clearly and accurately convey information in a timely manner. Timeliness is particularly important for messages used to provide user feedback. Designers should keep in mind that human performance can be improved if users are rewarded quickly and appropriately for good performance. For example, login procedures can be complex and annoying. In many systems, while users are trying to get connected to the computer they are more likely to be told that they have done something "illegal" or "improper" than to be rewarded for their good performance.

Providing good feedback messages is a means to keep users informed of where they are, what has been done, and whether an action was successful. Highlighting can be used as a means of providing immediate feedback to the user. For example, when a user selects an object on the screen, the computer could highlight that item. The user then knows that the computer has correctly accepted and interpreted the input.

Warm and Cold Messages

Anthropomorphism is the ascribing of human characteristics to nonhuman things, and anthropopathism is the ascribing of human feelings to nonhuman things. The obvious growth in the capabilities of the computer as a machine that can "think" has led to many comparisons with human characteristics. Engel and Granda (1975) suggest that computer responses should be strictly factual and informative and the designer should avoid making the machine appear human. There is a fine line between having the computer behave in a human-like manner and having the computer behave as a human.

Some computers are designed to be as friendly and warm as possible. For example, computers may address the user with such statements as "Hi, my name is PDP. What's yours?" and proceed to address the user by his or her first name throughout the remainder of the interaction. This may be perceived as being novel or funny at first, particularly by young or inexperienced users. But this type of interface is likely to become irritating as the user gains experience with the system.

There are, however, some instances where a computer should probably come across as warm, such as when interacting with children, particularly children having their first experience with a computer; and in hospital information systems where a computer is used to elicit information at the time of admittance.

Computers will tend to exhibit "personalities" particularly when the same person deals with the same computer and set of software for a long period of time. The "personality" exhibited will be a combination of the programmer's style and the unique characteristics of the hardware and software. Personality has been defined by Chaplin (1975) as "that which permits a prediction of what a person will do in a given situation." Each new computer-based system is unique and becomes predictable with use. In this sense computer-based systems will take on personality characteristics, and some will be seen as being warm and friendly and others as cold, and possibly hostile.

Computer Speech

Computers and associated electronic equipment can produce signals sounding remarkably like words produced by a human voice. In some situations, it may be preferable to have a voice output rather than a CRT display or printout. Voice output is most useful in situations where a message is simple, short and will not be referred to later. Also, computer speech may be preferred when the message calls for immediate action, vision is already overburdened, or the job requires the user to move about continually.

Voice output from the computer may seem more natural than reading from a CRT or from a printout, but it is not always the best way to present information. When used, computer speech should be broken into phrases or sentences in a way that clearly meets the requirements of users. For example, only short user requests and short system answers should be used in inventory type control systems. The length of speech output may not be so critical as long as the user can request that the output statement be repeated. In fact, after each computer speech output, the computer should provide the user the choice of responding with "wait," "go ahead," or "repeat." Finally, the user should be provided with a "backup" option to return easily to the next previous step in the program sequence.

Computer-based speech is very much a part of industry today. Its use began on a wide scale in the banking industry, where the largest number of speech output systems now exist (Hornsby, 1972). Another application that makes effective use of voice response capabilities is the telephone company's directory assistance. Other applications include the stock market (quoting current prices), the weather bureau (providing weather forecasts), and the medical profession (where some computers conduct interviews with patients using voice output).

FOR MORE INFORMATION

Engel, S. E. and Granda, R. G., *Guidelines for Man/Display Interfaces,* IBM Technical Report TR 00.27200, Poughkeepsie, New York, 1975.

Martin, J., *Design of Man-Computer Dialogues,* Englewood Cliffs, N. J.: Prentice-Hall, 1973.

Ramsey, H. R. and Atwood, M. E., *Human Factors in Computer Systems: A Review of the Literature,* ADA075679, 1979.

Shackel, B., (Ed.), *Man/Computer Communication,* Volume 1, Maidenhead, Berkshire, England: Infotech International Limited, 1979.

Shneiderman, B., *Software Psychology,* Cambridge, Mass.: Winthrop Publishers, Inc., 1980.

15

FORMS AND CRT SCREEN DESIGN

INTRODUCTION

Forms and cathode ray tube (CRT) screens have many characteristics in common. Designers who have successfully developed forms will most likely also be able to develop good CRT screens and vice versa. Many of the same considerations are important for both types of media. We will first review issues that relate to both forms and CRT screens, and then discuss some considerations specific to each.

One of the main reasons both forms and screens exist is to assist in collecting information, and they do this by *displaying* a question, prompt or message of some kind. Once an entry is made, they *display* both the question and the answer. For this reason, we will refer to both types of media as visual displays or simply displays.

DISPLAY DESIGN STEPS

The following steps lead to a systematic design of an effective display. We can divide these steps into two basic groupings. The first set of four steps include activities related to *data collection*. The second set of ten steps include activities related to *analysis* and *design*.

Data Collection

The systematic collection of information concerning the use and elements of a display is probably the most overlooked aspect of display design. Yet this is precisely the activity in which it is determined *what* goes on the display. Basically, the designer needs information that will help determine the proper type and size of display, and the compatibility of the new display with other displays in the system. The characteristics of the potential users are also very important. One sequence for the collection of display design information is listed below:

1. Determine specific requirements for the display.
2. Collect information relevant to proposed items and processing.
3. Determine characteristics of potential users.
4. Collect information on existing displays in the system or adjacent systems.

Analysis and Design

There are also some guidelines to help in the actual design of displays. The following ten steps for the design of displays assume that a designer has done an adequate job of collecting the information just discussed. Each of the following steps is important to the final product:

1. On the basis of data collection, determine the most appropriate grouping of items using grouping principles.
2. Develop headings or labels for groups and item names as required.
3. Determine groups of groupings where necessary.
4. Determine media to be employed (e.g., paper form, CRT display, etc.).
5. Attempt several alternative layouts and choose the one that will elicit the best human performance.
6. Determine a meaningful title for the display.
7. Develop instructions and/or other explanatory material to be placed on the display or to be used as separate instructions.
8. Determine physical characteristics of the display (e.g., type size, contrast, etc.).
9. Test the proposed layout on people representative of potential users.
10. Redesign as required.

GENERAL CONSIDERATIONS

Legibility means simply that the display can be read easily. Here, a knowledge of typefaces, letter sizes, spacing, flicker rates, and contrast all come into play. If the display is illegible it will fail to communicate and most of the other characteristics will become meaningless. Legibility may differ substantially depending on the light where a display is used. The design objective should be to make each display as legible as possible.

Readability means that the display communicates well and the reader understands precisely the meaning or purpose of what is presented. Readability is achieved in part from having the appropriate sequential order of instructions (based on a task analysis).

Accuracy is a two-sided coin. First, the display must be designed to *elicit* accurate information from a user. Instructions or headings that are vague or ambiguous may result in inaccurate entries. Secondly, the display must *transmit* accurate information.

A display has to be *compatible* with the user's knowledge and skill, with other activities being performed, and with the work environment in which it is used. Compatibility is an important factor in both performance and user acceptance. Systems have sometimes failed to be readily accepted simply because users refused to accept the displays. One way to cultivate initial user acceptance is to establish clearly user requirements early in the developmental process, and then to ensure that the new display adequately reflects these requirements.

Another important factor in gaining user acceptance is the *timeliness* of the data built into the display. Make information available to the user while it is still useful. Also, a good display should contain only *essential information,* and a display should present information that is *directly usable,* requiring minimal interpretation.

ADDITIONAL CONSIDERATIONS

Besides the general guidelines just presented, two more major areas require careful attention. These also apply to either a paper form or a computer-based CRT. The first concerns grouping items and the second concerns standardizing items. In addition to grouping and standardizing guidelines that apply to all displays, there are others that concern each of the two major types of displays separately—some for paper forms and some for CRT displays. These will be discussed later in the chapter.

Grouping

Grouping techniques are important in helping to organize information. There are four fundamental techniques for grouping data.

1. Sequence
2. Frequency
3. Function
4. Importance.

Sequence

Sequential grouping is based on the principle of grouping together items in the order they are transmitted or received. For example, if a designer is developing a visual display that must be used with information that always arrives, then the display should be designed so that the data can be entered in the same sequence. For output data, sequential grouping would dictate that the data be output in the order of use. The first information needed should be the first shown at the top of the display, the second item used should be the second one shown, and so forth.

This principle not only applies to item sequence, but also to the sequence of item groups. An example would be a process in which the user is required to verify all historical data before going on to other sections of the displays. The historical data should not only be grouped together, but be placed in a prominent position near the top of the display.

The natural order of data is another basis for sequential grouping. Other factors being equal, some sequences seem more natural than others. For example, name groups are usually alphabetical by last name, and number groups are usually placed in ascending order.

One method of establishing sequential groupings is by using a technique called procedural flow analysis. Closely analyzing the procedural flow determines what functions a display serves and what processes are performed on the data at each point in the flow of information. With this information the designer could elect to group together those items of information that need to be processed at about the same time. The order of the item groups would then follow the same sequence as the processing steps. The sequence of items within a grouping would also parallel the sequence of processing steps.

Frequency

The frequency of use technique is based on the principle that items used most often should be grouped together. If a designer develops a display completely on the basis of this technique, it would be, in effect, a rank order listing of the items according to their frequency of use. The most frequently used item would be at the top and the least frequently used at the bottom. For example, a preprinted mark-sense form for keeping a record of tool use might be designed so that the most frequently used tools are grouped near the top of the form.

Another application might be to arrange items within the groups established by the sequential grouping technique. After sequentially grouping the items, place items most frequently used at the beginning of each group.

One of the most common methods used to establish frequency of use is *link analysis* (cf. Chapanis, 1959). Briefly, link analysis is a technique for determining the connections between various items and/or procedures, and expressing these connections or linkages in terms of relative frequency. Using link analysis, a designer could not only determine and chart the *existence* of links, but also compute the *frequency* with which the various links occur. Of course this requires a final set of work modules so that all links are relatively stable. Through such an analysis a designer may find that item A is used 60 percent of the time and item B only 25 percent of the time. However, even more important, the designer may find that after item X is used, item A is used only 20 percent of the time, but item B is used over 95 percent of the time.

Function

A functional grouping technique may also be used. A designer may consider grouping items according to special needs of the activity being performed. For example, it may be advantageous to group all of the items that pertain to inventory in one location, and those related to requisition in another location. If the sequence or frequency of use are not too important, then layout based on functional relationships may be the best criterion for grouping. Items grouped on this basis should be identified as such.

Importance

One final technique is to group items according to how important they are to the success of the system. If a certain item or items are critical, then it may be best to place the critical items in the best position on the display so that they are not overlooked.

Grouping Tradeoffs

Grouping items on a visual display is a delicate process of elimination, weighting, and judgment. The designer should determine which of the techniques just discussed actually apply to the items. Of those that apply, determine how much weight each should carry in the final decision. And determine weighting on the basis of eliciting the best possible human performance. When determining the best grouping, *give precedence to human performance requirements* over any other requirements—for example, software or hardware requirements. After accomplishing these steps, perform tradeoffs in terms of applying one technique in one case and another technique in another case. The object, of course, is to arrive at an arrangement of items that has the highest probability of eliciting an acceptable level of human performance.

Standardization

Two kinds of standardization are important: standardization of items on a display, and standardization of displays. Although standardization is desirable, it should not take precedence over the grouping principles just described.

By standardization of items we mean that all items in the system with the same information should have the same name. They should also be coded the same way and be given the same field length (if appropriate). In many systems a large number of different people use the same items for a variety of different reasons. Frequently a large number of the items have more than one name. This, of course, could produce considerable confusion and lead to degraded human performance.

When using the same items on different displays, make an attempt to locate them in the same place on all displays. People tend to have good "location" memories. They can remember where an item is located even though they cannot recall the item itself. In addition, group those same items in the same groups and give them the same headings and the same basic design and style.

CRT DISPLAY GUIDELINES

Physical Characteristics

General

A display should be designed to suit the particular conditions of its use. For example, the designer should consider the viewing distance. The maximum viewing distance will influence the size of the details, such as scale markings and numerals, shown on the display. Usually designers assume about a 16-to-24-inch viewing distance from a user's eye to a CRT display. For displays with adjustment knobs, the maximum distance is generally set by reach limitations at about 28 inches.

Also suit the size of display details to the expected illumination level. CRT displays are usually hindered rather than helped by other lighting in the work station (Grether and Baker, 1972). Using general illumination with CRT displays requires careful design to minimize glare on the cover glass.

Another physical characteristic to consider is the viewing angle. How much the observers line of sight strays from being perpendicular to the display surface is the viewing angle. Buckler (1977) states that performance decrements begin to occur when the CRT screen is slanted between 19° and 38° from perpendicular.

A designer should keep in mind that each of the conditions involved in the design of CRTs that are presented here interact with all other conditions.

Statements that we can make regarding a particular condition are limited since the effect on the user of a given level of one condition depends on the levels of other important considerations. Even so, we will present one condition at a time and specify a range of levels for each. The following recommendations assume an ambient illumination of about 100 ft-c.

Luminance and Brightness

Luminance is the radiant intensity of a surface measured in millilamberts (mL). The luminance level of the pages in this book in a well-lighted room is about 50 mL; however, any luminance about 25 mL or so is usually acceptable, assuming adequate contrast. This level may be used as an estimate of the recommended luminance for symbols on CRT displays (Gould, 1968). Brightness is the subjective impression of luminance and typically is not uniform across a CRT screen. All displays should provide the user with the advantage of controlling luminance (brightness) by adjusting appropriate display controls.

Contrast

Contrast is the ratio of background luminance minus symbol luminance to background luminance plus symbol luminance. Howell and Kraft (1961) simulated CRT characters and found that there was little increase in character legibility when the contrast ratio was increased from 13 to 38. Gould (1968) recommends a contrast ratio of 30 (contrast of 94 percent) as preferred and 15 (contrast of 88 percent) as acceptable. Rouse (1975) reports that providing adequate contrast still seems to be a problem with many CRT displays as evidenced by the many dark CRT terminal rooms and the cardboard hoods taped to many CRT terminals.

Flicker

If a CRT is not regenerating fast enough, it appears to flicker. In general, a regeneration rate of 60 Hz, which is the rate of home television receivers, is probably sufficient to prevent the perception of disturbing flicker (Gould, 1968).

Resolution

The resolution required for a CRT display depends primarily on the viewing distance of the user and the application of the display. A designer should ensure that image quality is adequate by providing the level of resolution needed for accurate reading. Most important is that there should be an illusion of continuous characters, with individual scan lines or matrix dots imperceptible to the user. To provide this illusion, there should be at least 50 scan lines per inch (Rouse, 1975). The minimum acceptable character size has been determined in terms of scan lines and visual angle. Ten lines per character seems to be a good minimum for accurate detection of individual characters

(Gould, 1968). Buckler (1977) recommends a minimum dot matrix character height of seven dots (5 x 7 dot matrix), although nine dots (7 x 9 dot matrix) is preferred (Stewart, 1976). The angle subtended on the eye should be at least 12 minutes but 16 minutes is preferred.

Character Style

For CRT displays, Buckler (1977) notes that generation technique heavily influences character style. Low resolution could cause symbol distortion especially for dot matrix characters. Vartabedian (1971) found dot matrix character generation superior, both in performance and user preference, to stroke character generation. However, his study was conducted using only uppercase letters. In general, uppercase characters should be used for headings and titles, while lowercase characters should be used for text and variable information. Chamberlain (1975) and Wright (1979) suggest that a mix of upper and lowercase is best for readability. Use of only lowercase characters should be avoided. Each character generation method has its own set of constraints and distorting properties that need to be taken into account.

Color

The colors of the characters and the background should agree. In multicolor displays avoid confusion by using few colors and placing them far apart. For single-color displays, white, grey, and black are recommended, as well as yellow, orange, and green. Blue and red may strain the eyes more than the colors just mentioned because they place higher demands on the eyes' focusing mechanism. Also, people who are unable to distinguish one or more colors tend to have most trouble distinguishing green and red.

Evaluating Physical Characteristics

It is becoming more common for system designers to secure production model CRT display terminals rather than design their own. Before a CRT display is used it should be carefully evaluated to ensure that it possesses characteristics that will lead to an acceptable level of human performance.

The following list of human performance requirements can be used when making an initial evaluation of a video display terminal—a CRT/keyboard combination. Terminals that meet other important system requirements can then be compared from a human performance point of view. Each terminal can be evaluated on each of the human performance requirements. When a requirement is met the weighted value can be added to the total of other "met" requirements. The weights give an estimate of the relative importance of each requirement. The weights range from 1 to 5, with 5 indicating the most important requirements. At the end of an evaluation add the totals for the terminals and consider the one with the highest total as best meeting the human performance requirements.

CRT DISPLAY	*WEIGHT*
Location and Design - The CRT screen should be designed and located so that it may be easily read by users in their normal operating position.	5
Viewing Distance - A 16 to 24-inch viewing distance (eye to screen) should be provided. Design should permit the user to view the scope from as close as 12 inches.	4
Orientation - The CRT screen should be perpendicular to the user's *normal* line-of-sight. The angle of the CRT from the vertical should approximate 10°.	4
Reflectance - Displays should be constructed, arranged, and mounted to minimize or eliminate reflectance of the ambient illumination from the glass or plastic cover. Reflected glare should be minimized by proper placement of the scope relative to light sources, use of a hood or shield, or optical coatings.	4
Flicker - The CRT screen should have no discernible flicker and be free of any movement.	4
Screen Contrast - Contrast on the face of the CRT (with light characters against a dark background) should be between 3:1 to 15:1, recommended between 6:1 to 10:1 (ratio of brightest illumination to lowest illumination) in room ambient light levels of 70 to 100 ft-c. When dark characters are displayed against a light background, the luminance of the character background should be at least three times the luminance of the character.	3
Adjacent Surfaces - Surfaces adjacent to the scope should have a dull matte finish. The brightness range of surfaces immediately adjacent to the CRT should be between 10 percent and 100 percent of screen brightness. No light source (warning lights, mode indicators, etc.) under control of the vendor in the immediate surround should be brighter than the CRT characters.	1

Operator Controls - A limited range CRT brightness control and a contrast control should be provided for the user. All other controls—linearity, height, width, and focus—should be provided, but accessible only to maintenance personnel.	2
Cursor - The equipment should have a character underline, motionless cursor. The cursor should have the following characteristics: left, right, up, down.	5
Character Height - Characters should be at least 0.17 inches in height.	4
Character Width - Character width should be between 70 percent and 85 percent of character height.	3
Stroke Width - Stroke width-to-height ratio should be 1:8 to 1:10.	3
Character Spacing - Spacing between adjacent characters should be at least 18 percent of the character height.	3
Interline Spacing - Distance between lines should be at least 50 percent of character height. A distance of 66 percent is preferred.	3
Character Resolution - The number of scan (raster) lines per character should be a minimum of 10.	4
Character Illumination - Characters should be evenly illuminated.	1
Format - Editing and error correction information should be presented to the operator in a directly usable form. Requirements for transposing, computing, interpolating or mental translation into other units should be avoided.	3
Content - The information displayed on the CRT screen should be limited to that which is necessary to perform specific actions or make required decisions.	3

KEYBOARD

Key Size - The key tops should be square or slightly rounded and have a diameter of 0.5 inches.	4
Key Surface - The key tops should be concave and treated to minimize glare.	2
Key Displacement - Displacement should be a minimum of 0.125 inches to a maximum of 0.187 inches.	3
Key Resistance - Resistance should be in the range of 2 to 4 ounces.	3
Key Labeling - Key symbols should be etched to resist wear, and colored with high contrast lettering.	1
Key Separation - The adjacent edges of keys should be 0.25 inches apart.	4
Key Mounting - Keys should be securely mounted and firmly fixed in place to minimize horizontal movement.	2
Keyboard Angle - The keyboard should be inclined at an angle of 11° to 15° from the horizontal.	1
Highlighting - Functional highlighting of the various key groups should be accomplished through the use of color coding techniques.	2
Special Keys - Special keys should be an integral part of the keyboard. The distance between them and the alphanumeric section of the keyboard should not exceed 1 inch.	3

WORKSPACE

Work Surface Width - A workspace at least 30 inches wide and 20 inches deep should be provided to accommodate records being transcribed.	1
Work Surface Height - The work surface top should be 30 inches above the floor.	3
Copy Holder - A copy holder should be available that can be freely adjusted by the user, and has a surface area larger than the documents being transcribed.	3
Palm Rest - A palm rest should be provided for the key operator.	3
Leg Room - Unobstructed leg room should be provided.	2

LABELS

General Requirements - All controls should be appropriately and clearly labeled in the simplest and most direct manner possible.	1
Character Height - Character height should range from 0.10 to 0.20 inches.	1
Character Design - Characters should be uppercase black lettering with differences among symbols highlighted by emphasizing the salient features of each, e.g., straightness, roundness, lumpiness, angularity. Particular attention should be given to: uppercase D, uppercase O and zero; and lowercase L and 1.	2

Noise Generation - Noise generated by the video display terminal should be less than the ambient noise level of the environment before the device is installed (about 50 to 60 dB). Other factors, such as continued highpitched whir or hiss also should be eliminated from the equipment.	3
Heat Generation - Heat generated by the equipment should be dissipated in a way that does not expose the user to excess heat either directly or through an uncomfortable increase in ambient room temperature.	3
Ambient Light Level - Provision shall be made for the satisfactory viewing of the CRT in a room having an ambient light level from 70 to 100 ft-c.	3

Content Characteristics

General

The following section contains a set of guidelines to help designers to develop an effective interface between CRT displays and their intended users. These guidelines apply particularly to CRT displays; however, many of them, especially those dealing with consistency of presentation, information presentation, and labeling are also applicable to paper forms. The present set of guidelines represents an effort to collect and identify what is known or can be reasonably deduced from the present state of knowledge.

Designers are encouraged to apply these guidelines, but should recognize that other considerations may force tradeoffs in some instances. The guidelines are based on material presented by Engel and Granda (1975), Pew and Rollins (1975), and Smith (1980). One obvious weakness of these and similar guidelines having to do with human/computer interaction is that many of them are based on opinion, judgment, and accumulated wisdom, rather than on research studies. The main problem with this approach is that designers are given much advice, but not told the consequences of ignoring it.

Messages

Computers communicate to people through the use of messages (and as indicated in the previous chapter, people communicate to computers with commands). Messages can serve a variety of purposes. A message can be a prompt for more information, it can be a notification generated by an error condition, or it can be purely informational such as one that identifies the user's location in a program. No matter what its purpose, the message should be concise but clearly understandable to the user. Meaningless words, abbreviations, and codes appearing in messages are of little value and can provide considerable frustration for many users (particularly those who are inexperienced). Keep in mind, however, that a more experienced user may prefer simple (abbreviated) messages, as long as more detailed messages are available when requested. Users should be able to control the amount, format, and complexity of information displayed by a computer.

Messages should reflect the user's point of view, not the computer's, and should be strictly factual and informative. In addition, the computer should present information to the user in directly usable form. A user should not have to search through reference information (either written or computer-based) to translate messages received from a computer (that includes error messages). Avoid requirements for transposing, computing, interpolating, or performing other mental gymnastics. For example, the presentation of numerical data that must be scanned and compared with other items, probably could be presented best in graphic form. However, if data is presented in graphic form, users should also be able to look at the raw data as an option.

Display formats used for data entry should be compatible with formats used for CRT screen output, scanning, and review of the same data items. Where a display format perfected for data entry seems unsuited for data review, or vice versa, design a compromise format, taking into account the relative functional importance of data entry and data review.

Where data entry involves transcription from source documents, in a form-filling mode, the CRT screen layout should match (or be compatible with) paper forms. In a question-and-answer mode, the sequence of entry should match the data sequence in source documents. For inputs involving prompting, required entries should be prompted explicitly by means of displayed labels.

For data entries involving selection among displayed alternatives, implicit prompting can be provided by marking (brightening, inverse video, etc.) the selected alternative(s).

Implicit cues should be provided when using form-filling on CRT screens to supplement explicit labels. For example, underscores could be used to delineate each entry field. Field delineation cues should distinguish required (dashed underscore) from optional (dotted underscore) entries, and should indicate the maximum acceptable length of the entry. Where item length is variable, the user should not have to justify an entry either right or left, and should not have to remove any unused underscores.

Users should be supplied only with information that is essential for taking the next action or making a decision, and should not be overloaded with information. For example, if a user must be told how heavily loaded (backed up) a computer is so that a decision can be made whether or not to compile a program, the computer should not provide a number from 1 to 100, but rather a more meaningful message such as "Very Heavy," "Moderately Heavy," or "Light."

Most computer-provided explanations should be of the quick reference type. More detail can be provided by having a user key "HELP" or using a "HELP" function key. Complete information should be stored in a computer-based or document-based set of instructions. If it is available in the computer, a user should be able to access slightly more information by keying "HELP" a second time, and complete information (with examples where possible) by keying "HELP" a third time.

Ensure that the computer makes an appropriate and meaningful message response to each user command.

Not:

User command:	remove file-name
Computer message:	ready for next command

But:

User command:	remove file-name
Computer message:	file-name removed

Most messages and other text (prose) should be displayed using both upper and lowercase type. However, where attention-getting is needed, as in labels or titles, use uppercase only.

Not:	But:
ALL UPPER CASE TEXT	Normal reading is
IS HARDER TO READ	easier if
THAN A MIXTURE OF	the text is in
UPPER AND LOWERCASE	both upper and lowercase

Put frequently appearing messages in the same place on the screen. Frames should be about the same physical length so that users can depend on finding items in about the same spatial location. Important but infrequent events, such as error messages, may need some enhancement or highlighting to be recognized. Place such messages in the user's central field of view.

Labeling

Labeling is the act of placing a descriptive title, phrase, or word adjacent to a group of related items or information. Good labels provide a quick means of identification and can assist the user in rapidly scanning for an item of interest, or help ensure that an item is being entered in the proper field. Labels should be highlighted for ease of identification.

If a label is a code name or abbreviation, then that code must be meaningful to the users of the display. Where there is some doubt about the familiarity of the label, do not use it. Find a code that *is* meaningful or use complete words. Do not spend time forming labels from a restricted vocabulary. Develop meaningful labels using the complete set of alpha characters, numeric characters, and any special symbols that are meaningful to users. Develop labels that are clearly unique; having only slight differences between labels increases confusion.

Labels can be used to identify a single data item, a group of items, or an entire display. For example, each menu should have a label (title) that reflects the question for which an answer is sought. If a good, descriptive label for a menu cannot be easily determined, it could mean that the menu does not reflect a logical and consistent set of material and possibly should be redesigned.

Label every item. Do not assume that a user will be able to identify individual items because of past experience. Context plays a significant role. For example, 513-721-2345 may be recognized as a telephone number if it is seen in a telephone directory, but may not be recognized as such on an unformatted CRT screen.

Word labels distinctively for data entry fields so that they will not be readily confused with data entries, labeled control options, guidance messages, or other displayed material.

Where entry fields are distributed across a display, adopt a consistent format for relating labels to entry areas. For example, the label might always be to the left of the field; or the label might always be immediately above and left-justified with the beginning of the field. Such consistent practice will help the operator distinguish labels from data.

Where a dimensional unit ($, mph, km, etc.) is consistently associated with a particular data field, it should be part of the fixed label rather than have it entered by the user. Where data entry displays are crowded, adopt auxiliary coding to distinguish labels from data. The recommended standard is to display fixed, familiar labels in dim characters and data entries in bright characters.

Interframe Considerations

The following guidelines cover situations where a designer has a display made up of several frames in a series. For the most part, these guidelines reflect the idea that a designer should make decisions consistent with reducing the load on the user's memory.

A message should be available that provides explicit information to a user on how to move from one frame to another, or how to select a different display.

Not:	But:
Incorrect entry	Incorrect entry. Choose one of the following: Order Sell Change

The function to "roll" or "scroll" should refer to the displayed lines of information, *not* to the window. "Roll up 5 lines" should mean that the top 5 lines of data would disappear and 5 new lines would appear at the bottom. The window through which the data is viewed should remain fixed.

Items contained in a numbered list and described on "continue" pages (or rolled/scrolled lines) should be numbered relative to the first item on the first page. In addition, when the user is forced to scroll or page-turn through a multiframe display, the present maximum locations should be available on a visible portion of the screen.

Not:	But:
Page 3, Line 8	Line 63 of 157

In a hierarchy of frames with different possible paths through the series, a visible trail of the choices already made should be available to a user.

If at all possible, supply all relevant information on a particular topic on *one* display frame. Do not force users to remember data from one frame to the next, or to write down information that will be used for subsequent interactions with the computer.

Non-Textual Messages

These guidelines on data presentation refer primarily to non-textual information, including coded information such as part numbers, telephone numbers, scores on a series of tests, etc. The guidelines attempt to ensure that when information is presented on a screen, it will be directly usable. If codes are used, either as commands or data items, they should reflect the guidelines discussed in Chapter 16, Code Design.

Make use of illustrations to supplement explanations in text.

Strings of six or more alphanumeric characters that are not words should be displayed in groups of two, three, or four with a blank character between them. The grouping should be consistent for all strings of characters. If natural groupings of 2 to 5 characters exist, they should be used instead of artificial groupings (e.g., Social Security Number: 518-40-1087).

Not:	But:
ABBA423675A2	ABBA 423 675 A2
ABBD252389K4	ABBD 242 389 K4
ABCR862534M3	ABCR 862 534 M3
ABRG563487W4	ABRG 563 487 W4
ACGL190537S0	ACGL 190 537 S0

People seem to be able to scan for a certain item of information more quickly and accurately if a tabular format (above right) is provided.

Data items should be presented in some meaningful order if possible, for ease of scanning and identification. For example, put historical dates in chronological order.

Not:	But:
1066, 1975, 1492	1066 1914
1215, 1941, 1861	1215 1917
1914, 1865, 1945	1492 1941
1917	1861 1945
	1865 1975

Identical data should be presented to the user in a standard and consistent manner, despite what it looked like when originally input.

Not:	But:
3/5/50	3-5-50
03-05-50	
March 5, 1950	
1950: Fifth day, third month	

Number menu items starting with one, not zero. In counting, people start with one. Do not use meaningless leading zeros.

With words, alpha or alphanumeric codes, use vertically aligned lists with *left* justification. Sub-classification can be identified by indenting.

Not:	But:
CANCEL (CONTINUE,	CANCEL
END, HELP, SAVE)	CONTINUE
STATUS, SUBMIT	END
	HELP
	SAVE
	STATUS
	SUBMIT

With numeric codes use ***right*** justification, especially in tables.

Not:	But:
47321	47321
539	539
67	67
482645	482645

Avoid unnecessary punctuation. With abbreviations and acronyms it is usually best to omit punctuation. For example, Calif not Calif. and CPU not C.P.U.

Use complete words, not contractions or short forms of words, so that users are more likely to understand the sense of the message. For example, "not complete" rather than "incomplete," "will not" rather than "won't."

Be careful not to change traditionally accepted ("natural") formats.

Not:

Enter Social Security Number, address and name

But:

Enter name, address and Social Security Number

In addition, use standardized (familiar) formats when they exist. For example, some suggested standard data items for American civilian users are:

Telephone: (914) 444-0111

Time: HH:MM:SS, HH:MM, MM:SS.S

Date: MM/DD/YY

With these and all other messages, select a screen size that comfortably accommodates the items to be presented. Do not crowd items on a CRT display. Putting too much on the screen frequently leads to confusion, lengthened human processing time, and an increased error rate. If a larger screen cannot be justified, the layout of the screen should be carefully organized so that it does not confuse the user.

Organization

Organizing a screen can be done in a variety of ways, ranging from the use of arbitrary but consistent grouping, to the use of a systematically developed grouping based upon the guidelines discussed earlier in the *Grouping* section.

Probably the first consideration should be to standardize the placement of displayed information, including labels and data items, within a specific screen format, and to remain consistent. For example, functional areas should remain in the same relative display location on all frames. This means that certain areas of a screen should be reserved for certain types of information. One area may contain computer output only, another may be reserved for user input, a third for reference information, and another for "housekeeping" commands. Avoid breaking up the screen into too many partitions ("windows"). Numerous windows are confusing.

But using different screen areas in this way is only the first step. Some way should be used to help the user know that the screen has been systematically divided. There are several ways of helping a user understand and appreciate screen divisions. On a large uncluttered screen, each area (window) may be separated by blank spaces in sufficient quantity (3-5 rows and/or columns) to indicate clearly that unique information can be expected within the area surrounded by those blank spaces. On smaller and/or more cluttered screens, where extra unused spaces are not available, the user's understanding of the screen divisions, including different areas and/or items on the screen, can be assisted by any one of the following techniques: different colors; surrounding line types (solid, dashed, dotted, etc.); or different intensity levels.

When enumerating advantages, alternatives, disadvantages, etc., start each point on a new line.

Not:	But:
Advantages - easy to use, fast learning, minimum errors, high input rate	Advantages -- • easy to use • fast learning • minimum errors • high input rate

In presenting text on a small screen, there should be a maximum of 50 to 55 characters on each line. On larger screens, break up text into two (or more) columns of 30 to 35 characters per line. Separate columns by at least 5 spaces if the text is not right-justified and by 3 to 4 spaces if the text is right and left-justified.

Feedback to User

Feedback is critical in human/computer interactions, and comes most frequently from the CRT screen. One of the more frustrating aspects of interacting with a computer is sitting at the terminal and waiting for a response. Questions arise such as, "Is the computer still up?," "Is the terminal still connected to the computer?," "Did the computer understand my input?" A constantly renewed CRT message that indicates the computer is still working on the problem provides the user with the necessary assurance that everything is all right.

Highlighting

Highlighting refers to emphasizing some objects such as label, data item, title, or message on the screen. This emphasis can be accomplished by the following:

- increasing the intensity of an object (relative to others)
- displaying the item in a unique color
- flashing the object on and off (at two to three flashes per second with a flash duration of at least 50 msec)
- underlining the object
- presenting it in a different style or size font (if the object consists of alphanumeric characters)
- pointing to it with a noticeably large flashing object (such as an arrow)

- reversing the image of the object (such as going from light letters on a dark background to dark letters on a light background)
- making a shaded box around the item
- putting some graphics (such as a rectangle composed of a string of asterisks) around or near the object.

Whatever the specific technique, the main purpose of highlighting is to *attract attention.*

Highlighting can be used to provide feedback to the user. For example, when a user is presented with a menu containing a list of mutually exclusive options, and one option on the list is selected, those remaining in the list could be dimmed. This is particularly important when the screen is cluttered. Another common use of highlighting is to help the user detect an item of information. If a user is to perform an operation on some item on a display, highlight that item so that the user is rapidly directed to it.

Cursor

A movable cursor with distinctive visual features (shape, blink, etc.) should be used to designate position on a CRT screen. If position designation involves only selection among displayed alternatives, some form of highlighting might be used instead. The cursor should not obscure any other character displayed in the position designated by the cursor. Where fine accuracy of positioning is required (as in some forms of graphic interaction) the displayed cursor should include a point designation feature. For arbitrary position designation, the cursor control should permit both fast movement and accurate placement. Rough positioning should take no more than 0.5 seconds for a displacement of 8 to 12 inches on the display. In position designation, the displayed cursor should be stable and should remain where it is placed until moved by the user (or computer) to another position.

When data is entered primarily by designating position, as in selecting displayed alternatives, cursor placement should occur by a direct-pointing device (light pen) rather than by incremental stepping or slewing controls (keys, joystick, etc.). In selecting display alternatives, the acceptable area for cursor placement should be made as large as possible, including at least the area of the displayed label plus a half-character distance around the label.

When a data entry display initially appears, the cursor should automatically appear at the first character position of the first input field. In addition, display formats for data input should be designed to minimize user required actions for cursor movement from one entry field to the next. Sequential cursor positioning in predefined areas, such as displayed data entry fields, should occur through programmable tab keys.

Selector Pen

When designing selectable areas on the screen with a selector pen, keep the target as large as possible. Designate an area at least as large as the height and width of the alphanumeric character string so that any portion of that area will be the selectable area.

FORM DESIGN GUIDELINES

General

Forms are centrally important in most systems. If well designed, they aid the human/computer interaction. If poorly designed, they can take far longer than is desirable to complete and/or produce large numbers of errors. Designers may or may not be interested in the *time* people take to complete a form. For example, forms completed by people outside the system boundaries—by customers—may take considerable time to complete without having a negative impact on system performance. Customers complete these forms on their own time. The only potential negative effect on system performance is that if customers feel it takes too long to complete a form, they may not do it. Or they may do it hurriedly, making many mistakes. A well-designed form collects the necessary information in the shortest possible time and with the fewest errors.

The *accuracy* of information collected by having people fill in forms is important whether or not the information is collected within or outside the system boundaries. Mistakes made by people completing forms are as much the responsibility of the designer, as the form-filler. In fact, most errors can be attributed to designers.

A designer has the responsibility to develop forms that elicit accurate information in the shortest time. As Wright and Barnard (1975) pointed out, we are no longer bound by the traditional, arbitrary conventions for writing that have evolved over the past few hundred years. We now have the results of considerable research on many human performance related characteristics of form design. Much of these results already have been discussed. Other guidelines specifically related to form design are summarized in the following paragraphs as general rules or important considerations.

Instructions for Completing Forms

The instructions for some forms leave users confused. They cannot understand what is wanted. We will review some of the more common problems designers encounter when preparing instructions for using forms. Much of the following discussion is adapted from Wright and Barnard (1975).

General Considerations

In general, all instructions necessary to complete a form should appear *on the form.* The main exception to this rule is when the same individual fills in the same form over and over again. The instructions should include:

1. A brief description of why the form is used.
2. A sample of a completed form.
3. A complete list of valid codes to be used on the form (when codes are necessary).
4. Specific instructions for filling out the forms that cover all contingencies.
5. Specific instructions for transmitting the form.
6. Procedures for correcting errors.

Make the instructions as brief as possible and use clear, unambiguous words and sentences. Generally the instructions should follow the same sequence used for completing of the form. Caution the user, in turn, to complete the form in the same order as the instructions.

Placement of Instructions

Whenever possible, put all instructions on the front of a form. Place general instructions at the top of the form and specific instructions just before the corresponding item(s). Instructions that pertain to an action to be taken with the completed form should appear at the bottom of the form. Forms where instructions are presented and then have to be remembered until the appropriate section on the form is reached can generate numerous problems. In fact, Wright (1969) found that even the reading of a question between the reading of a set of instructions and making an entry is sufficient interruption to increase errors.

Error Correction Procedures

Spell out error correction procedures. This becomes especially important with machine-processed forms such as those used in mark-sense or optical-scan processes. Unless instructions are provided, the initiator may assume that he or she can simply cross out errors and put another mark in the correct column.

Testing Instructions

Always test instructions on a group of people representative of the group for whom the instructions are intended. Do not simply ask the people if they understand the instructions or if the instructions are clear; rather, ask them to perform using the instructions (with no help from the tester) and then evaluate their performance.

Additional Considerations

Other guidelines for instructions used with forms are discussed in the following paragraphs. Chapter 19 contains a more detailed discussion of some of these topics.

Headings

Appropriate headings can greatly benefit a reader (cf. Dooling and Lachman, 1971). There is good reason to expect similar benefits from the use of headings within forms. Headings appear to provide a context that assists understanding of both words and sentences. Some methods for displaying headings are better than others. For example, headings printed sideways to bracket several rows of questions all relating to the same topic, will be less effective than a heading written horizontally (Wright and Barnard, 1975).

Words

Familiar words are easier to read and remember (cf. Loftus, Freedman and Loftus, 1970). People also find it easier to think about and draw inferences from sentences using familiar words (Wason and Johnson-Laird, 1972). Familiarity seems to apply to abbreviations as well.

Wright and Bernard (1975) present examples where unfamiliar words are used on forms:

> *gainful occupation* when they mean *paid work,* or to *emoluments* when they mean *income.* One form had *obtains pecuniary advantage* rather than *gets money.* The notes accompanying a current rate demand have a section dealing with mixed *hereditaments* not connected to the public sewer. Yet hereditament is not even included as a dictionary entry in several pocket editions. (p. 215)

A designer should check words used on a form against their frequency of use using such sources as a study by Kucera and Francis (1967).

One difficulty with familiar words is that they may not mean the same thing to everyone. For example, "income" may mean income before any tax deductions to some people, to others it may mean after tax deductions.

Sentences

Short sentences seem to be more easily understood than long ones. Sentences with more than one clause, where the clauses are nested inside one another, are particularly difficult to understand.

One alternative to long sentences may be to convert them into a list. Wright and Barnard (1975) give an example of a long, unwieldly sentence that appeared in an instruction on a form:

> It should be noted that a rent allowance cannot be granted to a tenant of a local authority or to a person with a service tenancy or occupying a dwelling partly used for commercial or business purposes but he or she may qualify for a rate rebate.

Ignoring the fact that this was printed in all capital letters, a 47-word sentence is bound to be difficult to understand. They suggest that a sentence such as this might be more easily understood if written as a list:

> The following people cannot have a rent allowance but they qualify for a rate rebate:
>
> - tenants of a local authority
> - people with a service tenancy
> - people occupying a dwelling partly used for commerce or business.

Blumenthal (1966) and Hakes and Foss (1970) have shown that sentences with subordinate clauses are more easily understood if the clauses are introduced by relative pronouns (e.g., *which, that*) than if these pronouns are omitted. For example, people grasp the meaning of "The dog that the milkman found chased the cat" more readily than "The dog the milkman found chased the cat."

Active sentences are easier to understand and remember than the equivalent passive forms (Herriot, 1970; Greene, 1972; Barnard, 1974). For example, application forms should not instruct "The notes should be read" when they mean "Read the notes." Research suggests that the passive voice focuses attention on the agent (Ainsfeld and Klenbort, 1973) whereas the active is relatively neutral showing no strong differentiation between the two nouns.

Positive sentences are best. It generally has been found that sentences with negative elements are more difficult to understand than positive sentences (cf. Herriot, 1970; Greene, 1972). Words such as *not, except,* and *unless* all have negative elements. The form that said, "Do not delay returning this form simply because you do not know your insurance number" could as easily have said, "Return this form at once even if you do not know your insurance number" (Wright and Barnard, 1975).

Even when single negatives have a clear interpretation, performance will be better if an alternative affirmative can be used (Wason, 1961; Jones, 1966, 1968). People more easily understand the instruction to "Leave this box blank if you already receive a pension," than the negative version "Do not write in this box if you already receive a pension." Similarly, when people must answer Yes/No questions, evidence suggests that instructions to check, underline, or circle what *does* apply are more easily followed than instructions to delete what does *not* apply (Wright and Barnard, 1975).

Some of the studies examining the terms "more" and "less" have the greatest relevance to designing forms (Palermo, 1973; Wright and Barnard, 1975). These studies show that people most easily understand the term "more." Yet forms still tell applicants that they need not declare their interest from savings if the amount is *less* than $100.

Finally, designers should avoid double questions. Wright and Barnard (1975) pointed out that double questions such as, "Are you over 21 and under 65? Yes/No," cause difficulties for many people over 65 who appear to take each part of the question in turn. They want to answer "Yes" to over 21 and "No" to under 65. In one study, over one-third of the responses to this question were either incorrectly answered or not answered at all. The problem was solved by making it two separate questions.

Physical Characteristics

It does not take a Rembrandt to produce a thoroughly useful form. In fact, it is usually much better to spend time dealing with performance-related considerations than aesthetic ones. In short, having a truly operational form is better than having a pretty one. Designers should concentrate on performance issues and where possible, let others deal with nonperformance issues. If the designer can indicate precisely what he or she wants on the form and where, and can specify the physical characteristics of the form, then an artist or professional printer can take over and deliver a fine finished product.

Printers are governed by long established procedures and standards and precisely defined terms. If designers know the standards and terms, the chances of getting exactly what they want from a printer are very good.

Layout Aids

Form layout sheets for practically every imaginable application are available from equipment manufacturers, forms manufacturers, and printers. Do not overlook any of these sources of aids.

Many printers will provide a designer with a sample book showing various type sizes and styles. Another useful device that a printer can supply is a ruler. This is used in laying out typed forms to determine how many characters can be placed in a given space. The ruler also provides direct measurement in "points" for determining what point size type you will need.

Form Dimensions

If you have considerable freedom in determining the dimensions of a form, use a length and width that is standard in the printing industry. Figure 15-1 lists the standard lengths and widths for commonly used forms.

LENGTHS				
3-1/2	5-1/2	7	8-1/2	11

WIDTHS								
4-3/4	5-3/4	6-1/2	8	8-1/2	9-1/2	9-7/8	10-5/8	11
11-3/4	12	12-27/32	13-5/8	14-7/8	16	16-3/4	17-25/37	13-15/16

Figure 15-1. Standard Form Dimensions

Answer Space

By "answer space" we mean the space on the form that is designed to permit users to enter information. Both the size and location of answer spaces need to be considered. Answer spaces need to be big enough for a form user to comfortably enter all the necessary information.

In addition, if machines will print the entries (for example, a typewriter), both the vertical and horizontal spacing must conform to the machine's printing characteristics. Most vertical spacing for machines is based on the standards of 6, 8 and 10 lines per inch. In horizontal spacing for machine preparation, the requirements of the equipment will, of course, put some restrictions on the design of the form.

If entries are to be handwritten, then allow about 1/3 of an inch vertical height for each line. If vertical space is very tight on the form, this dimension can be reduced to 1/4 of an inch, but not smaller (Schaffer, 1980). In planning for horizontal spacing of handwritten entries, allow at least 1/6 to 1/8 of an inch per character or letter.

A second major consideration is that the answer space should be located relative to the question so that form users know exactly where to make each entry. A designer should put the answer space as close to the end of the question as possible. Large gaps between questions and answers are likely to lead to errors. Research has shown that even when people are simply copying information, the bigger the gap the greater the error (Conrad and Hull, 1967). These errors are most likely due to short-term memory limitations. Figure 15-2 shows some difficulties associated with matching questions and answers.

1. Do you use any part of the dwelling for commercial or business use? YES/NO

2. Do you pay rates directly to the city? YES/NO

3. Are you receiving supplementary benefits? YES/NO

4. Is your wife in paid employment? YES/NO

5. Are you an owner occupier? YES/NO

6. Are you related to landlord? YES/NO

7. Is the whole building let to you? YES/NO

8. Is the landlord responsible for internal repairs and decorations? YES/NO

Figure 15-2. Example of Form Layout with Large Gaps Between Questions and Answers. This increases the probability of error when people either forget answers or misalign questions and answers (Wright and Barnard, 1975).

A third consideration with answer spaces concerns whether or not to use boxes or tics. Some designers automatically provide answer boxes, requiring one character per box, without stopping to consider the effect on human performance. Several researchers (cf. Devoe, Eisenstadt and Brown, 1966; Barnard and Wright, 1976; Barnard, Wright and Wilcox, 1978; Apsey, 1978), suggest that subdivided answer spaces slow people filling out a form, and possibly may reduce the legibility of their responses. In fact, these studies show that writing one letter per box slowed both the writer and subsequent reader. Writing on lines where subdivisions are denoted by small vertical marks (ticks) may be the best technique for both writers and readers when delimiters are needed. But keep in mind that totally unconstrained printing (no boxes or ticks) may be the best approach in many systems (cf. Sternberg, 1978).

Leading

Look closely at this printed page or at any printed form. Notice that there is some white space between the bottom edge of the letters of one line and the top of the letters of the next line below it. This between-line spacing is called "leading," and on forms should always be 2 points or more (at least 1/36 inch). (Hartley and Burnhill, 1975).

Space Between Groups

Another kind of spacing that is important in forms design is the space between item groups and the space between headings and item groups. Many people feel that at least 12 points (1/6 of an inch) should be provided between groups of items to maintain the distinction among the groups. At least 6 points (1/12 inch) should be allowed on each side of a heading and 12 points between the head and the group.

For columns of item groups, designers should provide for item separations of at least 12 points (1/6 inch) if they do not use vertical separation lines. This spacing can be reduced slightly if a line is provided, but the 12-point minimum tends to be a good rule.

Margins

Margins of a form should provide at least 1/2 inch at the top, bottom, and sides unless a portion of the margin space is to be used for special purposes. An example of an exception would be a form designed for entries to be made in the margins for easy reference. Forms that will be bound may require considerably more than a 1/2 inch margin on the bound edge.

Typefaces

The most common typefaces used in forms design are: Gothic, Italic, and Roman. Gothic type is simple, squared off, and easy to read; it has no serifs. Italic types does have serifs and a distinctive slant causes it to stand out. Roman type has serifs but does not slant.

Gothic is one of the best all-round typefaces for readability. It can be compressed more than most types without losing its legibility. This is especially important if limited space requires the use of small 6-point type. Also, Gothic is an easy typeface to read when it is capitalized. This makes it desirable for headings. Italic type is hard to read in large amounts or long lines, but it is very good when a word or phrase needs to stand out. Roman style is considered best for large quantities of text. For this reason Roman type is commonly used for instruction sections of forms.

A given type style can be set in light, medium, or bold face. Bold refers to a heavy line type that stands out. It is another useful way to highlight a word or heading.

Type Size

The recommended type size for forms is eight to twelve points (Tinker, 1965; Poulton, 1969). A point is 1/72 inches or 0.0138 inches; twelve points is equal to about 1/6 inch. Smaller type sizes (six points or less) slow reading speed, and for some, of course, would not be read at all—resulting in omissions or errors. Table 15-1 is a useful reference for judging the readability of type sizes.

It has long been established that people read lowercase print more easily than uppercase (Starch, 1914; Tinker and Paterson, 1928; Tinker, 1965; Poulton, 1968). USE UPPERCASE PRINT ONLY TO CALL ATTENTION TO CERTAIN STATEMENTS. Otherwise use lowercase (with appropriate uppercase to begin sentences, of course) on forms.

Point size is 6, vertical spacing is 8 — A paragraph of running text illustrates different fonts and point sizes. Point size is 8, vertical spacing is 10 — A paragraph of running text illustrates different fonts and point sizes. Point size is 10, vertical spacing is 12 — A paragraph of running text illustrates different fonts and point sizes. Point size is 14, vertical spacing is 16 — A paragraph of running text illustrates different fonts and point sizes.	Point size is 18, vertical spacing is 20 — A paragraph of run... Point size is 24, vertical spacing is 26 — A paragraph of run...

Table 15-1. Point Size Comparisons

We have covered some of the most important physical characteristics of forms. The designer should be cautioned that many of the forms design guidelines just discussed have evolved over the years in the absence of good research. These general guidelines are probably acceptable for most applications; however, if a system demands an exceptionally high level of human performance, the designer may need to conduct studies to determine the best set of guidelines for that system.

FOR MORE INFORMATION

Cakir, A., Hart, D. J. and Stewart, T. F. M., *Visual Display Terminals,* New York: John Wiley and Sons, 1980.

Grover, D. (Ed.), *Visual Display Units and Their Application,* London: IPC Science and Technology Press Ltd., 1976.

16

CODE DESIGN

INTRODUCTION

This may be a good time to formally introduce myself: my name is Robert Bailey. My parents, as far as I know without the help of any coding research, devised a six-character code, "Robert," to go with the six-character code, "Bailey," the latter having been handed down through many generations. I found as I grew older, however, that many of my associates rejected the six-character code, "Robert," preferring the shorter code, "Bob."

One day I realized that there were some who even rejected my three-character code. To my auto insurance company, for example, I wasn't Bob or Robert; I was 950424F1130B. My driver's license code, a terribly long 15 characters, is B81811658863405. Blue Cross/Blue Shield also rejects the Robert and Bob codes and gave me one of their codes. I recently called the bank, and they wouldn't do business with me until they knew both my name code and my bank number code. Exxon, preferring my credit-card code, doesn't even know my name code.

For my employer's records, I found that I was no longer Robert Bailey but 270471. Further, I couldn't become 270471 until they verified that the government had me on file as 518-42-1887. The telephone company, not to be outdone, has given me a ten-character code which I am supposed to memorize. Aside from the phone number, the telephone company also has much more information, all in the form of codes, that must be directly or indirectly generated, processed, or updated to offer me telephone service (Table 16-1).

Table 16-1. Codes Associated with Customer's Telephone Service

Cable/Wire Number = 79	Low Binding Post = 105
Apparatus Number = 18	Cable/Wire Size = 404
Signaling Arrangement = SX	Building = 13
Type Set = HCK-BLK	Equipment Type = DLCID
Pair Number = 1511	Customer Code = 660
Building Suffix = A	Class of Service = A
Numeric Address = (201) 745-3621	Wire Center = 348
Unit = 3	Terminal Capacity = 24
Street Code = ABAB	Even/Odd/Both Indicator = E
Multiplicity = 3	Terminal Location Code = 16
Vert. Row = 130W - 020	Terminal ID Number = 650032-101
Class of Service = 2FR	Engineering Plate Number = 1475
Bunch Block Number = 31	Terminal Type = F
Section of Plant Length = 1562	Cable Number = 13
Complement Type = 3	USOC Code = CJJER
Connect tp Section of Plant = 136784	Distribution Frame Side = H
Facility Type = 2	Bridge Lifter Number = 190
Type of Plant = MUL	Section of Plant Number = 3381
Taper Code = 110100	Bay Relay Rack = 0103-02
Binder Color = 660	Distribution Frame Type = MDF
Wired Out of Limits = 1	Pair Present Condition = 13

CODES

Codes are supposed to be a shortened way of representing longer messages, and for that reason it is desirable to have each code convey as much information as possible. My name, Robert Bailey, has 12 characters. It is a very inefficient code because the 12 characters convey only one message—the identity of a single individual. We might, however, assume two other messages. The name Robert suggests that I am male, and both names together may suggest that I was born in the United States. Nevertheless, taking 12 characters to convey my identity does not use code efficiently. The post office's five-character ZIP code is an example of a more efficient code. The first character refers to a sectional center, the next two to a post office, and the last two to a delivery area.

As it is used here, a *code is a shortened representation of a word or group of words.* Codes are usually made up of all letters (alpha codes), all numbers (numeric codes) or a mixture of letters and numbers (alphanumeric codes). Some codes use special symbols that are not alpha or numeric—$, #, ;. Codes are commonly used to define people or items, such as equipment parts, locations, facilities, and so forth. Thus, rather than spell out California, we write "Cal"; rather than say that a person lives on the left side of the road, in the 23rd house from the corner, starting from where the brick barber shop is located, we say the person lives at 1444 Oakley Avenue; and rather than ask for the person who is 5'9," weighs 175 pounds, and has dark hair with hazel eyes, we ask for "Robert."

There are at least two reasons why words by themselves are not convenient. The first is that they are not necessarily unique, in situations where it is essential not to confuse one identification with another. There are many Adams, Baileys, and Smiths and even with the first and middle names added, there is no guarantee that two individuals will not have the same name. The second reason is that the full words can be very lengthy, especially if they must be unique. This is a handicap for human use, when the words must be read and copied by hand or typewriter. It is also a handicap for computer processing, particularly in systems characterized by large volume files.

Zipf (1935) observed that in a number of diverse languages, a word's frequency of use and its length were negatively correlated. Further, this relation appeared to be causal; as a word or phrase increased in frequency of use, it became shorter. Examples of this sort of word abbreviation (or code making) are plentiful in English. Consider for example, that *television* becomes the code *TV, cathode ray tube* becomes the code *CRT,* and *video display terminal* becomes *VDT.*

For numerous reasons, then, not the least of which is the advent of computers, communication takes place with information in a shortened form. There is no question that the use of codes has proliferated to the extent that virtually everybody uses them. Coded information can help a person communicate faster and more accurately.

TYPES OF CODES

There are two broad types of codes, *arbitrary* and *mnemonic.* Arbitrary codes aim to provide a unique identification. The numbers or letters of these codes hold little or no special significance to the user— examples are a Social Security Number, passport number or a ticket number. Mnemonic codes tend to convey information that is meaningful in some way to the user.

Arbitrary codes usually bear no direct relationship to the word or groups of words they are to represent. Most telephone numbers are arbitrary codes. Usually knowing a person's seven-digit telephone number or nine-digit Social Security Number tells little about the person.

Mnemonic codes are purposely designed to have some association with the word or groups of words they represent. Mnemonic codes frequently, though not always, consist of alpha characters and tend to be easier to learn and recall than arbitrary codes. A person's name is a mnemonic code and represents a shorthand way of referring to a person without a lengthy description of age, height, weight, type of nose, and color of hair. Many mnemonic codes are abbreviations of the words they represent—NJ for New Jersey, Chevy for Chevrolet or Fri. for Friday.

Certain arbitrary codes, even those that are all numeric, may become mnemonic codes if used frequently enough for an association to develop between the code and what it represents. Consider, for example, a system where each type of computer-detected error is assigned a specific error code. An omitted name may be error type 1, a misspelled name, error type 2; a wrong address, error type 3; and an incorrect telephone number, error type 4. After a relatively short period of time, the people working in this system will begin communicating by making statements like, "yesterday I corrected twenty-seven 3's and nineteen 4's." Translated, this means that yesterday 27 wrong address errors and 19 incorrect telephone numbers were corrected. In this case, *meaningless* arbitrary codes were converted to *meaningful* mnemonics.

CODING ERRORS

Errors made while using codes may be classified as *clerical* or *procedural.* Generally, clerical errors are related to code design, while procedural errors are more closely related to other aspects of system design. A procedural error occurs when a user selects and enters the wrong code. It can be an error related to:

1. The general principle of code usage, such as failure to encode a street address when operating procedures call for a code to be assigned

2. The information content of the message unit, such as entering the code for one equipment type when a different equipment type should have been entered

3. The formal setup of the message unit, such as entering the code in the wrong field.

A clerical error occurs when a user correctly selects the code, but incorrectly enters one or more characters while printing or keying. Clerical errors are not usually related to the information content of the code. Thus, the definition of clerical errors excludes omission of a whole code or substitution of one code for another. In general, when using codes about one out of every five errors is clerical (Bailey, 1975).

Character-level errors can be classified in many ways. Probably the most common classification scheme is shown in Table 16-2.

Table 16-2. A Common Classification Scheme for Character-level Errors

Category	*Example*	
	Correct	*Error*
Substitution	ABC	A*K*C
Transposition	ABC	A*CB*
Omission	ABC	AC
Addition	ABC	AB*B*C

A richer classification scheme, oriented toward the stage in cognitive processing where an error probably occurred, is shown in Table 16-3.

Table 16-3. A Classification Scheme for Character-level Errors Based on a Stage Model of Cognitive Processing (see Chapter 6)

Category	*Example*	
	Correct	*Error*
Perceptual		
Visual	ABCDE	AB*G*DE
Auditory	ABCDE	*H*BCDE
Intellectual and Memory		
Anticipation	ABCDE	AB*D*E A*C*CDE A*C*BDE
Perseveration	ABCDE	AB*B*DE AB*B*CDE ABC*A*DE
Movement Control		
Adjacent Key	ABCDE	*S*BCDE

This scheme suggests that character-level coding errors occur due to faulty processing in one of the three major cognitive stages: perceptual, intellectual, or movement control. Within each stage, different types of errors can occur. For example, in the intellectual (including memory) stage, it is suggested that there can be either anticipation or perseveration errors. Whether or not anticipation or perseveration errors actually do originate in the intellectual stage is still not clear. We have more confidence in identifying errors from the perceptual and movement control stages than the intellectual stage. Nevertheless, this classification scheme provides a more useful way to consider character-level coding errors than the older, purely descriptive method in Table 16-2. We will review briefly each of the categories.

Perceptual Errors

Perceptual errors probably occur while sensing, storing, or encoding stimulus information. We suggest the majority of errors in this stage will occur from misreading visual stimuli or mishearing auditory stimuli. Confusion matrices are the best available source of commonly confused characters. Numerous confusion matrices have been published showing which uppercase or lowercase English characters tend to be most frequently confused with one another. A list of visual confusions for uppercase letters and numbers taken from several studies is shown in Table 16-4 (Neisser and Weene, 1960; Owsowitz and Sweetland, 1965; Fisher, Monty and Glucksberg, 1969; Bailey, 1975).

Table 16-4. A Listing of Visual Confusions to Help Identify Perceptual Stage Errors

Response	*Stimulus*	*Response*	*Stimulus*
B	R	P	D
C	F	S	J
C	G	T	I
D	O	T	J
D	P	U	J
F	E	U	V
F	P	V	U
G	C	V	Y
G	Q	W	N
K	Z	X	Y
L	Z	Y	V
M	H	Y	X
N	H	1	I
O	C	2	Z
O	D	5	J
O	G	5	S
O	Q	6	G

The visual confusion "T for Y" is also an eligible combination, but is not included because it also qualifies as a movement control confusion. A list of auditory confusions taken from Conrad (1964) and Hull (1973) is shown in Table 16-5.

Table 16-5. A List of Auditory Confusions (Sound-alike Characters) to Help Identify Perceptual Stage Errors

A-J	C-E	E-V	H-8	N-A	T-2	4-5
A-K	C-P	F-S	I-R	O-A	V-3	6-8
A-L	C-T	F-X	I-4	P-Q	Z-7	
A-N	D-B	G-P	I-5	P-T	O-4	
A-O	D-E	G-Q	I-9	P-V	O-8	
B-C	D-T	G-T	J-2	Q-T	1-7	
B-D	D-V	G-U	K-N	Q-U	1-8	
B-E	E-G	G-V	L-O	Q-E	1-9	
B-G	E-P	H-S	L-R	R-4	3-8	
B-P	E-T	H-X	M-7	S-X	4-1	

Intellectual Errors

The second category in Table 16-3, intellectual, involves all the processes required for choosing a response, that is, translating perceptual information into action. Even though all three stages make use of long-term memory, the intellectual stage also makes considerable use of short-term memory. Intellectual errors are defined operationally as errors in the sequence in which required characters are entered. Two major types of errors are included—anticipation and perseveration.

Anticipations

Anticipation errors occur when a character is keyed before its proper time in a sequence. There are three subclasses of anticipation errors. In the first the erroneous—anticipated—character is one or possibly two spaces ahead of its position in the original stimulus. For example, ABCDE is keyed as AB*D*E, or ABCDE is keyed as AB*E*. The old classification scheme emphasized characters not entered between the last correct character and the anticipated character by calling this type of error an omission error. What was omitted was emphasized instead of what character was anticipated.

In the second subclass a character is again omitted, but the erroneous or anticipated character is correctly entered in its original character position—ABCDE but keyed as A*C*CDE.

A third subclass of anticipation errors is commonly known as character transpositions. Transposition errors exist generally, but not exclusively, when the positions of two adjacent characters are interchanged—ABCDE keyed as A*CB*DE. Exactly why transpositions occur in the human cognitive processing is still controversial. For example, Murdock and vomSaals (1967) argue that transpositions only occur after characters are stored, whereas Conrad (1965) believes that transpositions occur as responses are being executed.

Perseverations

Perseveration errors occur when a character is entered in its proper position and again in an incorrect position. There are three subclasses of perseveration errors. In the first subclass, the wrong character is entered immediately after the correct character. It is always a repeat of the first character (a "stutter") and an omission of a subsequent character—ABCDE keyed as AB*B*DE. In the second subclass, the persisting character again is entered immediately after the correct character. It is always a repeat of the first character, but without an omitted character. In fact, all of the original stimulus characters end up in their proper order in the response—ABCDE keyed as AB*B*CDE. The third subclass of perseveration errors involves a character perseverated in a character position at least two away from its original position. Again, all of the original stimulus characters are in their proper order, but not in their proper positions—ABCDE keyed as ABC*A*DE. Because an extra character shows up in the latter two subclasses of perseveration errors, these errors used to be termed "commission" or "addition" errors.

Movement Control

The third category in Table 16-3 is movement control errors. These errors occur on a keyboard when a subject attempts to depress a key for a properly perceived and translated response but hits the wrong key. These errors, by definition, result from typing a letter on the keyboard immediately adjacent to the one required by the stimulus—for example, by depressing a key to the left or right. A list of keyboard confusions is shown in Table 16-6. The character combinations of N for H, and U for J are also visual confusions. In addition, movement control errors may result, again by definition, from using the correct finger, but the wrong hand—should key an "E" but strikes an "I." We assume both of these error types occur after a stimulus is correctly perceived and the proper action is decided. In typing and keying tasks, these movement errors are commonly called "motor confusions."

ERROR CONTROL

Errors are costly; therefore, it is important to design codes that elicit the fewest errors. When a proofreader or computer detects errors, the point of origin is frequently obscure. This frequently requires costly and time-consuming analysis to determine the correct information. However, the most costly errors are usually those not detected; they can be proliferated in printouts and files and eventually could result in a general lack of integrity for the total system.

One of the best ways to eliminate coding errors comes through having well-designed codes. In addition, it is important to have adequately designed facilitators (instructions, performance aids, training) that support the use of codes.

Table 16-6. A Listing of Motor Confusions for a QWERTY Keyboard

Response	*Stimulus*	*Response*	*Stimulus*	*Response*	*Stimulus*
A	S	L	K	X	C
B	V	M	N	Y	T
B	N	N	B	Y	U
C	X	N	M	Z	X
C	V	O	I	1	2
D	S	O	P	2	1
D	F	P	O	2	3
E	W	Q	W	3	2
E	R	R	E	3	4
F	D	R	T	4	5
F	G	S	A	4	3
G	F	S	D	5	6
G	H	T	R	5	4
H	G	T	Y	6	7
H	J	U	Y	6	5
I	U	U	I	7	8
I	O	V	C	7	6
J	H	V	B	8	9
J	K	W	Q	8	7
K	J	W	E	9	0
K	L	X	Z	9	8
				0	9

Feedback on incorrect codes helps workers learn the correct codes. Feedback is most effective if given at the time of, or immediately after the error occurs. Thus, the best feedback comes from self-detecting errors or having them detected quickly by a computer. Keep in mind that if proofreading is the only checking of codes for errors a considerable delay in feedback to the user usually results.

Designers should know what general characteristics of codes enhance human performance. They will learn this by carefully considering issues related to the user, activity performed, sources of information, type of response required and type of format used. These general considerations and the specific guidelines that follow are derived from Sonntag (1971); Hodge and Field (1970); Jones and Munger (1968, 1969); Field, Hodge, Manley and Sonntag (1971); Hodge and Pennington (1973) and others.

USER CHARACTERISTICS

Codes should be designed for human use, and not to accommodate forms, software, or hardware. We have dealt with many user characteristics in great detail in earlier chapters. Other characteristics that are closely related to code use are reviewed below.

1. Users tend to make more errors working with codes that sound or look alike than with codes that do not. (Perceptual)
2. Users should make use of the code as soon after seeing or hearing it as possible. (Short-term memory)
3. If users have trouble speaking a code, they will have trouble rehearsing and remembering it. (Short-term memory)
4. Users prefer short codes (one to three characters). The longer the code, the greater user rate of error in copying tasks. Long strings of code characters become more manageable when they are grouped into three-character or four-character elements. (Short-term memory)
5. Users tend to make more errors at the end of short codes or to the right of center in longer codes. (Short-term memory)
6. Users work best with codes that are consistent with their expectations. Problems with retention may occur when codes are not mnemonics. (Long-term memory)
7. Users react more quickly and more accurately to familiar codes than to unfamiliar codes. (Long-term memory)
8. Users best recall codes when the context (environment) at the time of recall is similar to the context at the time the material was learned. (Long-term memory)
9. Users have more difficulty keying with the middle, and little fingers than with the index and ring fingers. (Movement control)

Except for activities where the same codes are frequently used and rely heavily on long-term memory, most of the critical design issues are associated with short-term memory. Short-term memory is used extensively in processing a vast number of look-up, copy, and keying tasks.

ACTIVITY CHARACTERISTICS

The activity helps to determine the type of codes. Frequently used codes should be committed to memory. For infrequently used codes, it is usually best to design a system where users look up codes in either a code book or a computer file. If users are employing a large number of different codes, and the code books or computer files are poorly designed and awkward to consult, users may attempt to guess the correct code and many errors are likely to result. In one telephone company study, users were correct only 73 percent of the time although they indicated that they thought they knew *all* the codes.

In some applications, imposing codes that are difficult to memorize may actually improve accuracy by requiring users to access code books rather than to rely on memory. This technique only should be used where accuracy is more important than performance rate. The fastest performance comes when people have memorized the codes.

RESPONSE CONSIDERATIONS

The design of an activity determines the way users will make *responses*. Responses may be verbal, printed, mark-sense, keyed, dialed, etc. Speaking is the most familiar and usually the easiest way of responding for most people. In computer systems some form of keying is very common. Certain types of codes elicit fewer responding errors. A small survey was made in one company to determine the frequency of different types of responses where codes are used. The results showed that in 12 work modules, 3 emphasized verbal responding, 8 emphasized handprinting, and 1 emphasized keying. A designer should determine the types of responses to be made with codes before developing the codes. Certain codes, for example, can be keyed with fewer errors than others.

FORMATS

Formatting, as applied to coding, refers to (a) how pages are formatted in a code book (b) how codes are formatted, and (c) how codes and code elements are arranged on a printed form or CRT screen. We will address guidelines related to the last two considerations later in this chapter.

A code dictionary or code book (whether printed or computer-based) should be formatted to reduce errors and to minimize time for use. Also, take into consideration the density of a page—performance appears to suffer with high density material.

Keep in mind that the format of a code book should be divided into two parts: one for encoding and the other for decoding. For example:

Encoding		*Decoding*	
Item	*Code*	*Code*	*Item*
Erasers	423	117	Pencils
Hole punch	605	183	Pens
Paper	321	321	Paper
Paper clips	719	412	Paper weight
Paper weight	412	423	Erasers
Pens	183	605	Hole punch
Pencils	117	719	Paper clips

GUIDELINES FOR CONSTRUCTING CODES

A designer should consider the following guidelines when developing a new set of codes. Many of these are adapted from a set of code design guidelines developed by the Bell Laboratories Common Language Department. If these considerations are appropriately applied, the probability of having code-related errors once the system is operational is reduced.

Secure Basic Information

Once the need for a code is established, the designer should secure the following information:

1. A description of the information to be coded
2. The purpose of the code
3. The total number of items the code is to encompass when the system is first operational
4. The growth potential of the coded items—the number of different items over the life of the system
5. The media on which the items are to appear—CRT, paper
6. The primary tasks performed by the people using the code
7. Accuracy requirements, including error detection and correction
8. The computer limitations that will be imposed on the code
9. Whether or not there is a code in use similar to the one to be designed and, if so, what that code is.

Establishing Rules

Some designers create code sets worthy of military intelligence. Some of these codes, unfortunately, are almost impossible to break without a code book. When designing systems, we should have just the opposite objective. Construct codes so that users, by knowing the rules for construction, can easily determine code meaning.

Develop a set of rules before designing a set of codes. The choice of rules used for developing a code set depend on various factors, including how and by whom the codes will be used. The rules, once clearly defined, should be followed consistently. A definite set of rules helps to provide users with a means of decoding——converting the code into the information it represents—and serves to facilitate both code learning and performance in use of codes. Encoding schemes that are not based on systematic and consistently applied rules may lead users to make incorrect code associations, thereby increasing the probability of error.

User Experience

Codes should be designed for the least skilled users within the expected population of code users. Codes that are designed for the newest users do not usually degrade the performance of more highly skilled personnel. Where the possibility exists that experienced users will have their performance penalized by codes designed for less experienced users, the designer should consider using two or possibly three different versions of the same code. In some computer systems, for example, less experienced people may prefer codes that are descriptive—"ed" for "edit", "sub" for "substitute", "wrt" for "write", whereas experienced users may prefer much briefer codes—"e," "s" or "w."

Preference vs. Performance

Most users eventually accept and use a set of codes as a second language, and become quite comfortable using even poorly produced codes. When this occurs, users sometimes have difficulty separating their preferences from their actual performance levels. Take user preference into consideration. However, give performance data higher priority if preference and performance do not agree. For example, many experienced users see no problem with using long codes, but performance tests consistently show that the proportion of errors increases as code length increases.

When asked to rank codes of 2 through 11 characters, users seem to show no preference for either short or long codes. Judgment of ease or difficulty of use seems to be based on other code characteristics. Performance data, on the other hand, give a clear indication that users have difficulty with codes that have seven or more characters. Users do rank those codes for which internal patterning is readily recognized as easier, which is consistent with performance data.

One telephone company study compared user code preferences and user performance. Three conclusions interest us.

1. Users showed high agreement about which codes were easy and which were difficult to use;

2. Performance data tended to support the idea that codes were *not* more difficult for different types of users;

3. Marked differences in performance found where different types of users used the same code sets appeared more related to the overall design of work modules rather than codes.

When users express dissatisfaction with a set of codes, usually it is because of inconsistent rules for code generation, lack of logic in codes (little relationship of code to meaning), difficult formatting requirements, or difficulties in finding codes in code dictionaries.

Most users believe that alpha codes are easier to work with than numeric or alphanumeric codes. Among the alpha codes, mnemonics tend to be rated higher than those which are non-mnemonic. On the other hand, performance data indicates that users make fewer errors with alpha codes only when the alpha codes convey special meaning, and the tasks for which they were used are complex.

The length and complexity of a dictionary definition appears to be associated with the judged difficulty or preferences for codes. Where the relationship between the code and its definition is not obvious, codes tend to be rated as more difficult. For example, the C911 code with a dictionary definition of "audible ring overflow" was rated as extremely difficult by all who had to use the code.

Uniqueness

The demand for code uniqueness means that codes close in meaning and auditorily or visually similar should not be used. A code should uniquely represent the element of information it defines. Keep in mind that where codes are verbally transferred there can be considerable confusion among those that sound alike. Failure to provide uniqueness leads to confusion and errors.

Meaningfulness

Meaningfulness refers to those attributes of a code which lead the user to readily associate the code with the item object, instruction, or action that the code represents. As indicated earlier, a meaningful code is known as a "mnemonic." Meaningfulness should be built into codes however and whenever possible.

Users tend to perceive codes more accurately and quickly if they are meaningful. Both complex and simple activities seem to benefit from meaningful codes. Characters that are grouped to provide meaningfulness, and that appear logical to the code user, elicit far fewer errors than characters that are drawn at random. Mnemonic alpha codes almost always produce better performance than numeric codes of the same length. Even in typing tasks, fewer errors occur when the text is meaningful than when codes made up of random letters are used. The reduction of errors when keying mnemonic codes appears related more to learning a code pattern than to learning the code's meaning. Certain alpha characters appear in combination much more frequently than others. Mnemonics also permit a significant decrease in keying errors where the task involves looking up a code in a code book and then keying.

The following attributes are related to the meaningfulness of codes:

1. Similarity to the English language. Part of the concept of meaningfulness is associated with the order of characters within a code

(cf. Howe, 1970). For example, a random code such as QVIL would be more subject to error than QUIL, since there is no general habit pattern in the English language using QV in sequence. While it is true that users can, over a period of time on the job, adapt to letter combinations that do not occur or occur infrequently in the English language, the design of code sets for the least experienced user dictates avoiding unusual combinations wherever possible. Codes that are pronounceable (i.e., rehearsable) tend to be more meaningful, more easily learned, and better remembered than those which are not pronounceable.

The material in Appendix C was prepared to assist designers in using common characters and character combinations. In each of the tables, the most frequently used letters, digrams, or trigrams are shown first, the second most frequent is second, etc. Designers should construct codes that contain frequently used characters and character combinations.

2. The number and type of different vocabularies—alpha, numeric, or mixed alphanumeric—used in constructing the code.

   All alpha codes such as "CRT" tend to be more meaningful than numeric or symbolic codes. Codes drawn from two vocabularies (for instance, the alphanumeric P38) are more meaningful than those drawn from three vocabularies (for example, the alphanumeric/symbolic MC*237$5).

3. The frequency of code use. Frequently used codes generally are considered more meaningful than moderately or infrequently used ones.

4. The context of code use. Codes that are placed consistently in the same location in a message tend to have higher meaning than those that appear randomly, since the location serves as an additional cue for code meaning. Codes that are part of a longer code phrase tend to have higher meaning than codes that stand alone, since cues from the total set contribute to the meaning of any one code in the set. This is similar to the idea that the meaning of a word may be more apparent in a sentence than when the word stands alone and out of context.

5. The similarity between the code and the full descriptor of the item, object, action, or instruction for which it stands. Codes that are similar to their basic words (abbreviations) that contain some of the same characters in the same order as they appear in the definition can be expected to have more meaning than those with characters that differ and/or appear in a different order.

6. The rule used to construct the code is known by users. Codes that are consistently derived by established rules tend to be more meaningful than codes with inconsistent or obscure design rules since the code user, when in doubt, tends to decode by application of the rule.

Alpha vs Numeric Characters

The two most popular character vocabularies from which codes are made are letters and numbers. Since it is usually easier to convey meaning with letter or alpha characters than with numbers, tasks requiring complex cognitive processing of a code should use alpha character codes as much as possible.

Frequently, alpha characters can be used to develop codes that have an easily associated meaning. Building meaningfulness into codes permits the code user to make maximum use of past learning. For many activities, meaningfulness may be the most critical characteristic in the code design.

Give numeric characters preference when designing codes for simple transfer tasks, for example, where a code is transferred from one form to another, or keyed from a form to a computer using a terminal. These are tasks where past learning may not be as critical. Numeric codes are also superior to random alpha codes in "listen-recall-write down" tasks as long as the user is not required to hold the code in memory for more than a few seconds (Conrad, 1957, 1959).

Gallagher (1974) pointed out that one advantage of using a combination of numbers and letters for codes is that they can be arranged in a pattern to mark off different fields without having to resort to other symbols such as hyphens or virgules. A code of the type:

B12A05P38

would create natural groups of B12, A05 and P38. It is likely, however, that the groups would be even better recognized if spaces were also used.

Hull (1975) compared the performance elicited by codes constructed using different code lengths, vocabularies and groupings. The codes that elicited significantly fewer errors were those that had the least characters (6 in this study), contained both letters and numbers and had no more than five characters in a group (e.g., 1 E2NVL or QVTEK 4). The most errors were elicited by eight-character codes made up of all uppercase letters that were ungrouped (e.g., JBDEBCJD). Codes that had eight ungrouped numeric characters (e.g., 64195897); and those that were constructed of nine alpha characters with consonants and vowels, grouped in a consonant-vowel-consonant format (e.g., DOL FIK TAG) elicited a moderate number of errors.

Meaningfulness and Frequency of Use

Where meaningfulness cannot be provided for all codes within a code set, preference should be given to those codes that will be used most. In a system that requires a large number of codes, an equal amount of meaning cannot be built into all codes. This problem has its origins in the high redundancy in the English language. For example, consider a situation where the words "subscriber," "substitute," and "suburban" all need to be encoded for a user. Probably the most meaningful code for all three is SUB. To keep the codes unique, the designer must either make the code longer or give up some of the meaningfulness. As a general rule, meaning should be stressed in the design of frequently used codes even if it means making them slightly longer.

Special Symbols

Codes should be composed of characters already in the user's vocabulary. Alpha and numeric characters are familiar to all workers. Even though many other symbols are available for building codes (e.g., $, !, #), most of them are not as commonly known by users. When special symbols are required, the most common ones should be given preference. They should have, wherever possible, the same functions or meanings that users would expect them to have—# means pound, ? means question, $ means dollar. When using a keyboard to input data, requiring a user to use the shift key for special symbols will tend to increase errors (Hammond, Long, Clark, Barnard and Morton, 1980).

Compatibility of Code Sets

The total set of different code types (all alpha, all numerics, etc.) developed for users of a new system should be taken into consideration. A tendency to independently make up different code types occurs when more than one designer develops a system. One designer may use only numerics, another only alphas, and still another may mix alphas and numerics with special symbols. Users tend to form habit patterns or preferences for the types of codes that they use most often. Where numeric characters dominate in a new system, more errors will be made against alpha characters than against numeric characters. Where alpha characters dominate, fewer errors tend to be made.

Where other factors do not provide a clear indication of the type of code to be used in a new system, consideration should be given to using the same type of codes already in use in the system being replaced. If that set of codes is not acceptable, a designer should at least ensure that the new code type does not severely conflict with codes being used in adjoining systems.

Limited Vocabulary

The vocabulary (characters used to make up codes) for a given code set should contain the fewest amount of different characters that are consistent with providing meaningfulness, unique coding and the ability to expand. The larger the vocabulary base from which a code is drawn, the more difficult it is to remember the characters that could be legally contained in a code. If the lowercase "L" and alpha "O" characters are not included, for example, and this is known by users, then symbols having the characteristics of a 1 and 0 have a better chance of being correctly interpreted as a one and zero.

If all codes are made up of only five alpha characters arranged in different ways, people and computers could immediately recognize any character other than these five as illegal. Codes made up using a single vocabulary (all alpha) are less subject to error than codes made up using multiple vocabularies (mixed alpha and numeric, or numeric and special symbols, etc.).

Code Length

Codes should be short, preferably containing six or fewer characters, consistent with providing meaningfulness, unique coding, and anticipated growth. Code length is one of the most important considerations in code design. In general, a long code is more difficult to use and elicits more errors than a short code. Users work best with codes of three to five characters. The more complex the task, the shorter the code should be. Studies clearly indicate that in complex tasks, errors increase substantially when codes contain more than six characters; in relatively simple tasks, errors increase rapidly when codes are longer than eight characters. There is at least one exception to the code length rule. One-character codes tend to elicit more errors—generally errors of omission—than do codes of two, three, or four characters.

Shorter codes elicit fewer handprinting and keying errors, and, as a bonus, require less memory storage in a computer. Even so, examples of long codes abound. Consider the New Jersey Division of Motor Vehicles code for a driver's license. It is *15* characters long. The New Jersey driver's license number, like many codes, is read, handprinted and keyed numerous times by a variety of different people, for example, by police when issuing a traffic ticket, court clerks while preparing dockets and collecting fines, banks and other places of business when trying to establish identification for cashing checks.

The number of possible 15-character combinations, with only the ten numbers zero through nine is 10^{15} or 1,000,000,000,000,000. But the New Jersey driver's license number also includes one character position for an alpha character, which increases even further the total number of possible 15-character combinations. When using all ten numeric characters in fourteen character positions and all twenty-six alpha characters in one character position, the number of possible different driver's license numbers is 2.6×10^{15} or 2,600,000,000,000,000.

New Jersey only has about four million licensed drivers, which means that a code of seven characters is all that is required. In fact, every person in the world, if we assume a population of about 5 billion, could be assigned his or her own number by using only a 10-character code. By using a 15-character code, unique numbers could be assigned to every man, woman, and child on earth, plus every person on over 700,000 other worlds with the same population as earth. The question is: From a human performance point of view, is it really necessary to use a 15-character code?

One of the first steps in developing any code set is to determine the code length. Once the length has been tentatively established, another early step is to determine how many codes of this length will be required for the life of the system. Depending on the nature of the system, the number of codes may range from two to a very large number. For example, only two codes would be required to respond to a computer's request to save a file, "yes" or "no." On the other hand, the number of unique Social Security Numbers required in the United States may someday exceed 500,000,000.

Once the code length and the total number of codes are determined, a third step is to consult either Table C-4 or Table C-5. These two tables show the minimum number of different characters necessary to develop a character vocabulary when the number of codes and the code length are known.

In Table C-4, for example, if the total number of codes required by the system is 20, and if each code is to contain only two characters, the number of different characters required to make these codes is five. Table C-5 is read in the same manner. For example, if the system requires 400,000 codes and each code is to contain four characters, the character vocabulary will require 26 different characters.

Tradeoffs may be required when selecting a character vocabulary. For example, if a system requires a total of 20,000 codes, and the original decision was to have a code length of four characters, the total number of different characters that would be required is twelve (Table C-5). The requirement of using twelve characters precludes using only numeric codes. The best performance may be achieved, depending on how the codes are used, by using codes that contain *five* characters (rather than *four* characters), and thus using only numeric characters.

Lengthening Codes Artificially

Since error potential increases as codes get longer, the addition of one or two character codes to enable computer detection of errors (e.g., using check digits) actually increases error potential. Cost tradeoffs between preventing errors and using the computer to detect errors should be considered before codes are artificially lengthened to provide for computer checks. The cost of computer detection of errors might exceed the cost of preventing the errors in the first place. Such tradeoffs should be carefully evaluated before doing something that is *guaranteed* to increase errors, for instance, artificially lengthening codes.

Growth Requirements for a Code Set

Few code sets remain the same size throughout their operational use. New equipment or modifications evolve, new instructions are written, and new locations and buildings are developed, all of which require expanding an existing code set. Designers must realistically estimate the requirements for new codes, and allow for expansion. When, for example, there is some doubt about whether to use a 6-character versus a 7-character code the best decision may be to go with the 6-character code, recognizing that another character can be added in 15 or 20 years (if needed). The costs for adding one character at some later date, may be far less than costs associated with detecting and correcting code-related errors over a 20-year period.

Visual and Auditory Confusion

As we pointed out earlier, characters that may be easily confused, either visually or auditorily, should be avoided. For example, confusion brought about by visual similarity of characters is very common for:

- The alpha character "O" and the numeric character "0"
- The alpha "I" and the numeric "1"

Visual and auditory confusion also can be a problem with entire codes. An example of similar visual codes is HAY and HAP, and an example of similar auditory codes is GOME and GOAN.

Visual confusions may occur even when copying from clear typescript, and to a considerable extent when copying from someone else's handprinting. The following information (Hull and Brown, 1975) show items (other then alpha O and zero, and 1 and I) highly likely to be confused when they are copied:

Z and 2
U,V and W
Y with 7 and 4
S with 5 and 8
G with 9 and 6
T with J and 7
Q with O and 2
D with 0

(adapted from "Reduction of copying errors with selected Alphanumeric subsets," by A.J. Hull and I.D. Brown. *Journal of Applied Psychology*, 1975, *60*, 231-237. Copyright © 1975 by the American Psychological Association. Reprinted by permission of the publisher and author.)

Hull (1976) suggests that the removal of O, I, Z, Y, S and G from an alphanumeric vocabulary would greatly reduce the incidence of letter-digit copying errors. In an all alpha vocabulary, U, Y, Q, and T or J, should probably not be used if it is possible to avoid them.

Studies using a noisy background to simulate poor listening conditions have shown that certain letters are easily confused with others. Gallagher (1974) summarized the results of Conrad (1964) for the 26-letter alphabet. Starting with the full alphabet, Gallagher totaled the stimulus errors in Conrad's matrix and estimated each successive letter with the highest number of errors. From the first set of totals this was C. The row and column in the matrix corresponding to C was then set to zero and the process repeated with C eliminated (see Figure 16-1). In the same way, Figure 16-2 summarizes the results of Hull (1973) for a combined vocabulary of alpha and numeric characters.

When a designer has full control over how codes will be visually presented, certain characters should *not* be used. However, when the designer

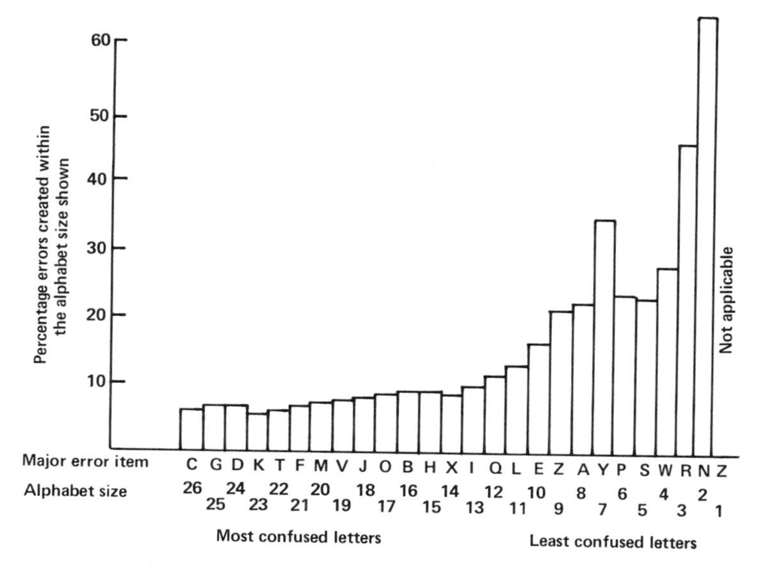

Figure 16-1 An Acoustic Confusion Ranking for Alpha Characters (adapted from a paper by C.C. Gallagher, in *Applied Ergonomics*, Volume 5, pp. 219-223, published by IPC Science and Technology Press Ltd., Guildford, Surrey, UK. The letter C created most errors. G created the most in an alphabet of 25, after C has been eliminated. D created most out of 24, with C and G eliminated; and so on through the alphabet.)

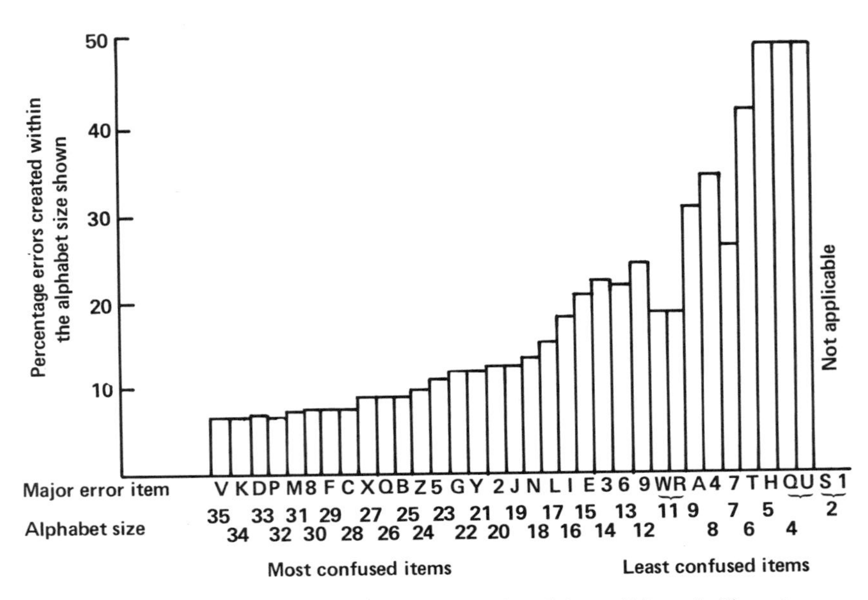

Figure 16-2. An Acoustic Confusion Ranking for Alpha and Numeric Characters Combined (adapted from a paper by C.C. Gallagher, in *Applied Ergonomics*, Volume 5, pp. 219-223, published by IPC Science and Technology Press Ltd., Guildford, Surrey, UK.)

does *not* have full control special techniques to aid in discrimination of perceptually similar (look alike) characters should be consistently employed. For example, ensuring that a "1" (one) is always perceived as a "1" and not a lowercase "L." This is the reason why it is so important to know how the codes will be used *before* they are constructed. There are different considerations depending on whether the codes will be handprinted, displayed on a CRT screen or read from a computer printout. Certain visual confusions are unique to each form of display.

For example, when handprinted, the alpha "Z" and the numeric "2" are often confused (Bailey, 1975; McArthur, 1965; Neisser and Weene, 1973). The accepted standard is to slash the "Z" (Ƶ) to differentiate it from the numeric "2." Figure 16-3 illustrates a proposed method for handprinting all 36 alphanumeric characters. Notice particularly the techniques used to differentiate between the I and 1, alpha O and numeric 0, S and 5 and Ƶ and 2.

There are certain characters made up of 5x7 or 7x9 dot matrices that are difficult to discriminate on many CRT screens. Those that tend to cause the most confusions are the alpha "O," alpha "D" and numeric zero; the alpha "I", lowercase "L" and numeric "1"; and the alpha "S" and numeric "5." If the CRT screen for the new system does not provide a high degree of differentiation for these and possibly some other characters, then the designer should avoid these characters in the new code set. If they are used, point out potential difficulties in reading these characters to users.

The same kinds of visual confusion problems result from using many computer printers. For example, one printer that is widely used presents a lowercase alpha "L" that is identical to the numeric "1." This same printer presents an alpha "O" that is slightly "fatter" than the numeric zero, but for all practical purposes, a user cannot tell the difference between an alpha "O" and numeric zero when reading a code. The same is true of lowercase "q" and "g."

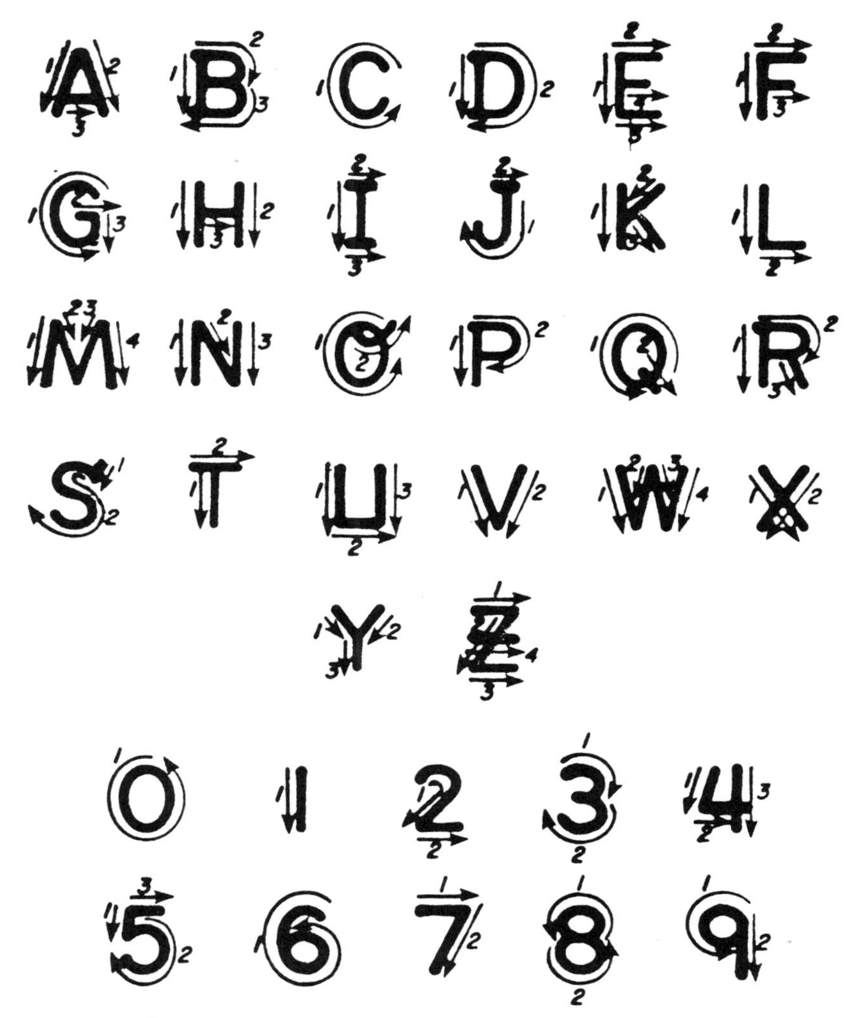

Figure 16-3. Proposed Handprinting Techniques to Aid in the Discrimination of Look-Alike Characters
(Note Particularly the special features of the alphas I, O, S and Z.)

We advise designers to carefully inspect devices (and even the hand printing of users) that will be generating characters in a new system, and try to identify characters that could cause confusions.

Consistency

Symbols should be used consistently for the same meaning or function. The use of like symbols for the same function should be consistent within the new system and with all other systems that a user is exposed to. For example, a designer could use either a dash or virgule as a separator in a code phrase—for example, month-day-year—but should not use a dash one time and a virgule the next: 11-9-39 versus 11/9/39.

Serial Position

It has been known for many years that errors are not equally distributed among all character positions. The characters in any sequence to suffer the highest incidence of errors varies with the length of the sequence in question. Where designers have options, they should use their knowledge of this serial position effect to reduce the negative effects of degraded performance. The "serial position" concept is discussed in more detail in Chapter 8.

Consider a task where users hear and then write down 6-character codes composed of two numeric (N) and four alpha (A) characters (NNAAAA). In this case most errors will probably occur in the fifth position. In general, with codes containing six to ten characters, the item second before the end of any sequence shows the highest error score (Howe, 1970).

There is another way of considering serial position. For example, in a simple "look up and write down" task using 3-character codes composed of two alphas and a numeric or two numerics and an alpha, there is usually more errors when the odd character is in the middle (e.g., code G2A had more errors than code GA2).

Formatting

Formatting of codes and code phrases should facilitate scanning for accuracy and completeness. Codes and code phrases with easily detected patterns are more meaningful and easier to scan. If users are accustomed to a certain pattern in codes, an inconsistent code pattern may create errors. For example, one group of codes observed in the same department had inconsistently structured code phrases involving dates. In one instance the date was *bracketed* by alpha characters: X10-10BET; in a second, the date was *followed* by an alpha character: 10-11-CA; and in a third, the alpha character prefixed the date X10-7. If these phrases had been arranged in a consistent pattern, users probably would have made fewer errors.

Code phrase format should follow the order in which the user will need the information. Codes that belong together to form a phrase or discrete part of a message should be grouped in a consistent manner. In addition, instructions should be organized so that they precisely match the sequence of codes in a code phrase. For example, the first instruction should relate to the first code, second instruction, the second code, etc. Code users frequently complain that when instructions do not follow this same order the users tend to misinterpret the instruction or miss a part of it.

Evidence suggests that formatting can affect the ability of users to detect their own errors. For example, a format that has exactly six spaces provided for a six-character code, would make it relatively easy to detect when only five characters are entered. Formatting also should facilitate scanning an output for error. When using a properly designed format, a user should be able to easily scan for errors in a check for internal consistency, for example, to ensure that a white cord is provided for a white phone.

Decisions on code formatting should take into account that the probability of error is reduced when related elements of information are grouped in one location. For example, in one work module, two of the three code elements related to closing an order were placed at the top of the message, and the third element was placed at the bottom. Thirteen percent of the coding errors made in this work module were omissions of the code at the bottom. In work modules where all three elements were located at the top of the message, no errors of omission were observed.

Grouping

When they set out to remember a series of numbers or letters, people break the series up into natural groups which enable them to remember the codes better. Grouping codes in a pattern encourages the best performance. A designer or user can do the grouping. When left to users, grouping may be synonymous with syllabication, such as dividing the code ALTECEP into AL, TE, and CEP. In other cases, patterning is associated with redundancy, such as mentally grouping 4LLL4 to three Ls between 4s. Users may also break codes by associating the parts with specific meanings, such as remembering BISCUS as BIS for "business" and CUS for "customer." There is no question that grouping codes into patterns facilitates remembering, but the ways of grouping vary widely with individuals. One telephone company study found that when a given grouping technique was defined and people instructed on that technique, those with instruction made significantly fewer coding errors than those without instruction.

It is certainly preferable for a designer to group codes rather than the user. The designer should take advantage of available information on grouping. For example, the familiar telephone number 2019816425 is a long code (actually a code phrase) containing three distinct codes. The first code (201) is the area code, the second code (981) a district within the area, and the third (6425) is a subscriber's number. Breaking code phrases into the individual codes facilitates the use of both the short-term and long-term memories. The telephone number, for example, is generally broken up as (201) 981-6425.

It is almost always preferable to use short codes, rather than long codes or code phrases. A code *phrase* is where two or more codes are used together. Even when symbols are used to separate individual codes there still can be confusion.

A considerable amount of work has been carried out in the area of grouping (cf. Klemmer, 1969). The most natural subgroup size is three or four units. Thorpe and Rowland (1965) report on a study where people were asked to speak codes containing nine numbers. They found that about 90 percent of people used three groups of three. At the same time the authors showed that the people who did not use the most common natural groups had many more errors.

Gallagher (1974) reports that where long lists of series are copied, more errors result when they vary in structure. For example, in the following two lists:

(a)	(b)
68 - 51381	685 - 138 - 1
924 - 1861	924 - 186 - 1
4 - 5921 -37	459 - 213 - 7

a list of type (a) will produce more errors than (b) (Ryan, 1969. Courtesy *Quarterly Journal of Experimental Psychology.*).

Abbreviations

Code design frequently requires rules for abbreviations since abbreviations are considered to be mnemonic codes. There are many methods for creating abbreviations. They range from the arbitrary elimination of characters, for instance, eliminating all vowels, to complex methods of reducing only certain syllables. We presented a set of rules for abbreviating computer commands in Chapter 14. With each method, the degree of meaning retained in the abbreviated term is different. The following guidelines may also aid designers create standard abbreviations for words or word combinations.

When abbreviating a single word:

- Determine the number of letters required in the abbreviated term.
- Remove endings such as -ed, -es, -er, and -ing from the word to be abbreviated. (Make sure that they are suffixes, as in "waxing," and not an integral part of the word, as in "spring.")
- Choose the first letter and last consonant of the word to be abbreviated as the first and last letter of the abbreviation. For example, Boulevard

would be B--D.

- Fill the remaining spaces with the consonants appearing in the word in the order in which they appear. Avoid the use of double consonants in the abbreviation. For example, a five-character code for boulevard would be BLVRD; and four character code, BLVD, and a three character code, BLD.

- If there are insufficient consonants, use the first vowel in the order in which it appears in the word. For example, a six-character code for boulevard would be BOLVRD.

The following guideline could be applied to abbreviating two-word groups. Divide the letters required in half and take half from the first word and half from the second word in the manner described for single words. If either word lacks sufficient consonants, use those in the longer word to make up the difference. For example, PROGRAM TEST would be PGRM TST.

Another guideline applies to abbreviating word groups containing three or more words. In general, when a term having three or more words must be abbreviated, it is best to use the first letter of each word. For example, Central Processing Unit would be CPU, and as soon as possible would be ASAP. If additional letters are required to improve readability, they should be taken from the last word in the manner described for a single word abbreviation. For example, PROGRAM RUN TEST would be PR TST.

Exceptions should be made only when special significance is achieved in selecting one letter or group of letters over another. These guidelines are meant only for the development of new abbreviations and should not be applied to existing standard abbreviations.

Abbreviations are also used in systems where a user needs to retrieve coded data when the codes are not known. This situation can be encountered when a customer has forgotten an account number or lost a credit card. The usual solution is to provide an alphabetical listing of the entire file, with code numbers, so that a person can look up the name and find the lost number. If the volume of such inquiries is high, however, the time required can defeat the whole purpose of using a computer for the system in the first place.

Wooldridge (1974) suggests that examples of systems with computer retrieval on narrative keys include medical and scientific abstract services, where the user wants to find all known references on the basis of key words; law enforcement systems using an alias or misspelled name, or phonetically spelled name, to locate possible matches; and musical royalty systems, where the match must be done on the basis of a song title. In most such systems, all approximations to the key are printed out for human inspection and decision making.

The Soundex system is one method of coding when the key is a proper name. The result is an abbreviation (of sorts) with one alpha character and three numerics, derived as follows:

1. Retain the first letter of the name.
2. Drop all letters A, E, I, O, U, W, H, and Y and one each of double letters.
3. For the next three letters remaining, assign the following numbers:

   B, F, P, V: 1
   C, G, J, K, Q, S, X, Z: 2
   D, T: 3
   L: 4
   M, N: 5
   R: 6

4. If less than three consonants are left, fill up with 0.
5. Ignore adjacent equivalent letters with the same number.

   Examples: MURPHY becomes M610
   WARRICK becomes W620 (RR treated as one letter; CK treated as one letter)
   ANDERS, ANDERSON and AMITER all become A536

Similar-sounding names will thus have similar codes. The Soundex system is only one of many coding systems for dealing with this type of narrative material.

17

SUPPORTING HUMAN PERFORMANCE

DOSAGE LEVEL EXAMPLE

Drug company researchers often face a problem that is very similar to one encountered by systems designers. Their problem is how to pinpoint the dosage of a drug at which an expected response does or does not occur (Dunnett, 1972). Antibiotics should be strong enough to destroy bacteria, but not poison the person who uses them. Pain killers should relieve headaches, but not induce palpitations.

Determining the proper dose level of a drug requires precise information about the effects of various quantities of that drug. Drug researchers first attempt to determine the amount of the drug that will produce a specific effect, i.e., they determine the drug's standard of potency. To establish this standard the researchers inject test animals (e.g., mice) of similar size and heredity with a known concentration of the drug being studied. A widely used, though arbitrary, choice is that the standard will be set as that concentration which will kill, on the average, half the animals tested.

For example, researchers may inject an amount of the drug *curare* into test animals and record whether a response (death) occurs in a given time period, say, 30 minutes. If they have chosen a dose that is too large, all (or almost all) the animals will die. If the dose is too small, few or none of the animals will die. The important question is how can they find the dose that, if increased, causes more than half of the animals, to die and, if decreased, causes fewer than half to die? The task would not be very difficult if all animals behaved in the same way, but, even in carefully selected animals, the amount of a drug required to bring about a response differs greatly from animal to animal.

One way to accomplish this goal is to test the same number of animals at each of a variety of dose levels (say, 1, 2, 4, and 8 mg). Table 17-1 shows a set of outcomes from such a procedure. Five animals were tested at each of four dosage levels, with the outcomes as shown. Twenty animals were required.

Table 17-1. Outcomes for Twenty Animals at Four Dosage Levels
(X Means Death; O Means Survival)

Dosage Levels	Outcomes				
8 mg	X	X	X	X	X
4 mg	X	X	O	X	X
2 mg	O	O	X	X	O
1 mg	O	O	O	O	O

One disadvantage of this procedure, other than it requires such a large number of animals, is if all these dosages are below the threshold of all the animals, the researchers will have learned nothing about the location of the threshold for these animals except that it is greater than the largest dose given, in this case, 8 mg. Or, if all animals at all levels respond, the researchers will learn only that the threshold is below the smallest dose, 1 mg.

Another, more efficient way of arriving at the proper dosage is to administer the drug to one animal and, depending on the response of the animal, increase or decrease the dose to the next animal. If the animal tested does not respond, the researcher increases the dose for the next animal. If the animal *does* respond, the researcher decreases the dose for the next animal.

For example, assume that the doses are 1, 2, 4, and 8 mg. The researchers begin by testing one animal at one of these levels, say, 4 mg. If the animal survives, a second animal is given the next higher dose of 8 mg. If the second animal does not survive, a third animal is given 4 mg (the amount given to the first animal). If this animal does not survive, the fourth animal is given 2 mg. In all, six animals are tested following this rule. If none of the 2 mg, and two-thirds of the 4 mg doses produce a response, the researchers are led to believe that the average threshold is somewhere between 2 mg and 4 mg. By using this strategy, they can obtain about as much information from six animals as they could from twenty animals using the strategy outlined in Table 17-1.

DETERMINING PERFORMANCE SUPPORT REQUIREMENTS

The problem of establishing dose levels closely resembles the system design problem of determining the proper "dosage" (amount and mixture) of instructions, performance aids, and training to include in a system. Too little tends to degrade human performance while too much wastes time, money, and other resources in their development. In one system, designers developed materials that required over five feet of shelf storage space. Little of the material was used to help human performance in that system. Ideally, the "dose" for each particular user will be only what is necessary for that user.

Selecting People

At one time, acceptable levels of human performance could be achieved simply by selecting a group of users with an adequate existing set of skills and knowledge. Most of the skills needed to perform activities already existed in a large part of the adult population. Work was simple and tasks were easy. Working on many assembly lines required a person to perform tasks that were readily understandable (within a few minutes) and relatively easy to carry out. Until recently designers have not had to worry too much about the lack of performance aids, poor training, or foggy instructions.

Facilitating Performance

Good written instructions, performance aids and training help to *facilitate* human performance. Selecting the appropriate people to work in a system is still important, but today's systems also require greater emphasis on perfecting *interfaces* and the performance *facilitators*. Both interfaces and facilitators should always be as well designed as possible. It is not acceptable for some to be better than others. With facilitators, however, there is the additional question of deciding how much of each type to develop.

The designer has almost total control over three performance facilitators: performance aids, written instructions, and training. One of the most basic differences between performance aids, detailed instructions, and training is the time that elapses between when information is presented and when the performance takes place. With a performance aid, for example, the user reads (or hears) the aid and then almost immediately begins performing. Also, written instructions tend to be used in a relatively short time after reading. With training, however, the information may be presented in January and not used until July.

In the development of every system, decisions must be made concerning when to use written instructions, performance aids, or training to produce an acceptable level of human performance. Some systems, usually through default, rely almost totally on verbal instructions passed from designer to supervisor to user, then from one user to the next, with little thought given to more efficient means. Designers who use this approach seem to assume that *all* essential information can and will trickle down from the designer to a user or set of users. Some designers even believe that once users finally figure out the intricacies of the software or hardware, the users will be able to prepare their own instructions, performance aids, and training materials. Cost-effective, reliable human performance cannot be obtained in this way, either in the short term or over the life of the system — particularly in a system that has a regular turnover of users.

DEVELOPING PERFORMANCE FACILITATORS

The design and development of human performance facilitators should be the responsibility of designers, *not* users. An acceptable level of *system* performance frequently depends on achieving an acceptable level of human performance. If the system designer assumes no responsibility (gives up control) for a major portion of the total system, how can he or she ever hope to achieve acceptable performance levels? Effective designers seek out and maintain control over as many critical system components as possible. They know that the system will be successful only if all major components are given adequate consideration. If the weakest component is human performance, and this performance causes a number of errors, user dissatisfaction, or high costs for training, then all the good design work in other areas could and usually does go unheralded and unappreciated.

Assuming that the *basic* human performance design decisions are adequate and the *interfaces* are well designed and developed, the only other major human performance consideration concerns the *facilitators:* instructions, performance aids, and training materials. Although many systems rely on all three methods to achieve long lasting and acceptable human performance levels, many successful systems use only one or two of these methods. Designers must decide the proper mix or balance. Too many performance aids and too few training materials may cause user confusion while too much reliance on training and very limited use of performance aids could result in costly development of training materials that will never be used.

In earlier approaches to system development, the general development sequence was design, document, and develop training. Performance aids if developed at all, were usually developed once the system was operational by users who were working with the system. Using this sequence, designers did not have the opportunity to decide whether training or instructions or performance aids were more appropriate for obtaining acceptable human performance.

This approach frequently resulted in the production of manuals overloaded with useless information that was not referenced on the job, and training courses that contained information that could be conveyed to workers in less expensive and less time-consuming ways. Often, written instructions and training courses covered the same tasks, with the only difference being one of emphasis, because they were developed by different designers. On the job, users found it so difficult to locate data in the mass of unnecessary information that they often dealt with these materials by storing them away and forgetting about them. When the information in these documents was needed, workers either asked someone for answers or guessed at their own solutions, rather than search for the answers elsewhere. This often contributed to human performance degradations.

In the development of some systems, large volumes of written instructions are still produced (usually *after* the software and/or hardware are ready for use). Designers who develop excessive wordy instructions in this way usually take little satisfaction in preparing the written materials, and users take even less satisfaction from having to read them. But "everyone knows" that all systems must have user manuals, so lengthy and frequently worthless manuals continue to be prepared.

There are few guidelines to help a designer decide when to use instructions, performance aids, and training. The best decisions are made when as much as possible is known about potential users, the activity to be performed, and the context in which the performance will take place.

After developing a minimal set of facilitators, the designer can test and, if necessary, revise the materials and then test the revised set. For this process to be most effective, the designer should begin by testing the smallest set of materials that could be considered adequate. If the user, who must be representative of the user population, is unable to perform adequately, the designer then can add more material to the set. If the user performs adequately, then a second person can try the material. If the second person does well, this may suggest that the materials are close to adequate. If the person does not do well, then this may mean that the materials are inadequate in certain areas. Revisions should be made and another test conducted.

One criterion of success could be that two (or three) people in a row do well with the materials. However, if the first person does well this may mean that the materials have been over-developed. To determine if this is the case, the second user should be given fewer materials. If the second person still does well, then the materials could be reduced even more. Using this process, the designer can find the optimum level of both quantity and quality for a set of performance facilitators. The users of the materials will benefit, and the designer will gain valuable feedback that he or she can use in developing future work modules.

How do designers who prepare instructions, develop performance aids and training, and write personnel-selection information decide what role each of these facilitator options will play in achieving optimum human performance? Making such a decision involves answering such questions as:

1. Should experienced personnel be selected, or should less experienced personnel be trained?
2. Will training be enough, or will a performance aid be needed?
3. Are step-by-step instructions needed even after training?

BALANCING FACILITATORS

Problems encountered in the past few years have clearly pointed out the need for a balanced use of facilitators that would produce more efficient and effective human performance. We will discuss a method proposed to balance the use of facilitators. It is based on approaches proposed by Wulff and Berry (1962), Harless (1969), Mager (1973), Joyce, Chenoff, Mulligan and Mallory (1973), Lineberry (1977) and Tannenbaum (1977). This approach asks designers to determine what means they think should be used to best facilitate human performance. This requires that each identified human activity be assigned the best means for facilitating its accomplishment.

To determine the best means to facilitate performance, the designer should be familiar with the factors involved in selecting the possible options of instructions, performance aids, training materials, and personnel selection. Then he or she should complete a systematic tradeoff of options for each work module. The following sections deal with these options.

SELECTING PERSONNEL

An obvious and often used way to achieve acceptable human performance is to select people who already know how to do the work. Personnel selection may be the *only* means used to facilitate performance. The availability of such people may make the development of training, instructions, and performance aids unnecessary. Generally, a designer can develop work modules at any level desired. If the only facilitator to be used is to select people who already have the necessary skill, then the work modules should be developed to reflect this decision.

The following are circumstances in which personnel selection may be the best option for facilitating performance:

1. An abundance of skilled people are available.
2. Skilled people are available and only a small number of people are needed to perform the activity, or the activity only will be performed for a short time.

3. The activity is so complex that developing and maintaining the facilitators would be more costly than assigning available people from other jobs with the necessary skills.

4. Skilled people are available and there are good reasons to start operating a system before other performance facilitators are developed.

DEVELOPING PERFORMANCE AIDS

Performance aids are devices that store information for immediate use.

They may be written or computer-based. A couple of familiar examples of performance aids are fingertip telephone indexes and frequently called telephone number lists placed under glass desktops. Sets of instructions also store information, but the information in a performance aid is usually characterized by having less to read and a quick-access format. There is no clear dividing line between performance aids and other types of instructions; in fact, a performance aid is a special type of written instruction.

The designer should consider opportunities to develop performance aids *before* looking for opportunities to develop lengthy instructions or training. This bias in favor of performance aids whenever possible is explained by the benefits that their use provides.

1. Performance aids are not as subject to problems of "forgetting" as are training or other types of written instructions.

2. Performance aids generally cost less to develop than instructions or training, and can be developed more quickly.

3. Performance aids are easier to revise when procedures change than lengthy instructions and the learning already established by training.

4. Performance aids generally shorten training time even when both are required.

To summarize, designers should use performance aids in preference to training or instructions whenever possible because they tend to deliver desired performance more reliably, cost less, shorten training time and are easier to revise.

It is usually best to develop a performance aid if:

1. Moderate speed and high accuracy are required at the time of performance. For example, accessing a frequently-called-telephone-number list before making a call.

2. Long procedures are required and there is much information to remember. The performance aid can provide "key words" or "pegs" for memory.

3. Written assistance is needed to make a decision.

4. Conversion of information is required, such as from yards to meters or square feet to square yards. A performance aid that lists conversions can be posted or carried in a pocket or purse.

5. The consequences of error are serious. For example, consider the performance of the crew of a commercial airliner in conducting various preflight routines. The possible consequence of degraded performance tends to rule out trusting the accurate performance of this task to memory.

6. The amount of behavior is large. For example, conducting a review of a new system to determine its adequacy. The amount of behavior involved in such a review indicates that the reviewer should be guided by a performance aid.

7. The behavior is performed infrequently. For example, a user is required to submit a report on cost effectiveness of various programs to management every other year. The infrequency of this performance points to the use of a performance aid for its accomplishment.

8. The performance methods are likely to change. For example, a means must be developed for users in a temporary system that is to be modernized with automated equipment next year. Adequate performance with the current equipment is required now, but the changes in performance anticipated due to automation may make performance aids the best approach.

9. Limited time is available. For example, people are hired on a short-term basis during peak sales periods to take orders from customers over the telephone. The time available for eliciting acceptable human performance indicates use of performance aids.

10. The required information is "hidden" in other system documents. Performance aids provide a way to summarize detailed information. A page number or command (for computer-based aids) can be included to find where more information is available.

DEVELOPING WRITTEN INSTRUCTIONS

Instructions may be contained in practices, user manuals, handbooks or a computer. The main question a designer must determine is whether or not a set of detailed instructions should be developed at all. Decisions at this point are not concerned with the form or format of the instructions.

It is best to develop instructions when:

1. A full description of the work is required (possibly including background information or information on where the work fits into the larger system).
2. Detailed information is required for activities performed infrequently, where retention of training information is unlikely.
3. Step-by-step guidance along with detailed reference information (such as that ordinarily provided to users in handbooks, formula tables, or computer programs) is required.
4. Complex visual presentations with detailed explanations are required; for example, graphs, flowcharts, schematics, and drawings.
5. Cross-referencing from one set of instructions to another is required, that is, information from one source must be matched with information from another source.
6. The activity is so complex that it is too difficult to learn and remember all the required information.

DEVELOPING TRAINING

There are many situations where performance should be based on learning situations that preclude the use of detailed instructions or performance aids. These situations are best identified by considering the following factors that indicate that providing training will help to achieve desired performance.

Training development is best when:

1. Skill levels must be attained and demonstrated before performance begins. For example, telephone installers should know their jobs *before* they try to install customers' telephones. In such cases, training is usually necessary.
2. On-the-job speed is so critical that there is no time to use performance aids or instructions. For certain critical and emergency activities, it is not desirable for workers to lose time looking through written material. It would be totally inappropriate, for example, for a member of a rescue squad to be reading about ways to stop bleeding while a victim lies suffering on the ground.
3. Activities are too complex to perform without training.
4. Workers need preparation—explanations, demonstrations, practice—to adequately use performance aids or instructions.
5. Training is the quickest way for users to achieve and maintain an acceptable performance level.

6. Training that is already available for other purposes will produce the desired performance if supplemented.

7. It is more cost-effective to train people who do not have the necessary skills than it is to select people who do.

8. Skills and knowledge constantly must be enhanced or updated using refresher training courses.

9. The performance is very frequent. For example, a user must conduct an inspection of computers on an average of three times per shift. The frequency of these inspections tends to support the use of training rather than performance aids or instructions.

10. The required performance is hard to communicate in written form.

11. The use of performance aids is impractical or inappropriate. For example, performance aids are usually not too useful for users who must perform troubleshooting under confining or poorly illuminated working conditions. Performance aids are usually not too practical for physicians conducting physical examinations or conducting routine office procedures. Most people want their physician to know what needs to be done without constant reference to a performance aid.

MAKING TRADEOFFS

Generally, training is more expensive to develop, provide, and maintain than performance aids or written instructions. Furthermore, while the cost of developing training materials is high, in many systems the most expensive elements of training are the combined costs of student salaries, training facilities, instructor salaries, and training administration.

The designer should be aware of the comparative costs of the various options. Generally, the development of detailed performance aids or instructions should be considered *before* training development is recommended. Furthermore, designing a system so that people can be *selected* with the necessary skills is frequently favored over developing performance aids, instructions or training.

There is some performance that is best facilitated by some combination of performance aids, instructions and training. Lineberry (1977) presented the following situation as an example:

> Performance involves on-the-spot troubleshooting of a massive multicolor printing press. The operator must monitor the material being printed at a rate of 30 yards per minute, recognize print defects, determine their cause, and make corrective machine adjustments to eliminate the defects. Corrective action is required several times per shift. Failure to eliminate print defects results in loss or downgrade of manufactured goods. (p. 91)

If we consider the performance required in this example against the facilitator options previously discussed, the consequences of error and amount of performance involved indicate the need for a performance aid. However, the frequency of performance and performance rate criterion point to heavy reliance on training. Probably the best solution is to provide a well thought out combination of training and a performance aid.

A designer should evaluate each task in each work module and determine the best way to achieve an acceptable level of performance. A worksheet, such as shown in Figure 17-1 may be used. When the worksheet is completed, every task should be accounted for and a recommended means for its achievement recorded.

WORK MODULE A	RECOMMENDATIONS			
Task Number	Personnel Selection	Performance Aid	Instructions	Training
1	X			
2		X		X
3				X
4			X	
5		X		

Figure 17-1. A Sample Facilitator Worksheet

18

SELECTION CRITERIA

DETERMINING AND REPORTING MINIMUM QUALIFICATIONS

Plato proposed a series of tests for the selection of the guardian of his ideal republic. From this, we can assume that he desired some control over who was selected to help ensure that the guardian would perform adequately. Similarly, for the same type of control, designers should propose guidance for the selection of people to work in their new systems. There is probably nothing quite as discouraging as developing a system for college-educated, experienced users, only to find that the people selected to perform in the new system are high-school graduates without experience.

Traditionally, designers have been almost totally excluded from the selection process. This is unfortunate and unfair. So much of determining whether or not there will be an acceptable level of human performance can be attributed to selecting the right people. Although most designers will never be able to actually select people, they should certainly provide basic selection information for those who will do the selecting. No one knows better than the designer the type of person the system was designed to accommodate.

DEVELOPING A STATEMENT OF MINIMUM QUALIFICATIONS

A designer should be especially concerned about two aspects of selection. First, an explicit and clear profile for each type of potential user should be prepared. The profile should consist of several detailed statements concerning the basic skills and knowledge a potential user should possess in order to use or begin learning how to use the new system. It is always important to specify the skills that the designer had in mind for each user *before* the user is exposed to additional training in the new system. In some systems with little or no training, these basic skills along with some written instructions or performance aids should be all that is needed to perform adequately.

This basic set of (before training) skills and knowledge will represent a *statement of minimum qualifications* or SMQ. An SMQ should be supplied by a designer for each work module. An example of an SMQ is shown in Figure 18-1.

STATEMENT OF MINIMUM QUALIFICATIONS

Work Module Title: Determine and Post Terminal Data

Work Module Number: 753-147

Work Module Description:
This work module describes two procedures for taking terminal and terminal complement data from the cable book and posting it to a form.

The person selected to perform this work module is assumed to have the following qualifications prior to training on this work module:

I. GENERAL BACKGROUND

Performance in the plant assignment office for a minimum of one working year with all assignment center records. Some of this experience should be in estimate assignment work.

II. SPECIFIC KNOWLEDGE AND SKILLS

Must be able to read Customer Cable Records to determine:

1. Terminal complements
2. Binding post numbers and where binding posts start and stop
3. Terminal addresses (e.g., terminal relative location, numeric address and street name)
4. Terminal type (distinguish among fixed binding posts, ready access with posts, ready access without posts, and stub types)
5. Presence of bridge lifters at terminal
6. Existence of station protection at terminal
7. Cables feeding each terminal
8. Originating, terminating and preferred counts on cross box terminals (need to know local methods of indicating counts)
9. Restrictions against terminals (e.g., insufficient loading, Midpoint load pairs, bridge gap, course gauge, and core pairs).

III. SPECIAL REQUIREMENTS

Must be able to score successfully on the Data Entry Qualification Test.

Figure 18-1. Example of Statement of Minimum Qualifications (SMQ)

PROPOSING TESTS

Second, the designer should provide ideas and suggestions on how to efficiently and effectively select people who meet a particular SMQ profile. That is, a designer should propose a means for *measuring* the differences between potential users. Using the suggestions provided by designers, management can better select the people who will most likely perform adequately in the new system.

As this book points out, people differ in many ways. A designer needs to keep this in mind when making basic design decisions, as well as decisions about interfaces and facilitators, and particularly when *specifying* the types of people needed to perform the system. The differences that exist from one person to another (in such things as education, experience and attitude) provide a foundation for identifying and selecting those people who *best match* the requirements for the activity.

We observe that people differ in such obvious physical traits as height, weight, or hair color. We are also well aware of differences in less obvious traits such as intelligence and personality. As we meet people we recognize these differences: some are dark, others light; some are heavy, others thin; some are bright, others dull; some are excited about their work, others are bored. Such simplified descriptions are usually of little value. They imply that each trait can be described adequately in terms of only two categories (for instance, tall-short, pleasant-disagreeable, superior-inferior). Such a classification may be convenient, but is not adequate for selecting people who match the requirements of a job.We need more discriminating classifications of people. Within any trait, the differences can best be discriminated by adding more and more categories. Well-constructed tests are used to make these finer discriminations possible.

If a designer can establish that the existence of a certain trait will help achieve an acceptable level of performance, then management should select people who possess that particular trait. For example, intelligence, as measured by an intelligence test, usually is related to performance in almost any kind of training activity. Generally, those with higher intelligence learn faster. Thus, the trait of "intelligence" is associated with performance in a training course, and an intelligence test may prove useful in identifying and selecting those people with a high probability of doing well.

The use of tests to assist in the systematic selection of people to fill various work assignments began in the United States about 1918, and has expanded rapidly ever since. Thus, for a period of at least 60 years, psychological research has produced an abundance of psychological tests that measure traits related to human performance. Many of these should be of interest to designers. From the assembly-line worker to filing clerk to top level management, there is scarcely a type of work for which some kind of test has not proven helpful. Unfortunately, few designers are ever exposed to available tests that may help in selecting the "best" person for a particular job.

If a trait is well understood, there are usually many equally satisfactory alternatives to measuring it. The problem becomes one of determining the right test for each selection question, and taking the necessary steps to ensure that the test does indeed help in selecting people who are potentially the best performers. There are published lists of available tests (Buros, 1961, 1974), as well as lists of critical reviews for many of the tests (cf. Buros, 1978). Many designers will not be able to spend a great deal of time reviewing and evaluating potential tests. Even so, they should be aware that these tests exist and take steps to help ensure that they are used (when appropriate) by management.

A word of caution is in order. Selecting a useful test is not easy. There are many times a designer who understands testing, and who understands the nature of the work to be performed, can propose a really useful test. Even testing experts, however, can make a wrong decision. A designer should be very much concerned with finding out *for sure* whether a test actually will help improve performance. This is called *validating* the test.

Test validity is concerned with how closely test scores are related to the performance that is being predicted. A test should be put to use only when there is clear evidence that it is valid. Unfortunately, some people pick tests on "looks" or using some other vague criteria, and do not take time to ensure that the test is measuring what they think it is measuring. Many such tests turn out to have little or no relationship to performance. Once you have selected a test, it may be necessary to acquire the help of a test specialist to ensure that the test is valid.

SPECIFYING SKILLS, KNOWLEDGE AND ATTITUDES

Decisions concerning this aspect of selection can be critical to the ultimate level of human performance obtained in the operation of a new system. The designer should specify the skills, knowledge, and attitudes required of users, and, where possible, the best tests for measuring these characteristics. The SMQ and tests can then be used to select the people to perform a job. This process is illustrated in Figure 18-2. Again, designers know best who they had in mind as potential users, and should take time to specify, in some detail, the types of people required for each work module.

MATCHING EXISTING PEOPLE

Frequently, new systems make use of people who are left over from the old system. In some cases few, if any, new people are hired until after the new system has been operational for quite some time. In the interim selection criteria should be used to match the skills, knowledge, and other traits of people from the *old* system with the activities to be performed.

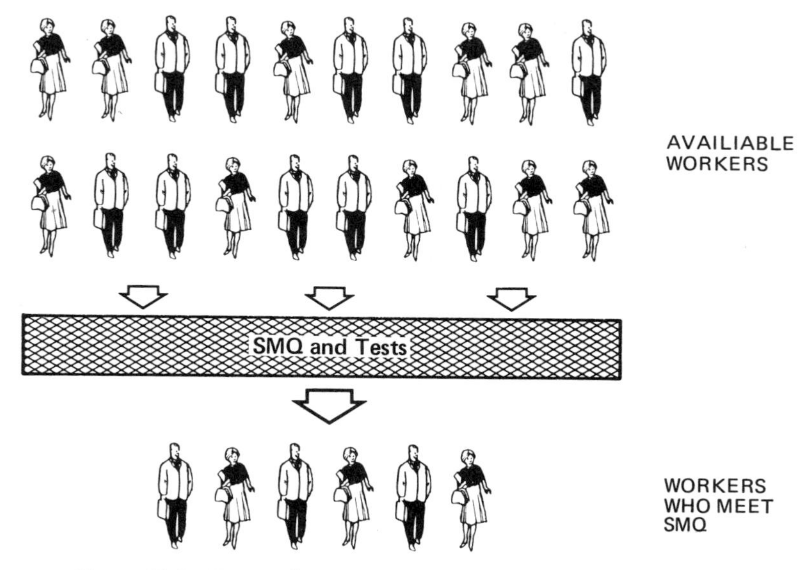

Figure 18-2. How an SMQ and Carefully Selected Tests Can Help Screen Potential Users

We need to make one final point. Selection is concerned with choosing one person from among many to perform a given activity. However, where people are already available, the emphasis should shift to choosing the one *activity* best suited to the strengths and weaknesses of a particular person. The primary question then is, "What activities will best use the individual's strengths?" Or on what activity will each individual perform best and feel the greatest satisfaction?"

If a designer is unable or unwilling to specify detailed selection criteria (including proposed tests) that may be helpful in selecting people, at the very least, he or she should provide a set of brief SMQs with the new system. Ideally, the entire selection process should begin with detailed statements of relevant characteristics of potential users that a designer had in mind when making design decisions. The more knowledgeable a designer is about important human performance-related skills, knowledge, attitudes, and other traits (as well as about ways to measure them), the more effective the selection process can be.

FOR MORE INFORMATION

Anastasi, A., *Psychological Testing,* New York: The Macmillan Company, 1976.

Cronbach, L. J. and Gleser, G. C., *Psychological Tests and Personnel Decisions,* Urbana: University of Illinois Press, 1965.

19

PRINTED INSTRUCTIONS

INTRODUCTION

The advanced state of technology ensures that no matter how well-designed the software or hardware in a system, people still may have difficulty performing in the absence of clear, accurate, and complete instructions. Experienced designers realize that even small changes in the way these instructions are written can have sizable effects on human performance. Instructions may be stored in a 3-inch thick user manual, on the face of a 2-foot square road sign, or on a 1/10,000 square-inch chip in a computer. It is the responsibility of the designer to develop and store instructions. Furthermore, the designer must ensure that when the stored information is required it can be accessed easily and used efficiently to assist a person in performing.

Laurence Saunders in *The Tomorrow File* gives an account of the use of instructions that, although fictional, applies to many real-life situations. In his society of the future a precisely timed sleeping pill or "hypnotic" called "Somnorific" is developed and marketed as an over-the-counter drug. Somnorific initially fails because after unwrapping the substance people fail to wait the required ten seconds before inhaling. Clear instructions to wait are on the wrapper. The solution of Saunders' human engineering team is to recommend that a more tenacious adhesive be attached to the product package. The adhesive sticks to users' fingers when they open the package and ten seconds pass before they can roll it into a little ball and completely flick it off.

We are constantly bombarded with instructions, and like the people who used Somnorific, we ignore many of them. One main purpose of a design is to provide instructions that people understand and use.

Many designers today must overcome a long-standing tradition related to the preparation, storage, and presentation of instructions. For years designers have felt that, given sufficient time, a person assigned a task could figure out what to do and how best to do it in the absence of good written instructions. More recently, attempts have been made at systematically preparing instructions for users but these still rely heavily on the user's intuition of what must be done and how it can best be accomplished. This is the case with many Christmas toys that must be assembled after purchase. Often there are no instructions provided or instructions are very brief. Or, if the product was imported, the instructions may have been developed in another language—for instance, Japanese—and then translated word-for-word rather than thought-by-thought into English.

The situation in many systems closely resembles the Christmas toy example. Instructions are provided as an afterthought. In these cases, designers must rely greatly on the past experiences of users. The designer who understands how the system must operate often acts as though users have that same knowledge.

Instructions surround us and assist us to perform many day-to-day activities outside the work context. For example, when driving an automobile people are constantly receiving instructions to control their speed, or to watch for an oncoming curve or children on the way to school. If while we are driving in an unfamiliar area someone in the car becomes very ill and needs medical assistance, we might follow signs showing where a hospital is located. In this case the instructions are on signs that are available but never noticed until a specific need arises. Another time we may want to visit a friend in an unfamiliar city. The friend may give us instructions to count six traffic lights, turn right, proceed until we find Main Street, and turn left until we find a house numbered 25, which is our friend's home.

VERBAL AND WRITTEN INSTRUCTIONS

Instructions usually are either verbal or written. Verbal instructions are prevalent when one person instructs another, such as a mother instructing a child or an air traffic controller instructing a pilot. Written instructions are seen on signs, in books and magazines, and more recently on computer controlled CRT screens.

It is important for a designer to consider carefully the exact information needed for an acceptable level of performance; to determine the best way of storing the information until it's needed; and to devise a method for presenting the information (when needed) so that it can be used as efficiently and effectively as possible. Verbal instructions are flexible, but the person who is giving the instructions may not consistently include complete information. Written instructions do not change over time. A designer who takes the time to ensure that the information is accurate and complete and then transfers it to a written page or a computer at least has the assurance that the information presented will be complete.

Written or verbal instructions are not merely an adjunct to a system, but an integral part that deserves careful consideration as early as possible in the design of a new system. Poorly designed instructions usually are reflected in both degraded human, and system performance.

MEANINGFUL INSTRUCTIONS

When a user does not understand an instruction, and consequently does not perform in accordance with that instruction, then the resulting degraded performance is the direct responsibility of the designer. Chapanis (1965b) reports on such a problem in a large building in Baltimore. This building has elevators on each floor. The wording of signs placed near the elevators is shown in Figure 19-1.

PLEASE

WALK UP ONE FLOOR

WALK DOWN TWO FLOORS

FOR IMPROVED ELEVATOR SERVICE

Figure 19-1. Sign Placed Between Two Elevators

Chapanis conducted a small study to determine what the signs meant to people and found that most people thought something like: "This must be one of those fancy new elevators that has something automatic. The elevator doesn't stop at this floor very often. If I want to get the elevator I'd better go up one floor or down two floors." And this is exactly what people were doing. Unfortunately after they had trudged up or down the stairs they found the same sign. A designer had developed a set of instructions that did not convey what he or she intended. Chapanis proposed some alternative wordings that would communicate the message more clearly. A long and short version of the same message is shown in Figure 19-2.

IF YOU ARE ONLY GOING

UP ONE FLOOR

OR

DOWN TWO FLOORS

PLEASE WALK.

IF YOU DO THAT WE'LL ALL HAVE

BETTER ELEVATOR SERVICE.

TO GO UP ONE FLOOR

OR

DOWN TWO FLOORS

PLEASE WALK

Figure 19-2. Alternative Versions of the Sign (adapted from *Words, Words, Words*, by A. Chapanis. Reprinted by permission of the Human Factors Society, Santa Monica, California.)

Chapanis also recounts finding the following warning in a military instruction manual:

> WARNING: The batteries in the AN/MSQ-55 could be a lethal source of electrical power under certain conditions.

However, next to the terminal referred to in the manual there was a piece of paper on which someone had printed in large red letters:

LOOK OUT!
THIS CAN KILL YOU

A soldier had cut through the wordy instruction that came in the manual and had extracted the heart of the idea. The words selected by the soldier were very clear and to the point.

Instructions should be written so they can be understood. Conrad (1962) tells of a tragedy that occurred during very cold weather in England. A small boy left school in the afternoon and did not return home. The next morning, a search was organized and he was found dead from exposure at a place nearby, where he often played. An inquest on the child's death disclosed that a search was not organized sooner because neither the boy's mother nor any of her neighbors knew how to use the public telephone at the end of the street. Conrad reviewed the instructions for using the telephone and was impressed by the difficulty of interpreting the instructions. He conducted a study to determine how well people actually could read and understand them. He found that only 20 percent of the people were able to use the telephone correctly. When another group of people received a shortened and rewritten version of the instructions, there was a marked increase in the number who could now perform the task.

Thus, even the simple use of a telephone is becoming more sophisticated. Not so very long ago, one merely lifted the receiver and responded to an inquiring voice with a two or three digit number. Now numbers with seven, ten or more figures are common. If something goes wrong in trying to place a call, the caller must know how to correct the problem. Many callers must now know how to dial direct, reverse charges, use a credit card, and charge to a third party number. Other telephone systems require people to know how to transfer calls, set up three-way calls, and hold calls.

However, the increased technological complexity of the telephone and other systems does not mean that the requirements placed on users of these systems should become more complex. If more complex systems continued to require more complex performance, we would soon find that the number of people that could perform with any system would become very limited. Usually, the more complex a system the harder a designer needs to work to ensure that the user population is able to adequately perform in the system. Having a clear and concise set of instructions is one way to help simplify performance requirements.

Problems with providing adequate instructions have been well documented for several years (cf. Wiedman and Ireland, 1965; Bailey, Steman and Kersey, 1975). In some cases documents did not give the user enough information. In other cases instructions contained omissions and were inaccurate and poorly organized. At times poorly designed tables of contents, indexes, section headings and cross-referencing schemes (or the omissions of these features) led to degraded performance. At other times information was at the wrong level of detail.

WRITING STYLE AND GRAMMAR

Thompson (1979) provides a short set of easy to understand guidelines for writing (p. 42):

1. Outline. Know before you start where you will stop.
2. Write to your readers' level. Remember your prime purpose—to explain something.
3. Avoid jargon.
4. Use familiar combinations of words. For instance, "We're going to make a country in which no one is left out," instead of "We are endeavoring to construct a more inclusive society."
5. Use "first-degree" (common) words.

First-degree words	Second/third-degree words
face	visage, countenance
stay	abide, remain, reside
book	volume, publication

6. Stick to the point. Use your outline to judge the value of each instruction.
7. Be as brief as possible. Tighter writing is easier to read and understand.
   - Present your instructions in logical order.
   - Do not tell people what they already know.
   - Look for the most common word wasters.

Word Wasters	Cut to
at the present time	now
in the event of	if
in the majority of instances	usually

   - Look for passive verbs you can make active. "George chopped down the cherry tree" (active), instead of "The cherry tree was chopped down by George" (passive).

- Look for positive/negative sections from which you can cut the negative.
- When you've finished, stop.

Strunk and White's *The Elements of Style* (1972) contain a more detailed, but also highly readable set of guidelines. Designers who are preparing instructions should have a good understanding of style and grammer. It is beyond the scope of this book to cover these areas in more detail. Table 19-1 lists some basic rules of good grammar, with examples cleverly included:

Table 19-1. The Fumblerules of Grammar

- Never split an infinitive.
- The passive voice should never be used.
- Avoid run-on sentences they are hard to read.
- Don't use no double negatives.
- Use the semicolon properly, always use it where it is appropriate; and never where it isn't.
- Reserve the apostrophe for it's proper use and omit it when its not needed.
- Do not put statements in the negative form.
- Verbs has to agree with their subjects.
- No sentence fragments.
- Proofread carefully to see if you any words out.
- Avoid commas, that are not necessary.
- If you reread your work, you will find on rereading that a great deal of repetition can be avoided by rereading and editing.
- A writer must not shift your point of view.
- And don't start a sentence with a conjunction.
- Don't overuse exclamation marks!!!
- Place pronouns as close as possible, especially in long sentences, as of 10 or more words, to their antecedents.
- Hyphenate between syllables and avoid un-necessary hyphens.
- Write all adverbial forms correct.
- Don't use contractions in normal writing.
- Writing carefully, dangling participles must be avoided.
- It is incumbent on us to avoid archaisms.
- If any word is improper at the end of a sentence, a linking verb is.
- Steer clear of incorrect forms of verbs that have snuck in the language.
- Take the bull by the hand and avoid mixed metaphors.
- Avoid trendy locutions that sound flaky.

(adapted from *On Language*, by William Safire, New York Times Books. Copyright © 1980.)

- Never, ever use repetitive redundancies.
- Everyone should be careful to use a singular pronoun with singular nouns in their writing.
- If I've told you once, I've told you a thousand times, resist hyperbole.
- Don't string too many prepositional phrases together unless you are walking through the valley of the shadow of death.
- Always pick on the correct idiom.
- "Avoid overuse of 'quotation "marks"'"
- The adverb always follows the verb.
- Last but not least, avoid cliches like the plague; seek viable alternatives.

WRITE FOR A SPECIFIC USER

It cannot be emphasized enough—prepare instructions for the people who will use them. Elegance of style, clarity of organization, and simplicity of presentation are wasted if instructions are prepared for someone other than the ultimate user. This becomes almost impossible if the designer does not have in mind a clear set of reader characteristics. *Designers must write for a specific audience.* Instructions that are acceptable for one audience may be totally unacceptable for another.

Occasionally, a set of instructions is intended for such a large audience that the only alternative is to reduce the instructions to the lowest common denominator—the least capable person who would be required to use them. But this should not become the standard way of approaching the development of a set of instructions.

DEVELOPING INSTRUCTIONS

Designers have considerable control over the quality of instructions that they develop. They begin with a precise definition of the target audience for each set of instructions. Designers should write instructions at the level that is most efficient, most comfortable and most acceptable to the target audience. If this is not done, users should return the instructions to the designer indicating that the instructions will not be accepted until they can be understood easily.

A designer preparing a set of instructions needs to know what specific human actions are desired and in what order these actions should occur. This information should be readily available from the task analysis. Once this is clearly defined, the designer can compose the instructions piece by piece so they will produce the required actions, and only those actions, in the correct order.

A designer should never release a set of instructions until they have been reviewed for completeness and performance tested—preferably by having people similar to the target group actually read the instructions and then do what they say (without any outside help). This kind of performance test would give the best predication of whether or not the instructions have actually elicited the kinds of human actions intended. For example, a person is given the assignment to develop and place a series of signs so that people attending a conference in an unfamiliar city will be able to find the conference headquarters. The wording is carefully planned, numerous signs are made and then placed at strategic locations. The true test of whether or not these signs will provide the necessary instructions for people as they enter the city is to have people actually follow the signs to find the conference center. If several people other than the designer are able to do this, then this particular set of instructions has accomplished its objective. The design process is not completed until the designer is assured that the instructions will elicit the appropriate actions. Thus, testing should not be thought of as evaluating the design. It should be an integral part of the design process that ends when a designer feels comfortable that the instructions are truly communicating.

Remember, developing written instructions for use in a system is not something that is done at the end of the design process. The development and design decisions associated with instructions should be made in parallel with other decisions relating to the configuration of a system. As the designer develops a procedure, he or she should ask, "Is this something I can readily communicate to a user with a written instruction?" The skill required to provide a good set of instructions is every bit as complex as that required to engineer the software or hardware of a system. If a designer finds that the necessary skills are unavailable he or she should request assistance in this vital area.

Keep in mind that instructions may range from one or two-word phrases on an equipment label to a full set of written instructions in a user manual. Some instructions can be very short (the familiar STOP sign); others require more extended description including illustrations and cross-referencing or indexing. Whatever form the instructions take, however, they should be developed as early in the design cycle as possible, and tested at the same time as other portions of the system.

GUIDELINES FOR PRESENTING INSTRUCTIONS

Procedural Guides

Instructions are prepared to provide the user with definitive information that will direct his or her immediate or future behavior. More often than not, an instruction tells a user *what* to do and *how* to do it. There are, of course, exceptions to this. Some instructions focus primarily on telling a user what to do with the assumption that the user already knows how to do what is required. Instructions always contain either an explicit or implicit notion of direction. Instructions usually attempt in one way or another to direct the thoughts and/or performance of a user.

Sometimes all that is required to instruct an individual is a word such as "stop," or a small group of words such as "Press button A to proceed," or even a sentence or two, "Do you wish to see a menu? If so, press the return key." Frequently, to have a full set of instructions, several pages of information must be prepared. These may be contained in a "procedural guide." More specifically, we can refer to it as a "work module procedural guide" because it contains the instructions necessary to enable acceptable performance of the work of a single or related series of work modules. The procedural guides are generally directed to a very limited and specialized audience. In fact, each procedural guide should be prepared for a specific type of user.

What a procedural guide contains is much different than the contents of the front page of the *New York Times*, an article in *Scientific American* or a novel. As we will discuss in more detail later, a work module procedural guide should convey information completely, accurately, concisely and objectively. The information contained in this chapter pertains equally to instructions of one or a few words, as well as much longer procedural guides of several hundred words. The instructional material could be stored in a large book or a computer, and read from a printed page or a CRT screen.

Numerous studies related to presenting instructions are available to designers to help them make good decisions. The guidelines that follow are taken primarily from Ronco, Hanson, Raben and Samuels (1966), Macdonald-Ross and Smith (1973), Wright (1977), Hartley and Burnhill (1977) and Dever, Friend, Hegarty and Rubin (1978). More detailed information can be found in each of these resources.

Typographical Features

Printed material has several typographical characteristics that affect legibility and should be carefully selected and evaluated by the designer. The characteristics that have received the most attention are size, case, style of typeface and use of capitals. Numerous studies have also been conducted to determine optimal spacing or leading, line length, and the paper-print contrast. Formats also have been studied in an attempt to enhance reading comprehension.

Legibility

Legibility has been studied under both favorable and unfavorable reading conditions. For studies conducted under favorable reading conditions, the aim generally has been to determine an optimal combination of typographical characteristics for normal reading. In studies conducted under unfavorable reading conditions, on the other hand, the aim usually has been to determine how small or how dimly illuminated a letter, word, or phrase can be and yet be correctly perceived. Both situations are of interest to designers. And in fact, a designer should carefully consider all possible situations where an instruction will be read. It is obviously not a good idea to design an instruction or set of instructions to be used with high illumination only to find that a good portion of the time, they are used with very low illumination. For example, do not design a highway sign that can be easily seen during the day but can be seen only with great difficulty during twilight or after dark.

Favorable reading conditions exist where the illumination is acceptable and other degrading influences such as vibration or viewing from long distances are absent. Unless stated otherwise, the conditions we will discuss will focus on legibility under favorable reading conditions.

Type Size

Three terms are useful when discussing the physical characteristics of type: point size, pica, and leading. A *point* is a unit of measure equivalent to 0.01384 inch (72 points approximate 1 inch). Points are generally used to indicate type size and interlinear spacing. In terms of letter dimensions, point size can be estimated by measuring from the top of the highest letter (a capital, or a lower case b for example) to the bottom of the lowest letter (such as a lower case p or y). The *pica* is another unit of length. One pica is equal to 12 points and 6 picas equal about 1 inch. Picas are used to specify typographical dimensions such as line width, column depth, and column spacing. *Leading* refers to spacing between lines and is measured in points. Type set with no leading appears crowded but is usually readable because the letters do not occupy all the vertical spacing available on the type body.

The type size in books and magazines usually ranges from 7 to 14 points with the majority being about 10 to 11 points. Probably the optimum range is from 9 to 11 points—sizes smaller or larger can slow reading speed (Tinker, 1963). Type size, leading, and line width tend to interact. The following table is adapted from Tinker (1963) and presents ranges of line widths and leadings for commonly used sizes of type. For each type size, legibility remains relatively constant within these ranges of line width and leading.

Point Size	Line Width Range (picas)		Leading (points)
6	14	28	2-4
8	14	36	2-4
9	14	30	1-4
10	14	31	2-4
11	16	34	1-2
12	17	33	1-4

For tasks that require locating specific words or headings in text—looking up a word in the dictionary, for instance—words set in 12-point boldface tend to be located faster than words set in 6-point boldface (Glauville and Dallenback, 1946).

Type Styles

Most typefaces or type styles in common use are about equally legible. However, some typefaces that have shown poor legibility are Cloister Black and American Typewriter (Tinker, 1963). In addition, italic print tends to retard the rate of reading. Text printed in all uppercase letters is read more slowly than text in lowercase letters. Boldface type is read as rapidly as ordinary lowercase type. Users prefer boldface type for emphasis. Finally, text presented in too wide an assortment of typefaces retards reading speed.

Line Widths

Line widths in common use fall in a relatively small range. In double column magazine printings the median is 17 to 18 picas, and for single column it ranges from 21 to 26 picas. Readers tend to prefer line widths of about moderate length (14 to 36 picas), and two-point leading. Ranked next are 1 point, 4 point, and last, no leading (Tinker, 1963). Leading is most effective in improving the legibility of smaller types (6 and 8 point) although the improvement is not generally sufficient to make them more legible than 10 or 11 point type set without leading.

Margins

Readers prefer ample margins and believe that wider margins increase legibility. However, studies have shown that text set with no margins are just as legible as text set with large margins. It is interesting to note that legibility may be adversely affected by page curvature such as exists near the inner margin of some books. This problem can be dealt with by using a larger inner margin. Readers tend to prefer multiple columns per page. Spacing between columns can be accomplished in a number of ways without adversely affecting legibility.

Print and Background Colors

For ordinary reading, black print on white background is more legible than white print on a dark background. In addition, black print on a white background is generally preferred by readers. However, the use of tinted papers with black print does not seem to impair legibility if the reflectance of the paper is high (70 percent or greater) and the type size is at least 10 point. Colored inks on colored papers can adversely affect legibility. Those combinations which provide good brightness contrast (e.g., a dark-colored ink on a light-colored paper) tend to be the most legible.

Unfavorable Seeing Conditions

Under unfavorable seeing conditions uppercase letters can be read more easily than lowercase letters when the two are equated in point size and/or in printing space occupied. Also, regular letter widths tend to be superior to narrower or condensed letters in both upper and lowercase characters. A stroke width 25 percent of letter width and spacing between adjacent letters of about 50 percent of mean letter width have been reported as optimal (Crook, Hanson, and Weiss, 1954).

Also, under unfavorable seeing conditions, legibility can be improved if the typeface is plain, such as the Gothic style or other styles that have relatively uniform stroke widths. For continuous text, lowercase letters should be used along with uppercase for headings. For ordinary reading distances the size of type should be at least 8 points. For conditions that are particularly severe, 11 or 12 point is preferable. A highly reflecting, non-glossy white paper and black ink should be used to maximize brightness contrast. Leading from 2 to 4 points should be used whenever page spacing permits, and margins should be reduced to allow the use of larger type sizes and increase leading if necessary. Finally, only high quality printing should be used since degradation of letter contours in the printing will affect legibility adversely (Ronco, Hanson, Raben, and Samuels, 1966).

Impact of Typographical Features on Performance

Designers should be aware that even after applying the best of all typographical characteristics, only small differences in performance usually occur. That the differences tend to be small is not surprising when we consider the refinement process in the preparation of printed pages that has taken place over the last few centuries. Even so, with the vast amount of reading of instructions that still must be done, any increase in speed of reading or accuracy in interpreting the meaning is worthwhile.

Finally, designers should constantly keep in mind where the reading of instructions will take place. Characteristics that are nearly optimal under favorable conditions still may lead to degraded performance under less favorable conditions. A maintenance manual that is well designed for use in a well-lighted office may be almost totally unreadable in an equipment room where the light is dim.

Language

When a word or concept does not mean the same to a designer as it does to a user, the intent of an instruction is altered and its effectiveness impaired. The specification and control of word and phrase meaning is therefore critically important to designers. Word meaning is related to such variables as word frequency, overall familiarity, and complexity.

The meaningfulness of words is usually defined in terms of their associative value. Meaningful words generally evoke greater imagery and understanding than less meaningful words. Within limits, the more frequently a word appears in a language, the more readily it is learned and the higher its associative value. And the stronger the associative connection between words, the faster learning takes place in many activities. Using words that are familiar facilitates the learning of instructions. A designer must exercise some caution, however, because words and phrases that are too common may change or lose their meaning through excessive repetition.

When possible, a designer should use words containing few syllables. However, it is better to use multisyllable words when they can convey the intended meaning more readily than words with fewer syllables. Abbreviations should be used sparingly and when an abbreviation is used, it should be the standard abbreviation and not one made up by the designer for use in a particular situation.

Sentence structure and style may also influence the effectiveness of instructions. Sentences must be complete, well-organized, unambiguous, and concise. In addition, a sentence should contain correct grammar and the appropriate placement of modifiers. Ideally, sentences should be relatively short (20 words or less) with appropriate punctuation.

As a general rule, *positive active sentences are the easiest to understand.* Introducing the passive or the negative creates problems, either by slowing people down or by causing them to make errors (Gough, 1965; Slobin, 1966; Broadbent, 1977). Consider the following instructions; the first is active, the second passive: "Flip the switch up to start the motor" and "To start the motor, the switch must be flipped up." Although these sentences mean the same thing, studies have shown that people understand the first one best.

Miller (1962) compared active sentences with their passive forms to determine which would elicit the best performance. On the average, people took 25 percent more time to understand simple passive sentences like "The small child was warned by his father," or "The old woman was liked by her son," than the corresponding active forms, "The father warned the boy," and "The son liked the old woman."

In Britain, public telephones have been changed so that people are required to dial their call, and when the person picks up the receiver at the other end it is signaled by a series of pips. At that point, the caller inserts money and begins the conversation. When the new approach was first introduced, the instructions were worded in the negative: "Do not insert money until number answers." Experience with this approach necessitated a change so that the caller was told in correct sequence and positive form what he or she should do, rather than what not to do.

There may be instances, however, when the use of negatives and passives is actually easier to understand than simple active sentences. For this reason, designers must consider carefully the appropriate use of active or passive and positive or negative and not rely exclusively on making all instructions in a positive active voice.

For example, a question is more likely to be answered correctly if it is in the same form as the sentence in the instructions: passive if the instructions are passive and active if they are active (Wright, 1969). To take a simple example, imagine a person who is holding an instruction book, while somebody else is repairing a copying machine. The person who is repairing the machine asks the other one "What is driven by the main gear?" The answer is more likely to be correct if the instruction book says "The spur wheels are driven by the main gear" than if it says "The main gear drives the spur wheels." Of course, it may be very difficult to predict the way in which a person is likely to ask a question of this nature except perhaps that they are more likely to use active sentences than passive ones (Goldman-Eisler and Cohen, 1970).

If the parts of a machine or a sequence are well known in advance, such that items or conditions are encountered one after another, it is desirable for the order of words that refer to them to agree with the order in which they are experienced. It is generally not a good idea to write an instruction such as "Before turning off the calculator, clear the memory" when one could say "Clear the memory before turning off the calculator." Thus, understanding is better if the sequence of words in the sentence corresponds to the sequence of events to be performed (Clark, 1971). From this we can establish a guideline that it is better for instructions to mention first the part or condition that is encountered first, even if this means making the sentence passive.

Designers must constantly evaluate to determine whether or not the person reading an instruction is likely to have an assumption or presupposition as to what the instruction will be asking them to do. If in fact they do have an assumption, then it may be a good idea to use a negative sentence to deny the assumption. This underlines the importance of considering not only the logical meaning of an instruction but also to think about the affect the instruction will have on the person at the very time the person will be reading it.

Remember, it is usually best to word instructions in an active or affirmative way, and to avoid negatives and passives. Yet, it is acceptable, and in fact preferable, to use a negative in some situations.

Sentence Formats

There are numerous special formats that have been studied in attempts to present instructions in other than the conventional use of phrases, sentences and paragraphs. Most notable of the special formats are patterning, spaced-unit and square-span. *Patterning* refers to the use of underlining, boldface type, uppercase or any other technique to emphasize certain words in order to stress their importance. An example of patterning using italics and boldface, is shown in Figure 19-3. Patterning provided by underlining certain words tends to increase the reading speed of people with high reading aptitude, while decreasing the speed of people with low aptitude (Klare, Mabry and Gustafson, 1955). Patterning provided by capitalizing the word or words that are most important in each sentence, tends to help people comprehend more (Dearborn, Johnston and Carmichael, 1951).

Now is the *time* for all GOOD men to come to the *aid* of their country.

Figure 19-3. An Example of Patterning

The *spaced-unit* format places extra spaces between thought units in a sentence. An example is shown in Figure 19-4.

Now is the time   for all good men   to come

to the aid   of their country.

Figure 19-4. An Example of Spaced-Unit Format

Square-span formats use the same principle of breaking sentences up into thought units. In addition, each thought unit is written on two lines in order to allow the reader to use the vertical visual span as well as the horizontal. An example is shown in Figure 19-5.

Now is	for all	to come to	of their
the time	good men	the aid	country.

Figure 19-5. An Example of Square-Span Format

In spaced-unit and square-span formats, units of relatively short length are superior to units of longer length. North and Jenkins (1951) compared spaced-unit, square-span and a standard format. They found that people could read faster and remember more with spaced-unit than with square-span or standard conventional formats. On the other hand, Klare, Nichols, and Shuford (1957) reported that spaced-unit and square-span formats tended to offer little improvement over the conventional formats in learning technical materials. Findings that show no difference between spaced-unit, square-span and other similar types of formats over the conventional format may be due to people's familiarity with conventional formats over the new formats. Klare, et al. (1957) for example found that the people initially read slower with the square-span format but people get faster as they use it more.

Frase and Schwartz (1979) studied the effect of segmenting technical documents into phrases. Segmentation was carried out based upon a method outlined by Johnson (1970). Johnson had adult readers divide written discourse into pause units. These units represent the places where one would stop to breathe, to add emphasis, or to add meaning to the text. When 50 percent of the readers agreed on a pause position then that position was accepted as a place where a new phrase would begin. An example of text segmented into these type phrases is shown in Figure 19-6 (Weiss, 1980).

Orienting a Map

When you read a road map that you have chosen
and are about to begin your comparison
of the map with the ground,
you must first orient your map.
To do this
you turn the map around
until it "agrees with the ground."
That is to say -
a road on the map must run in the same direction
as does the real road it represents;
a farm on the map
must lie in the same direction
as does the real farm,
and so on.

Figure 19-6. An Example of Text Segmented into Phrases

Frase and Schwartz found that text segmentation did produce a faster reading speed than the regular prose text. In a similar study, Weiss (1980) reported that 4th through 7th grade readers had significantly higher comprehension with the text segmented into phrases than with standard sentences.

Illustrations (Graphics)

As designers prepare instructions they must constantly ask whether the instruction can best communicate the needed information through the use of words or illustrations or both. Because most people are taught to communicate only with words from the time they begin handprinting, they are not familiar with the power obtained when illustrations and written information are combined.

In many situations illustrations can be used as the *primary* form of communicating information with words used as a secondary approach. Before using illustrations, a writer should consider whether they will add to the understanding of an instruction. If this is so, a designer has to decide what form of illustration will best convey the information. Once he or she determines the form of an illustration, a designer needs to decide on the number of illustrations needed to convey the information, the amount of material to be included in each, the appropriate size and proportions for each, and where they should be placed. Absolute size of an illustration usually makes no difference so long as the content is visible. Designers must also make decisions concerning the use of contrast, shading, and color in the preparation of illustrations.

Illustrations generally should not be included if they repeat in some graphical form what has been said adequately in words. Thus, the use of illustrations should depend on the illustration adding something to the meaning of an instruction. More specifically, illustrations should be used to show details that are difficult to describe verbally. For example, tables generally present detailed numerical information in a compact and organized way. Probably the best approach to incorporating illustrations in instructions is to determine the illustrations first and then write around the illustrations with whatever narrative is then needed.

Illustrations also can make instructions more interesting and serve to make a presentation of rather dull material more lively. Whenever an illustration is used the narration should provide directions on reading and interpreting it.

One of the most difficult problems with providing adequate illustrations in a set of instructions is determining the most appropriate illustration for a particular type of information. Tables work well when large amounts of specific numerical data must be presented.

Other types of illustrations include maps, photographs, drawings and cartoons. Maps could be included when distances are to be calculated, routes chosen or to instruct an individual on how to get from one point to another. Photographs are useful for providing an exact impression of an appearance of an object and to illustrate objects in three-dimensions. Drawings are helpful for showing the simple dimensions of an object and for emphasizing certain detail. If cross-sectional diagrams are used the designer should indicate from what part of the entire object the cross-section was taken. Exploded-view drawings are frequently used when providing instructions on assembly and disassembly of mechanical objects and for showing relationships among mechanical parts. Artistic drawings attract attention and may help motivate individuals to read the text. In fact, cartoons can be used to highlight points that call for special attention—such as warnings. They should, however, contain the idea to be communicated and not serve merely as decoration where they might detract from the main intent of a set of instructions.

Probably the most important consideration concerning illustrations is to keep them close to related textual material. If there is not room for an illustration on the same page as the discussion of it, it should be placed on the facing page. Illustrations should not be saved up and then presented as a set at the end of a text. The only exception might be when an illustration needs to be consulted throughout a set of instructions. It may be convenient to place such an illustration in an appendix at the end, if possible on a foldout.

A designer should seek to keep wording on an illustration to a bare minimum. In addition, illustrations should not be overloaded with data or have multiple subjects covered on a single display.

Symbols used on illustrations should be few, familiar, and precise. The symbols should be compatible with one another and consistently used throughout the instruction. Straight lines and dashes tend to be somewhat difficult to interpret as symbols, whereas arrows indicating movement and serrated edges indicating cutaways are two of the most easily interpreted symbols.

For critical instructions color could be used to highlight specific ideas or objects. However, when color is used for coding purposes, no more than eight different colors should be used. A designer should ensure that the illustrations will be viewed under acceptable lighting or else the colors may not be seen as the color they are supposed to represent. If colors are to be viewed in low level illumination, changes in color tend to be less when the colors are on a white background. The designer should also keep in mind when using colors for coding purposes that about 8 percent of the population will have some kind of color weakness or color blindness.

Contrast, shading, and texture can be used to advantage in presenting certain types of illustrations including tables, and tends to increase the attention volume of the illustration content. Designers should exercise caution in using shading or cross hatching, particularly when they cross each other or when their patterns are extremely divergent. Cross hatching textures that are fine enough not to impair legibility of overprinting are likely to be overlooked when an illustration is used with low illumination.

Tables, graphs, and photographs most effectively communicate when they are accompanied by good titles or captions. Titles should be placed above tables and below graphs or other figures. Tables and graphs particularly should be numbered and have titles that are descriptive of the content. In addition, words used in titles and labels of illustrations should be consistent with the words used in the text. Titles should be short, between 5 and 11 words tend to be remembered best. If a subtitle is used to help clarify the title, the subtitle should be placed beneath the title and have lettering that is smaller than the size of the title letters.

A more detailed discussion of guidelines pertaining to illustrations can be found in Macdonald-Ross (1977) and Macdonald-Ross and Smith (1977).

Tables

To design effective tables, use the following guidelines (Poller et al., 1981);

1. Ensure that the row and column information is complete, and that all relevant levels of a variable are displayed in terminology familiar to the user.
2. Wherever possible, avoid making the user interpolate or draw inferences to find the needed information.
3. Arrange columns from left to right, so that information the user seeks is read to the right of the item the user looks up.
4. Use typographic cues to make row and column boundaries clear, such as:
    a. Use vertical lines or spaces to clarify column separation.
    b. Use horizontal lines to separate table items and major table sections.
    c. When a table has seven or more lines, use a space or line separation after every fifth line.
5. Use a different typeface from that of the general text.
6. Provide a logical order for the information within the table.

In general, as the number of conditions presented in a table increases, the table gets harder to understand and use. Sometimes it is best to simplify a table by creating two (or more) smaller tables. User difficulties also can be reduced by designing tables to present all conditions on one axis, either horizontally or vertically. This presentation is superior to one in which the user must combine conditions from both axes.

Organization

The ways instructions are organized can obviously influence the speed and accuracy of their understanding. Introductions, headings, titles, topic sentences, summaries, can all be used to help enhance the effectiveness of instructions. The following suggestions, if followed, should help in obtaining a meaningful and orderly organization of instructions.

Organization of instructions depends on the potential readers of those instructions. Although there are a few basic rules to keep in mind, the designer should be willing to adapt any set of instructions to fit the reading habits of the intended audience.

The *title* of a set of instructions deserves careful attention. The title can either enhance or impede the memory of material. Effective titles can also help in look-up tasks. In addition, providing a well worded *introduction* helps to facilitate comprehension, learning, and retention of instructions. This introduction could contain an outline of what is to come and the most important facts to be learned. Also of assistance in helping to improve comprehension and retention of instructions are well thought-out section *headings* and good *topic sentences.* All of these "early" parts of a set of instructions help the reader to call up appropriate schema (relevant past experiences). That is, the reader is able to pull together any appropriate information already known concerning the instructions to be given. A designer should work hard to find key cues and good introductory statements that will enable people to quickly and accurately receive, comprehend, and retain instructions.

Where it is necessary for a set of instructions to be rapidly accessed, a table of contents, indexes, and tabs should be included. It may be useful for both the traditional hierarchical and the "key word" types of indexing to be included.

It is usually not possible for designers to anticipate all requirements for looking up material in a set of instructions. However, through a carefully conducted test of the new system the designer can get a good idea of the majority of words that will be used as key words.

There are many different approaches that can be used for presenting a set of instructions. For example, when introducing and describing new equipment, an effective approach is to begin with a nontechnical description followed by a semitechnical and finally by the most technical description required for an understanding of instructions. In other situations, an inductive approach, building up from the specific to the general, may be more useful to a reader. The latter approach is used frequently when instructions tend to be complex. Probably the most frequently used approach is a narrative one in which instructions are presented in a chronological sequence. The narrative approach is obviously very useful for procedural instructions, although it frequently does not allow a designer to show the relative importance of various items.

Material at the beginning and end of a set of instructions tends to be learned most rapidly and recalled more easily. This places great emphasis on a clearly stated beginning for a set of instructions and, possibly, a well-developed summary at the end (Reder and Anderson, 1979). Designers, then, should carefully consider what material is presented early, in the middle and toward the end of a set of instructions as they attempt to provide instructions that leads to an acceptable level of human performance.

Generally, paragraphs should contain between 70 and 200 words and should deal with a single logically presented idea. Shorter paragraphs tend to facilitate comprehension better than longer paragraphs.

Amount of Information

An instruction may be complex due to the large amount of information that a designer attempts to cram into it. Conrad (1962) found that when faced with a large mass of printed instruction, people simply "close up their minds" because it just seems too much to understand. In Conrad's study he rewrote a set of instructions, including only those things that were directly related to performing the task. The subjects were able to read and understand the rewritten instructions, which substantially improved human performance on that task. It seems that the greater the amount of information that must be processed within a given period of time, the poorer the performance. This unwanted effect may be intensified still further if relevant information is not clearly differentiated from irrelevant information. The designer, therefore, is well advised to simplify the presentation of instructions to their essential points, to avoid unnecessary detail, and to maximize the ability to identify and understand critical material.

Redundancy

Redundancy may be achieved in two ways—by repetition of the same material or by providing alternative ways of looking at the same information (e.g., with examples).

Concepts tend to be remembered best when an instruction is followed by an example, making the concept more concrete. The example should be presented as close as possible to the initial introduction of the concept. Examples increase the reader's ability to generalize the concept and understand its application to new situations. For the designer, this suggests that presenting information in terms of various frames of reference and providing numerous illustrative examples of critical points will aid greatly in the understanding of instructions. Positive examples are more effective than negative ones. The musical admonition to "accentuate the positive" seems to be the best rule for designers to follow. Remember, with a set of easy to understand instructions, simple repetition is likely to bore the user.

Selecting Information

To communicate instructions effectively, the designer must be very selective of the information available. Attempting to transmit *all* available information to a user would in most systems produce only confusion. A designer first should consider all available information; decide what information is relevant to the user; and then determine how best to organize the information. Unfortunately, there are some designers who feel that the best way to achieve an acceptable level of performance is to dump everything known about the task on an unsuspecting user.

Amount of Detail

Experienced users perform better when given more general instructions, while users with less experience seem to derive more benefit from very specific instructions. However, instructions that are too specific may impede generalization to other activities. In the same light, a designer must provide preliminary instructions that are neither too concrete nor too abstract. Concrete, specific instructions seem best suited to situations requiring precise performance of a set of well-ordered activities. More general instructions encourage response flexibility. A general type of instruction may be preferable in situations demanding originality or creativity on the part of a user.

Motivation

Instructions are most effective when the reader is motivated to learn and understand them. Readers who approach instructions with a definite intent to learn will remember the material better than those who do not. Even though designers generally have a limited amount of control over reader motivation, they should attempt to develop instructions that are interesting.

FORMATTING PROCEDURAL INSTRUCTIONS

A procedure is a set of steps performed to obtain a specific result. For example, instructions to install a piece of equipment, operate a machine, or fill out a form, are types of procedural information.

It is important to recognize that the purpose of procedural instructions is to cause the reader to do certain things, *not* to cause the reader to be trained or entertained. Good procedural writing is not the same as good narrative writing. A good set of instructions has little in common with a volume of Shakespeare; the different styles have different objectives.

In most situations, written instructions should *not* be presented as a series of prose paragraphs—the traditional text format. There are at least two other ways of formatting information that elicit better performance: numbered lists and flowcharts. A *numbered list* format describes the activity using steps numbered sequentially and arranged from top to bottom down the page. A *flowchart* portrays in graphic form, by use of symbols such as blocks and diamonds, a sequence of specified operations. The symbols are connected by lines or arrows (see Figures 19-7, 19-8, and 19-9).

The following discussion assumes the use of either numbered lists or flowcharts for procedural instructions. These guidelines were derived from Poller, Friend, Hegarty, Rubin, and Dever (1981).

Instructions for Logging In

Users will login with the login name assigned by their course manager. For simplicity, the passward for a given login will be the same expression as the login itself. A successful login will place the user in the appropriate student directory of the file system. During login, the unconditional execution of .profile will solicit responses to these prompts:

TERMINAL?

The valid responses are hp, ti, and dasi for Hewlett-Packard, Texas Instruments, and Data Access Terminals, respectively.

Figure 19-7. Example of the Traditional Paragraph-Style Format for Instructions

Instructions for Logging In

1. Obtain login name and password from course manager.
2. Login with login name and password assigned by course manager.
3. Wait for prompt word "TERMINAL?"
4. Use the following table to determine what response to make to the Prompt word "TERMINAL."

IF using a:	THEN respond:
Hewlett-Packard	hp
Texas Instruments	ti
Data Access	dasi

Figure 19-8. Example of a Numbered List Format for Instructions

Organization of Related Information

A fundamental principle of presenting procedures is organizing the document to prevent the separation of individual procedures and their related support information. In general, it is more effective to cluster information items related to achieving a common goal, enabling a particular step in a process, or explaining a given concept. This helps to ensure that related information will be read without the interruptions introduced by flipping pages, and consulting other documents.

Sequencing

Correct sequencing of procedural steps produces effective procedural documents. The ordering of events in documents must:

- be consistent with the sequence required by the nature of the work environment,

- be clear to the user of the document,
- be presented to minimize both writer and user errors.

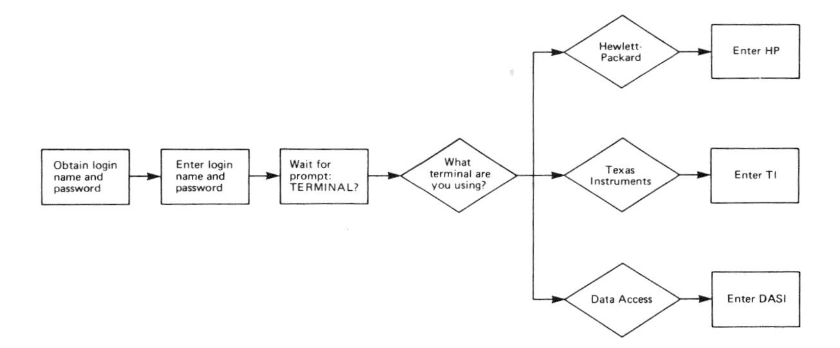

Figure 19-9. Example of a Flowchart Format for Instructions

The results of a good task analysis helps to specify the preferred order of actions and decisions in a given work environment. If the task analysis indicates that event A precedes event B, then place the step with the written instructions for event A before the written instructions for event B. Although common sense and psychological research dictate this, sometimes it is overlooked.

Overviews

Use overviews at the beginning of tasks to supply meaning to the work activities contained in a procedural document. They give users an overall idea of what activities are involved and how these activities fit together.

To write an effective overview:

1. Describe work activities in general terms.

2. Describe inputs to the work activities, or the conditions likely to be present when that set of procedures is applicable.
3. Briefly describe work activities to be performed in the set of procedures.
4. Describe the outputs or ensuing conditions that result from successful execution of the work activities.
5. Include admonishments (i.e., dangers, cautions, and warnings) relevant to the set of procedures to alert the users before they start the procedures.
6. Identify, when appropriate, the destination of the outputs or subsequent operations that result from correctly performing the procedures.

Operational Statements

An operational, or action, statement describes an *action* the user must perform. Both numbered list and flowchart formats can contain operational statements, with or without support information.

Use the following guidelines when writing an operational statement:

1. Start the operational statement with a verb. Do not use phrases such as "should be" or "must be" to indicate an action. Rather, state the action positively in an imperative sentence.
2. Place the object of the verb close to the verb.
3. Place relevant text after the object.
4. Use 20 words or less in the statement, if possible.
5. In list formats, underline key words (verb and object) with a continuous line. If boldface type is available, print key words (and intervening words) in boldface with no underlining.

Presenting Contingencies

This section describes rules for procedures that require the user to make a decision by choosing the appropriate answer to a question. In these situations, the performed *actions* are contingent on the *answer* the user chooses.

Decision Statements

Decision statements typically appear as questions and answers, or as conditional statements of the form "if ..., then ...." Even when these statements are not in question form, they always imply a question.

Write decision statements that ask exactly one question each. Look for the presence of the word "and" in the decision statement—this often indicates that you have clumped two or more questions together in one statement. For example, "Are the fire alarm and the water sprinkler on?" clearly contains two separate questions. Be sure to include all possible answers to the question asked by the decision statement. Also, state each answer explicitly. Do not use words like "else" or "otherwise" for one of the answers.

The possible answers to the question asked by the decision statement should be obvious to the user of the document. Nonobvious answers require the user to complete a series of operations or steps to determine the correct answer. If the answer is not obvious, include the procedural step(s) necessary to make the answers obvious before the decision statement.

Decision Display Structures

Decision display structures present decisions and the actions taken depending on the outcome of those decisions. Decision display structures include IF-THEN lists, IF-AND-THEN contingency tables, logic trees, and decision tables. These structures are required for the display of contingencies in list and paragraph-style formats.

The structure you choose depends on the number of decisions involved. The final form also depends on:

- whether or not the answers to the questions are obvious,
- the number of actions contingent upon the answers,
- the amount of support material required to carry out the operational statement(s) involved,
- logical limitations imposed by the page size, and
- characteristics of the user. (Many users have difficulty using decision tables effectively. For this reason, decision tables are a less preferred decision display structure).

IF-THEN Lists

Use the IF-THEN list when you are exactly *one* decision in a list format. Also, use it for flowcharts when one decision has more than seven answers. See Figure 19-10 for a specific example of an IF-THEN list format.

Use the following table to decide which action to take for different colors of traffic lights.

IF	THEN
light is red	stop.
light is green	go.
light is yellow	proceed with caution.

Figure 19-10. Example of IF-THEN List format

IF-AND-THEN Contingency Tables

Use the IF-AND-THEN contingency table when you have *two* consecutive decisions. Figure 19-11 shows an example of a contingency table.

Use the following table to decide which action to take for different colors of traffic lights and different traffic conditions.

IF	AND	THEN
The traffic light is red	-	stop.
the traffic light is green	-	go.
the traffic light is yellow	there is heavy traffic	go with caution.
	there is no heavy traffic	go.

Figure 19-11. Example of IF-AND-THEN Contingency Table

Logic Trees

The logic tree—a tree-like arrangement of decisions and the actions taken based on each combination of answers—directs the user from one decision to the next with connecting lines. Logic trees differ from flowcharts in several ways:

- No flowchart symbols are used.

- Actions occur only at the end of a sequence.
- Each item in the logic tree, except the first decision, has only one line leading to it.
- Logic trees do not extend longer than one page.

Use the logic tree for *three or more* consecutive decisions. Figure 19-12 contains an example of a logic tree.

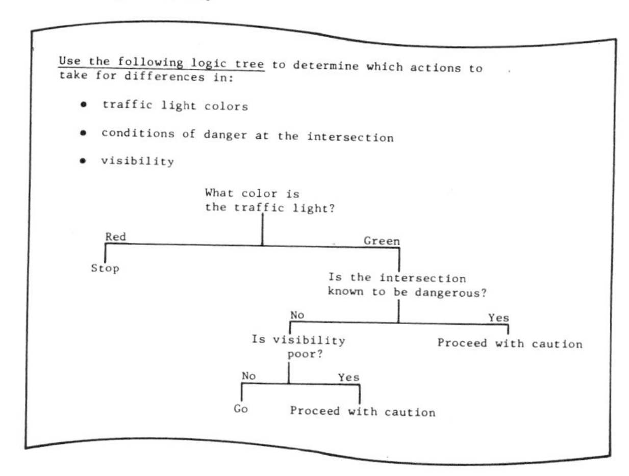
Use the following logic tree to determine which actions to take for differences in:

- traffic light colors
- conditions of danger at the intersection
- visibility

Figure 19-12. Example of a Logic Tree

Decision Tables

You can display several decisions compactly in a decision table. However, since many users have difficulty using decision tables, it is usually best to use them only if your intended audience has special training in reading decision tables.

Support Information

Support information explains, clarifies, or expands the meaning of operations in procedural documents. It enhances the user's ability to perform a procedure, but is not an operational (action) statement or decision statement. Although support information never directs the user to do a particular procedure, it does:

- alert the user to possible consequences of performed actions,
- increase the user's knowledge about a procedure,

- provide the user with feedback on machine, equipment, or environmental responses to procedures, or
- reference other sources of information that might be needed to do the procedures.

Be sure to place support information with the action or decision it supports. This prevents users from having to search for important information. In addition:

1. Do not include required actions or decisions in support statements.
2. Place support information in the sequence of procedures to correspond with the behavioral sequence.
3. Provide feedback to the user about the correct outcomes of actions, where the user may not know the correct outcomes. This feedback should *immediately follow* the action statement.

Check Statements

Check statements keep users on the right track and prevent them from performing incorrect actions. Users of procedural documents often make errors when the instructions tell them to go to a step other than the immediately following one. In some cases, the user continues to the immediately following (incorrect) step. In other cases, the user carries out the GO TO instruction, but goes to an incorrect step. Using check statements can reduce these errors. Place the check statement *immediately* before the action to which the user was directed by the procedure. See Figure 19-13 for an example.

1. Is the form completed?

IF	THEN
no	continue to Step 2 now.
yes	GO TO Step 4 now.

2. *Write in today's date* in section A.
3. *Write in your telephone number* in section B.
4. CHECK: You should be here only if you are working with a completed form.
5. *Mail* form to local administrator.

Figure 19-13. Example of a Check Statement

Admonishments

Admonishments warn a user of the possibility of personal injury, interruption of service to the customer, or equipment damage. Place the admonishment at the point in the procedure where it is needed, according to the task analysis. For example, if acid may leak out when a lid is unscrewed, then place the admonishments *before* the step that says, "Unscrew lid."

It is usually a good idea to place the admonishment in the overview of the procedure, as well as at the point where it is needed. Some users will use the overview incorrectly to bypass detailed procedures, and will completely miss the admonishment in the step-by-step procedures.

TRANSLATING FOREIGN INSTRUCTIONS

With international trade growing as rapidly as it is, there is a need for ensuring that instructions are still meaningful even after translation. Broadbent (1977) tells of an electric razor bought in England but manufactured in Holland. It was intended to be used worldwide and therefore carried a switch allowing it to be set for either a 220V or 110V supply. With the razor sold in England, a piece of paper was included which contained the instruction:

> "This razor should only be used on the 220V setting in the United Kingdom."

Broadbent observed that most people who saw the sentence probably figured that an equivalent statement would be "Outside the United Kingdom this razor should be used on the 110V setting." However, much of Europe has 220V and if a person obeyed the instructions he or she would certainly destroy this razor very quickly. A better phrasing would have been:

> "In the United Kingdom, this razor should only be used on the 220V setting."

This example is of particular interest because it was probably written this way due to differences between the Dutch and English languages. Well written instructions in one language does not necessarily mean they will be well written after translation into another language. The Dutch language does not allow a sentence to be written in the form just given. In Dutch one must say something like "The razor must in the United Kingdom be used only on the 220V setting." This however is incorrect English and may have led to the instruction being written as it was.

Byrne (1980) observed that translated instructions (or is it mistranslated instructions?) have provided a source of humor for years. There have been numerous mistranslated metaphors such as "watered-down male donkey" for "hydraulic jack."

Frequently, much more information is included than is necessary. Bryne (1980) provides a classic example from a translated set of repair instructions contained in an automobile manufacturer's guide: "The rear view mirror is fallen down resulting from breakage of the screw made of plastic in case of collision. And it is removed by loosening the screw. When reinstalling it, be careful not to damage the screw because it is made of plastic." Another foreign text reminds the mechanic about to check the condition of a vehicle's tires that "they are the primary points of contact with the road surface."

Probably the two most important considerations when instructions are to be translated are:

- determine the unique characteristics and special needs of foreign users,
- have the translation done by representatives of the foreign users themselves.

INSTRUCTIONS AS PROGRAMS

Hill (1972) has proposed that there are circumstances when ordinary English is too ambiguous to present written instructions well. He feels that the manner of presenting instructions should be much more formalized, to prevent misunderstanding. English seems to abound in examples of ambiguity, such as "You would scarcely recognize little Johnny now. He has grown another foot."

Hill suggests that we should be able to write instructions for people in programming languages, just as we do for computers. He illustrates this

"For best results wet hair with warm water. Gently work in the first application. Rinse thoroughly and repeat." Repeat from where? Surely the rule must be that, in the absence of other information we repeat from the first instruction. But this means that we have to wet the hair we have just rinsed! Let us use a little common sense and not bother with that. But the next instruction refers to the 'first application' and we cannot do that again, so perhaps logic tells us to leave that one out too. So the only thing left to repeat is 'rinse thoroughly and repeat' and now we are in a closed loop, and must continue rinsing our hair until aborted. (p. 308)

How much clearer it would be to say:

For best results:

begin
wet hair with warm water;
for *j* : = **1,2 do**
begin
gently work in application *(j)*; rinse thoroughly
end
end

Designers should be aware of situations where English ambiguities lead to degraded performance. Perhaps in the future many of our instructions will come to resemble computer programs.

COMPUTER-BASED DOCUMENTATION

The traditional ways of developing and distributing paper documents have not kept pace with the increased need for information in systems. With the ever increasing availability of computers with text editing and storage capabilities, a considerable amount of "documentation" is becoming computer based. The main advantage of computer-based documentation is the ease with which it can be updated. Paper documents are "frozen" when they are written. Since they are not easily changed, they must be complete when issued. This frequently leads to having large, cumbersome documents that contain more information than any one person ever seems to want or need. In contrast, computer-based documents can be easily changed. The requirement for having "everything for all people" included in the first place is not so great.

Well-organized computer-based documents can be as complete as paper documents, without containing as much information. Computer-based documents can be organized so that information is selectively displayed, reducing the need for redundancy. In addition, a record can be kept of what sections are accessed most frequently (or not at all), allowing the designer to make the necessary adjustments (add, subtract or change information) while the system is in use.

As Glushko and Bianchi (1980) observed, it is important not to overlook the desirable properties of paper documents—many of these advantages are not possible with computer-based documentation. For example, computer displays cannot produce pictures or graphics with the quality available in print. In addition, paper documents give the reader the option of underlining important points in the text, adding marginal notes, making "dog ears" or using paperclips to mark pages for later reference.

While designing a new system, Glushko and Bianchi (1980) conducted a detailed analysis of why decisions were made to use both paper and computer-based documentation. They suggest that even the most technologically advanced computer-based documentation will be surrounded by paper documents, at least for the foreseeable future.

The important message here is that designers should *not* automatically provide computer-based documentation, and then think that they have satisfied all documentation-related needs of users. Most systems probably require some mix of computer-based and paper documents that takes full advantage of the benefits provided by each approach.

READABILITY MEASURES

It is amazing that so many sets of instructions are written at levels too difficult for the majority of readers in the intended user population. Although long sentences using a difficult vocabulary with multi-syllable words may be preferred by certain readers, most users in systems prefer instructions made up of *short sentences* with familiar words of *one or two syllables.* Readability measures have been developed to assist designers in preparing instructions that make better use of word and sentence characteristics.

Readability measures provide a means for estimating the difficulty a reader may have with a set of instructions, whether paper or computer-based, because of the way they are written. Readability scores depend on the writing style rather than the content of a set of instructions. These stylistic features are under the control of the designer and range from the specific words a designer may choose to the way a set of instructions is organized into major topics.

A number of readability formulas have been developed over the last several years. The first true readability formula was published in 1923 (Lively and Pressey). However, the prototype of modern readability formulas was not published for another five years (Vogel and Washburne, 1928).

These early formulas and those developed since have continued in their attempt to quantify stylistic difficulty. Readability measures are most useful as *predictors* of reading difficulty. Faced with two texts covering the same subject matter, a designer can predict that the more readable text is likely to be more quickly read and easier to understand.

Even though readability measures attempt to *predict* reading difficulty, they do not always reveal *why* a collection of written material is difficult. This is because many important attributes of style that contribute to difficulty cannot be quantified. But, fortunately, many of these unmeasurable attributes are highly correlated with stylistic features that *can* be measured.

Klare (1974-1975) in a review of readability formulas, concluded that,

> ...as long as predictions are all that is needed, the evidence that simple word and sentence counts can provide satisfactory predictions for most purposes is now quite conclusive. (p. 98)

Before a designer accepts or rejects a set of instructions on the basis of its readability, a number of other things should be considered. For example, the instructions should be read to see if they make sense. A document classified as highly readable solely on the basis of a formula could be a disorganized disaster. In addition, it is possible that the readability of a document may be very important when it is used for step-by-step instructions or in training, but less important for a reference document. Consider, for example, that Coke (1976) found some evidence that readability was not as important when readers looked for specific information as it was when they had to remember that information.

The user and conditions under which a document is used should be carefully considered. Klare (1975) found that highly motivated readers were unaffected by readability in many cases. In circumstances where time is not crucial and readers are highly motivated, the readability of a document may be of less importance. Finally, other features of a document that are not related to content should be examined. A difficult document, according to a readability formula, might be made more readable by changing its format rather than its writing style (Frase and Schwartz, 1977; Hartley, 1977).

In using a readability measure to evaluate instructions, a decision has to be made as to what constitutes an acceptable level of readability for the material being prepared. This decision should be based on information about the reading abilities of the intended audience. As a general rule, it is better to write to a readability level that is *below* the reading skills levels of the intended audience (Kulp, 1976). This reduces the risk of developing a set of instructions that are too difficult for users. The reading ability of a group of potential users can be measured with a standardized reading test (cf. Nelson-Denny, 1973). A designer should avoid a large mismatch between the reading demands of a set of instructions and the reading skills of the intended user.

As used here, reading grade level is defined as a score on a standardized reading test. For example, a high school graduate who reads at an eighth grade reading level would have a score on a reading test about the same as that made by the average eighth grader.

If it is not practical to use a reading test, some idea of reading skill levels can be obtained by looking at the user's educational levels. In general, people with more education have better reading skills than people with less education. However, the *actual* reading ability of a person should not be confused with his or her educational level. A high school graduate does not necessarily read at the twelfth grade level. Coke and Koether (1979) found that for a group of over 200 craft-level telephone company employees with an average education of 12 years, 95 percent had reading test scores above the 10th grade reading level. In fact, the group as a whole averaged at the 14th reading grade level.

The readability measures that follow use only two features of text to predict readability: a measure of word difficulty, and a measure of sentence difficulty.

Several readability measures were evaluated by Coke (1978). Based at least partially on her recommendations, two formulas have been selected and will be presented. The first is the Kincaid (Kincaid, Fishburne, Rogers and Chissom, 1975), which uses average length of a word in syllables to measure word difficulty. The second will be the Automated Readability Index (Smith and Kincaid, 1970), which uses the average number of letters per word to measure word difficulty. Both use the average words per sentence as a measure of sentence difficulty.

Using these two formulas, readability calculations can be made with or without a computer. When a formula is being computed by hand, it is easier to use the Kincaid where syllables are counted rather than letters. When a formula is being computed by a computer, letters are easier to count than syllables, which is what the Automated Readability Index does.

Readability Formulas

To calculate the readability of a document using the Kincaid formula,* a designer should:

1. Select five or more 100 to 150-word samples,

---

* The process is almost the same for using the Automated Readability Index. The only difference is that with the Automated Readability Index the average number of letters per word is used rather than the average number of syllables per word. Using the average number of letters per word facilitates the use of this formula with computers. The Automated Readability Index formula is shown below for designers who want to computerize the readability calculation process:

  Reading Grade Level = 4.71 L + 0.5 W - 21.43

  where,

  L = average letters per word

  W = average words per sentence

2. Determine the average number of syllables per word in each sample,

3. Determine the average number of words per sentence in each sample, and

4. Apply the following formula to make the calculation:

Reading Grade Level = 11.8S + 0.39W - 15.59

where,

S = average syllables per word

W = average words per sentence.

Often it is not practical to use the entire set of instructions when making the needed counts. A common practice is to take a number of short selections from the document. The average readability of these selections is then used to estimate the readability of the set of instructions. Each selection should be from 100 to 150 words long. Selections should be taken from the beginning, middle, and end of a document. Enough selections (at least 5) should be chosen to obtain a good overall estimate of the readability level.

Abbreviations, acronyms and numbers can often be a problem when syllable counts are made because these words usually do not conform to the usual rules for determining syllables. When counting syllables by hand, Flesch (1949) has suggested dropping all words whose pronunciation is in doubt. Since most words in a set of instructions ought to be pronounceable, omitting a small percentage of the words should have little effect on average word length or the average number of words per sentence. This same reasoning holds for computer programs that make syllable counts. Coke (1976) has recommended that words that do not conform to a program's syllable-counting algorithm should be dropped from the analysis.

Usually the only difficulty encountered when determining the average number of words in a sentence is knowing for sure what a sentence is. Sentences commonly end with a period, question mark or exclamation point. However, a writer may separate complete thought units with a semicolon or other punctuation mark. In analyzing readability, a decision must be made whether to treat these units as separate sentences. Kincaid et al. (1975) suggests that these units be counted as sentences. But determining when such units have occurred can be a problem, especially when readability is analyzed with a computer. It is much simpler, and little accuracy is lost by defining a sentence as a set of words terminated by one of the punctuation marks commonly used to end sentences (Coke, 1976).

Readability Example

As an example of how to use the Kincaid readability formula, consider the following paragraph. It is a 127-word section from a much larger document.

> Readability formulas predict reading difficulty, but they do not reveal why a document is difficult. Many important attributes of style that contribute to difficulty cannot be quantified. These unmeasurable attributes are often highly correlated with stylistic features that can be measured. For example, a poorly organized document might contain a lot of long sentences because the writer is forced to reference topics he or she has or will discuss. In this case, the document would be classified as difficult by most readability formulas since average sentence length is often used in these formulas. But shortening sentences to produce a more "readable" document would not get at the source of difficulty. A readability formula should be used as a *predictor* of difficulty and not as a *diagnostic tool.*

This selection contains 230 syllables. The number of syllables (230) is divided by the total number of words (127), which produces an average syllables per word of 1.81 (S=1.81). The 127 words are contained in 7 sentences. The number of words (127) is divided by the number of sentences (7), which produces an average words per sentence of 18.14 (W=18.14). These results are then used to enter the formula:

$$\begin{aligned}\text{Reading Grade Level} &= 11.8\ S + 0.39\ W - 15.59 \\ &= 11.8 \times 1.81 + 0.39 \times 18.14 - 15.59 \\ &= 21.36 + 7.07 - 15.59 \\ &= 12.84\end{aligned}$$

The reading grade level for this one section is 12.84. A designer should take similar readability counts for at least four other sections (the more the better) and then average them. For example, assume that the following readability counts were taken for a document:

Section	*Readability Count*
1	12.84
2	11.03
3	10.65
4	13.48
5	14.72
6	11.64
7	9.87
8	13.52

The average of these counts, which is 12.22, will be the readability score or reading grade level for this document.

Assume that the designer knows that this particular document is to be used by people who read at about the fifth grade level. The designer found this out by having the potential users take a standardized reading test (cf. Nelson-Denny Reading Test, 1973). The designer then rewrites the document without changing its meaning. The section reviewed before now reads as follows:

> There are a number of formulas that predict how hard a document is to read, based on average word and sentence lengths. They may show *that* a passage is hard to read, but not *why*. Writing style cannot be measured, but some effects of style can be. For example, a writer may need to use long sentences because his writing is not well organized. The long sentences are the result of the problem, not the cause. The cure is not to make the sentences shorter, but to organize the material so that shorter sentences can be used.

This section now contains 97 words, 134 syllables and 6 sentences. Applying the Kincaid formula:

$$\begin{aligned} \text{Reading Grade Level} &= 11.8\ S + 0.39\ W - 15.59 \\ &= 11.8 \times 1.38 + 0.39 \times 16.17 - 15.59 \\ &= 16.28 + 6.31 - 15.59 \\ &= 7.0 \end{aligned}$$

The reading grade level is now down to 7.0, which is much closer to the fifth grade reading level of the potential users. It could be written on even a lower level if desired.

By comparing the readability of a document to the reading abilities of its users, a designer can determine whether or not a document will convey its information effectively. A designer should keep in mind that a readability formula is a useful *predictor* of the difficulty that a particular group of users may have with a document, and the formula does not tell how to make instructions more comprehensible. Incidentally, the average reading grade level for this book is 12.3.

WRITER'S WORKBENCH

An even more useful technique for evaluating written material has been developed at Bell Laboratories. This computer-based approach for assessing textual material is called the *Writer's Workbench* (Macdonald, Frase, and Keenan, 1980; Macdonald, Keenan, Gingrich, Fox, Frase and Collymore, 1981; Macdonald, Frase, Gingrich, and Keenan, 1982).

The Writer's Workbench is actually a system that consists of over 20 separate computer programs that evaluate and suggest improvements for written documents. The programs detect features that characterize poor writing, including problems with punctuation, spelling, split infinitives, long sentences, awkward phrases, and passive sentences.

Although the Writer's Workbench contains many short programs, users need to know only one or two short commands to start the editing process. For example, if a designer has the Writer's Workbench available, he or she can make it run by using a keyboard to enter the command "wwb" and the filename where the text is stored. Users also can run any of the 20 programs individually if they choose.

The Writer's Workbench output includes information such as the following:

1. Possible spelling errors. For example:

   cannto *condlusions* *operationel*

2. The computer prints any sentence that appears to be incorrectly punctuated and follows it by a proposed correction. For example:

   line 7:
   OLD: Very little information has come to light
   (there are no published reports.)

   NEW: Very little information has come to light
   (there are no published reports).

3. There is a listing of any double words that may be in the text. For example:

   line 8: for for

4. Sentences with possible wordy or misused phrases are listed, followed by suggested revisions. For example:

   line 7:
   [very] little information has come to light.

   Possible Alternatives

   omit "very"

5. The computer prints any split infinitives that it finds. For example:

   line 10: to easily determine

With most of the programs, a designer can secure more information if he or she does not understand the identified problem. For example, if the concept of "split infinitives" is not understood, the designer is instructed by the computer to type "splitrules." The computer then prints a brief overview of information concerning split infinitives:

> An infinitive is a verb that contains the word "to." Examples include:
>
> 1. to make
> 2. to be going.
>
> An infinitive is said to be split when a word or phrase occurs between "to" and the verb. Possible split infinitives include:
>
> 1. to often make
> 2. to soon be going.
>
> There is nothing ungrammatical about split infinitives; however, they are awkward. If the meaning of the phrase is clear without the split infinitive, by all means do not use it.

If a designer does not understand some of the basic punctuation rules, the computer will print a summary in response to the appropriate command. These are the rules the computer uses when evaluating a text. For example:

1. Periods and commas always go inside double quote marks.

   EXAMPLE: "I want to go to the fair," he said.

2. Semicolons and colons always go outside of double quotes.

   EXAMPLE: He knew what was meant by "hardcopy"; he didn't know about "software."

3. When a sentence is enclosed in parentheses, the period goes inside the closing parenthesis.

   EXAMPLE: (This is a sentence.)

4. If the words inside parentheses do not constitute a sentence, the period goes after the closing parenthesis.

   EXAMPLE: This is a sentence (but not this).

Users are able to adapt certain portions of the Writer's Workbench for their own needs. For example, if a designer is told that "dialog" is misspelled, but this is the spelling preferred by the writer (rather than "dialogue"), then this alternative can be added to the designer's "spell" file.

Other programs provide numerous statistics and editorial comments to help in critiquing the written material. One program, for example, provides a readability analysis much more detailed than the one discussed in the previous section. A typical output may look like the one shown here:

SENTENCE STRUCTURE

You have used more *passives* than is common in good documents of this type. A sentence is in the passive voice when its grammatical subject is the receiver of the action.

You can find all your sentences with passive verbs in them, by typing the following command: style -p filename.

You have used many more *expletives* than is common in documents of this type. Expletives are words that have no content. For instance, "it" and "there" are often used as expletives.

To find all the expletives, type the following command: style -e filename.

ORGANIZATION

You can use an *organization* program to look at the structure of your text. This program will format your paper with all the headings and paragraph divisions intact, but will only print the first and last sentence of each paragraph in your text so you can check your flow of ideas. To do this, type the following command: org filename.

STYLE STATISTICS

Readability grades:

(Kincaid) 15.6 (Auto) 15.2

Sentence information:

no. sent 13; no. wds 243
ave sent leng 18.7; av word leng 5.79
short sent (<14) 15%; long sent (>29) 0% (0)
longest sent 25 wds; shortest sent. 8 wds

Sentence types:

simple 46%; complex 38%;
compound 0%; compound-complex 15%

Word usage:

verb types as % of total verbs
to be 59% (17); aux 41% (12) inf 14% (4)
types as % of total:
prep 10.7% (26); conj. 3.3% (8); adv 5.3% (13)
noun 25.9% (63); adj 21.0% (51); pron 3.3% (8)

The Writer's Workbench provides other aids for designers who are preparing written instructions. Those that we show here should suggest some of the ways this tool can be used to help develop documentation that elicits acceptable levels of human performance.

FOR MORE INFORMATION

Dreyfess, H., *Symbol Sourcebook*, New York: McGraw-Hill Book Company, 1972.

Felker, D. B. (Ed) *Documentation Design: A Review of the Relevant Research*, Washington, D.C.: American Institute for Research, 1980.

Horn, R. E., *How to Write Information Mapping*, Lexington, Mass.: Information Resources, Inc., 1976.

Poller, M. F., Friend, E., Hegarty, J. A., Rubin, J. J. and Dever, J. J., *Handbook for Writing Procedures*, Indianapolis, Indiana: Western Electric Indiana Publication Center, Select Code 700-242, 1981.

20

PERFORMANCE AIDS

INTRODUCTION

A performance aid is a device or document containing information that a person uses to complete an activity. Performance aids are designed to be used *during* performance. They help by reducing the cognitive processing requirements of an activity, usually by reducing the amount of information to be remembered.

Using performance aids to help achieve acceptable levels of human performance is not a new idea. The Incas of the 14th century in Peru used a device called a "Quipu" (pronounced: KEE-POO). The quipu was used by runners who carried messages from town to town concerning the trade of livestock, goods, births, deaths, and other such information. This device was simply a colored length of rope with a series of knots in it. Each knot, depending on its shape and position along the rope, enabled the runner to recall information such as who sent the message, the subject of the message, and the number of items involved (Lanning, 1967).

Performance aids are not the same as training materials. Training materials cover performance in the *future,* whereas performance aids affect performance in the *present*. In other words, training materials are designed to increase the probability that a user will remember and apply the necessary skills and knowledge at appropriate times in the future. Each entry on a performance aid assists the user to perform at the time the performance is needed.

Performance aids also should not be confused with training aids or tools. Training aids are documents or devices designed to promote the *learning* of a particular skill, for future use. Performance aids, on the other hand, are designed to assist the performance of activities *after* the skills have been learned. Tools usually extend the user's physical abilities (e.g., a hammer or wrench) while performing. Performance aids extend the user's *cognitive* abilities, particularly the ability to remember, while performing.

Performance aids are only one means of ensuring that a user has the information needed to perform an activity adequately. People can be hired who already possess the appropriate knowledge and skills. Or who can develop the skills with training and other types of instructions. All of these methods can be, and often are, used in combination. Consideration for deciding which aids should be used are discussed in Chapter 17. Performance aids should be designed only under certain circumstances. Provide performance aids when needed, but be sure not to waste the expense of producing them when they serve no useful purpose.

The most significant benefits of performance aids are:

1. Reduction of errors (the user relies less on long-term memory)
2. Increased speed of certain task performances (reduced uncertainty can lead to faster responses)
3. Reduced training requirements (although users must be taught to use the performance aid, they do not have to learn and remember all the information contained in the aid)
4. Lowered minimum selection requirements (in many situations, a well-designed performance aid allows the work to be done by a person with fewer skills and less knowledge).

TYPES OF PERFORMANCE AIDS

There are two types of performance aids that are used frequently. Both assume that (a) the user has been trained (if necessary) on how to use the performance aid itself, and (b) the user possesses the skills and knowledge to use the information contained in the performance aid.

The first kind assists in the *memory of specific items* of information. For example, rather than memorizing the 15 items we need to buy at the food market, we prepare a written shopping list to use while at the store. In many similar situations, human performance is helped by using well-designed performance aids.

Table 20-1. Performance Aid Types

Type	*Characteristics*	*Examples*
Cue (signals or directs the user)	Directs the user's attention to specific characteristics of information, objects, or situations.	Use of arrows, circles, underlining, or other cueing devices on an illustration to point out the difference between two data recording forms. This helps the user tell one form from another.
	Provides a signal to act in a particular way without giving a complete directional statement.	A checklist of abbreviated directions—the user knows what the brief statements mean and acts as if they were detailed directions.
Associate (translates or converts information)	Provides a means of looking up further data related to a piece of information. The information the user looks up may be an elaboration of what is already known (as in a glossary) or a conversion of what is known (as in a square-root table).	A code book used for encoding and decoding data. A rate-conversion table.
Analog (displays abstract or relational information)	Provides a means of displaying information that cannot be represented directly. Usually, analogs provide a means of displaying the relationship between events in time, locations in space, or processes within a system.	A time-line chart used to show the relationship in time of processes and events. A diagram used to show the functional relationship of components in a system or device.
Example (shows sample or examples)	Provides illustrations, samples or classes of objects.	A sample of a properly completed form. Sample responses of a service representative to a customer's question.

The second kind of performance aid is one that provides step-by-step *guidance* for performing an activity or executing a set of procedures. A guidance type performance aid, familiar to many people, is the step-by-step assembly instructions which accompany an unassembled bicycle or model airplane. The bicycle or model airplane is assembled using the performance aid and a set of appropriate tools (e.g., screwdriver, wrenches, knife, etc.). Performance is facilitated because little or no special training is required, enabling the assembly to take place in a systematic way with few errors. We assume, of course, that both the work and the performance aid itself are well designed.

Different performance aids may provide different types of information. Four other types of performance aids are outlined in Table 20-1. Most performance aids will probably contain only one type of information.

IDENTIFYING THE NEED FOR PERFORMANCE AIDS

A 500-lb unexploded bomb is sticking out of the ground in an army camp. A nervous officer is shouting defusing instructions from a bunker to a terrified private next to the bomb 30 feet away.

Officer: "Turn the tailpiece counter-clockwise and lift it off slowly."

Private: "Yes, sir."

Officer: "You will have exposed two wires, one red and one black, these are to be cut."

Private: "Yes, sir."

The private carefully cuts the red wire and begins to cut the black one.

Officer: "Oh yes, cut the black wire *first* or it will explode."

In this situation, a well-designed performance aid may have been more helpful than step-by-step verbal instructions from the nervous officer.

A designer should carefully consider all activities to determine where performance aids can best help. Determining the need for performance aids usually requires the asking of numerous questions. As a minimum, designers need to know for each activity:

1. As precisely as possible, the type of person who will perform the task
2. How and when the task will be performed
3. Under what conditions or context the task will be performed.

A well done task analysis can answer all of these questions.

GENERAL CONSIDERATIONS

To design an effective performance aid, a designer should first determine the human/activity/context characteristics of the situation in which the performance aid is to be used. A designer can then make other design decisions in order to develop a performance aid that is as responsive as possible to user needs.

HUMAN (USER) CHARACTERISTICS

Earlier in this book, we briefly examined some general user characteristics. We will now discuss some user characteristics specifically related to using performance aids. There are at least three major considerations.

Reading Ability

The first concerns the user's reading ability. The user obviously should be able to read and understand the performance aid. For example, a performance aid to help prepare a Federal Income Tax return should be written clearly and at a level that is easy to understand (especially if the material in the lengthy preparation instructions is confusing). This applies no matter if the information is written on a form or on a CRT screen.

Effect of Past Experience

Second, a designer should determine if users have had (or will have) training and/or experience in the activity. Prior experience in the activity or similar activities may inhibit or promote performance in the new activity. For example, a person may have considerable experience dialing telephone numbers using a pushbutton phone, while the new activity may require the extensive use of a hand-held calculator with the keys located differently. This change may cause a degradation in performance due to the similarities and differences between the two types of keyboards. A new performance aid for using the calculator must take the previous telephone experience into account.

Consider another example where certain modifications had been made to a series of Air Force fighter planes. The pilots were retrained in the flight operation of the "new" aircraft, and after numerous "flights" in a simulator, the pilots were sent out to actually test fly the modified craft. The first jet began its takeoff. More than halfway down the runway, when it should have started to ascend, it was still zipping along the asphalt at an ever increasing speed. The plane never left the ground. It crashed through the base's perimeter fence, skimmed across a local highway, and plowed its nose into the ground.

Later investigation revealed that the direction of movement for one of the main controls had been reversed. With earlier planes, in which this pilot had put hundreds of hours of flight time, he would pull back to climb and push forward to dive. But in the new planes, pulling back on the stick, as this pilot had done, forced the plane down. Pushing forward would have given it lift. Lengthy retraining and a performance aid were not sufficient to displace all of the experience with earlier planes. The old performance behavior interfered with the new. The consequence is clear: New performance aids should be designed to accommodate, not counteract, old behaviors.

Resistance To Use

Finally, a designer should determine whether or not the user is likely to be opposed to using performance aids and if opposition exists, determine the reason. For example, certain people performing tasks consider themselves highly skilled, master technicians or experts. They may view a performance aid as a crutch to be used only by the less skilled. Also, people in particular tasks may consider any type of performance aid an inconvenience—many key operators do not like to interrupt their keying to access even the best designed performance aids.

Handicapped Users

A designer should never assume that the user population does not include people with handicaps. Such an oversight may lead to performance problems later.

For example, some calendar-type performance aids would be very difficult for wheelchair-bound users to reach and flip over easily. A flip file would be difficult for a user with a prosthesis, such as a hook, to manipulate. An auditory performance aid would be difficult to use if hearing impaired users could not adjust the volume to their own listening level. Designers who use color-coding in a performance aid must be aware that all users may not be able to differentiate among colors. Finally, not all users may be able to read a portable performance aid, such as a small card with tiny lettering.

This is not to say that all performance aids must be designed with the handicapped user in mind. However, a designer should be aware at the time a performance aid is being developed whether or not any of the intended users are handicapped and design the performance aid accordingly.

ACTIVITY CHARACTERISTICS

When considering the actual performance of the given activity in connection with a performance aid, the designer should ask:

1. Are the data in the performance aid unchanging (stable)? If not, how often will data be changed, and can the aid be easily updated?
2. Is the activity performed often? If so, will the aid be able to withstand frequent use?

3. Will the aid need cross-referencing for other closely related activities?

4. Will the aid require special postures or movements, so that its use may interfere with the activity to be performed?

The information content of a performance aid can be contained on a single page or divided into several pages called frames. Each frame should consist of all the information necessary to perform a single occurrence of the activity. If each occurrence of an activity requires the same information, the aid could consist of only one frame. However, the performance aid may consist of multiple frames if the information varies on separate occurrences of the activity.

An example of a single—frame performance aid is a card used by automobile assembly line workers to install a windshield. A dictionary or a telephone book are multiple frame performance aids. For each activity, the designer must determine whether a single or multiple frame performance aid is required.

When designing each frame, a designer should concentrate on how to make the information in the performance aid communicate clearly and effectively to the user. A designer should focus closely on how each piece of information should be presented, and in so doing attempt to optimize at least five quality standards.

1. Accessibility (the information must be arranged in a way that permits the user to find the information easily)
2. Accuracy (the information must be correct, and it must be presented in a way that ensures a correct and accurate response from the user)
3. Clarity (the information must be directly usable; no judgments or interpretations should be needed)
4. Completeness/conciseness (the aid must contain *all* the information the user needs, and *only* the information he or she needs)
5. Legibility (the information must be legible, even under the worst possible conditions where it may be used)

CONTEXT (ENVIRONMENT) CHARACTERISTICS

To be effective, performance aids must be suited to their context or work environment. The characteristics of the work area should be identified for each activity. Consider the following:

1. Is the illumination sufficient or is it abnormally low or high?
2. Will the activity be performed on a horizontal work surface (desk or table)?

3. Can the aid be placed on a wall, yet still be readily accessible and legible at a distance?
4. Will the aid be used with equipment to which it may be attached?
5. Will the task be performed at one location or several?
6. Will there be enough room at the workspace to accommodate the aid?
7. Will the work area be dirty? If so, a designer will need to provide a performance aid that can be kept clean enough to be read.
8. Will the work area be subjected to adverse conditions—heat, snow, or high humidity—that would require the performance aid to be constructed of special materials?

PERFORMANCE AID ON PERFORMANCE AID DEVELOPMENT

Morris Bolsky and Chris Yuhas at Bell Laboratories developed a performance aid on how to design good performance aids (Bolsky and Yuhas, 1975). They briefly outlined important considerations for planning, designing, organizing, and producing performance aids in systems. A slightly adapted version of their performance aid is shown in the following pages. It can be used by designers to see how convenient a well-designed performance aid can be.

WHAT IS A PERFORMANCE AID?

- A performance aid (PA) is any means of condensing and organizing how-to-do-it and/or reference information for *on-the-job* use. It can assume a wide variety of physical forms.
- A PA often assumes previous training—that the task is understood and the PA's purpose is to remind people of details. Sometimes, however, people may need to be trained in how to use the PA for performing the task.

WHY AND WHEN TO HAVE PERFORMANCE AIDS

- Because PAs are usually of compact/convenient size, they are more likely to be kept on the working area than are regular documents. Because they are clear, concise, and highly organized, they are easy to use. Thus they:
    - Aid memory/eliminate guesswork.
    - Save reference time.

- Improve efficiency/ speed, and accuracy/reliability of performance.
- Lower minimum selection requirements.
- Increase safety.
- Reduce training requirements.
- PAs are valuable for almost any task, but are especially needed if:
    - Task is critical.
    - Task is too lengthy, complex, or infrequently done to be remembered.
    - Other documents are less effective.

GENERAL CONSIDERATIONS

- *Analysis* - During system development, anticipate needs.
- *Observation* - Watch and talk with system users. Especially, look for "cheat sheets," notes pasted on equipment/wall or written in margins, etc.—these are candidates for PAs.
- *Consideration* - Picture yourself in the user's place. Consider HOW, WHEN, by WHOM and WHERE the PA will be used. Tailor it to *users'* (not your) skills, needs, and physical environment.
- *Participation* - Perform the task yourself under actual conditions.

PLAN

PREPARATION

- Familiarize yourself with:
    - How other PAs are designed. Talk with PA designers. Look in libraries. Take a course on PA development.
    - Services of Art/ Drafting. Documentation/Printing.
    - Materials/special designs. Study catalogs showing paper stock, colors; and special devices for PAs, e.g., all types/sizes binders, indexes, charts.
- Explore possible physical forms for the PA e.g., regular or pocket-size; permanent or loose-leaf binder; chart, single or multipage card; see-through overlay.

PURPOSE

- Get/study all the documents and PAs you can, on the task (and similar ones) you are designing the PA for.
- Consider including in the PA:
    - Procedures/Instructions/Checklists
    - Examples/Illustrations/Pictures
    - Reference Data/Tables/Formats
    - Conversion Tables/Formulas
    - Error Codes/Messages/Explanations/Correction Procedures
    - Trouble Symptoms/ Causes/ Correction Procedures
- Clearly establish the PA's purpose. Talk to your supervisor, and to users and their supervisors. Discuss orientation, content, level of detail, format, physical form, schedule, approvals.

TESTING/FOLLOW-UP

- After issuing first draft, talk with people to ensure you are proceeding correctly. Do this periodically. Hold design review.
- When prototype PA ready:
    - Send Xerox copies to many/all users, so they can use it at once, and also give you feedback.
    - Observe some users actually using the PA while performing the task. Ensure they interpret the PA correctly and thus perform the task accurately and efficiently.
- Follow up after distribution:
    - Put address on the PA right under Title, to which users are asked to send corrections/suggestions.
    - Personally *observe* PA's use, and *talk* with users.

DESIGN

ACCURACY

- Do not rely on documents alone for source information. Personally talk/check with experts.
- Personally proofread all drafts.

- Check final draft with source information and with experts.

PRECISENESS

- Be specific/explicit. Do not require interpretations/ assumptions. E.g., say "Model 33 Teletype," not just "terminal."
- When using words "it" and "they," ensure it is clear what these words refer to.
- Clearly label examples as such, so they won't be confused with actual formats.
- Do not muddle together discussions, with specific "how to" procedures or data.
- Opinions, assumptions, estimates do not belong in a PA.

CURRENTNESS

- If the system is changing often, design the PA for easy revision. Do not make it a bound notebook.
- Be sure to revise the PA as needed.
- Indicate changes from previous edition in a cover letter. Not on the PA itself, so as not to clutter the PA.
- Just below the title, include the date and system version.
- Identify names and editions of manuals on which the PA is based.

STRUCTURE

- Combine related information in one place. Refer to it if it is also needed elsewhere.
- Clearly distinguish or separate (in an appendix?) little needed instructions/data, from that needed all the time.
- Combine similar type information for similar type items. E.g., see example in FORMAT section, where information for 6 control cards is combined.
- Put procedural steps in numbered-list form.
- Put data to be compared or used together into columns. Columns should be close together and dots or dashes used to relate corresponding items (e.g., names in one column, phone numbers in the other).

- List data in different ways for different uses.
- Use extra spacing, horizontal, and vertical lines of differing widths, and perhaps color, to set off and highlight data.
- Use special symbols (e.g., ● or →) to indicate position and to direct attention.

CLARITY/SIMPLICITY

- Make the PA clear and simple to use. Also make it *appear* clear and simple at first glance, so people will *want* to use it.
- Use short words, sentences, paragraphs, sections.
- Use active voice/present tense.
- Be informal. Use *you, we,* etc. (Avoid using the *he* pronoun exclusively. Use *he/she* if pronoun unavoidable.)
- Do not feel constrained by rules of grammar. Use short phrases, even if they are not complete sentences.
- Use diagrams, checklists, lists, charts, etc., instead of or in addition to text.
- For large charts, put heading on both top and bottom, or both sides. Have thin lines to break it up into sections. Have a legend to explain symbols.
- Items in a list should usually have the same general sentence (or non-sentence) structure.
- Minimize cross-referencing. If needed, give page/section numbers.
- Use symbols, abbreviations, and terms familiar to users. Define them in one place at the start. And be consistent; if several forms are in use, stick to one. And if a term has several meanings, tell which you are using.
- Edit and re-edit. Also ask others, especially future users, to edit the PA.

COMPLETENESS

- Include enough identifying, instructing, and detailed information so that a trained person can use the PA without further explanation or references.
- The PA can also serve as a cue to more detailed information, by providing page references to manuals.

- For unusual or complex procedures that would be impractical to cover in the PA, refer to a manual or tell the user to ask for help. But do not just ignore them.

CONCISENESS

- Use as few words as possible to convey as much meaning as possible.
- Do not repeat the same information in more than one place unless there is *good* reason. If you *must* do so, use the same wording, list the items in same order, etc.
- Procedures to be used just once (e.g., how to get a computer job number) should not be in the PA.

DIRECTIONS

- Do not establish long/complex procedures for minor/simple tasks. Concentrate on the difficult/unfamiliar, not the obvious.
- Do not require uniformity where it is not essential. "Unity in essentials, freedom in incidentals."
- Clearly distinguish recommended from mandatory procedures.
- If a procedure is *really critical,* involving safety and/or high cost, use every means you and others can think of to alert users.

USABILITY

- Quality and strength of materials used for the PA should be in proportion to its importance, and length/stress of use.
- In all circumstances, consider the users' work space and if hands are free or not, to use the PA.
- A PA used with equipment may be pasted or otherwise attached to it.
- For terminals, consider an adjustable holder to keep the PA open at eye level, so the user can use it hands-free and without bending to see it.

ORGANIZE

TITLE

- Make the title clear, meaningful, functional. Not clever or cryptic.

- Identify the overall system, and the specific subsystem/task(s) covered.
- Typeset title in large print on the top, front.

AUDIENCE

- Tell *who* should use which parts of the PA, and *when*. Use text and/or chart.

CONTENTS

- Have a Contents section if the PA is multipart, and alphabetic and/or functional indexes. Put in page numbers.
- Put the contents/indexes on the outside, front of the PA, directly under the Title and identifying information. *NOT* on an inside page.
- Place page numbers in large type in a uniform place on each page. On reference cards, place on top, outside margin. If color used, print in color.

HEADINGS

- Use clear, meaningful, functional (not clever or cryptic) headings/subheadings.
- Typeset in large, bold type (larger type for main headings, etc.).
- If color used, print in color.

NOTATION

- Use a uniform notation for describing similar type items.

FORMAT

- Use a standard format to specify information for similar items.

EXAMPLES

- Use examples liberally.
- Place them as closely as possible to the text they illustrate.
- Use both *specific* examples (e.g., of individual computer control cards), and *overall* examples of entire task sequences (e.g., of control card decks).

- Consider annotating examples with explanations, using arrows, etc. to point or otherwise relate to the sections of the examples.
- Ensure examples are correct, clear, relevant, and do not contradict the text of the PA!
- A PA may consist mainly or even only of examples.

WRITE-IN SPACE

- Leave space above the Title for users to write in their name (in case PA is misplaced).
- Leave space within the PA for users to write in specific information needed while using the PA. E.g., their computer job number.
- Leave space in and at the end of the PA for notes, hints, etc.

PRODUCE

LEGIBILITY

- Ensure that text and diagrams will be clearly legible under all conditions of use. If the PA will be used by people in a hurry or under other stress, and/or under poor lighting or other unfavorable conditions, printing must be larger/bolder/clearer than usual.
- Proportional spacing improves legibility.
- Boldface type is more readable, especially if text is to be reduced in size.
- Use regular Roman type, not sans serif, which may cut legibility.
- Use regular lowercase (initial caps only) for text.
- Do not use all italics for text, nor for all caps headings. Italics, especially in all caps, are hard to read.
- Have all text and charts running the same way. Do not make readers have to keep turning the PA sideways.
- Write numbers in digit, not word form. Use commas or spaces if many digits, unless this would be erroneous (e.g., in format for computer control card).
- Have adequate spacing for margins, and between words, lines, paragraphs, and sections. Keep the spacing uniform.

- Blank space around an item emphasizes its importance. This can be used for critical information.
- Do not use wide columns of text. 3-1/6 inch column width is good.
- If color used, get expert advice on visibility, contrast, readability. E.g., bright paper colors suitable for posters are not suitable for text.
- A useful device for PA designers is a circular slide rule for calculating print reduction sizes. Borrow from Art/Drafting or buy one.

TYPING

- Have typist use a backing sheet while typing with heavy lines indicating the PAs dimensions.
- Sometimes it may not be possible to type what you asked for, due to size or other restrictions. Instead of leaving it to the typist to decide what to do, work with the typist to achieve optimum layout.

ORDERING

- Facilities and prices of printers vary. Get a few estimates. But do not sacrifice quality.
- Price per copy (especially for special physical designs) may go down considerably, the more that are made. Consider ordering more copies than initially needed, to avoid small, expensive reprints later. Consider:
    - Copies will be needed for new people coming on the job.
    - Some (all?) users may want more than 1 copy.
    - Other areas of the company may want copies.
    - Lost/damaged copies will have to be replaced.

PRINTING

- Specify type, weight, color of paper or card stock; type and colors for printing, final size, typesetting, etc.
- *Always* ask for a proof copy before final printing. Check *carefully*. If printing in more than one color, ask for a proof of the complete text, and of the text in each color except the main one. The complete text proof should be in the physical form as for the

final copy.

DISTRIBUTION

- Ensure that all who need the PA, get it!
- Make provision for new people to get it.
- Place reference copies in key areas.
- Have an address on the PA, under the title, where copies can be requested.
- List the PA in document indexes. Send copies to libraries.
- Send copies to other areas of the company that might use it, or might want to develop a comparable PA for their application.

PERFORMANCE AID EFFECTIVENESS ASSESSMENT

It is worthwhile to state again that once a prototype performance aid is developed it should be carefully tested. Designers should attempt to identify any physical characteristics that are incompatible with design requirements—for example, a card designed to fit under a piece of clear glass on a desk top when the user's desk top is constantly covered with books or papers. If there is a nearby free wall, a wall chart may be a better choice. A designer should also evaluate the performance aid for poor design such as heavy use of underlining, and excessive color coding. The designer should also evaluate the performance aid's legibility, accessibility of information, clarity, accuracy, completeness and conciseness, and compatibility with the work conditions.

The evaluation of a prototype performance aid should include discussions with several potential users. Users' opinions on the usefulness of the performance aid help to determine the users acceptance of it, and the likelihood of its being used appropriately on the job. User evaluations help to identify many flaws in the design of performance aids.

Finally, each performance aid should be tested in a simulated situation very close to the real one. Users should use the prototype aid as it will be used on the job. This will help a designer make necessary final modifications before full scale production of the performance aids begins.

21

TRAINING DEVELOPMENT

INTRODUCTION

Training is the systematic acquisition of *skills, knowledge,* and *attitudes* that will lead to an acceptable level of human performance on a specific activity in a given context. The primary purpose of training is to improve (change) the user in some task-related way, not merely to add to his or her store of knowledge. For example, when developing training for a new computer system, a designer could ensure that terminal-use *skills* are developed, that a general *knowledge* about the system is provided, and that users *attitudes* about the new system change to be positive.

The central goal of training is to achieve acceptable performance. Keep in mind, however, that training development is only one part of a much larger whole that encompasses a variety of human performance considerations (Silvern, 1961). Some designers mistakenly feel that once the hardware and software is developed, all that is necessary to ensure an acceptable level of human performance is a set of good training materials. This is simply not true. In systems that are *operational,* for example, many human performance problems (perhaps even the majority) have little if anything to do with training.

Training development should not be initiated automatically, casually, or haphazardly. An often quoted estimate for training costs is 200 hours of development time for each hour of training. For instance, it costs the government up to $750,000 to train a military pilot (Weintraub, 1978). This is a hefty investment for a population characterized by significant turnover. The annual bill for adult training in the United States exceeds the annual defense budget (Gilbert, 1976, 1977).

With such great amounts of money, time, and resources at stake, the development of training must be approached with considerable care. A few years ago a major communications company developed and presented a ten-week training course. The course was given to new employees who were hired to install and service equipment. A later analysis of the course indicated that at least five weeks of the course were being spent teaching irrelevant material. The course was redesigned as a two-week self-paced course. The redevelopment costs—though a substantial $350,000—paid off. Since nearly 2000 employees receive the training each year, reducing the course from ten weeks to an average of two weeks resulted in savings of over $4 million a year. Making the course self-paced enabled some students to finish the course early or, if necessary, take extra time. This company reported that in one six-year period, there was a total saving of $37,800,000 from this course alone. Perhaps even more important, a follow-up evaluation indicated that performance on the job had improved. This savings and improved performance could have been realized from the *beginning* if the initial course had been designed properly.

Consider another example where improved training reduced costs. Many telephone systems now have features such as:

Automatic Callback - Where you can be called back automatically when a busy telephone you are trying to call becomes free.

Call Forwarding - Where all of your incoming calls will ring at another telephone.

Threeway Conference - Where you can set up a threeway conversation without operator assistance.

Transfer - Where you can transfer a caller from your telephone to someone else's without operator assistance.

The problem here is to train new users to take full advantage of these features.

Originally this was done for business customers by having a telephone company instructor explain and demonstrate each feature to groups of 8 or 10 people seated around a conference table. Following a step-by-step description and demonstration of each feature, one or two people from the group would be given the opportunity to try out the feature. This procedure would be employed with different users so that each had an opportunity to practice a few of the features. Each of these sessions took about two hours. For the telephone company with customers having hundreds of telephones, this type of training proved to be very expensive.

One approach to reducing training costs focused around the use of "hands-on" training. Originally, it seemed that learning to use a complex set of features would definitely require "hands on" practice by users. Ellis (1977) and others conducted studies where half the people had training sessions that included an opportunity to practice, and half had training sessions with *no* practice. Ellis found that there was no improvement in user performance for this type of task (requiring little movement skill) when people were provided an opportunity to practice during training. Karlin (1977) estimates the savings realized by eliminating "hands on" training in this situation was about $2.5 million for one year.

Another approach taken by designers to reduce the cost of training was to develop a one-page card performance aid that described the procedure for using each feature. In the new training sessions, users were given an overview of the features, and the details were given in the performance aid (Ellis and Coskren, 1979). Again, training time was reduced, with no decrease in feature use or performance.

LEARNING VS. TRAINING

A distinction should be made between learning theories and training. *Learning theories* usually describe the conditions under which a behavior is acquired. Such theories are *descriptive*; they provide a theoretical base that can be modified for application in specific practical situations (Glaser, 1976). *Training theories* specify the most effective and efficient ways to obtain knowledge, skills or attitudes at identified levels and under particular conditions. Such theories are *prescriptive*. They suggest principles of instruction, criteria for learning, and the conditions that are likely to ensure that learning will take place (Burner, 1966; Rickards, 1978).

While inroads have been made in the direction of developing research-based guidelines for training development, the job has by no means been completed (cf. Klahr, 1976; Patterson, 1977; Siegel, 1967; Snelbecker, 1974; West and Foster, 1976). Considerable experimentation over the past hundred years has yielded only a few principles that have received broad acceptance (Hilgard and Bower, 1975). For example:

1. Keep the trainees *active* (skill can be best developed by doing, not just listening or reading).

2. Make use of *repetition* (practice makes perfect).
3. Make use of *reinforcement* (reward correct responses).
4. Have trainees *practice* in many different situations so that they are able to generalize.
5. *Organize* the presentation of information in some meaningful way.
6. Provide for learning with *understanding*.
7. Encourage *divergent thinking* (urge students to develop creative solutions and explore alternative solutions).
8. Consider the trainee's *ability to learn* (some people learn quickly, some learn slowly).

DEVELOPING SKILLS

Training courses usually help to develop new skills, knowledge, or attitudes. Probably the most critical to human performance is the development of *skills*. There are two basic ways to acquire a new skill. One approach is simply to have the person perform, and over a period of time the skill develops. Frequently, there is a model of some kind to imitate, but no specific instructions are given. Most of us learned to walk, talk, and ride a bicycle in this way. The second way is to have another person communicate in some more or less systematic way what is to be done (i.e., suggest a strategy). These instructions can be verbal, written, or contained on a CRT display.

For the most part, designers should not allow skills simply to evolve. Instructions should be provided to make the learning process as efficient as possible. The following discussion focuses on a skill development process that usually begins with and uses instructions.

Developing Skills with Instructions

Skills seem to reflect a set of internalized instructions (or plan) that were originally voluntary but that have become relatively inflexible, involuntary, and automatic. Once the internalized instructions that control a sequence of skilled actions becomes fixed through overlearning, they function in much the same way as instinctive behavior in animals. The conditions under which various skilled components are triggered, or released, is much the same in both cases.

By consciously following a good set of verbal or printed instructions, a beginner may achieve the same objective as the automatic following of internalized instructions by a skilled performer. In a sense the performance is the same. But the beginner's performance is carried out in a way that is voluntary, flexible, and communicable, whereas the expert's performance is automatic, inflexible and usually locked in. The development of skill appears to free the expert to automatically implement larger and larger behavioral units.

When a person sets out to acquire a new skill, particularly in a system, he or she usually begins with a set of instructions of some kind. But just having the basic strategy in verbal form does not mean that the learner can correctly perform on the first try. For example, when an individual learns to fly an airplane, a set of instructions like the following may be provided:

> To land this plane you must level off at an altitude of about ten feet. Then, after you have descended to about two feet, pull back on the elevators and touch down as you approach stalling speed. You must remember that at touch-down the control surfaces are less sensitive, and any gust may increase your airspeed. That may start the plane flying again, so be prepared to take corrective measures with the throttle and elevators. And if there is a cross-wind, lower the wing on the windward side, holding the plane parallel to the runway with the opposite rudder. (Miller, Galanter and Pribram, 1960, p. 83)

Those are the instructions for landing the plane. When skillfully executed they serve to get pilot and airplane safely back to earth. It is a short paragraph and could be memorized in a few minutes, but it is doubtful whether the person who memorized it could land a plane, even under ideal weather conditions. In fact, it seems likely that someone could learn all the individual acts that are indicated in the instructions and still be unable to land successfully. Even given the description of what to do, the trainee still faces the major task of converting the *knowledge* into *separate skills*, and the separate skills into an *integrated skill.*

Skill Development Stages

Dreyfus and Dreyfus (1979) present an example of stages through which a pilot passes as he or she develops the flying skill. First of all, the novice pilot focuses all attention on a list of memorized instructions (procedures) to be appropriately applied. In doing so, the pilot trainee is so absorbed in details that he or she is unaware of most surrounding events, and experiences little sense of actual flying.

With further experience the pilot trainee acquires the ability to recognize and learn the importance of such situations as being in the landing envelope and such sensations as acceleration, and characteristic sounds and vibrations. The pilot analytically determines the appropriate actions by applying rules, such as determining whether the aircraft is in the landing envelope, or returning to base when vibrations are abnormal. This intermediate pilot trainee begins to feel that he or she is flying the plane.

Finally, a pilot's repertoire of flying experiences becomes so extensive that each *whole situation* is recognized as similar to a previous typical situation, and this previous situation elicits remembered appropriate responses. Furthermore, associated with the memory of each of these past experiences are other associated experiences. For example, suppose that the current situation is a normal landing, and hence location in the center of the landing envelope is a crucial aspect. If the pilot perceives that he or she is very high in the landing envelope, the associated past experience suggests a "go around and try again" situation. Analytic, conscious control becomes almost completely bypassed and replaced by an automatic (skilled) response. The pilot now feels that he or she is flying.

Dreyfus and Dreyfus (1979) suggest that the same type of phenomena shows up whenever a person acquires a complex skill, be it highly intellectual like chess, or largely associated with movement control like tennis. For example, in chess a beginner uses instructions to learn simple rules, such as to trade pieces to maximize material balance (calculated by adding up the values of the individual pieces involved). In tennis, a player first learns several independent movements such as swinging the racket at the proper speed or transferring weight from one foot to the other while making a stroke.

With experience, a chess player learns to follow rules such as exploit a weakness on the king's side or avoid an unbalanced pawn structure. And in tennis an experienced player may be advised to use top-spin on a return shot.

When truly proficient, a chess master, immersed in the world of the game, immediately perceives the forces and tensions on the board as similar to those previously experienced in actual play or in the involved study of previous games. The world class tennis champion no longer thinks about shifting weight or using top-spin. Bypassing analytical control of performance, he or she enters into an almost unconscious series of movements that are appropriate responses to the other player's actions and further, leaves the player free to develop new offensive strategies as the game goes on.

Skill development, then, probably takes place for most activities roughly as a three-stage process. First, the performance is almost totally under conscious control. Second, the performance is under shared control—some activities require conscious deliberation and others are automatic. Finally, in the third stage performance is totally under automatic control, perhaps leaving the person free to engage in other performance or to monitor and improve the performance presently taking place under automatic control.

Developing Skills Without Instructions

Although it at first seems necessary to begin all training with a set of clear, precise instructions that will eventually become internalized as skills, this is not always possible. Almost no one, for example, can communicate the instructions for a new bicyclist to maintain balance. According to Miller et al. (1960), the underlying principle would not really be much help even if a person did know how to express it: "Adjust the curvature of your bicycle's path in proportion to the ratio of your unbalance over the square of your speed." For most people this instruction is almost impossible to understand, much less to do. In such cases, a designer uses the other option for building a skill, and simply has someone run along beside the bicycle, holding it up until the trainee "catches on."

A careful review of the tasks on which skills are to be developed should suggest which of the tasks require explicit instructions and which do not. Keep in mind that small children frequently acquire skills without first memorizing verbal descriptions of what they are supposed to do.

Another problem with relying too heavily on an initial set of instructions when developing skills is that even the finest set of instructions may be too general from a user's point of view. Training instructions in many situations tend to deal more with overall strategy than with movement-by-movement or thought-by-thought details required by the trainee.

Miller et al. (1960) observed that the general strategy provided by the designer usually says little or nothing about the activities of individual muscle groups. The designer knows these interrelated acts because he or she knows how to perform, but they are implicit, rather than explicit and communicable. Thus, designers are usually working from the *general* (actions that need to be done) to the specific when attempting to communicate training instructions, while the trainee is working from the *specific* (thoughts and muscle movements) to the general when trying to carry them out. The designer and trainee may not see the required performance in the same way.

General Versus Detailed Instructions

In fact, in many situations it is probably best for a designer to provide general information and let the trainee work out the best specific way of doing it. In this way a trainee is free to develop specific thought and movement patterns that best enable him or her to meet the performance requirements.

In some systems, however, designers have provided very detailed instructions. On an assembly line in a factory there may be a task that consists of, let us say, assembling three washers on a bolt. The analysis of this task into "micromotions" will specify the exact time at which each hand should move and the operation it should perform. For the left hand, the instructions may read "Carry assembly to bin," "Release assembly," "Reach for bolt," etc., while at the same time the right hand is instructed to "Reach for lock washer," "Select and grasp washer," "Carry washer to bolt," etc. For each of these motions a fixed duration is specified. This is about as near as anyone can come to *writing programs* for people that are as detailed as the programs we write for computers.

With instructions as specific as these, designers attempt to find a sequence of motions that achieves the result most efficiently, with fewest movements and in the least time. Following these rules, designers may be able to develop chains of responses that can be executed with very high efficiency. But, unfortunately, many users may object to being so tightly regimented. When people have time to develop the skill themselves, they are able to determine the interposed elements that produce the skill. Once the skill has been developed, alternative modes of action then become possible, and the person is free to make changes that can help to improve performance.

Developing Efficient Skills

We need to make one final note on skill development. Designers should seek to develop skills that lead to performance with the fewest errors and shortest time frame. Many skills tend to be developed in such haphazard ways that the resulting performance is inefficient. Most people perform numerous activities every day without considering whether they are performing in an efficient or inefficient manner. Morehouse and Gross (1977) provide an example that helps illustrate the problems we frequently encounter with inefficient skills:

> The first time you made a bed, you probably followed someone's instructions—your mother's or your sergeant's. Like a rat in a maze who finds a route to the food and then repeats that route over and over without searching for a better one.
>
> The next time you make a bed, observe how many times you go from one side to the other before you're finished. The first time I tried this experiment I was surprised to learn that I was making six trips around the bed. I decided to see whether I could make the bed in one trip. I could, but it was a pretty sloppy-looking bed. So I modified my objective, and found that in three trips I could often make a bed as neatly as I could in six.

(adapted from *Maximum Performance*, by Laurence E. Morehouse, Ph.D., and Leonard Gross. Copyright © 1977 by Laurence E. Morehouse, Ph.D., p. 48. Reprinted by permission of Simon and Schuster, a division of Gulf Western Corporation.)

To develop skills that produce efficient performance is no simple matter, but can be done with effort and by knowing what to do. Keep in mind that simply *passing out information* is not the same as training a user. Good training concentrates on *building skills,* and in many cases *changing attitudes.*

USING REINFORCEMENT

Studies on the systematic modification of behavior supplies additional information for training developers. The concept of behavior modification states that people will act according to a set of rules if those rules are reinforced in a direct, immediate, and consistent manner. The emphasis is on the interaction between what the person does to the world and how the world reacts (Margolis and Kroes, 1975).

Much of the information on behavior modification is grounded in the work of B. F. Skinner. Skinner proposed reinforcement as the key factor in learning. *Reinforcers* are positive or rewarding consequences that increase the likelihood of a certain set of performance behaviors recurring. Possible reinforcers include:

- Attention and praise (social reactions)
- Prizes
- Status
- Privileges
- Awards
- Money.

Punishers are negative consequences, such as:

- Reprimands
- Ridicule
- Fines
- Deprivation.

In training, the appropriate use of reinforcers enhances the possibility that the desired performance skills will be acquired and not forgotten. The use of punishers is usually inappropriate because their consequences are difficult to predict. Ignoring undesired performance is often aversive enough to decrease the likelihood of the undesirable performance being repeated.

Another relevant behavior principle is *modeling.* A person's performance can be modified by their observations of reinforcers or punishers received by others. People learn by watching and imitating others. An observer is more likely to imitate behavior performed by a person with prestige, particularly if the behavior was reinforced. Finally, if an observer sees a behavior being punished, he or she is less likely to imitate that behavior (Mager and Pipe, 1976).

A designer should consider the following when scheduling consequences for behavioral changes during training:

1. There should be sufficient opportunities for reinforcement.
2. Responses to correct performance behavior should be *positive, immediate,* and *consistent.*
3. Specific desired behaviors should be recognized and rewarded even if the overall or final performance is less than adequate.

4. Reinforcers that support *undesirable* performance behavior should be identified and removed.

MOTIVATING TRAINEES

In training, people will learn most effectively if they are motivated to attain a training goal. If people remain unmotivated, learning is unlikely to occur. Some motivational approaches available to the designer are:

- Rewards
- Competition
- Goal setting
- Feedback
- Praise
- Recognition.

Again, these motivational approaches should be built into the training materials while they are being developed. A design should not rely on future instructors to provide all the necessary motivation.

THE TRAINING DEVELOPMENT PROCESS

A systematic approach to training development enables designers to produce courses which are relevant, effective, and efficient. A *relevant* course has high validity, that is, the knowledge, skills, and attitudes users learn in the course match the job requirements. A course is *effective* when students who complete the training achieve acceptable scores on criterion examinations and case problems, thus indicating that they have achieved course objectives. An *efficient* course is "lean" and does not waste the time of the training developer, students or instructors. One of the main goals of training, as Butler (1972) points out, is to bring about the "greatest amount of change in performance capability in the shortest amount of time."

The training development process can begin as soon as a designer determines that training can help to ensure an acceptable level of human performance. Some major considerations in the training development process are listed below:

1. What are the specific performance requirements?
2. What performance requirements are to be met through training?
3. What skills, knowledge and attitude do trainees bring with them?

4. What skills, knowledge and attitudes should the trainees have at the end of training?

5. How can trainees best learn the needed skills, knowledge and attitude?

6. What assurance is there that the instruction will be effective?

7. Exactly what performance is desired?

8. What performance level is acceptable?

9. Is *training* a good way to achieve the needed performance level?

10. What are the training alternatives for ensuring acceptable performance?

Most experienced designers know that training cannot overcome problems caused by such things as inadequate function allocation, poor work design or interface, incomplete instructions or performance aids, or inappropriate personnel selection.

As mentioned earlier, training may not be the only or even the best way of achieving adequate human performance. Other options include personnel selection, instructions, and performance aids. Since training is expensive to develop, deliver and maintain, it is usually selected only after other approaches are ruled out.

Analysis

The foundation of any training development enterprise is the analytic work that precedes it, such as function allocation, task analysis, and work module design. There are no shortcuts.

Actual training development efforts cannot begin until the analysis work is complete. A critical part of the analysis deals mainly with identifying the skills, knowledge, and attitudes necessary for acceptable performance on a particular activity. Clearly, the better the analysis and initial design, the less likely that performance problems will develop.

The results of task analysis and work module design should lead to a detailed work description that includes:

1. An identification of the skills, knowledge, and attitude required to perform each task in the work description (Miller, 1972). This becomes a list of the skills, knowledge and attitudes that training must produce or enhance.

2. Well-developed statements of minimum qualifications (SMQs) for each work module. These will provide a good description of the target

population (e.g., the skills and knowledge they already possess). A well-defined target population provides the starting point for training development. The performance expected of that population at the end of the course is the finishing point. The content of the course is established by subtracting what trainees can do from what one wants them to be able to do (Mager, 1967). No matter how carefully a target population is specified, the analysis will always reflect a spectrum of abilities.

When the appropriate analyses are complete, the design portion of training development begins.

Training Design

Determining Objectives

Well-stated, detailed, and measurable objectives are the logical first step in the training design process. Objectives control subsequent course design by providing a framework for the rest of the development process. A good set of objectives should specify skills, knowledge, and attitudes not already possessed by the target population and that therefore, the course must produce or enhance. In addition, objectives should specify required performance levels of the target population at the completion of the course (Butler, 1972).

The two major types of objectives are:

1. Terminal or end-of-course objectives: These reflect the performance levels expected of the trainee upon course completion.

2. Lower-level, subordinate, or enabling objectives: These include all levels and/or kinds of performances required to reach the terminal or end-of-course objectives.

All objectives should lead to one ultimate goal: an acceptable level of human performance. The objectives should track closely with all levels and kinds of abilities required of the trainee.

A clear objective includes the following elements:

1. *A given* - a statement of conditions, limitations, aids, or tools that impact the performance involved

2. *An action* - a description of the overt performance expected of the student

3. *Criteria* - the standards for that performance (time allotted for performance, number of items, degree of accuracy).

The development of objectives is iterative; as development progresses and more is understood about the new system, the objectives should be adapted to reflect the new level of understanding. Objectives ordinarily should not be changed just because they are difficult to reach.

Developing Tests

Tests serve a number of purposes. In-class and end-of-course tests *measure performance*. They also *provide feedback* to trainees and instructors on the attainment of objectives. By providing feedback, the test results potentially affect both learning and motivation. By providing instructor feedback, test results suggest needed revisions.

As soon as all levels of objectives have been completed, the designer should begin to develop test items that match the objectives. Good test items depend on well-thought-out objectives.

There are several types of tests. Two of the most important are prerequisite and entry-level tests. *Prerequisite tests* are administered to ascertain whether a trainee has the skills and knowledge necessary to begin a course. *Entry-level tests* can help determine what material can be omitted from the course because it is already known by the trainee. Both kinds of tests reveal what need *not be* taught to ensure competent performance.

Two other important tests are those administered *during* a course (in-session tests) and at its *termination* (end-of-course tests). These tests measure student success in accomplishing the objectives.

A designer may discover that some objectives cannot be tested. If so, such objectives should be redone. An objective should be measurable. The designer must ensure that all objectives truly reflect the expected end-of-course performance.

When the designer has developed a clear set of *objectives*, and is satisfied that the *test items* measure the objectives, he or she should move on to the next step in the design process.

Developing Training Materials

A basic document outlining key elements of the course should be produced at this point. Ideally, it includes the following elements:

1. Introduction and course outline: a general description of the course, including the overall objectives, an analysis of the target population, and an outline of the lessons in the course.

2. Unit outlines: including objectives for each unit, specifications for training strategies (e.g., unit sequencing, practice strategies, review and feedback points, testing), and specification of materials and media.

3. Implementation materials, facilities, and procedures: including a description of administrative roles and materials, instructor/course manager roles and materials, trainee materials, audiovisual materials,

training aids, training facilities, instructor training requirements, and pre- and post-course procedures and materials.

This design work is done first on a broad preliminary basis and then in a more detailed fashion. When these efforts are completed, drafts of the material are ready for trial and the system designer moves to evaluating the course.

An important consideration is whether to develop group-paced or self-paced training. There are many variations or methods of each mode. For example, *group-paced* includes lecture, illustrated lecture, lecture/discussion, or discussion. The outstanding characteristic of the group-paced mode is that trainees constitute a group and move at the *same pace*.

The predominant characteristic of the *self-paced* mode is that each student proceeds at his or her own pace. In some self-paced courses, the trainee learns from previously prepared or "programmed" materials where:

- the materials provide for continual student activity (responding),
- frequent quizzes are provided for self-checking,
- materials are divided into modules,
- the student demonstrates mastery of a module before he or she moves to the next module.

In many self-paced courses a course manager or mentor is provided to carry out managerial and tutorial functions as needed. There are several variations of the self-paced mode. Figure 21-1 lists common methods of training classified under the two modes.

Instructor-led, group paced courses, may have small inefficiencies or gaps because of inadequate training development that can be overcome by a good instructor. Self-paced training, on the other hand, is often used by trainees working alone, with little support except for occasional assistance from a course manager or mentor. Since self-paced instruction materials must "stand alone" or bear the whole, or major, burden of delivering the training, their design and development must be particularly thorough and well done.

While the ideal self-paced training course is *effective*, it is not always most *efficient* in the use of available resources, and it is seldom as economical as traditional lecture-based training. Four conditions conducive to the development of self-paced training are learning situations where:

1. There is a need for quality-control of the output of instruction; the organization requires *identical responses* from all learners to standard work conditions.
2. The *scheduling constraints* (availability of trainees) for training do not permit the organization of group-paced classes of instruction.

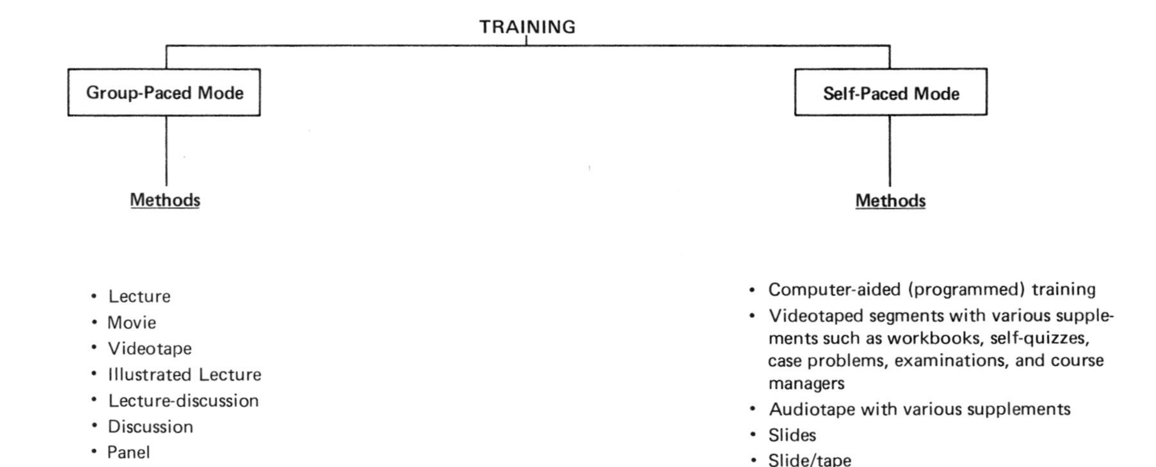

Figure 21-1. Some Representative Methods of Training used in Group-paced and Self-paced Modes

3. The *location constraint* of users does not permit the organization of group-paced classes of instruction.

Conditions *not* conducive to development of self-paced training are learning situations where:

1. The subject matter is yet too volatile (considerable changes are still being made to the new system).
2. The potential benefits, including number of trainees, do not exceed the break-even cost of developing and implementing self-paced training courses.
3. *Group* learning exercises are essential to simulate actual work conditions.

Evaluation: Tryout and Revision

Throughout the entire training development process, the course is tested. Two concerns are paramount for the designer:

1. Internal validity: Does the course elicit the type of performance identified in the objectives?
2. External validity: Will a person who completes the course perform at an acceptable level on the job? The final test of a course's effectiveness is a follow-up study of trainees after they are on the job.

Some of the things to consider when conducting *training tryouts* are:

1. Were *preliminary tryouts* done with first-draft materials?
    a. What were the results?
    b. What revisions were made?
2. Was another tryout done which included a *sufficient number* of trainees to establish confidence that the course is effective?
    a. Do the data summaries indicate high achievement (80 percent to 90 percent of objectives achieved)?
    b. Were indicated revisions made based on the tryout data?
3. Were the tryouts done with *representative users* of the actual target population?
4. Were the tryouts done in an *environment* that resembles as closely as possible the actual training environment?

Each unit of the course should be evaluated separately; then the entire course should be tested as a package. Trainee comments are a primary source of formal and informal feedback, not only during the tryout but throughout the life of a course. Each tryout will suggest revisions. A number of tryouts or revisions are inevitable before a finished product is ready.

Training courses should be kept up-to-date, even after they are trialed, revised, and in regular use. Designers should keep courses current by exploring changing needs in the user population and updating subject matter.

Conclusions

Designers who are developing training should keep several important principles in mind. First, training is a means of improving *human performance.* If a training package is to be effective, it must be based firmly on the analysis of the desired performance. Training is only one viable option that may be chosen to increase the likelihood of competent performance. Because the development, presentation and maintenance of training is expensive, it is not an option to be chosen lightly.

Second, when developing training, the designer should keep it *lean.* It seems to be simpler to add missing pieces to a course than to spend too much development time covering all aspects of everything.

Finally, whatever presentation strategies are chosen (for example, instructor led or computer based), it is good to remember that learning is, in a sense, *always self paced.* A designer must support the reality of people progressing at their own speed.

COMPUTERS AND TRAINING

A discussion of training development would be incomplete without a look at the impact of computers. Computers affect training in at least two ways:

- Their growing presence in homes, schools, and the work place creates a mushrooming need for training in their use.
- Computers may be used as instructional or learning tools in training design and delivery.

The first point is self-explanatory. Few of us remain untouched by computers and more of us are actual users. The second point, the utility of computers as instructional and/or learning tools, requires more discussion. All the power of computers is available to training developers.

To understand the potential of the computer for training, let's look first at some computer-related possibilities. These include:

- *The opportunity to provide demand scheduling:* A trainee can be scheduled for a course when/where he or she wants it by using a computer.
- *The opportunity to reduce training costs:*
  - — The use of the computer can reduce travel costs by providing training or segments of training at the trainee's home base (in fact, in a trainee's home).
  - — Time away from the job and its attendant costs are significantly reduced if on-site training is available.
  - — There is a reduction of such costs as those for instructor, facilities, or lab equipment.
- *The opportunity to provide more flexible training design:* Many training courses have become so complex that they can be difficult to control. The computer can perform many information management activities for the training system, thus freeing trainers to spend more time on creative design activities.

There are several names currently in vogue to describe the role of computers in training. For example, one might speak of "computer-based training," "computer-based instruction," "computer assisted instruction" or "computer-managed instruction." In the following discussion, it will be referred to it as *computer-based training* or *CBT*.

COMPUTER-BASED TRAINING

Advantages

There is little question that in many systems training needs can be best met using computer-based training (CBT). Much of the following discussion is adapted from by Medsker (1979). CBT can present explanations of course content, exercises with feedback, tests, and diagnostic information to trainees (Faust and O'Neal, 1977). Other advantages include having *interactive* training. Frequent responding and immediate, tailored feedback are important training advantages. CBT also enables considerable adaptation to individual needs.

Centralized control is an advantage because all users of a particular CBT course can depend upon receiving *exactly the same course*. If there is a human performance problem associated with training, it can frequently be corrected by making the right change in just the one course. With centralized course material, the necessary changes can be accomplished at all locations simultaneously. This reduces costly reproduction, storage, and distribution of print materials.

Training delivery can be made when and where it is needed. Trainees need not wait for a scheduled class. They can take a course or even portions of a course as work schedules permit. This helps provide training just before it is needed. In addition, many system users can work at their own job-related terminals.

CBT courses can be designed to monitor and record trainee progress and performance. A trainer at one location can monitor and direct the work of trainees at a number of locations. Data about the achievement of objectives and the efficiency of a course can be maintained and easily accessed by trainers. Although the original CBT courses can be developed by the system designer, it can be left to trainers in the new system to maintain and improve the courses.

Disadvantages

Of course there are also some disadvantages associated with using CBT. Expensive hardware and specialized software are required. Acquisition, development, and maintenance of the hardware and software may turn out to be much more costly and difficult than other forms of training. Computer down-time and slow or irregular response time may also prove to be a problem. Historically, one of the greatest disadvantages of CBT has been the high costs of terminals and computer processing time.

Another disadvantage, at least in some situations, is that designers require special knowledge and skills to make the best use of CBT. Because its special qualities are often not understood, CBT can be misused. For example, there is a tendency by less experienced designers to use the computer to simply move from one page of text to the next (i.e., an electronic page turner).

Designers should recognize that CBT course development often requires more time than do other types of course development—up to 200 or 300 hours per hour of instruction. However, estimates of CBT development time vary greatly, depending on how they are computed. Medsker (1979) suggests that factors that may affect development time calculations include whether a systematic design process has been used, whether time spent testing and revising lessons has been included, the developer's familiarity with CBT, his or her subject matter expertise, and the complexity of the CBT lesson format. Usually, the more sophisticated the software (i.e., the better it uses computer capabilities), the longer the preparation time. For systems that change fairly rapidly, long development time may not be feasible.

Evidence for Computer-Based Training

CBT is being used more frequently in new systems, either to deliver all the training, or as a supplement to training delivered in other ways (cf. Bright and Kerr, 1974; Smith, 1979). Existing CBT approaches range from courses to teach a user to operate a computer terminal, to courses used to train aircraft pilots to fly an airplane. Trollip (1979), for example, provides an evaluation of American Airlines flight crew training using a CBT simulator. He reports that when compared with more traditional simulator methods, CBT simulation results in more effective and less expensive training.

Studies of CBT at secondary school and college levels have produced mixed evidence in favor of CBT. When used in a variety of subject matter courses, CBT is about as effective as other delivery modes (Jamison, Suppes, and Wells, 1974). However, some studies do show a savings in student learning time.

Keep in mind that the quality of CBT projects tends to be quite variable. Imaginative and sophisticated applications of CBT produce the best results. It is not enough to simply have the computer present information and have a user respond while at a terminal. The novelty of using a computer soon wears off, and the basic problems associated with knowledge learning and skill building continue. Consider, as an example, the teaching of computer programming using CBT. Results of these studies tend to follow the pattern established in many other subject matter areas where there are usually no significant differences found between effectiveness of CBT courses and their alternatives (cf. Danielson and Nievergelt, 1975; Montanelli, 1977; and Gilbert, 1979).

Deciding When to Use CBT

To select the most appropriate means for delivering a training course (whether it be computer-based or not), the designer should consider the available and reasonable alternatives and match them to the needs of the trainee. Mager and Beach (1967) suggest some principles for selecting the best delivery approach. Briefly, they suggest choosing the approach that:

- Most closely approximates performance conditions on the job
- Causes the learner to perform in a manner closely approximating the performance called for on the job
- Requires frequent interaction by the learner.

Medsker (1979) suggests that CBT is a good candidate when individualized, self-contained training with low trainer involvement is appropriate. To help determine if this is the case, she suggests that the designer see if these conditions are present:

- Mastery of specific, measurable objectives by each learner is required
- Active responding by the learner (practice) is needed
- Job conditions are constant or predictably variable, which minimizes the need for on-the-spot adaptation to local conditions
- Long development time is acceptable
- Subject matter is fairly stable
- A large number of people will take the training
- Flexible scheduling of trainees is desired
- Instructors are expensive or difficult to find.

CBT is *not* desirable when:

- Training must adapt to rapidly changing subject matter or job conditions
- Courses teach interpersonal skills
- Limited time and/or resources exist
- Courses will not be needed over a long period of time.

Within the category of individualized, self-contained instruction with low instructor involvement, CBT is a particularly good choice if the system designer already possesses the required course development skills, including experience with CBT and programming ability or training on how to author a CBT course.

New Developments

Faust and O'Neal (1977) suggest that new developments with CBT are making its use even more attractive to system designers. Some of these improvements include:

- Easy-to-use ways of authoring new courses, instead of cumbersome authoring languages and procedures

- Computer-based guidance for the author during the CBT development process
- Great reductions in size and cost of hardware
- New types of display options: color, graphics, audio, motion, touch panel, voice recognition capabilities, computer speech, etc.
- Increased learner control, as opposed to program control, over the accessing of particular training components (e.g., practice, reviews, examples)
- Special function keys and other controls to assist movement through a CBT course.

FOR MORE INFORMATION

Briggs, L. J., *Instructional Design: Principles and Applications,* Englewood Cliffs, N. J.: Educational Technology Publications, Inc., 1977.

Butler, F. C., *Instructional Systems Development for Vocational and Technical Training,* Englewood Cliffs, N. J.: Educational Technology Publications, Inc. 1972.

Dick, W. and Carey, L., *The Systematic Design of Instruction.* Glenview, Ill.: Scott, Foresman & Co., 1978.

22

PHYSICAL AND SOCIAL ENVIRONMENTS

INTRODUCTION

All human activity takes place within some environmental context. The contexts in which people live, work, and spend their leisure hours are varied. Some environments expose users to physical extremes such as the temperatures near the equator or in the arctic. Other contexts include the weightless environment of outer space or the watery environment of the ocean.

The natural environment is no longer so natural, however. The light level, temperature, and sounds in many modern environments are predominantly human-made. In addition, the contamination and pollution of the natural environment places new demands on people's ability to resist toxic and irritating substances. On the other hand, technology frees people from being dependent on nature. For example, homes and offices help to protect most people from wide variations of heat and cold.

The context in which people work can enhance or degrade their performance. This does not necessarily mean that the environment always controls performance in totally predictable ways. Only extreme environments can do this. The environment is more like a stage on which we perform. Each stage has many and varied characteristics. Some of these will foster good performance; others will handicap performance. Sometimes the effects of the context are subtle and the impact is difficult to measure. At other times the effects are more blatant and have obvious impact, such as the noisy, vibrating environment of some helicopters, steel mills, and rock concerts.

Although the designer usually has some control over the environment, exceptions exist. A telephone line installer-repairer may have to work in the freezing wind high up on a pole, or in the intense heat and sun glare of the tropics. In such cases special steps have to be taken to provide appropriate clothing and other forms of shielding to enable adequate performance. Whether entirely human-made or natural, environment always seems to exert some influence over system users, and at times may become one of the most crucial factors affecting human performance.

AN EXAMPLE OF AN ALTERED ENVIRONMENT

Designers should be aware of how environment may affect task performance, and have sufficient understanding to design, prescribe, and maintain environments that reflect human needs and foster acceptable levels of human performance. Consider the following example.

Reliable Knitwear, Inc., is a medium-sized knitting mill that makes sweaters and sportswear. The company employs over 200 machinists, sewers, steam pressers, and maintenance personnel. Productivity is measured quantitatively (number of items produced) and qualitatively (percentage of seconds and rejects). In reviewing the books, the company's president saw that company profits were declining. She diagnosed the problem as resulting from excessive overhead costs. It seemed to her that it was costing too much to operate and maintain the plant. She announced to her employees that changes had to be made to cut down on overhead.

Her policy was instituted. Cooling and heating units in the steam press room were put on half-power. Lighting was reduced in storage areas, as well as in the knitting room itself. One of the two sewing rooms was closed and the workers all placed in the remaining room, making the room crowded and noisy. As a final economy measure, two of four maintenance people were dismissed.

Unfortunately, the company did not begin to save money. To start with, the overhead costs did not decrease nearly as much as she had estimated. Individuals turned lights back on and reset the controls on the heating and cooling equipment. In addition, production went down and the percentage of rejects and seconds increased. The workers began complaining; more began calling in sick. The production manager could not pinpoint the problem, and suggested that the company president call in a consultant. The president agreed.

After an investigation the consultant reported her findings and recommendations. The workers were not producing at their old rate because they felt abused by the company's policy of reducing the overall comfort level of the work environment. They felt the company did not care about them, only about profits.

But a negative attitude was not wholly responsible for low productivity. The poor lighting and the extreme temperatures in the press room seemed to be causing workers to make more errors and become fatigued sooner. The crowded and ill-maintained conditions in the sewing room caused further discomfort and also impeded the work process.

The president of this company had treated her workforce as if they were machines indifferent to the environmental conditions of the workplace. Her actions could make only a small difference to the overhead costs while bringing about a major problem resulting in a considerable loss of revenue for the company and the loss of the goodwill of many employees.

As this example suggests, the working environment is very important to human performance. There is a mistaken idea that designing acceptable environments consists of wrapping the worker in luxury and providing a soft reclining chair. Ensuring acceptable system environments is not a matter of pandering to the workers. Rather, it is a way of providing a work situation or context in which an acceptable level of human performance is most likely to occur.

THE GOALS OF ENVIRONMENTAL DESIGN

Performance Support

Just as an astronaut is provided a life-support system in order to live during space flight, so we can think of the context or workspace in which an activity is performed as a *performance support system*. The designer requires and expects a certain level of performance from users as they undertake and complete various tasks. The environment in which the tasks are performed has to support the required performance level, which takes into account meeting the basic physical, physiological, and psychological needs of users. For example, a driver should fit comfortably into his or her car and be within easy reach of the controls (physical needs). The automobile suspension should reduce vibration as much as possible, and the exhaust system must be functioning properly to avoid carbon monoxide poisoning (physiological needs). If the trip is a long one, a radio may alleviate boredom (psychological needs).

At the very least, environments should do no harm. That is, the environment should not drain off energies or be a point of dissatisfaction in itself. The environment should provide for the user's basic needs such as adequate light, air, temperature, seating, workspace, and safety. But more than that, the environment should *support* the performance of work so that work is easily organized, resources are readily accessible, and communication is facilitated.

Considering the environment as a performance-support system also includes aspects of the social environment. When designing an environment for a research group, one would wish, in general, to provide each person with his or her individual space, while at the same time encouraging interaction, group problem-solving, sharing of ideas, and a sense of group identity. While there are many considerations, one of the most important is placing the group members in physical proximity. This helps to encourage effective interaction.

Organizational Support

Not only should the environment support user performance, it should also transmit such messages to users as the nature of the organization, its structure, its values, and the nature of the work. People are very good at picking up these cues, and tend to behave and react accordingly.

Most people have been interviewed for employment. On entering the premises they receive numerous messages, conscious or otherwise, about the organization. A side-by-side interview with a casually dressed person who uses first names tells something quite different about the organization than an interview with a Dr. Comesky who is seated several feet away behind a forbidding desk. In addition, this formal interview may take place in a room where the wall coverings, window coverings, and furnishings echo the differing status of the personnel. It is hard to be informal in a very formal atmosphere, and it is equally difficult to behave formally in an informal atmosphere. Human performance is affected by surroundings. Consider, for instance, the difference in behaviors exhibited by the same people in a church, museum, and baseball park.

The issue is not that one organizational structure is better than another, or that one environmental design is better than another. The issue is rather that the environmental design portrays or reflects the organizational structure, and at least indirectly the kinds of performance behaviors that are acceptable in that particular environment or context. Designers should recognize that the environmental recommendations they make will help create an organizational image. Careful consideration of design issues associated with the environment is needed to ensure that the image is the one desired, that the performance elicited by the environmental cues is acceptable and desirable, and that the intended organizational hierarchy is reflected by the layout of the working environment.

A Normal Environment

A third consideration when designing a work environment is that all critical conditions should fall within a normal range. People are quite tolerant of environmental diversity, but not infinitely so. For any condition in the environment there is a normal range wherein people perform best. In environmental conditions outside that normal range, human performance may be degraded. The more abnormal the environment, the greater the probability of having degraded performance.

Guidelines have been developed that include information concerning workspace size, noise, lighting, and temperature. Special applications may require designers to consider additional aspects of the environment such as vibration, acceleration, toxic substances, electrical shock hazard, or radiation hazard. System designers should conform to available guidelines wherever possible. It is particularly important that designers recognize the environmental conditions outside the normal range, and take appropriate steps either to change the environment or minimize its effects on users. For example, a system design should specify warm clothes for working in severe cold, and salt tablets for working in heat.

Assisting Learning

A fourth consideration when designing a work environment is that it should function to help the user learn about the system. Where possible, environments should be designed so that they enhance the process of learning appropriate knowledge, skills, and attitudes. In some cases, the total work area helps to accomplish this—corporate libraries, training facilities, mock-ups, and simulators. On a smaller scale, environments can be designed to include the use of bulletin boards, safety posters, flip charts, and blackboards. Inclusion of these and other similar considerations help enhance the user learning process and, if done well, ultimately will assist in achieving an acceptable level of human performance. For example, the use of office-size blackboards or flip charts to convey information in group meetings is likely to facilitate communication and learning.

HUMAN-ENVIRONMENT INTERACTION

Someone once observed that people who chop their own wood get warmed twice. The human and the environment interact in some very interesting ways. Consider working in a cold environment (cf. Fox, 1967). The body responds in a variety of ways to conserve heat (for example, withdrawal of blood from the periphery) and to generate heat (for example, shivering). The person may also do exercises (such as slapping his or her arms), put on warmer clothing, limit exposure time to the cold conditions, or turn up the thermostat. People and their environments frequently interact in a chain of events. The environment impacts on people; people then make changes either to lessen the effects of the environment or to alter the environment. A designer should provide the user with as much control over a work environment as possible.

In the real world, the effects noted may be more or less pronounced depending on the nature of the work activity and the particular people involved. Delicate, fine work performed by sedentary, highly skilled people is more affected by adverse environmental conditions than less skilled, muscular work. In addition, highly motivated people usually tolerate inferior environmental conditions better than those who are less motivated.

These variations emphasize the interactive nature of the situation we are studying, and are in keeping with the model presented in Chapter 1. Human performance in any situation is a three-way interaction among the human, the work activity, and the environment. For example, studies show that heat affects Morse code reception (the more difficult task) more than Morse code transmission. Yet highly skilled telegraphers have been found to maintain their performance even in heat conditions which seriously diminished the performance of less skilled persons.

We must not take too lightly the potential effects of environmental conditions on human performance. The role of the designer is to understand the ranges of human adaptation to various environments and, most importantly, to design the environment to be a positive factor in promoting an acceptable level of human performance.

ADAPTATION AND THE OPTIMAL ENVIRONMENT

Poulton (1972) compiled a list of environmental conditions and resulting effects on performance, which is shown in Table 22-1. Designers should take care that the characteristics of their work environments fall within the ranges shown. An inappropriate amount of heat, light, noise, or vibration, for example, could result in degraded human performance.

Designers should not be misled by the apparent adaptation capabilities of people. The concept of adaptation refers to changes (usually cognitive changes) that take place in an individual to enable him or her to deal adequately with an adverse environmental condition. Adaptation may occur when people are exposed to a totally new environment or an existing environment that has changed considerably. There are limits to adaptation—limits imposed by physical, physiological, or psychological characteristics. When a person reaches the limits of adaptation he or she will experience stress, which in turn may result in degraded performance. Designers should consider the limits of human adaptation and design realistic environmental conditions that do not exceed these limits. In so doing, designers will be creating performance supportive environments.

Table 22-1. Environments Just Severe Enough to Produce a Reliable Deterioration in Performance (adapted from *Environment and Human Efficiency*, by E.C. Polton, 1972. Reprinted by permission of Charles C. Thomas, Publisher, Springfield, Illinois.)

Environmental Condition	*Condition just severe enough to degrade performance*	*Task*	*Kind of deterioration*
Heat	Air temperature 27° C (80° F)	Making knitted clothes	Reduced output
		Reading	Reduced speed and comprehension
		Making munitions	Increased accidents
Cold	Air temperature 13° C (55° F)	Tracking	Reduced time on target
		Making munitions	Increased accidents
Dim light	7 to 10 ft-c	Reading 7-point newspaper type	Reduced speed
		Reading 6-point italic type	Reduced speed
Glare	Depends upon the angle of the glare source, of the line of sight, as well as upon its brightness	Inspecting cartridge cases	Reduced speed
		Typesetting by hand	Reduced speed
Noise-Continuous	100 dB	Tapping 1 of 5 targets in response to 1 of 5 lights	Increased errors
		Threading photographic film through machine in dim light	Increased broken rolls and shutdowns attributed to the human
Noise- Intermittent	95 dB for 1 second	Reporting differences between pairs of cards	More omitted reports
	Noise varying randomly between 90 and 65 dB	Recording digits	Increased variability in rate of work

Environmental Condition	*Condition just severe enough to degrade performance*	*Task*	*Kind of deterioration*
Noise- Interference with speech communication	70 dB 600 to 4800 Hz	Speaking at 3 feet	Reduced comprehension
		Telephoning	
Vibration of human	0-0.02 inch at 19 Hz	Reading 12-point digits at 10 feet with 0-2 ft. L of illumination	Increased errors and reduced speed
Acceleration	3G upward (gravitational force 3 times the normal size)	Subtracting 3s starting with a number close to 100	Reduced speed
Weightlessness	1/6 G (gravitational force one sixth the normal size, as on the moon)	Tightening bolts	Reduced speed
		Joining connectors	Reduced speed
Carbon monoxide	Carboxyhaemogobin 8% in non-smokers	Adding columns of numbers	Reduced speed
Work after isolation	24 hours of isolation	Solving a difficult problem	Problem less often solved
		Thinking of uses for objects	Fewer uses thought of
Danger producing fear	Threat of injury - before emergency	Following complicated instructions	Increased errors

ENVIRONMENTAL COMPONENTS

A designer should think of the environment as having both a physical component and a social component. The physical environment includes such factors as noise, lighting, temperature, space, vibration, wind, air pollution, radiation, and gravity (weightlessness, acceleration). The social environment includes such considerations as social facilitation, conformity, privacy, personal space, territoriality, and crowding. It is usually true that designers have more direct control over the physical environment than the social environment. This by no means suggests, however, that a designer should not attempt to make design decisions to enhance the social environment.

PHYSICAL ENVIRONMENT

Human systems are vulnerable to extreme conditions in the physical environment. In aviation many accidents have been related to such things as poor visibility, wind, or thunderstorms. Automobile accidents frequently are associated with sun glare or icy roads. These conditions, however, seldom directly cause the accidents. Rather an environment with difficult and hazardous conditions may severely stress the human who is performing a particular activity. The environment can alter a person's ability to perform. Excessive cold or heat over long periods of time cause physiological changes—for example, frostbite, heat exhaustion—that not only lead to seriously impaired performance, but also may result in injury or even death.

Controlling the environment to make it more compatible with human physical, physiological and psychological limits has become commonplace. A person leaves an air-conditioned home to drive an air-conditioned car to an air-conditioned restaurant for dinner. Except for brief periods, the temperature is maintained at a comfortable level. Divers explore and do useful work in the oceans as a consequence of the development of underwater breathing gear. The use of fast, high-altitude aircraft in commercial aviation is possible only because the passenger cabins are maintained at a low altitude pressure. These are some of the many ways designers have controlled people's environment to enable adequate performance.

Noise

Noise may be defined as unwanted sound. This is not an entirely satisfactory definition because a sound wanted by one person may be unwanted by another person. Think, for instance, of a noisy party next door which is not considered noisy to the party goers; or the noise of a rock band to a rock fan versus the same sound to others. Thus, some sounds will be noisy because they are unwanted, intrusive, irritating, or distracting.

People constantly experience a wide variety of noises. These noises range from the gentle hum of a fluorescent light to the deafening roar of a jet at takeoff. Due to human adaptation and other similar phenomena, we tend to grow accustomed to most noises and pay little attention unless they become annoying. Even so, designers should be aware that despite the adaptation capabilities of the human, an inadequate environment still may result in performance problems.

The single most often voiced complaint in work environments is that there is too much noise and that it is distracting. The individuals most likely to be affected are those engaged in complex tasks requiring close, undivided attention and concentration (Glass and Singer, 1972; Finkelman, 1975).

As indicated earlier in the book, sound has at least two characteristics related to human performance: frequency and loudness. There is no single frequency at which noises produce harmful effects; however, low frequency noises, when sufficiently loud, can produce the most serious hearing impairments. On the other hand, high frequency noises usually interfere most with work and are considered the most annoying. Noises in the intermediate frequencies tend to interfere most with the intelligibility of speech.

Loudness is directly related to the mechanical pressure it exerts on the ear. Table 22-2 shows intensity levels of various common sounds, and their effects on people.

In general, a loud noise (90 dB or more) maintained over at least 30 minutes will cause degraded performance. High frequency noise (over 1200 Hz) has been found to interfere with performance more than noise in the lower frequencies. Intermittent and unpredictable noises are worse than steady sounds. When unpredictable, even low-intensity noise can be most annoying and distracting. Such noises are great attention grabbers and so become distractors. They may not affect the rate of work, but they can cause the worker to commit more errors.

The annoyance value of noise is related to its loudness, frequency, predictability, and the degree to which it is under the control of the hearer (Kryter, 1970; Glass and Singer, 1972). The more control a user has to start or stop a noise, the less the annoyance value of the noise.

However, the total absence of noise may also impair performance. Noise tends to have an arousal value with many people. Particularly for dull or uneventful tasks, it is important to have some sounds in the environment to help keep people alert (Warner, 1969; Eschenbrenner, 1971; Poulton, 1972).

The aim of a designer should be to have a noise level of about 50 dB in the environment. This is equivalent to the hum of voices and quiet machinery. This noise level is also information bearing in that it keeps the hearer apprised of what is going on in the environment, but it is not loud enough to be distracting.

Lighting

Although the sun, and sometimes the moon, provide adequate lighting outdoors, people require artificial light to perform activities inside. Lighting requirements range from the low, softly illuminated environment of a commercial airline cockpit to the bright surroundings of a hospital operating room. The lighting in both these situations has been designed to elicit acceptable levels of human performance.

The optimization of artificial illumination has long been a concern to researchers and designers. In particular, a great deal of attention has been given to work situations where there is too little light, low brightness contrast, or glare. The Hawthorne studies began in 1929 to determine the effects of

Table 22-2. Intensity Levels and Effects of Various Common Noises

Common Sounds	Noise Level (dB)	Effect
Carrier deck jet operation Air raid siren	140	Painfully loud (blurring vision, nausea, dizziness)
Jet takeoff (200 feet) Thunderclap	130	Begin to "feel" the sound
Discotheque Auto horn (3 feet)	120	Hearing becomes uncomfortable
Pile drivers	110	Cannot speak over the sound
Garbage truck	100	
Heavy truck (50 feet) City traffic	90	Very annoying
Alarm clock (2 feet) Hair dryer	80	Annoying
Noisy restaurant Freeway traffic Man's voice (3 feet)	70	Telephone use difficult
Air conditioning unit (20 feet)	60	Intrusive
Light auto traffic (100 feet)	50	Ouiet
Living room Bedroom Quiet office	40	
Library Soft whisper (15 feet)	30	Very quiet
Broadcasting studio	20	
	10	Just audible
	0	Hearing begins

illumination on performance. A short time later, Wyatt and Langdon (1932) and Weston (1949) found a deterioration of performance (reduced speed) in inspection and typewriting tasks when subjects performed these tasks under glare conditions. Tinker (1943, 1952) has shown decreases in human reading speed in dimly lighted environments.

As an example of the role that lighting can play in system functioning, consider the following. A leading manufacturer of men's suits had been losing profits steadily over a two-year period. A review of the company's books showed that returned merchandise costs were high. Approximately 10 percent of all suits had been returned by either dissatisfied retailers or individual customers. Further investigation revealed that most of the suits in question were made of dark-colored fabric. It was determined that the inspection and quality control personnel were not detecting defects in darker colored suits. It was discovered that the illumination level in the inspection area was high enough for workers to detect flaws in light-colored but not in darker colored fabrics. The situation was remedied by increasing the lighting. It was further found that by using lamps at an angle, certain types of fabric flaws could be seen much more clearly.

The above example underlines the importance of systematically considering every aspect of the work environment in systems design. With lighting, some considerations tend to be very subtle. Designers need to recognize that the illumination requirements of at least *some* visual tasks deviate from standardized recommended levels, and should be given individual consideration where possible.

Individual workers require different levels of lighting. A common mistake with both noise and illumination is to assume that all people are the same. Thus, we erroneously provide an average level of illumination to accommodate the average person. As mentioned earlier in the book, the average person does not exist. Guth and Eastman (1955) suggested that although it is perhaps idealistic to provide for individual lighting requirements in the workplace, it should be obvious that supposedly average illumination levels automatically provide insufficient lighting for a certain percentage of the work force. Designers should establish human performance requirements for each activity in their particular system, and then provide sufficient illumination levels to ensure the fulfillment of these requirements for *all* users.

Blackwell (1959, 1961, 1964, 1972) studied a variety of factors related to poor visibility and his work has served as the basis for specifying illumination levels for various work activities. A wide range of recommendations can be found in the I.E.S. Lighting Handbook. In general, research indicates that insufficient illumination, glare, reflectance, shadows, low brightness contrast, and flickering adversely affect human performance and should be controlled by system designers. Specific design recommendations are summarized in Tables 22-3 and 22-4.

Table 22-3. Recommended Illumination Levels (footcandles) for a Variety of Different Task Conditions

Task condition	Footcandles
Assembly	
Rough easy seeing	30
Rough difficult seeing	50
Medium	100
Fine	500
Extra fine	1000
Auditoriums	
Assembly only	15
Exhibitions	30
Social activities	5
Banks	
Lobby	
General	50
Writing areas	70
Teller's stations	150
Posting and keypunch	150
Conference rooms	
Critical seeing tasks	100
Conferring	30
Note-taking during projection	30
Control rooms	
Vertical face of switchboards	30
Bench boards (horizontal level	50
Emergency lighting, all areas	3
Drafting	
Detailed drafting and designing, cartography	200
Rough layout drafting	150
Hospitals	
Anesthetizing and preparation room	30
Autopsy and morgue	
Autopsy room	100
Autopsy table	1000
Fracture room	
General	50
Fracture table	200
Lobby (or entrance foyer)	
During day	50
During night	20
Medical records room	100
Nurses station	
General—day	70
General—night	30
Desk for records and charting	70
Obstetrical suite	
Labor room, general	20
Delivery table	2500
Recovery room, general	30
Patients' rooms	
General	20
Reading	30
Observation (by nurse)	2
Night light, maximum at floor	0.5
Examining light	100
Toilets	30
Pharmacy	
Compounding and dispensing	100
Surgical suite	
Operating room, general	200
Operating table	2500
Recovery room	30
Inspection	
Ordinary	50
Difficult	100
Highly difficult	200
Very difficult	500
Most difficult	1000
Library	
Reading areas	
Reading printed material	30
Study and note taking	70
Card files	100
Carrels, individual study areas	70
Circulation desks	70
Offices	

Reading poor reproductions, business machine operation, computer operation	150
Reading handwriting in pencil, reading fair reproductions, active filing, mail sorting	100
Reading handwriting in ink, intermittent filing	70
Reading high contrast or well printed materials	30
Conferring and interviewing	30
Corridors	20

Residences

Specific visual tasks	
Dining	15
Grooming, shaving, make-up	50
Food preparation and cleaning	150
Ironing	50
Sewing	
Dark fabrics	200
Medium fabrics	100
Light fabrics	50
Table games	30
General lighting	
Conversation, relaxation, entertainment	10
Passage areas, for safety	10
Areas involving visual tasks, other than kitchen	30
Kitchen	50

Sports lighting

Baseball

Major league	
Infield	150
Outfield	100
Municipal league	
Infield	20
Outfield	15

Basketball

College and professional	50
High school	30
Recreational (outdoor)	10

Football

Distance from nearest sideline to the farthest row of spectators	
Over 100 feet	100
50 feet to 100 feet	50
30 feet to 50 feet	30
Under 30 feet	20
No fixed seating facilities	10

Ice Hockey

College or professional	100
Amateur	50
Recreational	20

Ski slope	1

(adapted from Kaufman, 1972)

Table 22-4. Recommendations for Glare Reduction (Van Cott and Kinkade, 1972)

Direct Glare	*Specular Glare*
1. Avoid intense light sources within 60° of any commonly used line of sight.	1. Use diffuse light.
2. Use shields, hoods, and visors.	2. Use dull, matte surfaces instead of polished or glossy.
3. Use indirect lighting.	3. Arrange direct light sources so that the viewing angle to the work surface is not equal to the angle of incidence from the source.
4. Use several lower intensity luminaires instead of one very bright one.	

Temperature

Another condition of the physical work environment that system designers should be aware of is temperature. Extreme outdoor temperatures, while largely beyond the control of the system designer, can be compensated for by cooling equipment to control heat and proper clothing to counteract cold. Indoor temperatures, on the other hand, may be under the control of designers and should be modified to support an acceptable level of human performance. Like many other conditions in the physical environment, extreme temperatures may degrade the performance of work activities.

The temperature inside the workplace also may vary. Most workers can recall one time or another when it was hard to concentrate on their work or to perform for any length of time because of excessively hot or cold temperatures. Most people readily adapt to moderate fluctuations in environmental temperatures, but not to extremes. As we indicated earlier, the designer should recognize that there are *limits* to this adaptation process, and design the system environment so these limits are not exceeded.

Heat and Performance

The effects of heat on human performance has a considerable research history (Baron and Bell, 1976; Bell, Provins, and Hiorns, 1964; Griffiths and Boyce, 1971; Provins and Bell, 1970). Bell, Fisher, and Loomis (1978), in a review of laboratory studies dealing with the effects of heat on performance, report that some researchers find detrimental effects, some find improvements in performance, and others find no effects at all. They suggest that these results are best interpreted in the light of arousal theory discussed in Chapter 9. More specifically, laboratory research findings seem to suggest that effective temperatures (effective temperature is an index of perceived comfort level influenced by temperature and humidity) of about 90°F (32°C) increase arousal and tend to help performance of simple tasks, and degrade performance of complex tasks.

Additional research dealing with the effects of heat on performance has been conducted in actual job settings. Tichauer (1962), for example, studied the relationship between temperature and performance of cotton-pickers. He found that workers produce optimally between 75° and 80 F, whereas workers slowed down when the temperature was above 90°F. Wyon (1970) found reduced speed and comprehension in readers at temperatures greater than 80°F (27°C). Research with mineworkers (Wyndham, 1969) and students (Pepler, 1972) has reliably shown performance deterioration with increasing temperatures.

It seems that the actual level of heat necessary to degrade performance varies from situation to situation and from individual to individual. What is important from a system designer's point of view, however, is the fact that excessive heat does affect human performance. Crockford (1967) and Hill (1967) have discussed the problem of excessive industrial heat in factories—near furnaces or boiler rooms—and have suggested that workers be provided with an adequate intake of water and salt, wear protective clothing, have frequent rest breaks, and when new to the job, have adequate time to adapt to working conditions. Other possible ways to help alleviate the effects of heat include the use of increased ventilation, fans, and air conditioners where possible. See Figure 22-1 for recommended exposure times to heat.

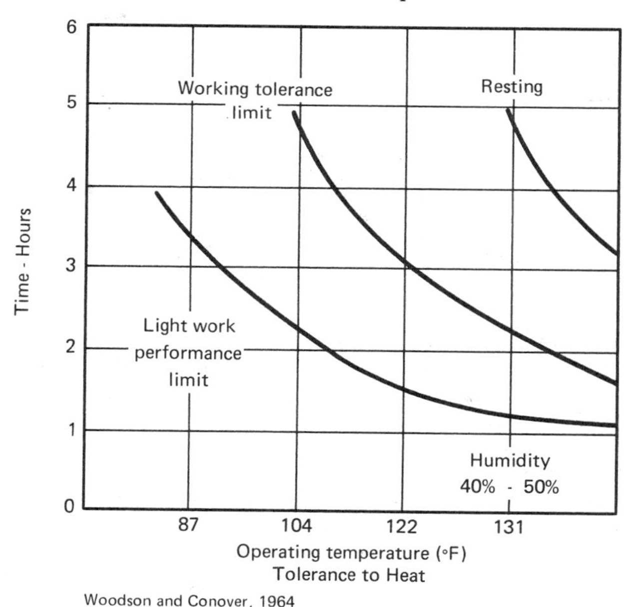

Figure 22-1. Recommended Exposure Time to Heat as a Function of Activity and Temperature

Designers also should consider the comfort levels of workers. Comfort levels usually are measured in effective temperature units discussed above. See

Figure 22-2 for a summary of comfort levels. It should be pointed out that comfort levels for feeling cool are generally below those at which performance deteriorates. However, a sustained experience of discomfort can lead to distraction, which may produce performance decrements.

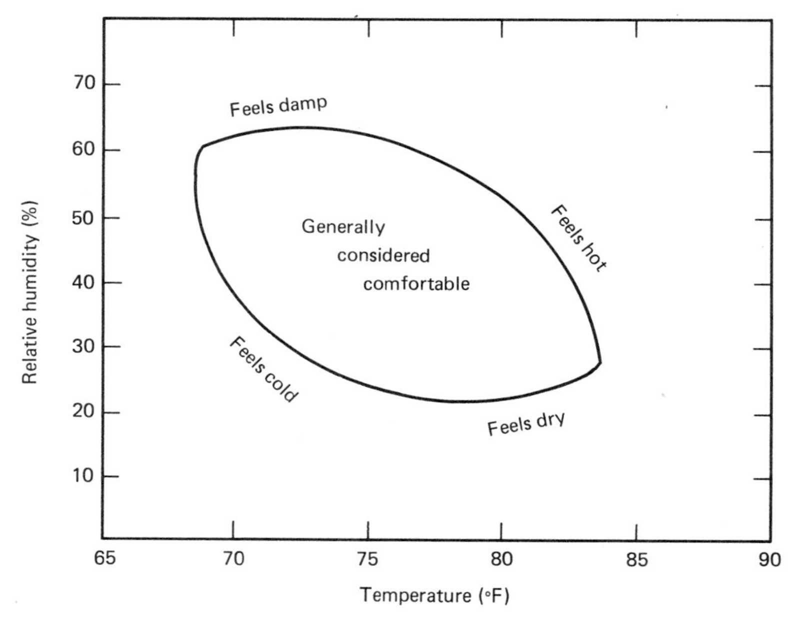

Figure 22-2. Comfort Zone as a Function of Relative Humidity vs. Temperature

Cold and Performance

It is relatively rare to find indoor environments sufficiently cold to seriously impair performance. However, some machines and storage areas require an ambient temperature below *65°F*. This creates a chilly environment for sedentary workers. If activities require the use of cold rooms for storage, people need to be clothed in ways similar to those specified for working in cold outdoor conditions. Energy conservation policies will most likely cause marginally cold conditions for many workers during winter months. It is important to remember that people vary widely in their ability to adapt to moderately low temperatures. For example, older people and those with poor vascular circulation could feel much colder than others.

Long exposure to cold temperatures as high as *55°F* may cause hands and feet to become chilled, and lose some of their strength, touch sensitivity, and ability to make fine manipulations (Fox, 1967). These losses increase with lower temperatures and longer exposures. The individual becomes progressively more awkward. In time, psychological abilities are affected—it becomes difficult to concentrate and memory lapses occur.

To guard against possible degradation of human performance in extremely cold temperatures, designers should make a variety of preventive decisions. First of all, they must consider the amount of protective clothing that is necessary when workers perform in environments of different severity. Secondly, selection procedures should be used to screen out individuals who cannot tolerate extreme cold. A third consideration would be to ensure that workers are given sufficient time to adapt to the environment. Finally, design decisions should be made that help guarantee that reasonable work and rest schedules are implemented.

Designers also should take steps to develop machine controls such as handles, pushbuttons, levers, or switches with oversized dimensions to make them easier to operate with gloved hands. In addition, they should consider entry and exit problems for people who may be slower and more awkward in their heavier clothing.

Workspace

Space allocation is particularly important to system designers in large growing companies. Although, ideally, the space allocation process is done on an objective basis, the end result frequently is that many workers end up with too little space and others with much more than they need. Often space allocation problems arise because the process does not take place on a functional basis (work requirements), but rather on the basis of seniority or status with the company. With increasing seniority and status in the company, workers are afforded larger offices just as they are afforded more pay and more vacation. This may start to degrade performance when employees are clustered so close together that they do not have adequate workspace to perform their work effectively.

Designers should assume the responsibility of specifying the minimum amount of workspace required for a user to adequately perform. For example, designers planning to place several individuals together in one area should be aware of the general requirement of a minimum of 70 square feet per individual. Some well-designed systems do not maintain an adequate level of human performance simply because the designer did not specify the necessary amount of work space.

Other Conditions of the Physical Environment

Aside from the major characteristics of the physical environment just discussed (noise, lighting, temperature, and space), designers should be aware of numerous other conditions that may affect human performance in certain systems. Vibration from heavy machinery, motor vehicles, and other equipment has been linked to performance decrements in tracking tasks (Lovesy, 1970), reading (Dennis, 1965), and tasks requiring considerable eye-hand coordination (Guignard, 1965). The effects of vibration can be reduced, for example, by selecting machines that vibrate less, incorporating spring suspensions where possible, and increasing letter and number size in displays for easier readability.

Anyone who has ever attempted to perform an outdoor activity on a windy day can attest to the fact that wind can be detrimental to performance. Poulton, Hunt, Mumford, and Poulton (1975) exposed a group of people to varying degrees of turbulence and found, as expected, that high wind (20 mph) and gustiness produced decreases in performance over a variety of tasks.

Wind has the greatest potential effect on performance when it occurs together with cold temperatures. The wind chill index (see Table 22-5) is important to designers who must calculate maximum exposure times and clothing requirements for system users performing work activities in cold environments. For example, telephone line workers making emergency repairs in $-4°F$ temperatures with a steady wind of only 15 mph are actually being subjected to a temperature that feels like $-37°F$. This type of environment will require shorter work periods, more frequent warm-up sessions, and better insulated clothing.

Table 22-5. Wind Chill Index

		Wind Speed			
		5 mph	*15 mph*	*25 mph*	*35 mph*
	32	29	13	3	- 1
Actual	23	20	- 1	-10	-15
Temperture	14	10	-13	-24	-29
°(F)	5	1	-25	-38	-43
	- 4	- 9	-37	-50	-52

Designers should be aware of the effects of air pollution (particularly carbon monoxide) on workers. Breisacher (1971), in a review of studies dealing with air pollutants, reports that exposure to relatively small amounts of these pollutants adversely affects human reaction time, attention, and manual dexterity. Wertheim (1967) exposed subjects to a level of carbon monoxide similar to what one would find on a moderately traveled freeway at rush hour. After 90 minutes of exposure, subjects evidenced a significant impairment on a time-judgment task. With heavier concentrations of carbon monoxide, less exposure time was required to produce performance decrements. Workers such as automotive mechanics, heavy outdoor machinery operators, and others exposed to even moderate levels of exhaust fumes or cigarette smoke (cigarette smoke contains relatively large quantities of carbon monoxide) are subject to the adverse effects mentioned over a period of time. Designers should provide proper ventilation. Reducing or even eliminating the pollutant at its source by selecting vehicles and machinery that emit less carbon monoxide is an even better solution.

Designers of some systems should be aware of the performance correlates of prolonged exposure to other environmental conditions, for example, weightlessness and acceleration in space capsules, and radiation in nuclear power plants. For a discussion of these topics, readers may refer to Poulton (1972).

Combined Environmental Stressors

When two or more adverse environmental conditions exist at the same time, the effect on human performance is not quite as easy to predict. The effect may be greater than the sum of the environmental conditions taken singularly, less than each individual condition, or equal to their sum. For example, Harris and Shoenberger (1970) found that the combination of excessive noise and vibration produced an additive decrement in a compensatory tracking task. Again, many of these effects are best understood by considering the arousal concept discussed in Chapter 9. A discussion of research on combined environmental conditions can be found in Poulton (1972).

Environmental Preferences and Aesthetics

Although little research has looked directly at the relationship between human performance and user preference for a given work environment, the issue of environmental preference itself should be important to system designers. If the work area or environment is not pleasing to users, it may lead to job dissatisfaction and degraded performance. Hertzberg's (1959) research suggests that a pleasing work area would not necessarily lead to greater satisfaction, but one that is nonpleasing would lead to dissatisfaction. S. Kaplan (1974), R. Kaplan (1974) and Wohlwill (1974, 1976) have investigated the concept of environmental preference. These researchers suggest that while people generally prefer environments that are complex, spacious, identifiable, coherent, and of smooth texture, individuals may have different levels of preference depending on past experience.

The issue of environmental colors and environmental aesthetics also has been of concern to researchers. Ertel (1973), for example, has shown that when school children were tested in rooms they thought were "beautiful" (light blue, yellow, yellow-green, orange) their IQ scores were significantly higher then when tested in rooms they considered "ugly" (white). Follow-up studies were conducted with similar findings.

THE SOCIAL ENVIRONMENT

Presence of Others

The user's social environment could include one or more co-workers, management people, customers and possibly even spectators. The total effect of these people on a user often impacts on human performance. As an example, consider one of the first research studies on the effects of the social environment, conducted in the late 1800s by Triplett. While examining the official records of bicycle races, Triplett noted that a rider's maximum speed was approximately 20 percent greater when he was paced by a visible multicycle. Desiring to learn more about the matter, he set up an experiment with children in the age range of ten to twelve, giving them the task of winding fishing reels. Alternating situations *alone* and *together,* he found that when working together 20 of his 40 subjects *excelled* their own solitary record, while 10 did less work and 10 were essentially unaffected. All in all, he concluded that the group situation must be thought of as producing greater achievement. He felt that the children's competitive spirit, together with the facilitative effects of others watching the competition, caused the increase in performance.

Since Triplett's early studies hundreds of other studies have been conducted on the relationship between human performance and the social environment. The general conclusion of investigators has been that the presence of others enhances performance of well-learned skills, but inhibits the acquisition of new skills, such as memorizing a list of nonsense syllables. Zajonc (1965) has suggested that the presence of others increases arousal in such a way that the dominant (or most probable) response becomes even more dominant.

The effect of others is more dramatic when *pressure* is placed on the individual to conform. People tend to feel uncomfortable when they are alone in their actions or out of step with the crowd. Two studies of interest in this regard are those conducted by psychologists Asch (1951) and Milgram (1965). Asch (1951) showed that the judgment of a subject was influenced by several other people who had first purposely made incorrect responses. That is, there was a tendency for the person to respond like other members of the group even when he or she thought the response was wrong. Many people follow directions from their supervisors and co-workers in much the same way.

Milgram (1965), in another study of conformity, instructed subjects to administer what they were led to believe would be a painful electrical shock to another person. They did this because they were expected (by the experimenter) to do so, even though it appeared to result in extreme pain to other people.

The implications for designers from these and other similar studies are relatively straightforward. Most people are strongly influenced by social pressure, competitive forces, and even the mere presence or absence of others in the workplace. Particularly in situations where important judgmental or evaluative decisions need to be made, designers should develop a system in such a way that decisions are made as independently as possible. For example, if an engineer needs to decide how much sheet metal to purchase for a given work project, he or she should make an estimate alone, and then individually consult with other engineers. Only after a decision is made should it be passed to a supervisor or others for approval. Such a procedure is preferable to a situation in which an engineer and two or three levels of management provide their views prior to the making of an initial decision. In the latter case, the engineer may be influenced against the best decision because some advisors (including members of management) gave their opinion first.

Personal Space and Privacy

People have very definite feelings about their control over privacy, being isolated or crowded, having a certain amount of personal space, and having a certain territory that is their own. For example, a person seated on a train or bus where there are many empty seats is likely to feel uncomfortable if another passenger chooses to sit next to him or her rather than in a vacant seat. Similar feelings are held by people in their workplace. Designers can improve human performance by understanding this aspect of the social environment.

Privacy

Altman (1975) has suggested that people are constantly struggling to obtain a certain desired level of privacy. People attempt to achieve a desired privacy level through the regulation of personal space boundaries and through claiming territory as one's own. There are three possible outcomes when a person attempts to regulate privacy:

- The achieved privacy is less than the desired level, in which case they experience crowding.
- The achieved privacy is more than the desired privacy level, in which case they experience isolation.
- The desired level of privacy is achieved.

Opponents of the open office or landscaped office concept, for example, point out that lack of privacy is a major factor behind employee discontent with such layouts. Restaurants, public rest rooms, and private homes all are designed in part with a consideration for privacy. Designers need to consider privacy when making decisions for their systems if they want to ensure an acceptable level of human performance.

Personal Space

The term "personal space" was first coined by P. Katz in 1937. The term has come to refer to an area with an invisible boundary surrounding a person's body. Encroachment of an individual's personal space may result in feelings of displeasure, anger, or withdrawal.

Hall (1966) investigated human spacing behavior in several different cultures, and developed a system of studying how people use space as a communications vehicle. Hall discussed four spatial distances or zones that are used in social interaction. The first zone, from 0 to 18 inches, is called the intimate zone and generally is used only for communication between intimate friends. The next personal zone extends from 18 inches to 4 feet and is used for contacts between close friends and everyday acquaintances. The third social zone, from 4 to 12 feet in distance, usually is used for impersonal and businesslike contacts, while the public zone extends from 12 feet out and is appropriate for contacts between politicians, actors, lecturers, and the general public.

Research on personal space indicates that the amount of personal space one maintains with another depends upon the individual and the activity being performed. In general, people find it unpleasant when they are forced to interact with others at inappropriate distances (either too close or too far) or when they must intrude upon the personal space of others. Unpleasant effects tend to be heightened when the other person is either a stranger or disliked, or when the encroachment takes place in a formal setting.

System designers should be concerned with minimizing personal and interpersonal problems which may lead to human performance degradation. They also should be aware of the consequences of inappropriate personal spacing behavior. Although designers cannot interfere with users' preferences as to how close or how far away they choose to interact with others, certain system design features—the size of a conference room or the distance between desks—should not inhibit an individual user from selecting the proper personal space for achieving the desired performance. Users should have the flexibility of choosing the level of personal space that they wish to maintain about them; designers should make decisions to accommodate such flexibility.

Crowding and Density

Crowding is closely related to personal space, and will be distinguished from density in accordance with a definition provided by Stokols (1972). Stokols defined density as a strictly physical variable—the number of people per unit of space. Crowding, on the other hand, is a subjective feeling of too little space. Density is a necessary though not sufficient condition for the feeling of being crowded. It is entirely possible, for example, for someone to be among many co-worker friends and not feel crowded at all.

Crowded working conditions are sometimes unavoidable. From a designer's point of view one of the most critical questions that can be asked about crowding is whether or not the crowding affects human performance. Performance effects of crowding are dependent on the task in question. More specifically, simple tasks do not seem to be affected by crowding (Bergman, 1971; Freedman, 1971) while more complex tasks may be adversely affected (Dooley, 1974; McClelland, 1974).

Territoriality

The concepts of personal space and crowding help predict how closely individuals like to interact in the environment. The concept of *territoriality* helps to predict with whom the interaction will take place. Territorial behavior usually is defined as a personalization or establishing ownership of a place by a person. The designer should keep in mind that defensive responses may occur when territorial boundaries are violated.

Human territorial behavior, at least in part, is a response or reaction to some need that people have to identify a place that they can call their own. Whether we are talking about a small child reminding a friend about whose sandbox they are playing in, or a computer operator who is quite particular about who steps into the computer room, the basic idea is still the same. One person lays claim to a particular location that is off-limits to unwanted others.

Studies suggest that the use of personal markers and objects—such as name plates—are effective in defining and maintaining territories. Territory holders seem to be more influential in their own territory as opposed to elsewhere. Based on these findings, it would seem to make good design sense to allow system users maximum flexibility in personalizing their given work area, and to provide boundaries or barriers to support individual territorial needs.

Isolation

While it is important to take into account the effect of crowding on human and system performance, the effects of isolation should also be considered. Freedman (1975) reviewed several isolation studies and concluded that most people perform adequately under isolated conditions. Freedman concluded that although most people can apparently endure seemingly unpleasant conditions of isolation for long periods of time with relatively few negative performance effects, they find little satisfaction in being so isolated.

FOR MORE INFORMATION

Poulton, E. C., *Environment and Human Efficiency,* Springfield, Illinois: Charles C. Thomas, 1972.

23

DATA COLLECTION

STUDIES AND TESTS

In developing systems concerned with human performance, there are two major types of data collection: studies and tests. *Studies* are used to help make good initial decisions, and *tests* are used to help in evaluating design decisions after they are made. Studies may be conducted at any time in the design process, even as early as when the decisions are being made as to the feasibility of developing a given system (i.e., whether a system should be developed or not). Tests are applied *after* decisions have been made and implemented, and require a "product" of some kind to evaluate.

The following three chapters present and discuss material helpful to designers when planning and conducting both studies and tests. This chapter deals primarily with using *questionnaires, interviews* and, *observation* to collect data.

GENERAL DATA COLLECTION CONSIDERATIONS

Systematic collection and analysis of data are important during the system development process. In fact, good data collection should serve as an important input to the design process. All too often, however, data collection is poorly planned and carried out (or not carried out at all), usually because designers are eager to "get on with the work." Not surprisingly, many design decisions in such situations range from disappointing to disastrous.

As used here, data collection refers to the acquisition of information that will serve as a basis for decisions within the system development process. As such, it usually includes the use of questionnaires, observation, performance testing (simulation), and comparison studies. The data collection process involves the full scope of activity from identifying a problem through the use of the data for making informed design decisions. This process can be divided into three major phases: planning, collection, and analysis.

A need for the collection of data may arise any time within the system development effort where there is insufficient information to support a necessary decision. When critical information is lacking, steps should be taken to acquire it. The only other alternative is for a designer to guess, which is a poor second choice. A wrong decision (or series of wrong decisions) could lead to degraded human performance over the life of the system.

Data collection can be expensive and involved, and only should be performed when it will help assure system success. Data should not be collected and/or analyzed for its own sake. Surveillance should be maintained over the impact of the data being gathered on system development decisions. At several points within the data collection process, evaluations should be made as to whether the results are impacting system development decisions. Any time it appears that the results of data collection are not having an impact on system development decisions, that particular data collection activity should be terminated.

The amount of effort put into data collection should be commensurate with the question being asked. In some situations, the alternatives are not of sufficient consequence to warrant data collection. On the other hand, if the best alternative decision is not readily apparent, and the selection is vital to having acceptable human performance, data collection activities and associated expenses might be the best way to go.

Certain data collection efforts, particularly comparison studies, should not take place without a thorough literature search to see if the information of interest has been collected by someone else. Therefore, a logical first order of business in many collection efforts is to attempt to discover if the needed information is already available. Appendix B describes several human performance resources where studies by others are reported.

Even though information of various types is needed throughout the entire system development process, it is usually true that the earlier in the system development process that data collection takes place, the more valuable the results will be. In other words, data should be collected when needed, without waiting for "a more convenient time." The later a requirement for data is acted upon, the less chance there is for the collected data to have a major impact on the overall system.

Traditionally, however, the early decisions tend to be the least supported factually. For example, early during system development users might specify an acceptable computer response time. The project itself gains a higher probability of success if a designer pauses at this point to gather factual data on the realistic costs of and need for designing and developing a response time at the level indicated in the statement of objectives.

The main objective of the data collection process is to provide information for the decisions required by the system development effort. For example, if one of the objectives of the system is to improve a specified process by 10 percent, then one objective of data collection could be to measure performance before and after the system is operational to show whether or not there has been a 10 percent improvement. If the 10 percent objective is not met, decisions can then be made for changes to the system that *will* ensure meeting the objective.

SAMPLING

There are at least three possible approaches when information is to be obtained about a population of people:

1. Examine all of the people in the population.
2. Examine a portion of the people, using objective (random) sampling procedures.
3. Examine a portion of the people, using some type of subjective (judgmental) sampling procedures.

Usually a designer is unable to gather data from everyone of interest. No matter whether data are collected using a questionnaire, performance testing, or as the result of a comparison study, the designer is usually forced to involve only a small portion or *sample* of all potential users. All people about which information is desired is usually called the *population.* Generally, designers deal with only a sample from the total population. Sampling lets us obtain information about the whole of something by examining only a portion of it.

To accurately say something about the population of users, it is necessary to ensure that the sample is appropriately selected so that it is truly representative of all possible users. Good sampling frequently depends upon the principle of "random selection." Random selection does not mean haphazard or aimless selection. Random selection implies selection governed wholly by the laws of chance, or in other words the selection of individuals for a sample that is completely independent of human decision. In drawing a sample of people from a much larger group, each person should have an equal chance of being selected. For example, if three people out of ten were to be used in a study, all ten people would first be assigned a number from 1 to 10. In a true random selection, each of the numbers 1 through 10 would have an equal chance of selection.

As the size of the sample increases and more of the population is included in the sample, the sample results become a better estimate of the population. However, this should not be taken as an argument for always using large numbers of people. Ideally, a designer would select a sample just large enough to comfortably represent the population. Determining the size of a sample is not always easy, and is best made in consultation with a statistics specialist. Hays (1973, p. 419) and many other psychology statistics books have good discussions of sample size considerations.

For different kinds of studies there are different types of sampling procedures, each of which has advantages. Three of the most commonly used types of sampling are:

- random sampling
- systematic sampling
- subjective sampling.

In selecting a sample, one of the first and most important steps is to decide what kind of sampling procedure should be used (again, consultation with a statistics specialist may be helpful). Each study has its particular characteristics that make one type of sampling more appropriate than others. To use random sampling, for example, when systematic sampling is more appropriate may add greatly to the cost of the study and/or produce less satisfactory results.

Random Sampling

Random sampling should be used when the entire population contains basically the same kind of people. In this procedure, the people included in the sample are drawn at random from the entire population. A common procedure for making this random selection is to first assign a unique number in serial order to each person in the population of interest, and then, using a table of random numbers, make a random selection of the serial numbers (and associated people) for inclusion in the sample.

An example of random assignment to different groups, using the random number table from Appendix A, is shown in Figure 23-1. Forty people were to be included in the sample. Ten were to be assigned to group A, ten to group B, ten to group C and ten to group D. Using the random number table, each person was assigned a different number as he or she reported to the study. The random number table can be read down each column, across the columns from left to right, or back and forth (from left to right in row one and then right to left in row two). The numbers in each five—character group can be read individually, or in groups of two, three or more.

As is shown in Figure 23-1, the designer decided in advance to use only the last two characters in each five-character group, and to move back and forth horizontally across the page. The first person to report was assigned the number 39, the second was assigned 33, the third was assigned 11, the fourth was assigned 9, etc. After all people were assigned a unique number (1 through 40), those with numbers 1 through 10 were assigned to group A, those with numbers 11 through 20 to group B, etc.

This particular set of random numbers was produced by computer. Generally, tables so prepared are not perfectly random but the imperfection is too slight to be of any practical consequence. A more extensive table of random numbers is shown in Table 1 of Appendix A.

Random Numbers

11339	19233	50911	14209	39594	68368	97742	36252	27671	55091
96971	19968	31709	40197	16313	80020	01588	21654	50328	04577
07779	47712	33846	84716	49870	59670	46946	71716	50623	38681
71675	95993	08790	13241	71260	16558	83316	68482	10294	45137
32804	72742	16237	72550	10570	31470	92612	94917	48822	79794

Figure 23-1. An Example of Random Selection Using a Random Number Table

Systematic Sampling

In systematic sampling, the people to be included in the study are selected from the population with a uniform interval between each person selected. For example, if a sample is to include one of every 100 people, the selection would be made by first drawing a person at random from the first 100 people in the population, and then selecting every hundredth person thereafter. This technique is useful when determining a sample to whom to send questionnaires. All that is required is a list of all people in the population of interest. For example, all people working in a particular system or at a particular work location.

Subjective Sampling

Any sampling procedure which does not follow precise statistical principles can be thought of as a subjective or judgmental sampling procedure. In subjective sampling, subjects may be selected by one or more persons who choose—hand pick—the particular people whom they feel best represent the population. This approach is used, for example, when selecting people for performance testing. Frequently, this kind of selection is unavoidable or is necessary because of cost considerations, but it has two limitations that a designer should keep in mind:

1. Subjective (judgmental) samples are subject to the personal bias of the individuals responsible for the selection.

2. The fact that the selection of people is not objective precludes the use of many statistical techniques.

DEVELOPING QUESTIONNAIRES

Types of Questionnaires

In the right applications, questionnaires can provide a large cross-section of data economically and relatively quickly. Questionnaires may be interviewer-administered or self-administered. Interviewer-administered questionnaires may be accomplished by face-to-face conversation or by telephone. Self-administered questionnaires are usually set up for the respondent to read the instructions, question himself or herself, and write the reply. Each type of questionnaire has its strong points and problems. There is no single best approach. Rather, it is a matter of fitting the technique to the circumstances.

The best results will sometimes be obtained by mixing types of questionnaires within a single activity. It would be entirely possible, for example, to use in-person interviews, telephone interviews, and self-administered questionnaires to collect data on the same set of system development questions.

Development of the questionnaire should not proceed until a decision is made on the type of instrument to be used. Some examples of questionnaire types related to data collection problems are shown in Table 23-1.

Table 23-1. Examples of Questionnaire Types

Data Collection Problem	*Questionnaire Type*	*Discussion*
Sample of 500 potential system users; simple five-question opinion questionnaire; three-day deadline.	Telephone interview	Tight time restrictions would probably rule out other two alternatives.
Sample population in the hundreds; multiple-part questionnaire; one-month survey time allocated.	Self-administered questionnaire	One-month study time and multiple response requirements make a self-administered questionnaire probably the best alternative.
Detailed information required of half-dozen members of upper level management.	Face-to-face interview	In-depth interview of busy executive particularly where so few are involved, is usually best handled face-to-face.
Attitude survey of 250 system users.	Face-to-face interview or Telephone interview or Self-administered questionnaire	Any one technique may be the best; there is not sufficient information given to choose with certainty, and additional details should be considered before one or more are selected.

Special care must be exercised in framing the questions for self-administered questionnaires. Because the self-administered questionnaire must play the role of interviewer, all the rules of interviewing apply. For example, terms must be defined, instructions must be clear, the questionnaire must be easy to fill in, and should look important to win the respondent's cooperation. In effect, the interviewing function is programmed into the content and makeup of the questionnaire.

Also, when self-administered questionnaires are used, the response rate is apt to be low unless some form of supervision is applied. Where large groups of people are to be sampled, as in an attitude survey among workers in a plant or an office, it may be best to bring the employees together in groups.

Questions

Once the type of questionnaire has been determined, the development of question content should begin. Designers must decide which questions are most likely to elicit information that will satisfy the objectives and requirements of the survey. An effective way to start assembly of questions is to simply write down as many possibilities as come to mind. Careful attention to exact wording is not important at this point. Emphasis on topics to be covered is more important. If the questionnaire is to include more than a few questions, it may be more convenient to write each on a separate card. This makes it easy to rearrange or eliminate questions as development of the questionnaire progresses. Once all questions are listed, the final selections and wording of questions should be considered. As with other steps in the data collection process, pretesting is important for individual questions, as well as the full set of questions.

Wording of Questions

Here are several rules or guides that will help in the wording of questions:

1. *Use words as simple and as familiar as possible.* While technical language assists clarity if the respondents are equipped to deal with the technical terms, it can be a liability. Thus, if there is any doubt about the technical understanding of respondents, use only simple words.

2. *Avoid "loaded" terms.* Terms may have an "emotional" value and can mean more to most people than their simple dictionary definitions. The use of such words can bias the results of a data collection effort.

3. *Avoid use of words that suggest the desired response.* "You do read a newspaper every day, don't you?" almost forces the respondent to say "yes." Responses to such a question would be worthless.

4. *Be careful to word questions so they do not embarrass respondents.* Questions that make the respondent appear poorly educated or inadequate at his or her job can produce distorted responses, and in some cases outright refusal to participate in the survey. For example, don't say "only a clerical worker."

5. *Do not place too great a burden on the respondent's ability to remember.* Ask only for details that you are reasonably certain the respondent will know. If more detailed information on specifics is desired, another type of data collection may be indicated, for instance, observation, mechanized measurement, or file searches.

6. *Anticipate the context of the responses.* For example, if the respondent is asked for an average, be sure that "average" means the same to all respondents. Otherwise, define the term in the question context.

7. *Use words that have precise meaning.* To illustrate the wide range in meaning that some words have, consider the study reported by Simpson (1944).

Is "almost never" almost always never? Which is more frequent: "frequently" or "often"? And how often is "often"? We use some words imprecisely. Worst of all, it is difficult to say just how imprecisely. Certain words are frequently ambiguous. Sometimes (occasionally?) we need to be reminded of that ambiguity. Thus, 100 students from an introductory psychology course at the University of Minnesota completed the following questionnaire:

What Do These Words Mean To You?

Below is a group of words which we use to indicate differing degrees of "oftenness" with which events tend to happen. Obviously, some of the words mean different things to different people. We wish to determine what each word means to you.

For instance, if "almost never" indicated to you that a thing would happen about 10 times out of a hundred, you should mark in the space before the expression "10." If it means about 1 time out of 100 to you, you should put "1" in the space before the expression. Simply indicate how many times out of 100 you think the word indicates an act has happened or is likely to happen.

1. Almost never
2. Always
3. About as often as not
4. Frequently
5. Generally
6. Hardly ever
7. Never
8. Not often
9. Now and then
10. Occasionally
11. Often
12. Once in awhile
13. Rarely
14. Rather often
15. Seldom
16. Sometimes
17. Usually
18. Usually not
19. Very seldom
20. Very often

The following results suggest some precise data about imprecision:

Word	*Median*
Always	100
Very often	87
Usually	79
Often	74
Rather often	74
Frequently	72
Generally	72
About as often as not	50
Now and then	34
Sometimes	29
Occasionally	23
Once in awhile	22
Usually not	16
Not often	16
Seldom	9
Hardly ever	8
Very seldom	7
Rarely	5
Almost never	2
Never	0

A closer examination of individual responses shows that use of most of these words is not consistent. One person's "rarely" is another person's "hardly ever." "Often" and "rather often" have the same medians. "Rather" is rather meaningless. "Very seldom" is very seldom less than "seldom." The results seem to indicate that people are exceedingly consistent about being exceedingly imprecise.

Types of Questions

There are two main types of questions:

- *Open-ended* questions are unstructured and permit free response.
- *Closed-ended* questions are structured, offering only specifically stated response alternatives which can be easily tallied.

Open-ended questions are characterized by little or no restriction on the range of responses. This kind of question is particularly useful when opinions or attitudes are being probed, or when it is difficult or undesirable to suggest possible answers to the respondent. Open-ended questions are often used in the exploratory phases of a study as a method of establishing alternates for final questions which can be tallied quantitatively. In such situations, the open-ended question helps establish final wording in terms used by the respondents themselves.

The following is an example of an open-ended question and some typical replies used in an interview where users are asked about the time they spend each day handling computer-detected errors.

Question: On a typical day, about how much time do you spend handling computer-detected errors?

Replies:

1. "Hardly any time at all." (Depending on whether the respondent liked or disliked this part of his or her job, such a response could mean almost anything, two hours, twenty minutes, five seconds.)
2. "Twenty per cent of my time."
3. "Four to five hours."
4. "It varies."
5. "Hard to say."

Obviously, most of these responses would be virtually useless because of the lack of definite information. Thus, in addition to being an example of an open-ended question, this is also an example of a poorly designed and/or misapplied question. The question is poor because if fails to indicate the kind of answer required. Better responses might have been obtained with a closed-ended question.

Assuming the respondents were qualified to answer, a more appropriate application for open-ended questions might be as follows:

> What could be done to reduce the time it takes
> to handle computer-detected errors?

While open-ended questions can be a great help in the early phases of a study, the unstructured responses may present problems in tabulation and analysis.

Closed-ended questions can take several forms, all carefully structured for quantitative tabulation and analysis. For example:

> On a typical day, about how much time do you spend handling computer-detected errors? Less than 1 hour _____ 1-5 hours _____ More than 5 hours _____.

A point to remember, however, is that the question should not ask for more detailed information than the respondent is likely to have. While the answers actually will be estimates or guesses, the detail will imply a degree of accuracy that does not exist.

Another form of closed-ended question asks the respondent to evaluate alternatives or check lists according to rating scales. These scales may be numerical or verbal. When such check-list questions are used, allowances must be made for biases that result from typical respondent patterns. For example, if a list contains more than six or seven items, the respondents may tend to concentrate on the first few alternatives and pay less attention to the others. Such a survey could erroneously lead to conclusions favoring the earlier items. For example, a questionnaire may have a question like the following:

> Would you please indicate how each of the following words describe your feelings about the majority of customer complaints? If the word describes your feelings very well, give it a rating of 7. If the word does not fit your feeling at all, give it a 1. If the work falls between the extremes, please choose that number which best describes where you think it belongs.

	1	2	3	4	5	6	7
Justified	___	___	___	___	___	___	___
Reasonable	___	___	___	___	___	___	___
Unreasonable	___	___	___	___	___	___	___
Too demanding	___	___	___	___	___	___	___
Should be ignored	___	___	___	___	___	___	___

An interview's questionnaire may have a similar question:

As I mention each of these words, please tell me how well it describes your feelings about the majority of customer complaints. Would you say each word describes your feelings very well, fairly well, not too well, or poorly?

	Very Well	Fairly Well	Not Too Well	Poorly
Justified	____	____	____	____
Reasonable	____	____	____	____
Unreasonable	____	____	____	____
Too demanding	____	____	____	____
Should be ignored	____	____	____	____

One way to offset such bias is to prepare more than one version of the checklist to be used in the survey. The simplest method for doing this is to print two lists, one in reverse order of the other. Then, when the results are tabulated and combined for analysis, the bias factor is somewhat balanced.

This simple solution, however, has limitations when the list is long, or when the alternatives are complex and require a lot of thought on the part of the respondent. The reverse-order method could still tend to bias the results against the items that appear toward the middle of both lists. In such cases, four orderings of the items could be printed, increasing position balance for alternatives. To illustrate, assume that there are 12 items in a response-selection list. The items below are grouped by quarters so each item appears only once in each quarter. This will help to average out any possible effects of the order of presentation on the responses.

	ORDER			
QUARTER	*A*	*B*	*C*	*D*
I	1	12	6	7
	2	11	5	8
	3	10	4	9
II	4	9	3	10
	5	8	2	11
	6	7	1	12
III	7	6	12	1
	8	5	11	2
	9	4	10	3
IV	10	3	9	4
	11	2	8	5
	12	1	7	6

Figure 23-2. An Example of Question Alternatives Grouped by Quarters

Organization

The sequence and interrelationship of the questions and their parts is also an important part of questionnaire design. Elements should be sequenced so that they carry the respondent through the questionnaire in a logical, easy manner. Organization of a questionnaire actually involves at least three separate levels of activity:

1. Logical ordering of questions according to topical groups
2. Logical sequencing of questions within topic groups
3. Ordering of response alternatives within individual questions.

Organization of the questions into logically ordered, topical groups is done so that all the questions covering a specific area are asked while the respondent is thinking about that subject. Topics should be ordered so they start with easier, more readily discussed subjects and lead gradually into more difficult areas. For example, in conducting a survey of teenagers about telephone use, starting with questions about how the telephone is used in dating or personal questions about grades in school could hamper cooperation. However, starting with neutral questions such as those about frequency of use and length of use could help establish credibility. Later in the interview the more personal questions are less likely to be a problem.

The order of the topics in the questionnaire should be developed so that ideas flow easily from topic to topic. When the questionnaire switches from one topic to another, a statement should be inserted that announces that the change is being made. Such statements are usually simple bridges: "Now I'd appreciate getting a few answers from you on the subject of . . . ." or "Let's turn to another subject."

Arrangement of the questions into logically ordered sequences within the topic groups is important because illogical question order could confuse respondents. To be successful in capturing the interest and cooperation of the respondent, both the topic groups and the questions themselves should follow some logical pattern.

In addition, question order can have a significant effect on the bias of responses. To illustrate, consider the two questions below that might be included on a public opinion survey about telephone service charges:

A seven percent increase in charges for telephone service seems

___ reasonable at this time

___ too high

___ unacceptable

___ I'm not sure

Do you consider telephone service

___ a good value

___ reasonable

___ too expensive

___ no opinion

If the questions were presented in this order, the response to the first could bias the selection of alternatives on the second. It would be more logical to put the second question first. This situation also illustrates an important general principle about presentation sequencing: It is usually best to begin with the general and proceed to the specific. The second question could be referred to as an "establishing question." It indicates an interest in the subject of telephone service charges and helps respondents establish a general attitude framework in preparation for a response to a specific situation.

Another device for channeling respondent attention logically is the "filter question," an item designed to establish respondent qualification for participation in a survey response sequence. An example of a filter question on a service charge questionnaire follows:

Do you have a telephone in your home?

___ Yes

___ No

Since a rate increase would affect telephone users only, it might be decided not to seek responses on the service charge questions from persons giving a "no" response to the filter question. A filter question could then be followed by a "branching instruction," which directs respondents to questionnaire sections that they are qualified to answer and away from areas for which they have disqualified themselves. For example, a branching instruction on the filter question about whether the respondent has a home phone might direct attention to another question aimed at finding out why the respondent does not have a phone. In such a case, a separate branching instruction would be needed to direct respondents who do have phones around the question on reasons for non-use.

For self-administered questionnaires, it is best to minimize the use of filter questions and branching instructions. If used on self-administered questionnaires, instructions to respondents must be very explicit. The same principles apply to the sequencing of response alternatives within individual questions.

Questionnaire Length

As part of developing a questionnaire, a certain target length should be set. The questionnaire length relates to a number of factors within the data collection effort:

1. The amount of time respondents can reasonably be expected to devote to a questionnaire
2. The available time designers can devote as data collectors
3. The actual data requirements for the study
4. The costs and available funds for the study.

These factors, obviously, must be considered in the development of a questionnaire. After questions have been "polished" and sequenced effectively, the constraints listed above are brought to bear in determining the final design of the questionnaire and the plan for using it.

Layout

The physical layout of a questionnaire depends in part on whether it is to be self-administered or administered by an interviewer. In both cases, the important thing is to make the instructions clear to the person using it. Although this requirement is more critical for self-administered questionnaires, the same general rules apply:

1. The instructions on the questionnaire should be as complete as possible, even when the survey is conducted by an interviewer.
2. The typeface for the instructions should be distinct from the one used for the questions. A common procedure is to use italics or all capital letters for the instructions.
3. Questions should be listed and numbered in an orderly manner so that the reader is easily led from one question to the next. Any complicated arrangement invites the omission of parts of the survey and/or errors in the recording of responses.
4. Where the question sequence calls for skipping questions under certain conditions, the branching instructions must be clearly written and, where possible, supplemented by arrows or other devices to assist the reader.
5. Provide adequate space for recording all responses. This is particularly true for open-ended questions. If the space is too small, only partial responses may be recorded.
6. All questionnaires should include entry and exit statements. These are to assist the interviewers or to enlist the cooperation of the respondent in the case of self-administered instruments. The following are simple examples of entry and exit statements:
    a. "Hello, my name is ______________. I'm involved in the design of a new system that I think you will find very useful. We're conducting a survey on computer usage in your department and I would like to ask you a few questions."
    b. "Thank you very much for your help."
7. Ensure compatibility with techniques to be used for tallying, tabulating, and analyzing results.

Testing

Before a questionnaire is used for full-scale collection of data, it should be pretested on selected, reasonably typical groups of respondents. If the questionnaire is to be administered through an interview, the pretest should be done by experienced interviewers who can pinpoint trouble spots.

Particular attention should be paid to such points as:

- Are the instructions clear and complete?
- Do the respondents understand the questions?
- Are the response categories clear and distinct?
- Can the interview or questionnaire be completed in the planned time?
- Is there adequate space for entry of responses?
- Is the questionnaire usable in all respects?

When a self-administered questionnaire is pretested, a small group of actual respondents should complete the questionnaires. This affords the opportunity to truly evaluate real-world aspects of the questionnaire. Some of the most important features of question evaluation are listed below. As a first step in testing, a designer should carefully evaluate each question on a questionnaire against the following criteria.

For all questions consider:

1. Is the issue fully defined?
2. Does each question reflect a clear understanding of the issue?
3. Is the issue meaningful to potential respondents?
4. Is there reason to suspect that the issue being dealt with by the questionnaire still is not sufficiently well understood to all potential respondents? Consider ways of changing certain questions or eliminating the uninformed.
5. What is the stage of development of the issue? It may be a mistake to ask a closed-ended question on a topic if the opinion of many respondents is till hazy on the subject. Conversely, if opinion is well crystallized or falls into definite patterns, using an open-ended question may be a waste of time.

For each open-ended question consider:

1. Is it necessary to ask open-ended questions? Remember that the analysis of thousands of verbal replies represents a good deal of work.
2. Which questions, if any, can be converted to a closed-ended type? If the different points of view on the issue are generally well known, then present them as alternatives (multiple-choice) rather than leaving them to the respondents to articulate in their various ways.
3. Are open-ended questions sufficiently directive? This type question can be too broad and leave respondents free to give answers from every direction and in every dimension. By carefully wording each question,

however, the answers can be confined to a particular frame of reference.

4. Do respondents know the number of ideas expected on each question? If one idea is accepted from one and five from another, it will be difficult to tell whether respondents are being weighted according to their ability to articulate or their weakness of conviction.

5. Is a probe question needed? If it is desired to extract all possible thoughts on the subject, it may be advisable to add a probe question.

6. Are check boxes needed? Even though the question is in an open-ended form, it may be possible to provide check boxes for the answers. This is especially likely if asking for amounts or figures.

For two-way questions consider:

1. Are all alternatives included?

2. Are alternatives complementary? In some cases, however, it is wise to take account of the realities of the situation rather than to use the literal opposites as complementary.

3. Are the choices mutually exclusive? If they cannot be made so, an answer box should be added for a "Both" category.

4. Are implied alternatives avoided? No fault can be found with stating the alternative while some harm may result from leaving it to be carried by implication.

5. Are "negatives" detailed enough? The phrase "or not" may not be enough to convey the full implications of the negative side.

6. Are "Don't know" or "No opinion" answers provided?

7. Is a reasonable middle-ground position available for respondents? If so, it must be decided whether to allow respondents this option.

8. Are double-choice type questions "better-worse" or "now-then" avoided?

9. Are fold-over questions used? Many two-way questions are easily converted to the fold-over type in which an expression of opinion and its intensity can be obtained in case a designer is interested in both.

For multiple-choice questions consider:

1. Are choices mutually exclusive?

2. Are choices exhaustive? That is, none of the alternatives should be overlooked. If combinations of the alternatives are possible, those combinations should be included.

3. Are some choices intentionally restricted? It may be all right to restrict the choices, but this restriction should be stated in the question.

4. Are choices balanced? The choices should be well-balanced within a realistic framework. The number of alternatives presented on one side or another of a central point can affect the distribution of replies.

5. Is it clear if respondents are expected to express one choice or more than one?

6. Are lists provided for respondents? If a questionnaire is used with an interview, give respondents a card list if the question has more then three alternatives.

7. Are numbers listed in logical order?

It is important for designers to put themselves in the position of the person responding to the questionnaire. The following items should be considered:

1. Avoid the appearance of talking down or otherwise insulting the intelligence of respondents.

2. Do not make questions sound stilted.

3. Do not use slang.

4. Do not try to be "folksy."

5. Help the respondents; do not confuse them.

6. If there can be any possible question about the meaning, restate it.

7. Keep away from wordings that result in ambiguous answers. A "Yes" that means "No" is worse than a "Don't know."

8. Each question should be specific without being elaborate.

9. Fine distinctions are often not understood by respondents.

When evaluating the use of particular words, the following items should be considered:

1. Use as few words as necessary. Most questions can be asked in twenty words or less.

2. Use simple, frequently used words.

3. Familiar words are the most useful, but only if they do not have too many meanings in context.

4. When a word with more than one or two syllables is used, make sure it is a familiar word.

5. Ensure that each word actually does have the intended meaning. Use a dictionary to check what other meanings the word might have that could confuse the issue.

6. If it is to be used in an interview, make sure the word has only one pronunciation. Also, consider the possibility of homonyms (pear and pair).
7. If a synonym is used, make sure that it actually is synonymous with the idea at hand.

When evaluating a question by an interviewer, the following items should be considered:

1. Misplaced emphasis can be minimized by underscoring the words that should be emphasized.
2. To reduce "jumping the gun" on the part of respondents, hold back the alternatives until all conditions have been stated.
3. When used in interviews, eliminate unnecessary punctuation because a pause may be taken as the end of the question.
4. With interviews, indicate correct pronunciation of difficult words.
5. Tongue twisters have no place in interview survey questions.
6. With interviews, spell out all abbreviations so that an interviewer will say them correctly.
7. Instead of the indefinite "how much?" approach, indicate the denomination in which the answer is desired—for example, percentages, dollars, miles, pints.

CONDUCTING INTERVIEWS

We use the term "interview" here to describe an activity aimed at gathering data person-to-person. We assume that the interview is structured by a questionnaire and/or special instructions to the interviewer. We assume further that respondents have been advised in advance of the interview activity and its subject area and that they have agreed to cooperate.

There are two basic types of interviews used for collecting data:

1. *Face-to-face interviews* - In these, the interviewer meets with the respondent and presents the questions as they appear on a questionnaire in the manner prescribed by the interview instructions, and records the responses. Interviews administered in person offer the highest potential in response rates and usually the greatest amount of information.
2. *Telephone interviews* - The differences between face-to-face and telephone interviews go far beyond a simple substitution of a phone call for a personal visit. Telephone interviewing generally involves use of shorter, simpler questionnaires and reduces the opportunities for the interviewer to use personal interaction to encourage cooperation from the respondent.

Face-to-Face Interviews

When time, budget, and other considerations permit, the face-to-face interview offers the best chance of collecting the most complete and usable information because:

1. Through persuasion, interviewers can elicit a higher rate of response and can encourage more thoughtful answers to the questions.
2. By making needed explanations interviewers can ensure that the respondent is answering the true meaning of the questions.

In a face-to-face interview, the interviewer can note possible distracting or disturbing effects on the respondent. For example, other employees could be distracting a respondent to the point where the answers might not be valid. The interviewer could note and possibly control this in a face-to-face interview, but may never know about it if using the telephone.

The face-to-face interview, however, also involves the highest costs for data collection. Therefore, with this approach it is important to weigh relative costs and results. Telephone interviews remove a measure of interaction and sensitivity between interviewer and respondent. However, they are more quickly conducted and usually cost less.

Response rates on telephone interviews tend to be lower than for face-to-face interviews. It is easier for respondents to terminate a telephone interview prematurely, and telephone interviews in themselves are easier to avoid. For example, other people can "protect" a potential respondent by indicating that the desired respondent is not in or is too busy.

Telephone Interviews

There are at least two special requirement that apply to telephone interviews. Unfortunately, these constraints restrict the volume and scope of data that can be gathered.

1. In general, they should be short, certainly no longer than 15 minutes.
2. The information requested must be readily available. If the respondent has to search files, contact assistants, or take other special steps to get information, the telephone interview is apt to produce either an incomplete questionnaire or one that is answered with guesses.

Telephone interviewing offers other important advantages:

1. The telephone has a psychological insistence that reduces waiting time and speeds the interview. For example, it is quite natural for a respondent to let someone sit in his or her office while taking a few minutes to answer the phone and complete an interview. If the interviewer had gone in person, it may have meant a wait until the respondent finished with the

first visitor.

2. Interviews by telephone are more economical than personal interviews because they take less time, and may also eliminate travel, time, and expense.

3. In some respects, telephone interviews are easier to control. In part, this is because interviewers can work from a central point where they can be supervised.

Selecting and Training Interviewers

In the majority of cases, a designer will develop and administer a questionnaire without the assistance of others. However, there may be times when it is necessary to have help from others, particularly when administering a questionnaire to a large number of people through the use of interviews.

Interviewer selection and training are basically the same for both face-to-face and telephone interviews. The same data collection skills are involved, except that on the telephone the emphasis is on voice contact rather than personal interaction. The *selection* of good interviewers is an important consideration. As a general guideline, people conducting the interviews should be as impartial as possible.

All interviewers should be *trained* for handling the required questions. Depending on the background of interviewers and the complexity and/or scope of the data collection project, training requirements could range from a brief discussion of background and the questionnaire, to a full-scale course on interviewing in general and the subject of the questionnaire in particular. As part of the training program and to assist interviewers when interviewing, written instructions, and possibly a performance aid, should be prepared and included on or with the questionnaire.

Keep in mind two specific training objectives. First, before asking a question, an interviewer should have enough training to be able to explain the meaning of each question and the purpose of the survey. Secondly, an interviewer must be able to pronounce correctly and define every word in the questionnaire.

It is vital to the success of a survey that the interviewers have a basic understanding of the subject and that they be trained to explain the questions. No matter how well the questions are written, it may be expected that a certain number of respondents will require additional explanation or clarification. Also, there will always be a few respondents for whom certain questions are inappropriate. For these respondents, the interviewer must make decisions about how the questionnaire will be completed. In these situations, the better an interviewer understands the objectives of the survey, the better the decisions he or she can make. The competence and decisiveness of the interviewer will also influence the respondent's attitude about the survey.

Interviewing Techniques

When conducting the interview, the first step is to enlist the respondent's cooperation. This can be a problem in some circumstances, but it is an important part of the interviewer's job. For example, special consideration may have to be given to workers with limited time or to employees who might not want to cooperate but were instructed to do so by management.

Probably the best way to obtain respondent cooperation is by asking for it—making a polite request for assistance in solving a problem. Also, in cases where respondents may be concerned about jeopardizing their positions, guarantees of anonymity may clear up difficulties. No matter how it is enlisted, the respondent's cooperation is important in getting the interview started and in maintaining rapport during the interview.

In addition, an interviewer should make certain that the respondent is qualified to answer the questions in the survey. For example, if the survey were to be conducted among clerical supervisors, the conversation might begin as shown below.

Interviewer: "Hello. Are you the supervisor?"

Respondent: "Yes, I am."

Interviewer: "Could you please help me out by answering a few questions for a survey?"

Right from the start the interviewer should maintain an attitude of friendly neutrality, and should make sure that the respondent fully understands the instructions and the questions. The interviewer should say all words in full, avoiding acronyms, abbreviations, slang, or ambiguous words. She or he should not be afraid to ask the respondent if he or she understands. On the other hand, the interviewer should be especially careful to avoid giving the impression of talking down to the respondent. One good device for repeating a question or an instruction without casting any reflection on the respondent's ability to comprehend is to say: "This is a long question/instruction. I'll read it twice." Or: "I have been instructed to read this twice."

At times, interviewers meet people with whom they share common interests, background, opinions, etc. In these cases it is easy to get drawn into lengthy conversations that may ultimately change the interviewer's neutral posture with the respondent and affect the outcome of the interview. The best policy in these cases is for the interviewer to finish the questionnaire before mentioning or discussing items of interest.

An important part of interviewing is probing for answers. Consider the following:

1. Probes should never be argumentative or challenging.

2. When questions are open ended—when they ask for an unspecified number of answers without providing a checklist of response alternatives—it is up to the interviewer to give the respondent adequate time and to encourage further responses using such probes as, "Anything else?"

3. When the respondent appears to be unsure of an answer or when the interviewer has reason to suspect that the truth is not being told, the probe should, again, be used. Examples: "Could you explain that response?" "Shall I repeat the question?"

As with other aspects of a data collection activity, interviewing should be tested with a small, representative segment of the total sample before the entire collection effort is launched. This provides an opportunity to evaluate the questionnaire, the training, and the capabilities of interviewers before full-scale collection is undertaken.

OBSERVATION

Some types of data pertinent to system development cannot be collected through interviews or self-administered questionnaires. Data collection through observation is used in situations where the information to be gathered is of an operational nature. For example, if a designer wanted to know how many employees come to work late or leave early in an existing system, it would not be a good idea to ask the individuals themselves. In such a situation, it would be far more reliable to run a tally of time cards. Or, if time cards were not available (or not accurate enough), it might be best to have a trained observer stand at the entrance and make a count.

In another typical case, it might be necessary to learn the number and type of clerical functions performed per hour. Asking employees on the job might not help much. They might be prone to exaggerate reports on their work productivity. In this instance, it would probably be more effective to collect this information through observation. Under similar circumstances, if a study were aimed at determining peak loads of work, on-the-spot observation might be necessary.

Data collection by observation requires careful planning. A plan for taking samples must be developed to determine how many situations will be studied and which ones are to be selected. Such considerations as whether the population should be represented by a few people conducting all activities, or more people conducting a few selected activities, must be taken into account. In addition, preliminary work should be done to isolate the specific operations that are important to the study. It is easy in observational work to fall into the trap of making observations on too many activities.

Keep in mind that one of the major problems in the use of observation is that the known presence of an outside observer can change the actions of the people being observed. To minimize these effects, advance arrangements should be made, supervisors briefed, and employees advised of what is being done and why. Usually, depending on the particular circumstances, workers become used to an observer after a short time and their usual habits and work patterns take over.

There is some value in having a designer make the needed observations, thus receiving first-hand information on the question of interest. However, when this is not possible other people with the necessary skills should be found and trained to make the observations.

Data collection through observation follows the same general rules in the design and formatting of collection forms as in the preparation of questionnaires. Information on these forms should be sequenced according to the work flow of the operation under observation. With observation this is particularly critical because the observer may have to make entries in cadence with job performance cycles. Therefore, the layout of the form should be keyed to the constraints that the observer might find. Entry spaces on the forms should be grouped to conform to performance patterns of the work being observed.

Stop watches and push-button counters are used when observations must be time-related or volume-related. Typically, such observation is done with special clipboards that hold the collection form and the watch or counter in convenient working proximity. Collection instruments and observation activities should be tested to be sure the methodology works and the data gathered is usable.

OTHER DATA COLLECTION TECHNIQUES

There are numerous special techniques that have proven useful in collecting human performance information. These include articulation testing (discussed in Chapter 13), link analysis, critical incident techniques and others. Chapanis (1959) provides a good discussion of these methods. As an example of these special types of data collection, we will briefly discuss *link analysis*.

Link analysis is an observational technique for determining the relative association among system components. The results of a link analysis usually are expressed in frequencies.

Consider, for example, that five people are sharing office space. They have a need to talk briefly with one another concerning their work throughout the day. The question is, "How can a designer arrange their desks in the new system so that those people who need to talk most to each other are closest together?" The existing arrangement of their desks is shown in Figure 23-3.

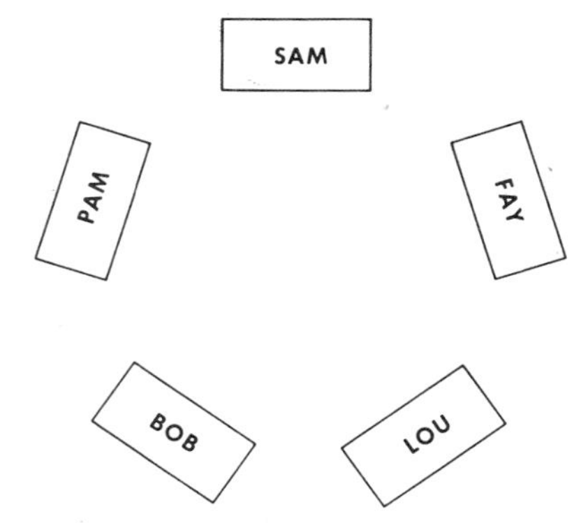

Figure 23-3. The Existing Arrangement of Desks

To collect the necessary information to make an informed design decision, the designer spent a day observing who talked to whom. Each time two people exchanged information the event was recorded. At the end of the data collection period the designer had the information shown in Figure 23-4.

1. Pam and Fay	15. Lou and Bob	29. Fay and Lou	43. Sam and Lou
2. Pam and Fay	16. Pam and Fay	30. Bob and Fay	44. Fay and Sam
3. Fay and Lou	17. Fay and Bob	31. Sam and Bob	43. Lou and Bob
4. Sam and Lou	18. Fay and Pam	32. Bob and Sam	46. Sam and Lou
5. Fay and Lou	19. Pam and Sam	33. Fay and Lou	47. Sam and Pam
6. Lou and Bob	20. Pam and Fay	34. Fay and Bob	48. Pam and Lou
7. Sam and Lou	21. Fay and Lou	35. Fay and Lou	49. Fay and Pam
8. Pam and Fay	22. Lou and Bob	36. Sam and Lou	50. Bob and Pam
9. Sam and Pam	23. Fay and Pam	37. Lou and Fay	51. Sam and Lou
10. Fay and Bob	24. Lou and Pam	38. Bob and Fay	52. Lou and Fay
11. Fay and Lou	25. Sam and Pam	39. Pam and Sam	53. Fay and Pam
12. Pam and Sam	26. Pam and Sam	40. Sam and Pam	54. Fay and Lou
13. Fay and Pam	27. Pam and Bob	41. Bob and Lou	55. Bob and Fay
14. Fay and Pam	28. Bob and Lou	42. Lou and Pam	56. Lou and Sam

Figure 23-4. Exchanges of Information in a One-Day Period

The information in Figure 23-4, which the designer assumed was typical of all days, was converted to a frequency count of *links* between workers and is shown in Figure 23-5.

The designer used the frequency count of links between workers shown in Figure 23-5 to change the seating arrangement. The seating shown in Figure 23-6 takes into account communication links, and places closer together those people who communicate with each other most often. Of course some tradeoffs were needed to come up with the best arrangement. But the seating diagram in

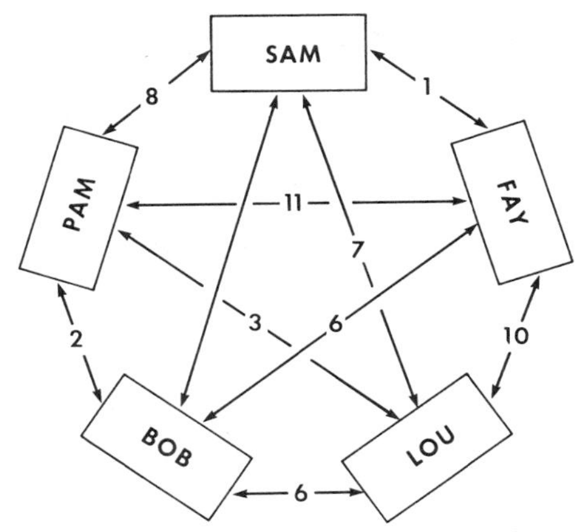

Figure 23-5. A Frequency Count of Communication Links Between People

Figure 23-6 enables about two-thirds of all communications to take place desk-to-desk without the need to talk across the room. With the old arrangement, less than half of the communications were desk-to-desk.

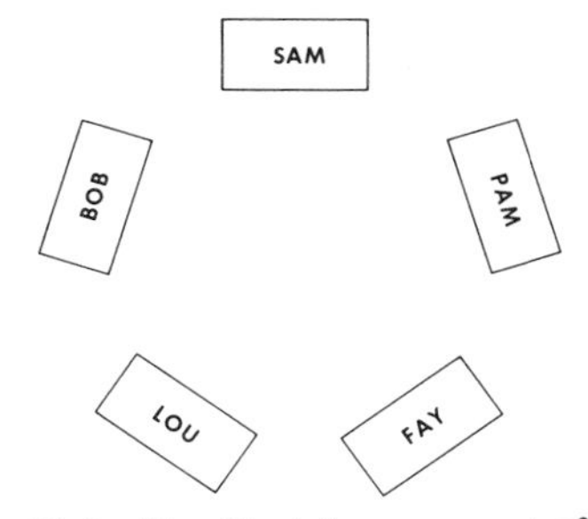

Figure 23-6. The Final Arrangement of Desks

Link analysis can be used in a wide variety of different situations. However, it is used primarily to help determine the best layout and arrangement of people and machines in systems. Keep in mind that the technique does not take into account how long a user spends with each link or the quality of the interaction.

FOR MORE INFORMATION

Chapanis, A., *Research Techniques in Human Engineering,* Baltimore: The Johns Hopkins Press, 1959.

Labow, P. J., *Advanced Questionnaire Design,* Cambridge, Mass: Abt Books, 1980.

Meister, D. and Rabideau, G. F., *Human Factors Evaluation in System Development,* New York: John Wiley and Sons, Inc., 1965.

24

PERFORMANCE TESTING

INTRODUCTION

Human performance testing should begin as soon as feasible and continue as needed until the system is completed.

MacPherson vs. Buick Motor Company in 1916 was a landmark case in determining liability suits. The plaintiff had been injured when a wheel collapsed on the auto he was driving. He had purchased the car from a dealer who had bought it for resale from the Buick Company. Suit was brought against Buick.

Buick based its defense on the fact that while it had sold the car to the dealer, it had no contract with the injured party and thus was not liable. The dealer in turn claimed that it was not liable because it had not manufactured the car. The judge ruled that the Buick Motor Company was indeed responsible. He stated that the manufacturer had a duty to inspect its products for defects, and that failure to do so constituted negligence (Hammer, 1972). A system designer has the same responsibility to his or her customers. Releasing a system that elicits degraded human performance could constitute negligence on the part of the designer.

The liability analogy can be taken even further. Over the years, the courts have greatly reduced the need for an injured person to show negligence on the part of a manufacturer, assembler, or retailer. The courts have held that since an injured person generally does not have the technical capabilities necessary to prove that negligence existed, the burden of proof is on the manufacturer to show that negligence did *not* exist. The injured person can recover damages if he or she can reasonably show that a product had a defect in design or manufacture when it left the manufacturer. The manufacturer then has to prove that he or she was not negligent, or prove that the user was negligent, that is, the product was misused. The burden of proof has shifted from the user to the designer.

In liability cases, negligence is defined as the failure to exercise a reasonable amount of care so that injury or property damage does not occur. It is possible that in the foreseeable future the definition could be expanded to include not only injury or property damage, but also situations where a designer's failure to exercise a reasonable amount of care leads to substantially degraded human performance and unreasonable operating costs for users.

Developing a new system, like most any new product, is a dynamic and complex process. This type of development activity involves elements of invention and creativity. The designer rarely duplicates even portions of systems already in operation. Systems under development are systems that have never been built before. To some extent, there is even some uncertainty about whether they *can* be built. As evidenced by the large number of negligence cases that crowd our courts, the risk of making incorrect design decisions is high. Designer errors can lead to extensive user errors.

ACHIEVING ACCEPTABLE HUMAN PERFORMANCE

Testing as it will be discussed here is conducted to help ensure acceptable quality for the human performance aspects of a system. There are three broad areas where using tests should be of interest to designers (cf. Hansen, 1963).

Quality of Design

Two products or systems may have the same functional use but be very different in design. One product may be made of cheap plastic with ill-fitting parts and an obviously short life expectancy. The other may be well designed using strong, durable materials, have parts machined to close tolerances, and give all indications of being able to withstand extended use. The same is true with systems. One may have been developed in too short a time by designers who had little understanding of how the system would be used or who the users would be. Another could have been very systematically developed, with consideration given to its present and future users from the time functions were initially allocated until the system was actually delivered. The quality of design of the second product is superior to the first.

The quality of design is greatly influenced by application, cost considerations, and customer demand. The quality and reliability standard applied to electronic equipment for a moon probe are much more stringent than those employed for an ordinary commercial calculator. As a matter of fact, frequently the quality of mass-produced items is purposely made marginal to stimulate both the repair and spare-parts activities of the business and replacement sales of the product. This also may be true with some systems where the completion of one system signifies the beginning of a second or third generation of the same system.

Care must be employed in the design process to avoid both underdesign and overdesign. Testing can help in arriving at the best design quality. For example, when a tolerance of ± 0.01 inch suffices for a machined part, there is no need to specify the more costly tolerance of ± 0.001 inch. Likewise, if an accuracy level of 95 percent is acceptable, there is no need to spend the extra time and money to increase the accuracy to 99 percent.

Quality of Conformance

Quality of conformance is the extent to which a product or system complies with its specifications or requirements. A product manufactured to specification and in conformance with the control limits of the production processes should satisfy the customer if the specifications have correctly translated the customer's requirements. The same is true with the systems that have a clear set of human performance requirements. By meeting or conforming to the requirements levied in the beginning, the system should satisfy its user. This assumes, of course, that the user was involved when the human performance requirements were established, and that the requirements have not been changed during development to accommodate design difficulties. Obviously, if no human performance requirements are levied, the quality of conformance cannot be measured.

Quality of conformance considers such areas as defect prevention, defect finding, and defect analysis and correction. When related to errors in computer systems, this refers to the prevention, detection, and correction of errors. A clear and measurable set of human performance requirements for a computer system should include these three aspects of error handling, and do so in a way that can be tested.

Another purpose of human performance tests is to discover whether a piece of equipment or a system conforms to some standard. Standards for a piece of equipment or a system can cover a wide range. At one extreme, the standard may be a precise specification of human performance. For example, "data must be keyed with an accuracy of 99.5 percent." At the other extreme, the standard may be so general that it is difficult to test. For instance, "the design of controls and displays must conform to good design practice."

Quality of Performance

The quality of performance of a product or a system is determined in large measure by its *quality of design* and *conformance*. Maintaining both at a high level should result in a high quality of performance. If the quality of design is poor or if there is a lack of conformance to the human performance requirements, the quality of performance will be adversely affected. Also keep in mind that even the highest level of conformance to requirements will be useless if the basic design is poor.

Early testing of individual work modules will provide designers with insights into design areas that require changes. System testing and trial testing help to "fine tune" and further improve performance quality. A carefully planned testing program will evaluate the adequacy of the design, conformance to requirements, and quality of resulting human performance. If the testing is begun early, design decisions relating to these three areas can be made that result in an acceptable level of human performance.

Another reason why testing throughout all phases of systems development is helpful, is that it enables detection of design deficiencies before they become "cast in concrete." Once the detailed design of a system has been completed, major design changes become expensive and time-consuming.

Formulating a concise set of rules to guide a designer through all the intricacies of human performance testing is difficult. Human performance tests conducted as part of a program of system development cover a wide variety of products and processes. Such performance tests include analyses of:

- the basic design,
- human interfaces with software and hardware, or even other people,
- workplaces,
- entire systems,
- training materials, performance aids, instructions, and/or selection procedures.

Several aspects of testing that are most useful to designers in the development of systems are discussed below.

THE TEST PROGRAM

Many system test programs place primary emphasis on the evaluation of software and/or hardware components of a system. Any information gained about user performance is usually a by-product of these tests, and tends to be of little value. A good human performance test program, on the other hand, will have a series of *tests developed specifically to evaluate human performance*. In some situations, one or more pretest is also helpful. If a pretest evaluation is included, there are at least four stages that should be included in most programs: pretest evaluation, work module test, system test, and on-site trial.

Pretest Evaluation

Pretest evaluation consists of *design verification* of early human performance products through the use of checklists and possibly some low-level simulation techniques.

Pretest evaluations should begin as soon as there are products to evaluate—outlines of work procedures, preliminary human/computer interfaces, rough draft of written instructions. Even materials that are not in final form can be reviewed. Early detection of deficiencies makes it easier to make changes. All materials associated with a work module should be evaluated before work module tests are conducted.

Designers should *not* stop testing once the pretest evaluations have been completed. Regardless of how good the various human performance materials may appear, experience has shown that we cannot merely ask someone to look over a completed work module, interfaces or facilitators, even in a semi-controlled environment, and expect a reliable determination of their effectiveness. Such evaluations have tended to be unreliable, even when evaluators are human performance specialists (Bailey, 1974; Rothkopf, 1963).

Work Module Test

This part of the test program usually consists of *simulation* testing individual components and groups of components (subsystems) as they are being developed. The primary purpose of this test is to determine whether the product or the process elicits the necessary level of human performance to meet the requirements established for it. When deficiencies or weaknesses are discovered, there is an opportunity for redesign and for retesting of the altered components.

Work module tests consist of well-controlled simulations of individual work modules (or sets of related work modules). A complete work module usually consists of a statement of minimum qualifications (SMQ) that describes the potential user, a set of instructions, performance aids, training materials, and the necessary interface items—for example, forms, CRT masks, or keyboard. The work module test has people that are typical of potential users perform in a simulated work situation. The assumption is that a person's performance on this test is representative of his or her performance on the job.

The SMQ is used to select test subjects. After potential users have been trained using only work module materials, a test session is conducted where these same users perform a representative sample of the actual tasks in the new system. Work module tests should be designed to represent the on-the-job environment as much as possible. Human performance data are collected during the test sessions and provided to designers in detailed test reports.

The work module test concept originated during the early development of large computer-based systems at Bell Laboratories (Bailey, 1970; Bailey and Kulp, 1971, Bailey, 1972). Since its inception the work module test approach has been successfully used in the development of numerous computer-based systems (Martin, 1974, 1975, 1979). These test programs have led to substantial cost savings.

In one work module test program, for example, 130 tests were conducted over a 20 month period (Martin, 1975). The test program used 448 subjects. The original tests elicited an average error rate of 13.1 percent. Based on recommendations in test reports, designers made changes to the basic design, interfaces and/or facilitators. Retests showed that errors were reduced by 35 percent.

System Test

System test consists of performance testing the entire system in a simulated work situation. The system test should not be performed until all human performance materials and interfaces have been developed and previously tested as work modules.

This facet of the testing program focuses on the integration of human, software, and hardware components. Some of the human performance objectives of these tests are:

- To determine whether the total system is operable without a lot of "hand-holding."
- To determine how well the overall (integrated) system design elicits an acceptable level of human performance.
- To determine if the system as a whole is capable of meeting human performance requirements.
- To re-evaluate instructions, performance aids, and training.

Trial

Trial is the final step in a good human performance test program. Even though it is valuable to observe human performance levels in a recently operational system, few substantial design changes can be made at this point. Indeed, a designer, if aware of the results, applies most of the information gained from such analysis of an operational system in the *next* system he or she develops. Usually, the system test, not the trial, is a designer's *final opportunity* to make meaningful changes. If the human performance test program consists only of pretest evaluations and trial, the chances are good that few (if any) changes will be made to improve human performance.

Trial is usually conducted in a real operational situation. One of the main aims is to determine if the total system actually meets performance objectives under normal operating conditions. This part of the test program is more concerned with the use of the system rather than its design. Some human performance objectives for this phase of testing are:

- To determine serious operational difficulties with the system (less serious ones are usually overlooked) and to identify ways of making improvements. Improvements at this point tend to be very costly.
- To determine the effects of human performance on system performance and system performance on human performance. The intent is to discover any defects in the system design that may substantially degrade human performance.
- To evaluate the "real world" adequacy of selection criteria, instructions, performance aids, and training.
- To determine if the system can be operated by the number and types of personnel assumed by designers.

PREDICTING HUMAN PERFORMANCE

The principal purpose of each of the first three tests just discussed (pretest evaluation, work module test, system test) is to act as a predictor, a way of forecasting what will happen when the system is actually operational. When we conduct a pretest evaluation or a test, we expect that the outcome will enable us to make valid statements about human performance in the future. However, the test techniques we have discussed differ greatly in their predictive power. Some are relatively good indicators, others give results that must be interpreted with great caution.

The variable that largely determines the predictive power of a test is the closeness or *fidelity* with which the test conditions reflect those in the real world.

Tests try to predict performance in the real world. The most direct way of finding out about the performance of people in the real world is to observe them working in the real world. When developing totally new systems, this is not possible until a system is being trialed. And even then the very process of making observations means that we may distort conditions and thus lose something in fidelity. People may not be as spontaneous or natural when they are aware that they are being observed. If inexperienced observers actually interfere with users, the amount of distortion and the loss in fidelity may be considerable. Unfortunately, predictions from observations made during trial may have high fidelity, but come so late in the development process as to be of little value.

Both system tests and work module tests may have less fidelity than observations made during the trial of a system, primarily because they are conducted under artificial conditions. System tests and work module tests cover a wide range on the fidelity scale. Some are highly realistic, while others may be so abstract that they bear little resemblance to the real world they are supposed to represent. The lower the fidelity, the more difficult it is to make accurate predictions.

Pretest evaluations have little or no fidelity. Rather, their intent is to review materials for completeness, accuracy, etc. Therefore, except in cases with extreme deficiencies in the basic design, interfaces or facilitators, it is very difficult to predict user performance in the real world from pretest evaluation results.

The fidelity of test conditions is, unfortunately, directly related to cost and inversely related to the ease with which meaningful changes can be made to a system. The more we find out about human performance in a system and the less uncertainty we have concerning this performance, the harder it is to make substantial changes to the system that could improve the performance. For example, the least expensive method is pretest evaluation, but it also gives the least information about changes that can be made to improve ultimate human performance. On the other hand, observations during trial give fairly good indications of performance, but most meaningful changes will be very costly. A highly realistic simulation is usually expensive, but it also improves on the ability to predict future human performance, and to identify deficiencies early enough that less costly changes can be made. The higher the fidelity of a simulation, the closer its cost to that of a real-world system. Even so, in some systems it is cost-effective to spend whatever is necessary on testing to reduce the even higher costs associated with redesign.

PREPARING FOR A TEST

Planning

Because people are so different and the activities they are expected to perform vary so widely, human performance tests tend to be much more complicated than most engineering or physical tests. People are constantly changing. They learn, they become bored, and they are influenced by what happens to them while performing. To get dependable results in the face of incessant change, the designer has to use some special techniques. Some of these techniques are discussed in Chapter 25. Designers should be aware that certain precautions are necessary in order to have a meaningful test.

Selecting Test Subjects

The designer should decide who will serve as test subjects. Two kinds of decisions are involved: the *number* of people to be tested and the *kinds* of people to be tested.

You should use enough test subjects to get *reliable results*, but not so many as to unreasonably increase the cost of the test. If reliable results can be obtained with five test subjects, it is wasteful to test fifteen. However, there is no easy way of deciding in advance how many test subjects are needed for a particular test. It is not uncommon to conduct work module tests or systems tests with four or six test subjects.

The second decision is an easier one. People used as subjects should represent as closely as possible the eventual users in the real world. A designer should make every effort to find people that are truly representative of future users as test subjects, even though this may involve considerably more trouble and expense.

To a large extent the goodness of human performance tests depends on how well the designer has been able to select test subjects that have the characteristics of the people who will ultimately work in the system. Some of the human characteristics that are most important include:

- Age
- Sensory characteristics—visual acuity, auditory acuity, color perception
- Responder characteristics—body dimensions, strength, handedness
- Cognitive characteristics—general intelligence, problem solving ability
- Motivational characteristics—cooperativeness, initiative, persistence
- Training and experience—level of general education, amount of specialized training, and specialized experience.

For example, if test subjects have an average age of 23 and most have college educations, while the user population has an average age of 47 and most have high school educations, the performance of the younger and more educated test group may show little resemblance to that of the actual user population.

Determining Test Items

A careful selection of test items helps ensure that *all important conditions are tested*. If selected, the test items can measure the ability of design, interfaces, and facilitators to achieve an acceptable level of human performance. A detailed set of human performance requirements—such as maximum acceptable error rate, processing time, training, and user satisfaction—should accompany each task. It is usually not practical to include one or more examples of every system input, output, and performance characteristic in a test. Therefore, an important part of developing a human performance test is deciding which features are most important and should be included in the test. The *validity* of any test is determined to a considerable extent by how successfully the designer has been able to identify and include representative test items.

Simulating the Context

A good test should take into account the context in which users will be performing once the system is operational. The designer may need to include a full range of conditions, considering both the physical and social environments. Telephone messages that can be easily heard in a quiet setting may be difficult to hear in a busy office. Trucks which can be easily driven in temperate climates sometimes cannot be used in the arctic region where drivers are hampered by several layers of protective clothing, thick insulated shoes, and bulky gloves. Electronic gear that might be suitable for moderate climates may be impossible to handle under a hot desert sun.

If a large range of contextual or environmental conditions exists, the designer must select a representative sample to include in the test. At the very least, test conditions should include values near each extreme and one in-between. For example, if a system is designed to operate in illumination levels ranging from bright sunlight to semi-darkness, tests should be conducted in both bright sunlight and semi-darkness as well as some illumination level in-between. If a representative subject performs satisfactorily at both extremes of an operating condition and at some point in between, it is generally safe to conclude that the same level of performance can be expected in the real-world environmental conditions. If human performance is degraded at one extreme, however, it may then be worthwhile to make tests at other values to discover at what point the performance begins to deteriorate.

Every test has some environmental conditions that may distract a person from adequately performing an activity. Frequently, these conditions are so common the designers ignore them. For example, having several children and a dog in an automobile may affect a driver using a car telephone, yet a designer may overlook these distracting conditions during testing.

The designer should make every attempt to ensure that the test subjects, test items, and test environment are all representative of the real world.

Reporting Test Results

All tests should result in some sort of summary and analysis. If there is statistical treatment of results, it should be relatively simple, involving no more than some summary tables containing averages and other easy to understand descriptive statistics. As an example, consider collecting and reporting results associated with errors in a computer system. Errors can be very costly. In fact, the costs to detect and correct clerical errors can be as much as $10 per error or even higher (Bailey, 1973). Simulation testing (particularly work module testing) can help identify the design deficiencies that may lead to excess errors.

Accuracy Objectives

To determine if a system that is being developed will meet overall accuracy objectives, each field of data should be assigned an *accuracy objective* based upon the data's criticality to system success. The determination of these accuracy objectives should be made *before* work module testing begins. For example, the Telephone Number data file may need to be 99.8 percent accurate, whereas the Terminal Location data file may be required to be only 96.5 percent accurate. To increase development and operational costs by trying to make the Terminal Location file more accurate is unnecessary. However, it may be of vital concern and very cost effective to do whatever is necessary to maintain the Telephone Number file at a 99.8 percent accuracy level.

Test results can then be reported in terms of the impact that human performance has on the accuracy of each specific data item file. For example, a test may show that requiring an operator to collect a customer's telephone number and to enter the telephone number on a form introduces about 2 percent error. Another test may show that a second handling of the data, transcribing the telephone number from form A to form B, adds another 1 percent error, and that a third handling, keying the data, adds still another 1/2 percent error. For every 100 telephone numbers being added, then, an average of three or four are incorrect (even after computer edits and validations). Thus, with a continual stream of new additions to the Telephone Number file, the accuracy of the file, could be expected to drop from a required 99.8 percent to 97.0 percent or lower. Faulty human performance adding 3 to 4 percent error means that the existing combination of basic design, interfaces, and facilitators do *not* elicit the necessary level of human performance to meet accuracy objectives.

A good test report should show not only that the accuracy objective is not being met, but also suggest reasons why it is not being met. By knowing that the accuracy objective for any data file is not being met, the designer can make the necessary changes to prevent errors from occurring.

Statistical Reporting

An example of the statistical results section of a test report is shown in Table 24-1. Almost all computer systems *inherit* many errors. These are errors that are made before the data flows into the new system being designed. These errors must be considered when projecting accuracy levels, and are shown in column A. Perhaps the most significant information to the designer is that contained in the Distance from Objective (DFO) column which must be either zero or higher for the data items to meet predetermined accuracy objectives. Remember, if each data item does not meet its accuracy objectives, it will be very difficult for the system to meet its performance objectives.

Table 24-1. An Example of Position Package Test Statistical Results

DATA FIELDS	A Inherited Error	B Estimated Accuracy Before Test (100%-A)	C Testing Error Rate	D Projected Accuracy-- No computer Validations	E Projected Accuracy-- with Computer Validations $(100\%-\frac{A+C}{2})$	F Accuracy Objective	G Distance From Objective (E-F)
Telephone Number	1.5%	98.5%	2.3%	96.2%	98.1%	99.0	-0.9
Terminal Location	5.4	94.6	3.6	91.0	95.5	95.0	0.5

Most workers use more than one data item at a time. In fact, data fields are frequently output in groups of five, ten, or more data items. For example, in the telephone company, a telephone installer usually requires the information in at least eleven different data fields, besides the new customer's name and address, to complete the installation of a new telephone. If any one of the data items is wrong, the installer must call the office and wait while correct data is generated. This could take several minutes or even hours. Obviously, the number of telephones installed in one day is dependent upon the accuracy of the information the installer is given.

Cost-Effective Performance

System performance is frequently measured in terms of cost-effectiveness. As far as errors are concerned, the system is cost-effective only if it produces information at an *accuracy level* at which users can perform their functions with an acceptable level of interruption and delay. The acceptable accuracy level may be extremely high (such as 99.8 percent for telephone number data) or much lower. The overall cost to purify the data, however, must be compared to the costs associated with using data containing a certain amount of error. The latter costs include those incurred because of work completion being delayed (e.g., a telephone installer searching for a non-existent address), customer irritation (e.g., a customer refusing to pay a bill because calls are listed that were not made), and the impact of errors on other data (e.g., an incorrect telephone number associated with a correct name and address may result in a customer not being billed for calls). It may be, for example, that a telephone installer must complete 95 out of every 100 installations, rather than 81 out of 100, in order to complete the installation process with the least overall cost to the telephone company. If so, then significant system changes may be required to improve the accuracy of the data. This usually means that human performance must be improved in some way.

User Acceptance

System performance can also be measured in terms of user acceptance. Experience has shown, for example, that users have great difficulty accepting a new system that produces less accurate information than that available in the system it replaced. For example, if the accuracy level of the data in the new system is not high enough to carry out telephone installations at the same rate as the old system, then some substantial changes must be made to the new system so that more accurate information is available to the user. Conducting meaningful tests helps a designer determine fairly early in the development process if the desired accuracy level will be obtained once the system is operational.

EXAMPLE OF A WORK MODULE TEST

The following example illustrates in more detail the kinds of decisions a designer must make when planning and conducting work module tests. The Loop Maintenance Operations System (LMOS) is a work module performed as part of the job of a Repair Service Attendant (RSA). This work module consists of a basic design, written instructions, CRT masks, training materials, and a performance aid. An RSA receives calls from telephone company customers when telephones or telephone lines need repairs. They record the customer's explanation of the problem and provide the trouble-report information to the individuals who make the repairs.

When contacted by a customer, the RSA first enters the customer's telephone number into a computer using a QWERTY keyboard and the computer responds with a copy of the customer's file on the CRT screen. After verifying name and address with the customer, the RSA enters a coded description of the customer's trouble, determines call-back information, information concerning access to the telephone, and gives the customer a date when the trouble will be repaired. The RSA report is then electronically transmitted to the appropriate test and repair bureaus.

Work module materials were prepared and eight representative test subjects selected. After completion of a four-day training class, the test subjects were given the five-hour work module test. The distribution of each type of trouble call used for the test corresponded closely to the actual distribution of calls received in the real world. A special computerized data base was used for the test. This test data base contained fictitious information for ninety-four customers, as well as appropriate CRT masks.

Simulated customer calls were made to each test subject by trained testers. Testers, as customers, used the same prepared script for each call. Each tester made a predetermined number of calls, and attempted to exactly duplicate the form and content of each call for each subject. Testers and RSAs were in separate rooms, and communicated using special telephones.

In general, the simulation proceeded as follows:

1. The customer would indicate a call attempt by pressing a button, which made a clicking sound in the earphones of the test subject (RSA);
2. The RSA responded by stating, "Telephone Repair Service, may I have the telephone number you are reporting, please?"
3. The customer gave the telephone number;
4. The RSA entered the telephone number, waited for the customer's record to appear on the CRT screen, and then verified the customer's name or address;
5. The RSA then requested a description of the trouble and asked any necessary questions;
6. The customer made only those responses contained in the script; otherwise, the response was, "I don't know;"
7. The RSA then asked for call-back and access information, and gave a commitment (date and time for repairing the telephone);
8. The customer, who acted either cordial or irate during the call, then made an appropriate closing comment and the call was terminated.

Subjects were tested in groups of two, three, or four, depending on the number of subjects available for each test. There was one tester for each subject. The testers rotated through the subjects. For example, customer 1 would talk with RSA 2, RSA 1, RSA 3, and then return to RSA 3 or RSA 2 with another call. This procedure helped to reduce the possibility of subjects and testers anticipating each other's remarks or behavior. It also required RSAs to wait for calls on occasion, or to get several calls in rapid succession. Both situations are typical of the real-world.

Information on ten different human performance measures was collected for each subject during the customer-RSA interaction. This required each tester to determine when errors occurred (usually when RSAs neglected to ask for or provide certain information), and to keep a record of the time spent by each RSA on each call. Information on other measures was gathered after the test session by comparing what the RSAs actually entered into the computer with what they should have entered. Based on these results, designers made changes to the system that improved human performance.

VALIDITY OF WORK MODULE TESTS

How valid is the work module test technique? That is, does the work module test technique produce a measure of human performance that reliably predicts future job performance?

Recall that a work module test is used because we assume that it will reflect what would happen in the real world if work module packages were released immediately, without further revisions. Validity is concerned with how closely work module test scores represent actual job performance.

Two different ways of considering work module test validity can be considered:

- *content validity*—the degree that the work module test includes a representative sample of all tasks, conditions, or situations that could have been included
- *predictive validity*—the extent that the results in a work module test are related to, or associated with, the results of on-the-job measures. When conducting work module or system tests, predictive validity is of most interest.

Simply because a work module test has high content validity, as judged by the test developers, or high face validity, as judged by the subjects, does not necessarily mean that the test will also have high predictive validity. This has been true for many types of psychological tests in the past (Guion, 1965).

Comparisons of human performance speed and accuracy scores for the LMOS work module just discussed are shown in Table 24-2 (Bailey and Koch, 1976). Note that the work module test provided fairly accurate predictions of on-the-job performance. With error data, the predictions were best for the first week or two of work. For the speed data, the predictions were best after three or four weeks on the job.

Table 24-2. Summary of Work Module Test and On-the-Job Monitoring Results Combined for All New RSAs (N=8)

Performance Measure	*Work Module Test*	*Week 1*	*Week 2*	*Week 3*	*Week 4*
Average time per Customer (seconds)	109.5	143.6	126.5	123.5	114.2
Overall error rate	0.17	0.15	0.17	0.18	0.22

As we discussed earlier, the closer a work module test represents real-world conditions the better will be the predictive value of the results. This assumes, of course, that the test contains and exercises a reasonably good sample of the actual activities to be performed on the job. We cannot overemphasize that the greater the fidelity of the test, the better will be the estimate of performance in the real world.

25

CONDUCTING COMPARISON STUDIES

COMMON SENSE VS. FACT-FINDING

Too often designers rely on common-sense guessing to solve human performance problems. Although useful at times, common sense should not be relied on solely. Consider that just 80 years ago it was only common sense that human flight was impossible. Common sense also seems to dictate that hearing in a very noisy environment is unlikely without the use of a microphone or earphones:

> Nearly everyone knows that boiler factories are very noisy and that workmen in boiler factories have difficulty hearing and understanding speech because of this noise. Does it seem sensible that you can improve the intelligibility of speech by putting earplugs in the workmen's ears? Research proves...that...workmen can hear better with earplugs!
> (adapted from *Research Techniques in Human Engineering*, by A. Chapanis. The Johns Hopkins Press. Copyright © 1959, p.7.)

Product design has, for years, depended on common sense. This has resulted in products with many different shapes, sizes, and colors. A good example is the control-burner linkages on stoves. The best design causes the fewest burns and least damage. Using only common sense, which configuration in Figure 25-1 is best?

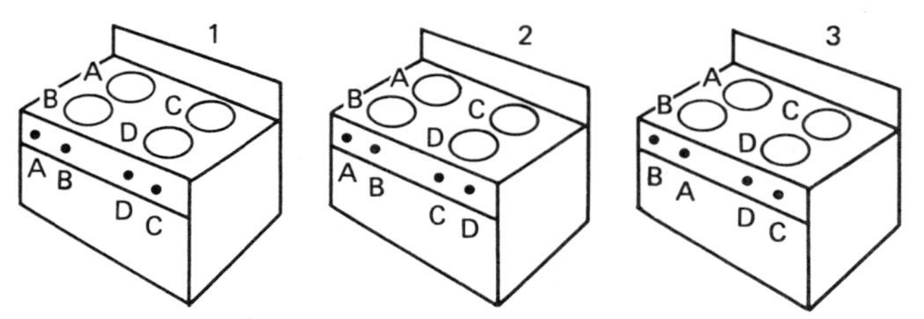

Figure 25-1. Control-Burner Linkage Designs
(Courtesy Johns Hopkins University Press)

While all three elicited numerous errors, configuration one had the fewest (Chapanis, 1959).

Designers tend to make decisions they feel are obvious. Unfortunately, basing design decisions related to human performance on truths that seem obvious may result in degraded performance.

Designers also rely on two other, usually unprofitable, approaches to decisions. First, the idea that something is true because someone says so. The "someone" could be a member of management, or a well-known authority in the particular area. Many designers depend on authorities for much of their knowledge. However, an authority's judgment should only be accepted if the authority can also provide factual evidence. Remember, Aristotle was considered an irreproachable authority for hundreds of years after his death, even though undeniable evidence contradicted his views. Authorities are useful and certainly can aid the design process, but they should not be used as an easy substitute for conducting a meaningful study.

The third approach to design decisions is following habit or tradition. Continued and unwavering belief in something does not make it true. Taking the time to conduct a study to compare the traditional approach with a suggested new one will result in a better design.

Reliable information for critical decisions should come only from careful studies. Designers can conduct many of these studies without the help of other specialists. The techniques covered in this chapter will help designers use methods that are more reliable than common sense guessing.

STUDY DESIGN AND STATISTICS

Measuring People

Human complexity makes the study of human performance difficult. When you measure a desk's height you can assume the measurement will not change. Even after measuring a single human dimension—speed of performance on a difficult task—you may find it often changes. The simplest human measurements are of length of limbs, weight, or height—and these can vary with time. Height can vary up to one inch during the day, and people tend to get shorter after the age of 60. Also, the measuring devices and techniques themselves may give different readings from one measurement to the next.

Most human performance studies attempt to measure cognitive (brain) functioning in terms of either an attitude or some aspect of performance. This includes how long it takes to learn or to perform an activity; how accurately an activity can be performed; and how satisfying the activity is. These are much more difficult measurements than height or weight because people do not perform consistently and available measurement devices are imperfect.

Artificial Performance

A study must have reliable results—that is, it must be possible to duplicate the results of the study. A good study takes into account that the situations being evaluated are deliberately created, and that the people performing in them usually behave in unnatural ways.

We can easily study the natural performance of microbes through a microscope or galaxies through a telescope, but the minute we turn to look at people performing, we find that they make adjustments to accommodate the person doing the evaluating. When human performance is studied, some elements of the situation tend to be somewhat artificial. Thus, a designer must work hard to make a study situation as *real-world* as possible, and recognize, also, that the resulting performance may be slightly different once a system is operational.

This problem with some studies, was dramatically recognized in the Hawthorne studies conducted several years ago (Roethlisberger and Dickson, 1939). Although questioned since that time, the so-called "Hawthorne effect" (the tendency of people to work harder when experiencing a sense of participation in something new and special) occurs frequently enough to make almost all human performance studies somewhat artificial.

Establishing Relations

Human performance studies look for the existence of relations: A certain performance aid is related to fewer errors, certain training materials are related to having shorter training time, certain activity characteristics relate to greater job satisfaction, or a certain activity configuration relates to shorter manual processing times. In all these relations we assume cause—the performance aids cause greater accuracy, training materials cause shorter training time requirements, activity configuration causes greater job satisfaction. That one causes the other is sometimes questionable, but establishing that one is closely related to the other is the essence of a good human performance study.

When things go together in some systematic way, a relation occurs that usually can be stated statistically. Most of the relations that interest us concern the effect design decisions have on the four standards: errors, processing time, training time, or job satisfaction. Appropriately using statistics will help determine the best way to optimize the critical variables. For example, a designer must decide which of two training approaches will result in a lower on-the-job error rate. A study measuring the errors elicited by the two different approaches will help the designer choose the best approach. The time invested in such a study often results in a payoff of greatly improved human and system performance.

Average and Range

Probably the most common statistic used to describe a set of study results is the *average*. The average is determined by adding up all the scores and dividing the result by the number of scores. For example, the average of the scores: 1, 2, 3, 4, 5 is 3 ($1+2+3+4+5 = 15$: $15/5 = 3$).

But to know the average of a group of scores only tells part of the story. Frequently, we are also interested in the "scatter" or "spread" of the separate scores around the average. This can be indicated for any group of scores by reporting the *range*.

Consider a situation where two groups are involved in a study to see if using a color CRT for a map reading task results in fewer errors than using a black and white CRT. The average number of errors for the "color" group is 23.6, and the average errors for the "black and white" group is 23.9. So far as the averages are concerned there is no meaningful difference in the performance of the two groups. However, the "color" group's errors ranged from 3 to 41 (3, 10, 15, 20, 24, 28, 33, 38 and 41), while the "black and white" group's errors ranged only from 20 to 28 (20, 21, 21, 23, 24, 24, 27, 27, 28).

This difference in range indicates that the "color" group is more *variable* than the "black and white" group. This greater variability in one group could mean that the color CRT was, in fact, affecting the error performance of people differently than the black and white CRT, but in a way that did not show up in the average number of errors. By taking a closer look at the individual error scores of people in the " color " group, we see that the color CRT tended to elicit the fewest errors from some subjects in the study, but it also elicited the most errors from other subjects. This could be important information when evaluating the two types of CRTs.

Consider another performance situation where we have two basketball players who each average 22.5 points a game. Suppose that one is consistent (exhibits low variability) and almost always scores from 20 to 25 points a game. The second player, on the other hand, is erratic (exhibits high variability); this player scores over 30 points in some games, and fewer than 10 in others.

There are important performance differences between these two points throughout the basketball season. The *differences* between the two players is best indicated by the range and not the averages.

Thus, a rough guide to the spread or variability of a set of study results (scores) can be given by indicating the range of scores, i.e., by noting the low and the high score. You can see why it is usually a good idea to report results giving both the *average* and the *range*.

Establishing Controls

In most studies conducted to aid design decisions, one object or method is compared with other objects or methods. No matter how limited a designer's background in statistics, numerous types of comparison studies can be conducted if a few rules are closely followed. The most crucial feature of a comparison study is the need to *control* conditions that might cause an ambiguous test. When planning and conducting studies, we achieve acceptable levels of control in many ways.

One of the main question we pose is "How much of the observed performance can be attributed to differences in the objects or methods being compared (e.g., performance on two CRT displays) and how much of it can be attributed to other sources?" A well-developed study tries to minimize the effects of "other" sources by:

1. Ensuring that all subjects perform the same activity in the same context
2. Conducting the study in a place with no unwanted distractions (unwanted distractions are those that one would not expect to occur in a real-world situation)
3. Using a consistent (standardized) set of instructions
4. Ensuring that subjects are as similar as possible to one another and to the larger group they represent.

Controlling all conditions is usually very difficult when using people as subjects. Consider a situation where we must decide which of two CRTs to use. Both appear acceptable and some people favor one; some the other. We decide that the most important performance measure is how quickly people can read the material on one CRT versus the other. We select *a person* to view the material on both CRTs and time the subject with a stopwatch. We find that the person can read the material on CRT A in 18 seconds and the same material on CRT B in 15 seconds. These results are shown in Figure 25-2.

TIME FOR ONE SUBJECT TO READ

CRT A: 18 seconds

CRT B: 15 seconds

Figure 25-2. A Comparison of CRT Performance Using One Subject

Based on these results, CRT B seems superior, but we feel that the study was too brief and extend it to include *two* people. The second subject reads the material on CRT A in 16 seconds and CRT B in 19 seconds. For both subjects, the *average* time to read the material on CRT A is 17 seconds and the average time on CRT B is also 17 seconds. Figure 25-3 shows the average difference between CRTs is 0, while the average difference between subjects is 3. Since the difference between subjects is substantially greater than any difference between CRTs, we have little confidence in using these study results to select the better CRT.

	Subject 1	Subject 2	Difference Between Subjects	Average Time to Read CRT's
CRT A	18	16	2	17
CRT B	15	19	4	17
			2 | 6	

Average difference between subjects= 3 seconds

Average difference between CRT's= 0 seconds

Figure 25-3. A Comparison of CRT Performance Using Two Subjects

Assume that two other subjects are asked to participate in a study of two other CRTs and they provide the data shown in Figure 25-4.

	Subject 3	Subject 4	Difference Between Subjects	Average Time to Read CRT's
CRT C	18	20	2	19
CRT D	15	13	2	14
			2 | 4	

Average difference between subjects= 2 seconds

Average difference between CRT's= 5 seconds

Figure 25-4. A Comparison of CRT Performance Using Two New Subjects and Two Different CRTs

In this case, the average difference between the two subjects (2) is much less than the difference found between the CRTs (5). Thus we have more confidence that the difference found between the CRTs is reliable. By measuring the hard-to-control difference between people and comparing it with the difference between objects or methods, we have an idea of how reliable our results are.

Using Statistics

When preparing a study it is important to try to provide a means for determining whether or not the difference between subjects is *more* or *less* than the measured difference between objects or methods. Modern statistical techniques help in making this determination (cf. Hays, 1973). Four useful statistics for designers are the t-method, F-method, b-method and chi-square method that will be discussed later.

Avoiding Bias

A second consideration when preparing to conduct a study is to *avoid systematic bias.* Look at Figure 25-5 and determine what can be concluded from this study.

		Time to Read (seconds)
Test 1:	Subject 1 reads CRT A	19
Test 2:	Subject 2 reads CRT A	18
Test 3:	Subject 1 reads CRT B	16
Test 4:	Subject 2 reads CRT B	15

Figure 25-5. A Comparison of CRT Performance Using Four Tests

The person who prepared this study made a mistake. In each case the subjects always read from CRT A first and CRT B second. There may have been a *transfer* effect; that is, the experience with CRT A may have affected performance on CRT B. We know the subjects always were slower reading CRT A than CRT B, but we do not know if this is because CRT A was always read first or because CRT A is more difficult to read. In this case, a systematic bias was included in the study.

Reducing Transfer Effects

It is important to be aware of the possible existence of *transfer effects* in any study. People change in the course of a study. Experience gained while performing changes them, even if only slightly. Positive transfer occurs when the trial performed second benefits from the practice gained in the first. This could have occurred in Figure 25-5. Negative transfer occurs when people have an experience in the first example trial that interferes with performance in the second. Two ways of controlling both these transfer effects is to counterbalance or randomize. In the example shown in Figure 25-5, the performance could be *counterbalanced* by having subject 1 perform on CRT A first and CRT B second, and subject 2 perform on CRT B first and CRT A second.

Another approach, particularly useful when dealing with a larger number of subjects, is to *randomize* the order of presentation for all subjects. Some would see CRT A first, others CRT B first. Besides having a random order of presentation, each subject's assignment within a study group can also be randomized. Consider the following example. An advertisement appeared in the company newspaper asking for twenty people to participate in a study. The person conducting the study decided in advance that the first ten people to volunteer would be in one group and the second ten to volunteer would be put into the second group. Twelve people volunteered the first day of the ad, while the other eight took up to two weeks to respond. Unfortunately, the study was related to motivation, and the motivation of the people who responded first was different than the motivation of those who responded later. By assigning the first ten to one group and the second ten to a second group, the person conducting the study systematically biased the results before the study had been conducted. In this case, after all twenty had volunteered, the subjects should have been randomly assigned to the two groups. Exactly how this can be accomplished will be discussed in the next section.

Determining Groups of Subjects

The groups in a comparison study are inevitably composed of people who differ in a variety of ways, such as manual dexterity or ability to make decisions. To reduce the chance of these differences between people affecting the final results, it is important to use participants that all have similar traits. This increases the chances that the results will be due to true differences in the objects or methods being compared (e.g., CRTs), as opposed to basic differences of people in the groups.

The groups in question might be composed of those who will perform on CRT A and those who will perform on CRT B. If a total of ten subjects is to be used, five could be assigned to view CRT A (group 1) and five to view CRT B (group 2). To ensure that the groups are approximately equal in all relevant characteristics, a designer should use one of three techniques when assigning subjects to different groups. These techniques are *random groups, matched groups* and *same-subject groups*. In the first two, different subjects are assigned to each group while in the latter, the same subjects are used in all groups.

Random Groups

In random assignment each subject has an equal opportunity to be assigned to each group. One simple way to assign group members to one group or another is by flipping a coin. Another quick and easy method is to put the names of all subjects into a hat and draw them out one at a time. The first name drawn is put in group 1, the second name in group 2, the third in group 1, etc.

In some instances, it is convenient to use a random number table like Table 1 in Appendix A. Consider another case where you are conducting a study using 30 subjects to test which of two performance aids is best. Pick any line of random numbers in the table.

Part of the first line in Table 1 is shown below:

11339 19233 50911 14209 39594 68368

Next, combine adjacent pairs of numbers to form two-digit numbers. (If there are nine or fewer subjects, use one-digit numbers; if 100 or more, use three-digit numbers.) Then pick the first occurrence of any values that fall in the 1 to 30 range (shown in bold below) and write them next to a subject's name until all subjects have been assigned a unique number.

11 33 91 92 33 50 91 11 42 **09**...

Subject	A	B	C	D	E	F...
Number	11	9	23	27	15	

Finally, put subjects 1 through 15 in group A and those with numbers 16 through 30 in group B.

Matched Groups

The second technique used to make groups equal in subject characteristics is to match the individuals assigned to the different groups. Matching can be achieved by measuring subjects in terms of their performance of some task, such as a typing speed test. The subjects are assigned so that each group is approximately equal—each group has about the same average typing speed. One way to accomplish this is to match each subject in group one with a subject in group two. For example, ten subjects who are participating in a study of how fast people can read document style A versus document style B could be matched on reading speed scores obtained from a standardized reading test. Assume that the available subjects and reading test scores are as follows:

Subjects	Reading Test Score
1	100
2	115
3	95
4	125
5	80
6	140
7	135
8	75
9	115
10	120

The subjects can then be matched and assigned to groups according to the test scores as shown below:

Group 1		Group 2	
Subject	*Score*	*Subject*	*Score*
1	100	3	95
2	115	9	115
8	75	5	80
7	135	6	140
10	120	4	125
Average	109	Average	111

One form of the t-method we discuss later requires the use of such matched pairs.

Same Subject Groups

The third and final technique is to have each subject participate in all groups. In this way, we keep to a minimum the differences between subjects (but we have to be very careful about *transfer* effects).

This method cannot be used in all studies because it does not always make good sense to have the same subjects involved in two or more conditions. For example, teaching the same person to use a new computer system by training them with two different methods. We also must avoid subject fatigue or boredom. Remember, people change as they spend time participating in a study.

Reliable Differences

After subjects are assigned to groups, the study completed, and the data collected, the results may be analyzed *statistically* to determine the *reliability* of differences. Statistical methods are most useful when there seems to be an important difference, but it is difficult to tell if the difference in performance is large enough to be reliable. If it is reliable, the same results should occur if the study were done again using different people with backgrounds similar to the original group. It is not always necessary to use a statistical method. If, for example, it is evident that, say, ten out of ten subjects performed with fewer errors using CRT A than CRT B, then a statistical method is not necessary.

Whether or not a difference is reliable depends on the probability that the cause of the difference is chance. In this book, we consider a finding reliable if the observed difference would have occurred by chance less than once in 20 studies (i.e., less than 5 percent of the time). The aim is to accept as many genuine differences as possible, without accepting chance differences. Remember that even though a set of results is found to be reliable, there is still a small possibility that they are due to chance. Stating that a difference is reliable does not indicate its size, only that if the study were repeated, we would probably get the same result.

Selecting a Statistical Method

Consider the following when attempting to decide which of the four statistical methods discussed in the following sections to use:

- Whether the study data are reported as *averages* of group performance or as *frequencies*—numbers of subjects in a given category.
- The *number* of different groups or conditions.
- The way that people are *assigned* to the different conditions being studied (random, matched, same-subject).

Table 25-1 can help in deciding the appropriate test for a set of data. An example of how to use this logic tree is discussed below.

Assume that a designer wants to compare the number of errors made on a given activity in crowded versus isolated conditions. Ten people are selected as subjects, and randomly assigned to one of the two conditions (crowded or isolated). The designer conducts the study and ends up with the average number of errors made under each condition.

There are three consecutive decisions to be made. First, are the study results reported as averages or frequencies? In this case they are reported as the *average* number of errors. The second decision has to do with the number of conditions evaluated. There were *two* (crowded and isolated). The third and final decision concerns how the subjects were assigned to each condition. They were *randomly* assigned to one group or the other. Following these branches on the logic tree leads to the conclusion that the t-method is the appropriate statistical method.

In some situations more than one method can be used. In these cases, designers can use the method they prefer.

Table 25-1. Logic Tree for Selecting the Most Appropriate Statistical Method

How are results reported?	How many conditions?	How were subjects assigned?	Method
Averages	2	Randomly	t-method (A)
Averages	2	Matched	t-method (B) or b-method
Averages	2	Same subject	t-method (B) or F-method or b-method
Averages	3 or more	Randomly	F-method (A)
Averages	3 or more	Same subject	F-method (B) or b-method
Frequencies	2	Randomly	chi-square
Frequencies	2	Matched	b-method or chi-square
Frequencies	2	Same subject	b-method or chi-square
Frequencies	3 or more	Randomly	chi-square
Frequencies	3 or more	Matched	b-method
Frequencies	3 or more	Same subject	b-method

STATISTICAL METHODS

The following sections introduce and demonstrate the application of statistical methods for determining whether the differences between groups are large enough to be considered reliable. The designer does not need an in-depth understanding of statistics to use these techniques. If none of these methods seems appropriate, or difficulties are encountered, the designer should consult a statistical specialist.

t-Method

The t-method is a commonly used technique that compares the difference between the averages of two groups to determine if there is a reliable difference between them. The number of subjects in each group should be the same. The following two examples illustrate two different ways of applying the t-method. The first example is used when subjects are *randomly assigned.* The second example is for when subjects are either *matched* or in *same subject* groups. The formulas are slightly different so be careful to use the correct one. Remember, the correct formula depends on *how the subjects were assigned* to the different groups.

Randomly Assigned Subjects (Example 1)

A designer wants to find out if different methods of training result in reliably different numbers of errors on an end-of-course performance test. He randomly assigns 24 subjects to two groups. He teaches group 1 using *videotapes* and group 2 using *lectures* by highly experienced workers. At the conclusion of the training, he obtains the following error scores:

Group 1 (videotape)		Group 2 (lectures)	
Subject	*Score*	*Subject*	*Score*
1	27	13	24
2	28	14	28
3	20	15	31
4	27	16	27
5	30	17	28
6	28	18	23
7	19	19	31
8	25	20	29
9	31	21	26
10	24	24	33
11	23	23	29
12	27	24	28

The following formula is applied to the data. The formula will become more meaningful with experience. For the time being, it is only necessary to carefully follow the computational steps shown after the formula.

$$t = \frac{A_1 - A_2}{\sqrt{\left[\frac{\Sigma X_1^2 - \frac{(\Sigma X_1)^2}{N_1} + \Sigma X_2^2 - \frac{(\Sigma X_2)^2}{N_2}}{(N_1 + N_2 - 2)}\right]\left[\frac{1}{N_1} + \frac{1}{N_2}\right]}}$$

where:

$A_1 =$ the average of the first group of scores

$A_2 =$ the average of the second group of scores

$\sum x_1^2 =$ the sum of the squared score values of the first group

$\sum x_2^2 =$ the sum of the squared score values of the second group

$(\sum x_1)^2 =$ the square of the sum of the scores in the first group

$(\sum x_2)^2 =$ the square of the sum of the scores in the second group

$N_1 =$ the number of scores in the first group

$N_2 =$ the number of scores in the second group

The following computational steps are used to calculate t:

Step 1. List the error data by group as shown above. No particular order is necessary within the groups.

Step 2. Add the scores in group 1 to obtain $\sum x_1$.*

$$27 + 28 + 20 + 27 + \ldots + 27 = 309$$

Step 3. Square every score in group 1 and add these squared values to obtain $(\sum x_1^2)$.

* Note that the $\sum$ symbol means "sum of" and the subscript 1 refers to group 1.

$(27)^2 + (28)^2 + (20)^2 + (27)^2 + (30)^2 + (28)^2 +$

$(19)^2 + (25)^2 + (31)^2 + (24)^2 + (23)^2 + (27)^2 =$

$729 + 784 + 400 + 729 + 900 + 784 + 361 + 625 +$

$961 + 576 + 529 + 729 = 8107$

Step 4. Count the total number of subjects in group 1. Call this value N_1.

$$N_1 = 12$$

Step 5. Square the value obtained in Step 2 ($\sum x_1$) to obtain the value $(\sum x_1)^2$.

Divide this squared value by N_1.

$$\frac{(\sum x_1)^2}{N_1} = \frac{(309)^2}{12} = \frac{95481}{12} = 7956.75$$

Step 6. Subtract the value obtained in Step 5 from the value in Step 3.

$$\sum x_1^2 - \frac{(\sum x_1)^2}{N_1} = 8107.00 - 7956.75 = 150.25$$

Step 7. Add the scores in group 2 to obtain $\sum x_2$.

$$24 + 28 + 31 + \ldots + 28 = 337$$

Step 8. Square every score in group 2 and add these squared values to obtain $\sum x_2^2$.

$(24)^2 + (28)^2 + (31)^2 + (27)^2 + (28)^2 + (23)^2 + (31)^2 +$

$(29)^2 + (26)^2 + (33)^2 + (29)^2 + (28)^2 =$

$576 + 784 + 961 + 729 + 784 + 529 + 961 + 841 +$

$676 + 1089 + 841 + 784 =$

9555

Step 9. Square the value obtained in Step 7 ($\sum x_2$) to get $(\sum x_2)^2$. Then divide the squared value by N_2.

$$(\sum x_2)^2 = (337)^2 = \frac{113569}{12} = 9464.08.$$

Step 10. Subtract the value obtained in Step 9 from the value obtained in Step 8.

$$\sum x_2^2 - \frac{(\sum x_2)^2}{N_2} = 9555 - 9464.08 = 90.92$$

Step 11. Add the values obtained in Steps 6 and 10.

$$150.25 + 90.92 = 241.17$$

Step 12. Divide the value obtained in Step 11 by $(N_1 + N_2) - 2$.

$$\frac{241.17}{(12+12)-2} = \frac{241.17}{(24-2)} = \frac{241.17}{22} = 10.96$$

Step 13. Multiply the value obtained in Step 12 by $\frac{1}{N_1} + \frac{1}{N_2}$.

$$10.96\ (\frac{1}{12} + \frac{1}{12}) = 10.96\ (.08 + .08) =$$

$$10.96\ (.16) = 1.75$$

Step 14. Take the square root of the value obtained in Step 13.

$$\sqrt{1.75} = 1.32$$

Step 15. Calculate the average $(\bar{x}_1)$ for group 1 by taking the sum $(\sum x_1)$ from Step 2 and dividing by N_1.

$$A_1 = \frac{\sum x_1}{N_1} = \frac{309}{12} = 25.75$$

Step 16. Calculate the average for group 2 by taking the sum $(\sum x_2)$ from Step 7 and dividing by N_2.

$$A_2 = \frac{\sum x_2}{N_2} = \frac{337}{12} = 28.08$$

Step 17. Subtract the average of group 2 from the average of group 1.

$$25.75 - 28.08 = -2.33 = 2.33$$

Note: For computational purposes, only the absolute difference of the average is important — the negative sign can be ignored.

Step 18. Divide the value obtained in Step 17 by the value obtained in Step 14. This yields the t value.

$$t = \frac{2.33}{1.32} = 1.77$$

Step 19. To determine whether the t value is reliable (i.e., whether the two groups are reliably different) the degrees of freedom (df) must be computed. The df is equal to $(N_1 + N_2) - 2$. In this example, $(12 + 12) - 2 = 22$.

Consult the t-method table (Appendix A, Table 2) to find the t value. In this case the tabled t is 2.07.

Because calculated t (1.76) is smaller than the tabled t (2.07), the designer must conclude that the difference between the two group averages is *not* large enough to suggest that one training method is better than the other, i.e., the results are not reliable.

Matched or Same Subject Groups (Example 2)

The following is an example using matched subjects; this technique also can be used with studies using the same subjects in each condition. A designer wanted to know if a telephone at each user's desk would increase the number of customer transactions completed each week. Two groups were selected by matching twelve customer representatives according to experience and then assigning them to two groups. The average years of experience in both groups were about the same. Two people in group 1 shared a telephone, while those in group 2 had their own telephone. The two groups were assigned to work in different buildings. The following t-method formula was used:

$$t = \frac{A_1 - A_2}{S_D}$$

where

$A_1 = average\ of\ group$ 1

$A_2 = average\ of\ group$ 2

$S_D = the\ standard\ difference\ determined\ by$:

$$S_D = \sqrt{\frac{N\sum D^2 - (\sum D)^2}{N^2\,(N-1)}}$$

where

N = the number of pairs
D = the difference between each score in group 1 and the related score in group 2.

Step 1. List the data in the form of a table, with the matched subjects side-by-side. For example, in the table below subjects 1 and 7 both had less than one year experience; subjects 2 and 8 each had about one year experience, etc. The order within groups is very important.

Group 1 (shared phone)		*Group 2 (own phone)*	
Subject	*Number of transactions*	*Subject*	*Number of transactions*
1	1	7	3
2	2	8	5
3	1	9	5
4	1	10	5
5	2	11	4
6	1	12	5

Step 2. Obtain the difference between each pair of scores to obtain the Ds.

Groups	*D*
1-3 =	-2
2-5 =	-3
1-5 =	-4
1-5 =	-4
2-4 =	-2
1-5 =	-4

Step 3. Square each D and add these squared values to obtain $\sum D^2$.

$\sum D^2 = (-2)^2 + (-3)^2 + (-4)^2 + (-4)^2 + (-2)^2 + (-4)^2 =$

$4 + 9 + 16 + 16 + 4 + 16 = 65$

Step 4. Count the number of pairs of scores. Call this value N.

$$N = 6$$

Step 5. Multiply the value obtained in Step 3 ($\sum D^2$) by N.

$$(\sum D^2)\ (N) = (65)\ (6) = 390$$

Step 6. Obtain the sum of the differences from Step 2 ($\sum D$). Square this term $(\sum D)^2$.

$$(\sum D)^2 = (-19)^2 = 361$$

Step 7. Subtract the value obtained in Step 6 from the value obtained in Step 5.

$$(\sum D^2)\ (N) - (\sum D)^2 = 390 - 361 = 29$$

Step 8. Take the value N from Step 4 and square it. Multiply this value by N - 1.

$$N^2 (N - 1) = (6)^2 (6 - 1)$$

$$= (36) (5)$$

$$= 180$$

Step 9. Divide the value obtained in Step 7 by the number obtained in Step 8.

$$\frac{29}{180} = .16$$

Step 10. Obtain the square root of the number obtained in Step 9.

$$\sqrt{.16} = .40$$

Step 11. Obtain the average of group 1 by adding the scores in column 1 and dividing by N.

$$A_1 = 1 + 2 + 1 + 1 + 2 + 1 = 8$$

$$\frac{8}{6} = 1.33$$

Step 12. Obtain the average for group 2 by adding the score in column 2 and dividing by N.

$$A_2 = 3 + 5 + 5 + 5 + 4 + 5 = 27$$

$$\frac{27}{6} = 4.50$$

Step 13. Subtract the average of group 2 from the average of group 1.

$$1.33 - 4.50 = -3.17 = 3.17$$

(*Note*: the negative sign can be ignored.)

Step 14. Divide the value obtained in Step 13 by the value obtained in Step 10 to obtain the t value.

$$t = \frac{3.17}{.40} = 7.93$$

Step 15. Degrees of freedom (df) = N - 1
= 6 - 1
= 5

In order to say that the availability of a telephone is related to a difference in performance, the calculated t for 5 degrees of freedom (df) must be 2.57 or larger (see Appendix A, Table 2). In this case the obtained t of 7.93 is much larger than the tabled t. The difference then *is reliable*. Therefore we can conclude that having access to one's own telephone is related to better performance. Having one's own telephone seems to lead to more completed transactions (average = 4.5) than does sharing a phone (average = 1.33). Again, there is a slight possibility, about 5 percent, that the differences in number of completed transactions is due to chance rather than telephone availability.

F-Method

The F-method provides a means to compare two or *more* groups simultaneously to help decide if there are reliable differences among them. Usually, if you have only two groups the t-method is preferred.

Determining if differences among groups are reliable by computing an F value and consulting a table is similar to the t-method procedure. If the F value is large enough to be reliable, it means that real differences probably exist among the groups and can be expected to reliably occur again under similar circumstances. Again, having an equal number of people in each group is important.

Randomly Assigned Groups (Example 1)

A designer wishes to see if people make fewer errors according to the type of training they receive on an activity. The three training types are formal classroom training, on-the-job training where they receive some assistance by those already working, and "pickup," where they receive no help and learn the activity through trial and error. The number of errors made during a one-week period is collected on a sample of 30 people who train with these three methods.

To calculate F, you use the formula shown below. Again, it is not necessary to understand what the formula means. It will become familiar with use. Simply follow the calculation steps.

$$F = \frac{\dfrac{Z - \dfrac{T^2}{N}}{C - 1}}{\dfrac{\left[S - \dfrac{T^2}{N}\right] - \left[Z - \dfrac{T^2}{N}\right]}{N - C}}$$

The meaning of the symbols C, N, S, T, and Z will be described in the calculation steps that follow:

Step 1. Start with three columns of error scores; one column for each condition (training type).

Type of Training

Formal	*On-the-Job*	*"Pickup"*
20	21	29
22	24	28
25	26	29
21	22	31
19	20	34
17	19	32
18	20	21
24	29	35
28	28	29
19	20	28

Step 2. Now add the scores for each column and then the total of each column together. This total is T.

$$20 + 22 + 25 + 21 + 19 + 17 + 18 + 24 + 28 + 19 = 213$$

$$21 + 24 + 26 + 22 + 20 + 19 + 20 + 29 + 28 + 20 = 229$$

$$29 + 28 + 29 + 31 + 34 + 32 + 21 + 35 + 29 + 28 = 296$$

$$213 + 229 + 296 = 738 = T$$

Step 3. Square the T value obtained in Step 2.

$$T^2 = (738)^2$$

$$= 544644$$

Step 4. Square each score and then add these squared scores for all groups together. This squared quantity is S.

$(20)^2 + (22)^2 + (25)^2 + (21)^2 + (19)^2 + (17)^2 +$

$(18)^2 + (24)^2 + (28)^2 + (19)^2 =$

$400 + 484 + 625 + 441 + 361 + 289 + 324 + 576 + 784 + 361 = 4645$

$(21)^2 + (24)^2 + (26)^2 + (22)^2 + (20)^2 + (19)^2 +$

$(20)^2 + (29)^2 + (28)^2 + (20)^2 =$

$441 + 576 + 676 + 484 + 400 + 361 + 400 + 841 + 784 + 400 = 5363$

$(29)^2 + (28)^2 + (29)^2 + (31)^2 + (34)^2 + (32)^2 + (21)^2 +$
$(35)^2 + (29)^2 + (28)^2 =$

$841 + 784 + 841 + 961 + 1156 + 1024 + 441 + 1225 + 841 + 784 = 8898$

$$4645 + 5363 + 8898 = 18906$$

$$S = 18906$$

Step 5. For each column square the sum obtained for each in Step 2. Also, count the number of scores in that column. Call this n. Divide the column squared value by n.

Column 1 = 213

$$(213)^2 = 45369 \qquad n = 10$$

$$\frac{45369}{10} = 4536.90$$

Step 6. Repeat Step 5 for each column and then add the results from the different columns. This quantity is Z.

Column 2 = 229

$$(229)^2 = 52441 \qquad n = 10$$

$$\frac{52441}{10} = 5244.10$$

Column 3 = 296

$$(296)^2 = 87616 \qquad n = 10$$

$$\frac{87616}{10} = 8761.60$$

$$Z = 4536.90 + 5244.10 + 8761.60$$

$$= 18542.60$$

Step 7. Count the total number of scores on the entire table. Call this quantity N.

Step 8. Use the results of Steps 3, 4, and 7 to calculate $S - \frac{T^2}{N}$

$$18906 - \frac{544644}{30}$$

$$18906 - 18154.80$$

$$= 751.20$$

Step 9. Use the results of Steps 4, 6, and 7 to calculate $Z - \frac{T^2}{N}$.

$$18542.60 - \frac{544644}{30}$$

$$18542.60 - 18154.80$$

$$= 387.80$$

Step 10. Subtract the quantity in Step 9 from Step 8.

$$751.20 - 387.80$$

$$= 363.40$$

Step 11. Count the number of columns on the table. Call this C.

$$C = 3$$

Step 12. Divide the quantity from Step 9 by the degrees of freedom, C-1.

$$\frac{387.80}{3-1} = \frac{387.80}{2} = 193.90$$

Step 13. Divide the quantity from Step 10 by the degree of freedom, N-C.

$$\frac{363.40}{30-3} = \frac{363.40}{27} = 13.46$$

Step 14. To find the value of F, divide the quantity from Step 12 by the quantity from Step 13.

$$F = \frac{193.90}{13.46}$$

$$= 14.41$$

Step 15. Use Table 3 in Appendix A.

a. Find the number C-1 (2). Use this number to read *down* the table.

b. Find the number N-C (27). Use this number to read *across* the table.

c. Where the two (column and row) *intersect*, read the tabled F value. In this case it is between 3.39 and 3.32.

Step 16. If the calculated F is equal to or greater than the tabled F, the difference among the groups is reliable; if the calculated F is less than the tabled F, there is *no* reliable difference among the groups or groups or columns.

Since the calculated F is larger than the tabled F in this example, we can conclude with some confidence that there are real differences among the groups (i.e., the differences are not due to chance). By inspecting the results, we see that the most errors occurred with the subjects who were trained using the pick-up method.

Same Subjects in Each Group (Example 2)

A designer wishes to determine whether illumination level will affect a satisfaction score measured by a questionnaire. She selects nine people from a group of employees and has each person work for one month under three different levels of illumination. To help reduce transfer effects, three of the people begin in the lowest light level, three in the middle, and three in the highest. Each month the people are assigned to different light levels. The average satisfaction score for each person in each group is shown in Step 1 below. This example uses three groups, but the same technique could be used with two groups, four groups, or more. The important consideration is that the *same* subjects perform in each condition.

Because each person performs in each group, a slightly different F-method formula is used. The formula when using the same subjects is shown below:

$$F = \frac{\frac{Z - G}{C - 1}}{\frac{[G + S] - [Y + Z]}{(C - 1)(R - 1)}}$$

The meaning of the symbols C, G, S, R and Z will be described in the calculation steps that follow.

Step 1. Start with a listing of the subject's scores separated by columns into the illumination levels to which they belong.

Subjects *(Employee)*	Illumination *Level A*	Illumination *Leval B*	Illumination *Level C*
1	18	20	19
2	17	21	14
3	23	20	27
4	28	23	26
5	16	17	15
6	15	19	17
7	19	20	18
8	22	23	21
9	24	25	22

Step 2. Add the scores for each column. Then add the totals of the columns. This total is T.

$$18 + 17 + 23 + 28 + 16 + 15 + 19 + 22 + 24 = 182$$

$$20 + 21 + 20 + 23 + 17 + 19 + 20 + 23 + 25 = 188$$

$$19 + 14 + 27 + 26 + 15 + 17 + 18 + 21 + 22 = 179$$

$$T = 182 + 188 + 179 = 549$$

Step 3. Square the T value obtained in Step 2.

$$T^2 = (549)^2$$

$$= 301401$$

Step 4. Count the number of columns on the table. Call this C. Count the number of rows on the table. Call this R.

$$C = 3$$

$$R = 9$$

Step 5. Divide the quantity T^2 (from Step 3) by the product of C times R (obtained in Step 4). Call this quantity G.

$$G = \frac{T^2}{(C)(R)} = \frac{301401}{(3)(9)} = 11163.00$$

Step 6. Square each score and then add these squared scores for all groups together. The squared quantity is S.

$(18)^2 + (17)^2 + (23)^2 + (28)^2 + (16)^2 + (15)^2 + (19)^2 + (22)^2 + (24)^2$

$= 324 + 289 + 529 + 784 + 256 + 255 + 361 + 484 + 576 = 3828$

$(20)^2 + (21)^2 + (20)^2 + (23)^2 + (17)^2 + (19)^2 + (20)^2 + (23)^2 + (25)^2$

$= 400 + 441 + 400 + 529 + 289 + 361 + 400 + 529 + 625 = 3974$

$(19)^2 + (14)^2 + (27)^2 + (26)^2 + (15)^2 + (17)^2 + (28)^2 + (21)^2 + (22)^2$
$= 361 + 196 + 729 + 676 + 225 + 289 + 324 + 441 + 484 = 3725$

$$S = 3828 + 3974 + 3725 = 11527$$

Step 7. Refer back to the total obtained for *each column* in Step 2. Square this value. Then divide the squared value by the number of scores (n) in each column.

Column 1 = 182 $\qquad$ $n = 9$

$$(182)^2 = \frac{33124}{9} = 3680.44$$

Step 8. Repeat Step 7 for each column and then add the results across different columns. This quantity is Z.

Column 2 = 188 $\qquad$ $n=9$

$$(188)^2 = \frac{35344}{9} = 3927.11$$

Column 3 = 179 $\qquad$ $n = 9$

$$(179)^2 = \frac{32041}{9} = 3560.11$$

$$Z = 3680.44 + 3927.11 + 3560.11 = 11167.66$$

Step 9. Referring back to the table of scores, add the scores for each row.

$$18 + 20 + 19 = 57$$

$$17 + 21 + 14 = 52$$

$$23 + 20 + 27 = 70$$

$$28 + 23 + 26 = 77$$

$$16 + 17 + 15 = 48$$

$$15 + 19 + 17 = 51$$

$$19 + 20 + 18 = 57$$

$$22 + 23 + 21 = 66$$

$$24 + 25 + 22 = 71$$

Step 10. Square each of these row totals and add the results. Then divide the quantity by C (from Step 4). Call this quantity Y.

$(57)^2 + (52)^2 + (70)^2 + (77)^2 + (48)^2 + (51)^2 + (57)^2 +$

$(66)^2 + (71)^2 =$

$3249 + 2704 + 4900 + 5929 + 2304 + 2601 + 3249 + 4356 + 5041 =$

34333

$$Y = \frac{34333}{3} = 11444.33$$

Step 11. Use the results of Steps 5 and 8 to calculate Z - G.

$$11167.66 - 11163.00 = 4.66$$

Step 12. Use the results of Steps 5, 6, 8 and 10 to calculate $[G+S] - [Y+Z]$

$[11163 + 11527]-[11444.33 + 11167.66]$
$= 22690 - 22611.99$

$= 78.01$

Step 13. Divide the quantity from Step 11 by the degrees of freedom, C-1 (C is from Step 4).

$$\frac{4.66}{3-1} = \frac{4.66}{2} = 2.33$$

Step 14. Divide the quantity from Step 12 by the degrees of freedom (C-1) (R-1) (C and R are from Step 4).

$$\frac{78.01}{(3-1)(9-1)} = \frac{78.01}{(2)(8)} = \frac{78.01}{16} = 4.88$$

Step 15. To find the value of F, divided the quantity from Step 13 by the quantity from Step 14.

$$F = \frac{2.33}{4.88} = 0.48$$

Step 16. Use Table 3 in Appendix A:

a. Find the number C-1 at the top of the table. Use this number (2) to read *down* the table.

b. Find the number (C-1) (R-1) on the left side of the table. Use this number (16) to read *across* the table.

c. Where the two (column and row) *intersect*, read the tabled F value (3.63).

Since the calculated F value (0.48) is less than the tabled F value (3.63), there is no reliable difference between the different illumination levels on human performance.

b-Method

Still another means for determining the reliability of a set of data is the b-method. With this method a designer can determine whether one item is reliably better than from one to four others. Consider the following example.

A designer wants to know which of four different terminals is preferred by a group of clerks. Eleven clerks agree to evaluate each of the four terminals (A through D) and to indicate which one they prefer. For example, subject 1 preferred terminal C, subject 2 preferred B, etc. The results of their evaluation are shown below:

Most Preferred Terminal Type

Subject	A	B	C	D
1			X	
2		X		
3			X	
4			X	
5	X			
6				X
7		X		
8			X	
9	X			
10			X	
11			X	

Terminal C was preferred by six of the subjects, while Terminals A, B, and D were preferred by two or fewer subjects. The question is, "How many times must Terminal C be selected in order to confidently conclude that the preference for Terminal C is *not* due to chance?" Note that we are using *frequencies* and not the averages used with the t and F methods.

By looking at Table 4 in Appendix A for eleven subjects and four categories, we see that six or more votes for a specific terminal is required for there to be a reliable difference. Thus, the results from the study are *reliable*. If another group of clerks with similar backgrounds were asked the same questions, the result would most likely be the same.

The same method can be used with *three* different groups. Consider a situation where a designer would like to determine which of three sets of computer commands elicit the fewest errors for experienced users. The three sets of commands use varying degrees of abbreviation: set A requires that the command be keyed in full (e.g., "remove"), set B requires an abbreviation in consonants (e.g., "rmv"), and set C requires only the first letter of the command (e.g., "r"). Thirty people with experience in command languages and terminals were used as subjects. The people were matched according to terminal-use experience and divided into ten groups. A set of simulated problems were given to each of the groups. Each person only used one command type. The number of errors made by each person in each group is shown below:

Errors Made While Using Each Command Type

Group	Command Type A	B	C
1	18	**15**	25
2	19	**17**	20
3	23	**12**	19
4	19	20	**18**
5	15	**11**	17
6	18	21	**15**
7	**16**	18	21
8	19	**15**	26
9	27	**12**	19
10	13	**11**	15

The results indicate that in 7 out of the 10 groups, command type B elicited the fewest number of errors. Entering Table 4 in Appendix A for 10 matched groups and three categories we find that at least 7 "hits" are necessary for the findings to be considered reliable. Had command type B elicited the fewest errors less than seven times, the chances of the findings being due to chance would have been quite high.

This technique can be used with the same people performing in all categories or with different people in each of the categories. If different people are used, make sure they are *matched* as carefully as possible. In the last example it was essential that all subjects in each group had about the same amount of experience using terminals and command languages. If they had not been carefully matched, the results would have been meaningless.

If *ties* occur on potential "hits," discard *all* of the responses for that subject or group. If this is done, make sure to reduce the number of subjects or groups used when entering Table 4 in Appendix A.

One final example. Consider a situation where a designer wants to determine which of two visual displays is best for a speedometer in an automobile. One display is a digital readout, and the other is the more traditional dial type.

Several different studies are used. In one study, seventeen people take turns driving an automobile. They perform a variety of driving tasks, including accelerating, decelerating, turning, etc. Each person performs twice, using a similar set of tasks but with the digital display one time and the dial the other. Half use the dial first and the other half use the digital display first. At pre-arranged times during the test session a beeper sounds. When the beeper is heard each driver responds by glancing at the speedometer and verbally reporting his or her speed. The designer is interested in knowing how quickly, on the average, they start to tell the speed. The results are shown below:

Average Time To Begin Reporting Readout
on Each Display (seconds)

	Display Type	
Subject	A	B
1	2.5	**2.3**
2	3.2	**3.1**
3	2.1	2.1
4	3.6	3.7
5	4.7	**4.6**
6	**2.1**	2.5
7	1.9	**1.7**
8	3.2	**3.0**
9	3.6	**3.4**
10	**2.8**	2.9
11	**2.6**	2.9
12	3.1	3.1
13	4.5	**4.2**
14	3.2	**3.0**
15	**2.8**	2.9
16	2.7	**2.4**
17	2.5	**2.3**

Note that the average times for subjects 3 and 12 were the *same* for both types of displays. These will be discarded, leaving 15 subjects. The majority of people did better with Display B, so Table 4, Appendix A is entered using 15 subjects and 10 hits for Display B. According to the table, at least 12 hits are required for these findings to be considered reliable, but only 10 are shown in these data. On the basis of this study we would have to conclude that there is *no reliable difference* between the two display types.

Because they are *averages,* these same data could be evaluated using an F-method. In some cases the same data could be evaluated using a t-method, F-method or b-method.

Chi-Square Method

In previous sections, particularly using the t-and F-methods, groups were compared in terms of the *amount* of a given characteristic. It often happens, however, that the data consist only of the frequencies falling into any one of a number of categories. That is, the data are in terms of frequencies (number of people), and we want to compare these observed frequencies with some expected frequencies. Usually, these problems consist of counting the number of people who prefer a particular alternative. The question asked by a designer dealing with frequencies is, "Can the differences between the observed frequencies and the expected frequencies be attributed to chance, or not?" The method used to evaluate such questions is the chi-square.

The chi-square is calculated by dealing with both observed frequencies and expected frequencies. The observed frequencies are the results obtained in the study. The expected frequencies are either calculated or assumed to be equal to each other.

If the differences between the observed frequencies and the expected frequencies are small, chi-square will be small. The greater the difference between the observed and expected frequencies, the larger chi-square will be. If the differences between observed and expected values are so large collectively as to occur by chance only 5 percent or less of the time, we conclude that the differences between the observed and expected frequencies are reliable. Make sure that each person is only represented once. Also, keep in mind that the method is only accurate if *expected* frequencies are 5 or more in each cell (the more people included in the study the better).

One Row Frequency Table (Example 1)

The simplest example of the chi-square is to use it with a frequency table consisting of one row.

Consider an example where a designer develops three alternative functional configurations for a new system—manual only, manual and computer, and computer only. Wondering whether or not they all have equal potential for success, he asks a random sample of 28 potential users which of the three configurations they would prefer. In evaluating his results, he used this chi-square formula:

$$\text{chi-square} = \sum \frac{(O-E)^2}{E}$$

where

$$O = observed\ frequencies$$

$$E = expected\ frequencies$$

Step 1. List the frequencies (number of people preferring each alternative) in one row.

Functional Configuration

Manual Only	*Computer and Manual*	*Computer Only*
6	15	7

Step 2. List the numbers from Step 1 as the *observed* frequencies in the table below. Determine the *expected* frequencies by dividing the total of observed frequencies by the number of observations ($\frac{28}{3} = 9.33$). This suggests that we expect about the same number of people to prefer each of the three alternatives.

	Manual only	Computer and manual	Computer only
Observed	6	15	7
Expected	9.33	9.33	9.33

Step 3. For each column subtract the expected frequency from the observed frequency.

Observed	6	15	7
Expected	9.33	9.33	9.33
	-3.33	5.67	-2.33

Step 4. Square the results from Step 3.

$$(-3.33)^2 = 11.09$$

$$(5.67)^2 = 32.15$$

$$(-2.33)^2 = 5.43$$

Step 5. Divide the quantities in Step 4 by the expected frequencies

$$\frac{11.09}{9.33} = 1.19$$

$$\frac{32.15}{9.33} = 3.45$$

$$\frac{5.43}{9.33} = 0.58$$

Step 6. Chi-square is determined by adding together the quantities in Step 5.

$$\text{Chi-square} = 1.19 + 3.45 + 0.58 = 5.22$$

Step 7. Calculate the degrees of freedom (df) by subtracting 1 from the number of columns (C).

$$df = C - 1$$

$$= 3 - 1$$

$$= 2$$

Step 8. Use the degrees of freedom to enter Table 5 in Appendix A.

Consulting the chi-square table for 2 degrees of freedom, we find that the tabled chi-square is 5.99. Since the obtained value of 5.22 is smaller than this value we conclude that there is no reliable difference between the preferences for the three configurations. Although on the surface it may appear that the computer-and-manual configuration is preferred over the other two, the difference is not large enough to preclude the finding from being due to chance.

Multi-row Frequency Table (Example 2)

A designer has the option of using either toggle switches, selector switches, or push buttons on a new terminal. She wonders if she should use different switch types for different job categories. She took an opinion poll using a random sample of 200 employees. She asked them to indicate their job category (clerk, technician, or manager) and the switch they preferred. Is there evidence to suggest that the three groups differed significantly in their preference for switches?

Step 1. Arrange the data into a contingency table as shown below. These are the *observed* frequencies.

Type of Switch

Category	*Toggle*	*Selector Switch*	*Pushbutton*
Clerks	20	25	15
Technicians	25	30	25
Managers	15	20	25

Step 2. Add the numbers across each row of the table to obtain the row sum.

$$\text{sum of row 1} = 20 + 25 + 15 = 60$$

$$\text{sum of row 2} = 25 + 30 + 25 = 80$$

$$\text{sum of row 3} = 15 + 20 + 25 = 60$$

Step 3. Add the numbers down each column of the table to obtain the column sum.

$$\text{column 1} \quad 20 + 25 + 15 = 60$$

$$\text{column 2} \quad 25 + 30 + 20 = 75$$

$$\text{column 3} \quad 15 + 25 + 25 = 65$$

Step 4. Add the sum of each column together to obtain N.

$$N = 60 + 75 + 65 = 200$$

Step 5. Calculate the *expected* frequency for each cell by multiplying the row and column totals obtained in Steps 3 and 4 and dividing by N.

$$\text{row 1 x column 1:} \quad \frac{(60)(60)}{200} = \frac{3600}{200} = 18.00$$

row 2 x column 1: $\frac{(80)(60)}{200} = \frac{4800}{200} = 24.00$

row 3 x column 1: $\frac{(60)(60)}{200} = \frac{3600}{200} = 18.00$

row 1 x column 2: $\frac{(60)(75)}{200} = \frac{4500}{200} = 22.50$

row 2 x column 2: $\frac{(80)(75)}{200} = \frac{6000}{200} = 30.00$

row 3 x column 2: $\frac{(60)(75)}{200} = \frac{4500}{200} = 22.50$

row 1 x column 3: $\frac{(60)(65)}{200} = \frac{3900}{200} = 19.50$

row 2 x column 3: $\frac{(80)(65)}{200} = \frac{5200}{200} = 26.00$

row 3 x column 3: $\frac{(60)(65)}{200} = \frac{3900}{200} = 19.50$

These are the *expected* frequencies.

	Type of Switch		
Category	*Toggle*	*Selector*	*Pushbutton*
Clerks	18.00	22.50	19.50
Technicians	24.00	30.00	26.00
Managers	18.00	22.50	19.50

Step 6. Fill in the first two columns of the following table using the observed frequencies from Step 1, and the expected frequencies calculated in Step 5. Then subtract the expected from the observed, and square the difference.

Observed Frequencies	*Expected Frequencies*	*Difference*	*Difference Squared*
20	18	2	4
25	24	1	1
15	18	-3	9
25	22.50	2.50	6.25
30	30	0	0
20	22.50	-2.50	6.25
15	19.50	-4.50	20.25
25	26.00	-1.00	1.00
25	19.50	5.50	30.25

Step 7. Take the squared differences from Step 6 and divide each by their *expected* frequency.

$$\frac{4}{18} = 0.22 \quad \frac{6.25}{22.50} = 0.28 \quad \frac{20.25}{19.50} = 1.04$$

$$\frac{1}{24} = 0.04 \quad \frac{0}{30} = 0 \quad \frac{1.00}{26.00} = 0.04$$

$$\frac{9}{18} = 0.50 \quad \frac{6.25}{22.50} = 0.28 \quad \frac{30.25}{19.50} = 1.55$$

Step 8. Chi-square is found by adding together the quantities in Step 7. Chi-square $= 0.22 + 0.04 + 0.50 + 0.28 + 0 + 0.28 + 1.04 + 0.04 + 1.55 = 3.95$

Step 9. Find the degrees of freedom by multiplying the number of rows minus 1 (R-1) times the number of columns minus 1 (C-1).

$$df = (R-1)(C-1)$$

$$= (2)(2)$$

$$= 4$$

Step 10. Entering Table 5 in Appendix A with 4 degrees of freedom, the tabled chi-square is 9.49.

The calculated chi-square value of 3.95 is less than the tabled chi-square. Thus, we must conclude that at least for the different switch types used in this study there is probably no reliable difference in the preferences for different controls by different types of employees.

Two-by-Two Frequency Table (Example 3)

A designer wishes to know if there is a difference between experienced and inexperienced employees in their preference for two types of documents. She selects a random sample of employees and asks them to choose between Type X and Type Y documents. The obtained responses are shown below. Is there evidence to suggest that the two groups differ in their preference for documents?

Number of experienced users preferring Type X = 30

Number of experienced users preferring Type Y = 12

Number of inexperienced users preferring Type X = 15

Number of inexperienced users preferring Type Y = 19

A slightly different chi-square formula can be used for this problem:

$$chi\text{-}square = \frac{N(AD - BC)^2}{(A + B)(C + D)(A + C)(B + D)}$$

where

N = total number of responses
A, B, C and D = number of responses in each cell

A	B
C	D

Step 1. List the frequencies in a two-by-two contingency table.

	Document Type	
	Type X	*Type Y*
Experienced	30	12

Inexperienced	15	19

Step 2. Calculate the total number of responses (N) by adding the number of frequencies in each cell.

$$N = 30 + 12 + 15 + 19 = 76$$

Step 3. Match the cell designations in the box (A, B, C or D) to the quantities in the two-by-two contingency table. Then multiply A times D and B times C.

$$(A)(D) = 30 \times 19 = 570$$

$$(B)(C) = 12 \times 15 = 180$$

Step 4. Subtract the quantities in Step 3 from one another, then square the result and multiply by N (from Step 2).

$$570 - 180 = 390$$

$$(390)^2 = 152100$$

$$N \times 152100 = 76 \times 152100 = 11559600$$

Step 5. Add A + B, C + D, A + C and B + D.

$$30 + 12 = 42$$
$$15 + 19 = 34$$
$$30 + 15 = 45$$
$$12 + 19 = 31$$

Step 6. Multiply together the quantities in Step 5

$$42 \times 34 \times 45 \times 31 = 1992060$$

Step 7. To find chi-square, divide the quantity in Step 4 by the quantity in Step 6.

$$\textit{chi-square} = \frac{11559600}{1992060} = 5.80$$

Step 8. For two-by-two contingency tables the degrees of freedom is always 1.

Step 9. Consult Table 5 in Appendix A. Since the calculated 5.80 value exceeds the tabled value of 3.84 we conclude that there is a reliable difference in the preference for these documents.

FOR MORE INFORMATION

Blommers, P. J. and Forsyth, R. A., *Elementary Statistical Methods in Psychology and Education,* 2nd Edition, Boston: Houghton Mifflin, 1977.

Hopkins, K. D. and Glass, G. V., *Basic Statistics for the Behavioral Sciences,* Englewood Cliffs, N. J.: Prentice—Hall, 1978.

Wright, R. L. D., *Understanding Statistics: An Informal Introduction for the Behavioral Sciences,* New York: Harcourt, Brace, Jovanovich, 1976.

APPENDIX A

Statistical Tables

Table 1. Random Numbers

11339	19233	50911	14209	39594	68368	97742	36252	27671	55091
96971	19968	31709	40197	16313	80020	01588	21654	50328	04577
07779	47712	33846	84716	49870	59670	46946	71716	50623	38681
71675	95993	08790	13241	71260	16558	83316	68482	10294	45137
32804	72742	16237	72550	10570	31470	92612	94917	48822	79794
14835	56263	53062	71543	67632	30337	28739	17582	40923	32434
15544	14327	07580	48813	30161	10746	96470	60680	63507	14435
92230	41243	90765	08867	08038	05038	10908	00633	21740	55450
33564	93563	10770	10595	71323	84243	09402	62877	49762	56151
84461	56618	40570	72906	30794	49144	65239	21788	38288	29180
91645	42451	83776	99246	45548	02457	74804	49536	89815	74285
78305	63797	26995	23146	56071	97081	22376	09819	56855	97424
97888	55122	65545	02904	40042	70653	24483	31258	96475	77668
67286	09001	09718	67231	54033	24185	52097	78713	59510	84400
53610	59459	89945	72102	66595	02198	26968	88467	46939	52318
52965	76189	68892	64541	02225	09603	59304	38179	75920	80486
25336	39735	25594	50557	96257	59700	27715	42432	27652	88151
73078	44321	77616	49296	55882	71507	30168	31876	28283	53424
81747	52244	38354	47800	48454	43304	14256	74281	82279	28882
47772	22798	36910	39986	34033	39868	24009	97123	59151	27583
54153	70832	37575	31898	39212	63993	05419	77565	73150	98537
93745	99871	37129	55032	94444	17884	27082	23502	06136	89476
81676	51330	58828	74199	87214	13727	80539	95037	73536	16862
79788	02193	33250	05865	53018	62394	56997	41534	01953	13763
92112	61235	68760	61201	02189	09424	24156	10368	26527	89107
87542	28171	47150	75523	66790	63963	13903	68498	02981	25219
37535	48342	48943	07719	10407	33748	93650	39356	01011	22099
95957	96668	69380	49091	90182	13205	71802	35482	27973	46814
34642	85350	53361	63940	79546	89956	96836	81313	80712	73572
50413	31008	09231	46516	61672	79954	01291	72278	55658	84893
53312	73768	59931	55182	43761	59424	79775	17772	41552	45236
16302	64092	76045	28958	21182	30050	96256	85737	86962	27067
96357	98654	01909	58799	87374	53184	87233	55275	59572	56476
38529	89095	89538	15600	33687	86353	61917	63876	52367	79032
46939	05014	06099	76041	57638	55342	41269	96173	94872	35605

APPENDIX A

Table 2. Values of t

df	t value
1	12.71
2	4.30
3	3.18
4	2.78
5	2.57
6	2.45
7	2.37
8	2.31
9	2.26
10	2.23
11	2.20
12	2.18
13	2.16
14	2.15
15	2.13
16	2.12
17	2.11
18	2.10
19	2.09
20	2.09
21	2.08
22	2.07
23	2.07
24	2.06
25	2.06
30	2.04
40	2.02
60	2.00
120	1.98
over 120	1.96

Note: In order to conclude that there is a reliable difference between the two groups, the calculated t must be greater than the tabled t.

Adapted from Fisher and Yates, 1963. Courtesy Cambridge University Press.

APPENDIX A

Table 3. Values of F

degrees of freedom	1	2	3	4	5	6	7	8	9	10
2	18.51	19.00	19.10	19.25	19.30	19.33	19.35	19.37	19.38	19.40
3	10.13	9.56	9.28	9.12	9.01	8.94	8.89	8.85	8.81	8.79
4	7.71	6.94	6.69	6.39	6.36	6.16	6.09	6.04	6.00	5.96
5	6.61	5.79	5.41	5.19	5.05	4.95	4.88	4.82	4.77	4.74
6	5.99	5.14	4.76	4.53	4.39	4.28	4.21	4.15	4.10	4.06
7	5.59	4.74	4.35	4.12	3.97	3.87	3.79	3.73	3.68	3.64
8	5.32	4.48	4.07	3.84	3.69	3.58	3.60	3.44	3.39	3.36
9	5.12	4.26	3.86	3.63	3.48	3.37	3.29	3.23	3.18	3.14
10	4.96	4.00	3.71	3.48	3.33	3.22	3.14	3.07	3.02	2.98
11	4.84	3.48	3.59	3.36	3.20	3.09	3.01	2.95	2.90	2.85
12	4.75	3.89	3.49	3.26	3.11	3.00	2.91	2.85	2.80	2.75
13	4.67	3.81	3.41	3.18	3.03	2.92	2.83	2.77	2.71	2.67
14	4.60	3.74	3.34	3.11	2.96	2.85	2.76	2.70	2.65	2.60
15	4.54	3.62	3.29	3.06	2.90	2.79	2.71	2.64	2.59	2.54
16	4.49	3.63	3.24	3.01	2.85	2.74	2.66	2.59	2.54	2.49
17	4.45	3.59	3.20	2.96	2.81	2.70	2.61	2.55	2.49	2.45
18	4.41	3.58	3.16	2.93	2.77	2.66	2.58	2.51	2.46	2.41
19	4.38	3.62	3.13	2.90	2.74	2.63	2.64	2.48	2.42	2.38
20	4.35	3.49	3.10	2.87	2.71	2.60	2.61	2.45	2.30	2.35
25	4.24	3.39	2.99	2.76	2.60	2.49	2.40	2.34	2.28	2.24
30	4.17	3.32	2.92	2.69	2.53	2.42	2.33	2.27	2.21	2.16
40	4.08	3.28	2.84	2.61	2.45	2.34	2.25	2.18	2.12	2.08
60	4.00	3.16	2.76	2.53	2.37	2.25	2.17	2.10	2.04	1.99
120	3.92	3.07	2.68	2.45	2.29	2.17	2.09	2.02	1.96	1.91

Note: In order to conclude that there is a reliable difference between groups, the calculated F must be greater than the tabled F.

Adapted from Hays, 1973.

APPENDIX A

Table 4. Minimum Number of Hits Required to be Reliable, Based on the Number of Subjects or Categories.

Number of Subjects or Matched Groups	Number of Categories 2	3	4	5
5	5	4	4	4
6	6	5	4	4
7	7	6	5	4
8	7	6	5	5
9	8	6	5	5
10	9	7	6	5
11	9	7	6	5
12	10	8	6	6
13	10	8	7	6
14	11	9	7	6
15	12	9	8	7
16	12	10	8	7
17	13	10	8	7
18	13	11	9	7
19	14	11	9	8
20	15	11	9	8

APPENDIX A

Table 5. Values of Chi-Square

df	Chi-square Values
1	3.84
2	5.99
3	7.82
4	9.49
5	11.07
6	12.59
7	14.07
8	15.51
9	16.92
10	18.31
11	19.68
12	21.03
13	22.36
14	23.68
15	25.00
16	26.30
17	27.59
18	28.87
19	30.14
20	31.41
21	32.67
22	33.92
23	35.17
24	36.42
25	37.65
26	38.89
27	40.11
28	41.34
29	42.56
30	43.77

Note: In order to conclude that there is a reliable difference between groups, the calculated chi-square must be greater than the tabled chi-square.

Adapted from Fisher and Yates, 1963.

APPENDIX B

Human Performance Engineering Resources

Books that contain information related to the material discussed in each chapter are listed in the "For More Information" section at the end of each chapter. Other good information resources are professional societies, journals, abstracts, and reports.

Professional Societies

Several professional societies concerned with human performance engineering are listed below.

Primary Societies

Human Factors Society
Box 1369
Santa Monica, California 90406

Ergonomics Society
4 John Street
London WC1N 2ET

American Psychological Association (APA)
Society of Engineering Psychologists (Division 21)
1200 17th Street, N.W.
Washington, D.C. 20036

Other Societies

American Institute of Industrial Engineers
25 Technology Park
Norcross, Georgia 30092

Society for Information Display
654 N. Sepulveda Blvd.
Los Angeles, California 90049

Society for Technical Communication
815 15th Street, N.W.
Washington, D.C. 20005

National Society for Performance and Instruction
1126 16th Street, N.W.
Washington, D.C. 20036

Institute of Electrical and Electronics Engineers
Systems, Man and Cybernetics Society
345 East 47th Street
New York, New York 10017

Acoustical Society of America
335 East 45th Street
New York, New York 10017

Optical Society of America
2000 L Street, N.W.
Washington, D.C. 20036

Journals

There are numerous journals published by professional societies or privately that contain information on human performance engineering. These journals generally are available in libraries. The journals of primary interest include:

Applied Ergonomics

Ergonomics

Human Factors

International Journal of Man-Machine Studies

Journal of Applied Psychology

Other journals include:

Human Information Processing

Journal of Experimental Psychology:
Human Learning and Memory

Journal of Experimental Psychology:
Human Perception and Performance

Journal of Motor Behavior

Perceptual and Motor Skills

Quarterly Journal of Experimental Psychology

Human/Computer Interface

AFIPS Conference Proceedings

Communications of the Association of Computing Machinery

Computer Graphics

Information Display

Journal of the American Society for Information Sciences

Written Instructions

Journal of Educational Psychology

Technical Communication

Training

Journal of Instructional Development

Performance and Instruction

Training

Training and development

Selection Criteria

Personnel Psychology

General

Journal of Industrial Engineering

Organizational Behavior and Human Performance

IE Industrial Engineering

Abstracts

Human performance engineering information is summarized in the following abstracts:

Ergonomics Abstracts (1969 - present)

Psychological Abstracts (1927 - present)

Science Abstracts - Series C

NTIS Reports

A relatively large number of reports are not published in journals or summarized in the above abstracts. One good source of these reports, particularly "AD" reports, is the:

National Technical Information Service (NTIS)
5285 Port Royal Road
Springfield, Virginia 22161

APPENDIX C

Code Design Table

Table 1. A Frequency Listing of the 26 English Letters Based on a Sample of 20,000 English Words

Letter	*Percent of Total*
E	13.3
T	9.8
A	8.1
H	7.7
O	6.6
S	6.1
N	6.0
R	5.9
I	5.1
L	4.5
D	4.3
U	3.1
W	2.9
M	2.5
C	2.4
G	2.2
Y	2.1
F	1.8
B	1.6
P	1.5
K	1.1
V	1.0
J	.2
X	.1
Q	.1
Z	.1
	100.0

Adapted from Mayzner and Tresselt, 1965

Table 2. A Listing of 200 Digrams Based on 20,000 English Words (Adapted from Mayzner, Tresselt and Wolin, 1965)

Digram	*Frequency*	*Digram*	*Frequency*	*Digram*	*Frequency*	*Digram*	*Frequency*
TH	3774	CA	368	IE	189	EW	106
HE	3155	NO	349	FR	188	EF	103
AN	1576	LO	344	EM	187	WN	103
ER	1314	YO	339	TR	187	FT	102
ND	1213	KE	337	EC	181	AP	100
HA	1164	OO	336	CK	178	NA	100
RE	1139	EL	332	AM	177	BL	98
OU	1115	LA	332	SU	175	GR	98
IN	1110	TO	331	EV	172	NC	98
HI	824	SH	328	PL	169	PI	97
OR	812	IL	324	SS	168	GI	96
AR	802	AI	322	HT	165	DS	95
EN	799	AY	319	IV	165	GA	94
AT	785	RS	318	MI	165	HR	91
NG	771	ET	316	CT	154	EP	90
ED	767	RI	309	FE	154	RU	89
ST	754	AC	308	TT	154	BR	88
AS	683	IC	304	YE	152	IO	88
VE	683	US	299	PO	149	OI	88
EA	670	CO	298	DI	148	AU	87
AL	664	GE	289	NS	148	EX	87
ES	630	LD	289	UG	148	UM	87
SE	626	MO	289	AK	146	FF	86
ON	598	RA	289	FA	145	IK	86
WA	595	GH	288	RY	145	MU	85
LE	591	CE	285	AB	144	TW	84
TE	583	WE	285	PA	142	DR	83
IT	558	PE	280	AG	138	KN	82
LL	546	UN	278	OP	138	LU	81
ME	530	LY	276	DO	137	YS	81
NE	512	IR	272	BA	136	NL	80
RO	504	WO	264	OV	136	OF	80
UT	492	ID	260	GO	135	BI	78
HO	487	TA	259	NI	135	MP	77
IS	484	BU	256	RD	133	HU	75
WH	472	IM	255	TU	132	TL	75
EE	470	TI	252	EI	127	LT	74
FO	429	UL	247	KI	127	CR	71
OM	417	BO	240	OK	123	RL	71
BE	415	AV	233	LS	121	UE	71
OT	415	IG	233	TY	121	FL	69
CH	412	OL	218	OD	119	RR	69
UR	402	SI	214	NY	115	PU	68
OW	398	SO	213	UC	114	AF	67
MA	394	TS	209	PR	110	CI	67
LI	390	FI	205	VI	109	OB	67
AD	382	SA	196	CL	108	QU	66
NT	378	OS	195	SP	107	OA	65
DE	375	RT	190	DA	106	RM	65
WI	374	EY	189	RN	106	UI	65

Table 3. A Listing of 200 Trigrams Based on 20,000 English Words
(Adapted from Mayzner, Tresselt and Wolin, 1965)

Trigram	*Frequency*	*Trigram*	*Frequency*	*Trigram*	*Frequency*	*Trigram*	*Frequency*
THE	2565	WER	119	ACE	80	EAT	65
AND	959	ATE	118	AID	80	ERY	65
ING	526	HOU	114	IND	80	HOW	65
HAT	479	OVE	114	URE	80	NIN	65
THA	438	NOW	112	COU	79	OSE	65
HER	414	WHO	112	LEA	79	RES	65
HIS	354	OUN	111	TUR	79	STE	65
FOR	353	COM	110	IDE	78	TIM	65
YOU	326	EVE	110	TIN	78	ION	64
WAS	304	HIN	109	ART	77	OOD	64
ALL	270	OUG	109	EAS	77	PLE	64
THI	259	USE	109	EST	77	RIE	64
ERE	255	ERS	108	VEN	77	WIL	64
ITH	238	AKE	107	HEM	76	NGE	63
ARE	228	MOR	107	LON	76	THR	63
WIT	227	WAY	106	ANT	75	TTE	63
OUT	225	INT	103	END	75	CHA	62
VER	221	STA	101	LED	75	HES	62
OUR	209	ABO	99	MEN	75	SHO	62
ONE	205	HIC	99	YEA	75	TEN	62
EAR	197	UND	99	HEA	74	DAY	61
AVE	194	AIN	98	PLA	74	ILE	61
NOT	191	ICH	98	ACK	73	MOS	61
OME	191	OWN	97	ARD	73	NEW	61
TER	179	OLD	96	GET	73	ONL	61
BUT	178	UST	96	ROU	73	ACT	60
HAD	173	ONG	95	SAI	73	BEC	60
GHT	163	WOR	94	ARS	72	LAS	60
IGH	161	BOU	93	LES	71	ANG	59
ORE	153	AME	91	PER	71	ICE	59
HAV	147	AST	91	UCH	71	ROV	59
ILL	146	CAN	91	AYS	70	ECT	58
OUL	145	HAS	89	EIR	70	EET	58
IVE	143	OST	89	RED	70	FIN	58
MAN	143	WOU	89	CAR	69	STO	58
SHE	143	ANY	88	HEI	69	AIR	57
ULD	143	KIN	88	SOM	69	EVE	57
OTH	140	WHA	88	THO	69	ITT	57
ENT	139	REE	87	NTO	68	MAR	57
FRO	138	BEE	86	TOO	68	ACH	56
HEN	133	IKE	86	AGE	67	EED	56
HEY	131	TED	86	CAM	67	RSE	56
WHE	131	ELL	85	NCE	67	TLE	56
ROM	130	LOO	84	ORT	67	IRS	55
EEN	128	OOK	84	CAL	66	ITE	55
HAN	128	SEE	84	DER	66	OIN	55
REA	128	EAD	83	FTE	66	PRO	55
UGH	128	LIK	83	IME	66	WAN	55
HIM	126	ITS	81	LLE	66	ADE	54
WHI	125	KED	81	NLY	66	DOW	54

APPENDIX C

Table 4. The Number of Different Characters Necessary to Develop a Set of Codes

Use this table when code length is between one and four characters and the total number of codes required by the system is between two and one-hundred

Total Number of Codes *Required by the System*	Desired Code Length *(Number of Characters Per Code)*			
	1	*2*	*3*	*4*
2	2	2	2	2
3	3	2	2	2
4	4	2	2	2
5	5	3	2	2
6	6	3	2	2
7	7	3	2	2
8	8	3	2	2
9	9	3	3	2
10	10	4	3	2
11	11	4	3	2
12	12	4	3	2
13	13	4	3	2
14	14	4	3	2
15	15	4	3	2
16	16	4	3	2
17	17	5	3	3
18	18	5	3	3
19	19	5	3	3
20	20	5	3	3
21	21	5	3	3
22	22	5	3	3
23	23	5	3	3
24	24	5	3	3
25	25	5	3	3
26	26	6	3	3
27	27	6	3	3
28	28	6	4	3
29	29	6	4	3
30	30	6	4	3
40		7	4	3
50		8	4	3
60		8	4	3
70		9	5	3
80		9	5	3
90		10	5	4
100		10	5	4

APPENDIX C

Table 5. The Number of Different Characters Necessary to Develop a Set of Codes

Use this table when code length is between two and ten characters and the total number of codes required by the system is between one hundred and 900,000)

Total Number of Codes *Required by the System*	Desired Code Length *(Number of Characters per Code)*								
	2	*3*	*4*	*5*	*6*	*7*	*8*	*9*	*10*
100	10	5	4	3	3	2	2	2	2
200	15	6	4	3	3	3	2	2	2
300	18	7	5	4	3	3	3	2	2
400	20	8	5	4	3	3	3	2	2
500	23	9	5	4	3	3	3	3	2
600	25	9	5	4	3	3	3	3	2
700	27	9	6	4	3	3	3	3	2
800	29	10	6	4	4	3	3	3	2
900	30	10	6	4	4	3	3	3	2
1,000		10	6	4	4	3	3	3	2
2,000		13	7	5	4	3	3	3	2
3,000		15	8	5	4	4	3	3	3
4,000		16	8	6	4	4	3	3	3
5,000		18	9	6	5	4	3	3	3
6,000		19	9	6	5	4	3	3	3
7,000		20	10	6	5	4	4	3	3
8,000		20	10	7	5	4	4	3	3
9,000		21	10	7	5	4	4	3	3
10,000		22	10	7	5	4	4	3	3
20,000		28	12	8	6	5	4	4	3
30,000			14	8	6	5	4	4	3
40,000			15	9	6	5	4	4	3
50,000			15	9	7	5	4	4	3
60,000			16	10	7	5	4	4	4
70,000			17	10	7	5	5	4	4
80,000			17	10	7	6	5	4	4
90,000			18	10	7	6	5	4	4
100,000			18	10	7	6	5	4	4
200,000			22	12	8	6	5	4	4
300,000			24	13	9	7	5	5	4
400,000			26	14	9	7	6	5	4
500,000			27	14	9	7	6	5	4
600,000			28	15	10	8	6	5	4
700,000			29	15	10	8	6	5	4
800,000			30	16	10	8	6	5	4
900,000				16	10	8	6	5	4

APPENDIX C

Table 6. The Number of Different Characters Necessary to Develop a Set of Codes

Use this table when code length is between five and ten characters and the total number of codes required by the system is between one million and one billion)

Total Number of Codes *Required by the System*	Desired Code Length *(Number of Characters per Code)*					
	5	*6*	*7*	*8*	*9*	*10*
1,000,000	16	10	8	6	5	4
2,000,000	19	12	8	7	6	5
3,000,000	20	13	9	7	6	5
4,000,000	21	13	9	7	6	5
5,000,000	22	14	10	7	6	5
6,000,000	23	14	10	8	6	5
7,000,000	24	14	10	8	6	5
8,000,000	25	15	10	8	6	5
9,000,000	25	15	10	8	6	5
10,000,000	26	15	10	8	6	6
20,000,000	29	17	12	9	7	6
30,000,000		18	12	9	7	6
40,000,000		19	13	9	7	6
50,000,000		20	13	10	8	6
60,000,000		20	13	10	8	6
70,000,000		21	14	10	8	7
80,000,000		21	14	10	8	7
90,000,000		22	14	10	8	7
100,000,000		22	14	10	8	7
200,000,000		25	16	11	9	7
300,000,000		26	17	12	9	8
400,000,000		28	17	12	10	8
500,000,000		29	18	13	10	8
600,000,000		30	18	13	10	8
700,000,000		30	19	13	10	8
800,000,000			19	13	10	8
900,000,000			20	14	10	8
1,000,000,000			20	14	10	8

REFERENCES

ASHRAE, *Handbook of Fundamentals*. New York: American Society of Heating, Refrigeration and Air Conditioning Engineers, 1972.

Adams, J. S.,

— Injustice in social exchanges. In L. Bertiowitz (ed.), *Advances in Experimental Social Psychology*, Vol. 2, New York: Academic Press, 1965.

— Toward an understanding of inequity. *Journal of Abnormal and Social Psychology*, 1963, *67*, 422-436.

Alden, D. G., Daniels, R. W., and Kanarick, A. F., Human factors principles for keyboard design and operation - A summary review. *Honeywell Systems and Research Center Technical Report*, March 1970.

Alluisi, E. A. and Morgan, B. B., Jr., Engineering psychology and human performance. *Annual Review of Psychology*, 1976, *27*, 305-330.

Alluisi, E. A. and Muller, P. F., Jr., Rate of information transfer with seven symbolic codes: motor and verbal responses. *WADC Technical Report*, 1956, 56-226.

Altman, I., *The Environment and Social Behavior*, Monterey, Cal.: Brooks/Cole, 1975.

Anastasi, A., *Psychological Testing, New York:* Macmillian, 1963.

Anonymous, Project TRENDS: A study of outside plant engineering documentation, Whippany, N. J.: Bell Laboratories Technical Documentation Department, 1978.

Apsey, R. S., Human factors of constrained handprint for OCR. *IEEE Transactions on Systems, Man, and Cybernetics*, April 1978, 292-296.

Asch, S. E., Effects of group pressure upon the modification and distortion of judgements. In Harold Guetzkow (ed.), *Groups, Leadership and Men*. Pittsburgh, Pa.: Carnegie Press, 1951, 177-190.

Atwood, G. E., Experimental studies of mnemonic visualization. Ph.D. dissertation, University of Oregon, 1969.

Averbach, E. and Coriell, A. S., Short-term memory in vision. *Bell System Technical Journal*, 1961, *40*, 309-328.

Averbach, E. and Sperling, G. A., Short-term storage of visual information. In Cherry, C. (ed.) *Symposium on Information Theory*. London: Butterworth, 1961.

Backman, G., Korperlange und Tageszeit, Upsala Lakar. Forhandl., 1924, *28*, 255-282.

Baddeley, A. D., *The Psychology of Memory. New York:* Basic Books, Inc., 1976.

Baddeley, A. D. and Patterson, K., The relationship between long-term and short-term memory. *British Medical Bulletin*, 1971, *27*, 237-242.

Bailey, R. W.,

— Testing manual procedures in computer-based business information systems. *Proceedings of the 16th Annual Meeting of the Human Factors Society*, October 17-19, 1972.

— Human Error in computer-based data processing systems. *Bell Laboratories Talk Monograph*, 1973.

— Handprinting errors in data systems. Presented at the 83rd Meeting of the American Psychological Association, August 1975.

Bailey, R. W., and Koch, C. G., Position package test validation study (LMOS: Houston). Private Communication, August 1976.

Bailey, R. W. and Kulp, M. J., Methodology for testing the position packages in BISCUS conversion. Private Communication, April 1971.

Bailey, R. W., Stemen, J. W., and Kersey, T. E., Human Error in broadband operations. Private Communication, May 1975.

Baramack, J. E. and Sinaiko, H. W., Human factors problems in computer-generated graphic displays. *Institute for Defense Analysis*, AD636170, 1966.

Barber, T. K., The effect of "hypnosis" on learning and recall: A methodological critique. *Journal of Clinical Psychology*, 1965, *21*, 19-25.

Barnard, P., Presuppositions in active and passive questions. Paper read to the Experimental Psychology Society, 1974.

Barnard, P. and Wright, P., The effects of spaced character formats on the production and legibility of handwritten names. *Ergonomics*, 1976, *19*, 81-92.

Barnard, P., Wright, P., and Wilcox, P., The effects of spatial constraints on the legibility of handwritten alphanumeric codes. *Ergonomics*, 1978, *21*, 73-78.

Baron, R. A. and Bell, P. A., Aggression and heat: The influence of ambient temperature, negative affect, and a cooling drink on physical aggression. *Journal of Personality and Social Psychology*, 1976, *33*, 245-255 (a).

Bartlett, F. C.,

— *Remembering*. Cambridge, England: Cambridge University Press, 1932.

Fatigue following highly skilled work. *Proceedings of the Royal Society*, 1943, *131*, 247-257.

— The measurement of human skill. *British Medical Journal*, 1947, 835.

Beach, B. H., Expert judgement about uncertainty: Bayesian decision making in realistic settings. *Organizational Behavior and Human Performance*, 1975, *14*, 10-59.

Beard, R. R., and Wertheim, G. A., Behavioral impairment associated with small doses of carbon monoxide. *American Journal of Public Health*, 1967, *57*, 2012-2022.

Bell, C. R., Provins, K. A., and Hiorns, R. F., Visual and auditory vigilance during exposure to hot and humid conditions. *Ergonomics*, 1964, *7*, 279-288.

Bell, P. A., Fisher, J. D., and Loomis, R. J., *Environmental Psychology*. Philadelphia: W. B. Saunders Company, 1978.

Benson, D. F., and Greenberg, J. P., Visual form agnosia. *Archives of Neurology*, 1969, *20*, 82-89.

Beranek, L. L.,

— *Acoustic Measurements*. New York: John Wiley & Sons, Inc., 1949.

— Revised criteria for noise in buildings. *Noise Control*, 1957, *3* (1), 19.

Bergman, B. A., The effects of group size, personal space, and success-failure on physiological arousal, test performance, and questionnaire response. Ph.D. dissertation, Temple University, 1971.

Blackwell, H. R.

— Development and use of a quantitative method for specification of interior illumination levels on the basis of performance data. *Illuminating Engineering*, 1959, *55*, 317-353.

— Development of visual task evaluations for use in specifying recommended illumination levels. *Illuminating Engineering*, 1964, *59*, 627-641.

— A human factors approach to lighting recommendations and standards. *Proceedings of the Sixteenth Annual Meeting of the Human Factors Society*, 1972, 441-449.

Blades, W., *Shakespeare and Typography*, 1892. In H. B. Wheatley. *Literary Blunders*. London: Elliot Stock, 1909. Reprinted by Gale Research Company, Detroit, 1969.

Blaiwes, A. S., Formats for presenting procedural instructions. *Journal of Applied Psychology*, 1974, *59*, 683-686.

Blankenship, A. B., Memory span: A review of the literature. *Psychological Bulletin*, January 1938, *35* (1), 1-25.

Bliss, J. C.,

— *Vision, IRE Transaction on Information Theory*. 1962, *8*, 92.

— *Yearbook of Science and Technology*. New York: McGraw-Hill, 1966, 357-360.

— In T. D. Sterling (ed.), *Visual Prosthesis - The Interdisciplinary Dialogue*. New York: Academic Press, 1971, 259-263.

Bliss, J. C., Hewitt, D. V., Crane, P. K., Mansfield, P. K., Townsend, J. T., Information available in brief tactile presentations. *Perception and Psychophysics*, 1966, *1*, 272-283.

Blum, M. L., and Mintz, A., Re-examination of the accident proness concept. *Journal of Applied Psychology*, 1949, *33* (3), 195-211.

Blumenthal, A. L., Observations with self embedded sentences. *Psychonomic Science*, 1966, 453-454.

Bobrow, D. G., Problems with natural language communication with computers. *IEEE Transactions on Human Factors in Electronics*, 1967, *8*, 52-55.

Boies, S. J., and Gould, J. D., User performance in an interactive computer system. Proceedings of the Fifth Annual Conference on Information Sciences and Systems, 1971, 122.

Boies, S. J., User behavior on an interactive computer system. *IBM Systems Journal*, 1974, 2-18.

Bolsky, M. I. and Yuhas, C. M., *Performance Aids for Greater Productivity: How to Plan, Design, Organize and Produce Them.* Private Communication, August 1975.

Book, W. F., The psychology of skill: with special reference to its acquisition in typewriting. University of Montana Publications in Psychology; Bulletin No. 53., Psychological Series No. 1, 1908.

Boring, E. G., *A History of Experimental Psychology*. New York: The Century Company, 1929.

Bower, G. H.,

— A multi-component theory of the memory track. In K. W. Spence and J. T. Spence (eds.), *The Psychology of Learning and Motivation: Advances in Research and Theory*. Vol. 1, New York: Academic Press, 1967, 299-325.

— Analysis of a mnemonic device. *American Scientist*, 1970, *58*, 496-510.

— Mental imagery and associative learning. In L. W. Gregg (ed.), *Cognition in Learning and Memory*, New York: John Wiley & Sons, Inc., 1972, 51-88.

Bransford, J. D. and Franks, J. J., The abstraction of linguistic ideas. *Cognitive Psychology*, 1971, 331-350.

Braunstein, M. and Anderson, N. S., A comparison of reading digits aloud and keypunching. *IBM Technical Report RC - 185*, November 1959.

Brayfield, A. H. and Crockett, W. H., Employee attitudes and employee performance. *Psychological Bulletin*, 1955, 396-424.

Breisacher, P., Neuropsychological effects of air pollution. *American Behavioral Scientist*, 1971, *14*, 837-864.

Bright, R. L., and Kerr, E. G., Computer-aided instruction is feasible now. *Technical Education Reporter*, 1974, *1* (2), 60-67.

Broad, D. J., Basic directions in automatic speech recognition. *International Journal of Man-Machine Studies*, 1972, *4*, 105-118.

Broadbent, D. E.,

— *Perception and Communication*. New York: Pergamon, 1958.

Differences and interaction between stresses. *Quarterly Journal of*

— *Experimental Psychology*, 1963, *15*, 205.

Language and ergonomics. *Applied Ergonomics*, 1977, *8.1*, 15-18.

Brown, J., Some tests of the decay theory of immediate memory. *Quarterly Journal of Experimental Psychology*, 1958, *10*, 12-21.

Bruner, J. S. and Minturn, A. L., Perceptual identification and perceptual organization, *The Journal of General Psychology*, 1955, *53*, 21-28.

Bryden, M. P., Tachistoscopic recognition of non-alphabetical material. *Canadian Journal of Psychology*, 1960, *14*, 78-86.

Bryden, M. P., Dick, A. O., and Mewhort, D. J. K., Tachistoscopic recognition of number sequences. *Canadian Journal of Psychology*, 1968, *22*, 52-59.

Buckler, A. T. A Review of the Literature on the Legibility of Alphanumerics on Electronic Displays. ADA 040625, May 1977.

Bugelski, B. R., Kidd, E., and Segmen, J., Image as a mediator in one-trial paired-associate learning. *Journal of Experimental Psychology*, 1968, *76*, 69-73.

Buros, O. K. (ed.),
— *Tests in Print II; An Index to Tests, Test Reviews and the Literature on Specific Tests,* Gryphon, 1974.
— *Mental Measurements Yearbook.* Gryphon, 1978.
Buros, O. K., Peace, B. A. and Matts, W. L. (eds.), *Tests in Print, a Comprehensive Bibliography of Tests for Use in Education, Psychology, and Industry.* Gryphon, 1961.
Butler, T., Domangue, J. and Felfoldy, G., Using human estimation to measure work time. Private Communication, May 1980.
Byrne, W. J., Imported instructions, or "the almighty switch." *Technical Communication,* 1980, First Quarter, 12-13.
Carbonell, J. R., On man-computer interaction: A model and some related issues. *IEEE Transactions on Systems Science and Cybernetics,* 1969, 16-26.
Carbonell, J. R., Elkind, J. L., and Nickerson, R. S., On the psychological importance of time in a time-sharing system. *Human Factors,* 1968, *10,* 135-142.
Cardozo, B. L., and Leopold, F. F., Human code transmission: letters and digits compared on the basis of immediate memory error rates. *Ergonomics,* 1963, 134-141.
Carter, J. W., Some select physiological, anthropometric and human engineering data useful in vehicle design of space flight operations, DSP-TM-260, February 1960.
Cazamian, P., Round table discussion on the social factors in ergonomics, introducing report. *Proceedings Fourth International Congress on Ergonomics,* Strassburg, 1970.
Chaffin, D. B., Some effects of physical exertion. Western Electric/University of Michigan Research Report, 1972.
Chambers, R. M., Acceleration. In P. Webb (ed.) *Bioastronautics Data Book.* Washington, D.C.: National Aeronautics and Space Administration, 1964.
Chapanis, A.,
— *Research Techniques in Human Engineering.* Baltimore, Maryland: The Johns Hopkins Press, 1959.
— *Man-Machine Engineering.* Monterey, Cal.: Brooks/Cole, 1965a.
— Words, Words, Words. *Human Factors,* 1965b, *7,* 1-17.
— Prelude to 2001: Exploration in human communication. *American Psychologist,* 1975, 949-961.
Chapanis, A., Garner, W. R., and Morgan, C. T., *Applied Experimental Psychology.* New York: Wiley, 1949.
Chaplin, J. P., *Dictionary of Psychology,* New York: Dell Publishing Co., Inc., 1975.
Chase, W. G. and Simon, H. A., Cognitive Psychology, In W. G. Chase, Ed., *Visual Information Processing,* New York: Academic Press, 1973.
Cherry, C., *On Human Communication.* John Wiley & Sons, Inc.: New York, 1961.
Clark, E. V., On the acquisition of the meaning of before and after, *Journal of Verbal Learning and Verbal Behavior.* 1971, *10,* 266-275.
Clarke, A. C., *2001: A Space Odyssey.* New York: The New American Library, 1968.
Coke, E. U.,
— Are Bell System Practices difficult to read? Measurement of the readability of a sample of BSPs. Private Communication, July, 1976.
— Reading rate, readability and variations in task-induced processing. *Journal of Educational Psychology,* 1976, *68,* 167-173.
— Readability and the evaluation of technical documents. Private Communication, October, 1978.
Coke, E. U. and Koether, M. E., The reading skills of craft and technical management employees--estimates from two samples of students. Private Communication, May 1979.
Collins, C. C. and Bach-y-Rita, P., Transmission of pictorial information through the skin. *Advances in Biological and Medical Physics,* 1973, *14,* 285-315.
Coltheart, M. Visual information processing. In Dodwell, P. C. (ed.), *New Horizons in Psychology 2.* Harmondsworth: Penguin, 1972.
Condry, J., Enemies of exploration: Self-initiated versus other-initiated learning. *Journal of*

Personality and Social Psychology, *35*, 459-477.
Conrad, R.,
— Errors of immediate memory. *British Journal of Psychology*, November 1959, *50* (4), 349-359.
— The design of information. *Occupational Psychology*, 1962, *36*, 159-162.
— Designing postal codes for public use. *Ergonomics*, 1967, *10*, (2), 233-238.
— Acoustic confusions and immediate memory. *British Journal of Psychology*, 1974, *55*, 75-84.
Conrad, R. and Hille, B. A., Memory for long telephone numbers. *Post Office Telecommunications Journal*, 1957, *19*, 37-39.
Conrad, R. and Hull, A. J.,
— Copying alpha and numeric codes by hand. *Journal of Applied Psychology*, 1967, *51*, 444-448.
— The preferred layout for numeral data-entry keysets. *Ergonomics*. 1964, *11*, 165-73.
— Input modality and the serial position curve in short-term memory. *Psychonomic Science*, 1968, *10*, 135-136.
Conrad, R. and Longman, D. J. A., Standard typewriter versus chord keyboard - An experimental comparison. *Ergonomics*, 1965, *8*, 77-88.
Conrad, R. and Rush, M. L., On the nature of short-term memory encoding by the deaf. *Journal of Speech and Hearing Disorders*, 1965, *30*, 336-343.
Considine, K. CRT Technology. *Conference Record of 1976 Biennial Display Conference*, Institute of Electrical and Electronic Engineers, 1976, 80-87.
Corcoran, D. W. J., Carpenter, A., Webster, J. C., and Woodhead, M. M., Comparison of training techniques for complex sound identification. *Journal of the Acoustical Society of America* , 1968, *44*, 157-167.
Cornsweet, T. N., *Visual Perception*. New York: Academic Press, 1970.
Craig, R. L. and Bittel, L. R., *Training and Development Handbook*. New York: Mc Graw-Hill Book Co., 1967.
Crawford, R. P., *Techniques of Creative Thinking*. New York: Hawthorn, 1954.
Crockford, G. W., Heat problems and protective clothing in iron and steel works. In C. N. Davies, P. R. Davis, and F. H. Tyrer (eds.), *The Effects of Abnormal Physical Conditions at Work*. London: E. and S. Livingston, 1967.
Crook, M. H., Hanson, J. A., and Weiss, A., The Legibility of type as a function of stroke width, and letter spacing under low illumination. *WADC Technical Report*, 1954, 53-40.
Crossley, E., A systematic approach to creative design. *Machine Design*, March 6, 1980.
Crossman, E. R. F. W., A theory of the acquisition of speed-skill. *Ergonomics*, 1959, *2*, 153-166.
Cureton, T. K., *Physical Fitness of Champion Athletes*. Urbana: University of Illinois Press, 1951.
Curry, E. T., Fay, T. H. and Hutton, C. L., Experimental studies of the relative intelligibility of alphabet letters. *Journal of the Acoustical Society of America*, 1960, *32*, 1151-1157.
Damon, A., Stoudt, H. W., and McFarland, *The Human Body in Equipment Design*. Cambridge, Mass.: Harvard University Press, 1966.
Daniels, G. S. and Churchill, E., The "average man"? WCRD-TN-53-7, Aero Medical Lab., Wright Air Development Center, Wright-Patterson AFB, Ohio, 1952.
Daniels, G. S., Meyers, H. C. and Worrall, S. H., Anthropometry of WAF basic trainees, WADC-TR-53-12, 1953.
Danielson, R. L., and Nievergelt, An automatic tutor for introductory programming students. *SIGCSE Bulletin*, 1975, *7*, 47-50.
Davenport, C. B. and Love, A. G., *Army Anthropology*. Washington, D.C.: U.S. Government Printing Office, 1921.
Davies, I. K., Get immediate relief with an algorithm. *Psychology Today*, 1970, *3*, 53-54.
Davis, G. A., *Psychology of Problem Solving: Theory and Practice*. New York: Basic Books, Inc., 1973.
Davis, G. A., and Houtman, S. E., *Thinking creatively: A guide to training imagination*. Wisconsin Research and Development Center for Cognitive Learning, University of Wisconsin, 1968.
Davis, R. M., Man-machine communication. In C. A. Cuadra, Ed., *Annual Review of Interscience*

and Technology, Vol. 1. New York: Interscience Publishers, 1966, 221-254.

DeGreene, K. B., Ed., *Systems Psychology*, New York: McGraw-Hill, 1970.

Dearborn, W. F., Johnston, P. W. and Carmichael, L., Improving the readability of typewritten manuscripts. *Proceedings of the National Academy of Sciences*, 1951, *37*, 670-672.

deCharms, R., *Personal Causation*. New York: Academic Press, 1968.

Deci, E. L., *Intrinsic Motivation*. New York: Plenum Publishing Corp., 1975.

Dempster, W. T., Space requirements of the seated operator. *Wright Air Development Center Technical Report*, 1955.

Dennis, J. P., Some effects of vibration upon visual performance. *Journal of Applied Psychology*. 1965, *49*, 245-252.

Deutsch, D., Interference in memory between tones adjacent on the musical scale. *Journal of Experimental Psychology*, 1973, *100*, 228-231.

Dever, J. J., Friend, E., Hegarty, J. A., and Rubin, J. J., Designing and presenting work procedures. Private Communication, June 1978.

Devoe, D. B., Alternatives to handprinting in the manual entry of data. *IEEE Transactions on Human Factors in Electronics*, March 1967, 21-32.

Devoe, D. B., Eisenstadt, B., and Brown, D. E., Manual input coding study. RADC-TR-66-476, September 1966.

Dick, A. O.,

— Relations between the sensory register and short-term storage in tachistoscopic recognition. *Journal of Experimental Psychology*, 1969, *82*, 279-284.

— Visual processing and the use of redundant information in tachistoscopic recognition. *Canadian Journal of Psychology*, 1970, *24*, 133-141.

— Visual information processing and the memroy trace. Unpublished Ph.D. thesis, University of Waterloo, 1967.

Dick, A. O. and Loader, R., Structural and functional components in the processing, organization, and utilization of tachistoscopically presented information. Technical Report 74-5, Center for Visual Science, University of Rochester, 1974.

Diffrient, N., Tilly, A.R. and Bardayjy, J.C., *Human Scale 1/2/3*, Cambridge, Mass.: MIT Press, 1974.

Doherty, W. J., The commercial significance of man-computer interaction. *Man/Computer Communication*, Vol. 2, Maidenhead, Berkshire, England: Infotech International, 1979, 81-94.

Donders, F. E., Die Schnelligkeit Psychischer Processe. *Archiv Anatomie und Physiologie*, 1868, 657-681.

Dooley, B. B., Crowding stress: The effects of social density on men with "close" or "far" personal space. Ph.D. dissertation, University of California at Los Angeles, 1974.

Dooling, D. J. and Lachman, R., Effects of comprehension on retention of prose. *Journal of Applied Psychology*, 1971, 216-222.

Dreyfus, H. L. and Dreyfus, S. E.,

— The scope, limits, and training implications of three models of aircraft pilot emergency response behavior. Report No. ORC-79-2, February 1979.

— The psychic boom: flying beyond the thought barrier. Report No. ORC-79-3, March 1979.

Dreyfuss, M., *The Measure of Man: Human Factors in Design*. (2nd ed.). New York: Whitney Library of Design, 1967.

DuBois, P. H., A test-dominated society: China 1115 B.C. - 1905 A.D. In A. Anastasi (ed.), *Testing Problems in Perspective*, Washington D.C.: American Council on Education, 1966.

Duncan, J. and Konz, S., Legibility of LED and Liquid-Crystal Displays, *SID Journal*, 1976, *17*, 4 , 180-86.

Duffy, F., Cave, C., and Worthington, J., (eds.), *Planning Office Space*. New York: Nichols Publishing Company, 1978.

Dunnett, C. W., Drug screening: the never-ending search for new and better drugs. In J. M. Tanur *Statistics: a Guide to the Unknown*. San Francisco, Cal.: Holden-Day, Inc., 1972.

Ebel, R. L. and Noll, V. H., Handprinting. *Encyclopedia of Educational Research,* 1969.

Edwards, W., Dynamic decision theory and probabilistic information processing. *Human Factors,* 1962, 59-73.

Edwards, W., Phillips, L. D., Hays, W. L. and Goodman, B. C., Probabilistic information processing systems: Design and evaluation. *IEEE Transactions on Systems Science and Cybernetics,* 1968, *4,* 248-265.

Egan, J. P., Articulation testing methods. *Laryngoscope,* 1948, *58,* 955.

Ekstrand, B. R., To sleep, perchance to dream (about why we forget). In C. P. Duncan, L. Sechrest, and A. W. Helton (eds.), *Human Memory: Festschrift in Honor of Benton J. Underwood.* New York: Appleton-Century-Crofts, 1972, 59-82.

Ellis, S. H. An investigation of telephone user training methods for a multiservice electronic PBX. In Proceedings of the eighth international symposium on human factors in telecommunications. Harlow, Essex, England. Standard Telecommunication Laboratories, 1977.

Ellis, S. H. & Coskren, R. A. New approach to customer training. *Bell Laboratories Record,* 1979, *57,* 60-65.

Engel, S. E., and Granda, R. E., Guidelines for man/display interfaces. *IBM Technical Report,* TR00.2720. Poughkeepsie Laboratory, December 1975.

Engen, T. and Ross, B. M., Long-term memory of odors with and without verbal descriptions. *Journal of Experimental Psychology,* 1973, *100,* 221-227.

Ericsson, K. A., Chase, W. G. and Faloon, S., Acquisition of a memory skill. *Science* 1980, *208,* 1181-1182.

Ertel, H., Blue is beautiful. *Time,* September 1973.

Eschenbrenner, A. J. Jr., Effects of intermittent noise on the performance of a complex psychomotor task. *Human Factors,* 1971, *13* (1), 59-63.

Faust, G. W., and O'Neal, A. F. Instructional science and the evolution of computer assisted instructional systems. *Electro Conference Record,* 1977, *34* (2), 1-6.

Festinger, L., *A Theory of Cognitive Dissonance.* Evanston, Ill.: Row Petersen, 1957.

Field, M. M., Hodge, M. H., Manley, C. W., and Sonntag, L., Guidelines for constructing human performance-based codes, Private Communication, April 1971.

Finkelman, J. M., Effects of noise on human performance. *Sound and Vibration,* 1975, *36,* 26-28.

Firing, M., *The Physically Impaired Population of the United States.* San Francisco, Cal.: Firing and Associates, 1978.

Fisher, D. F., Monty, R. A., and Gluckberg, S., Visual confusion matrices: Fact or artifact? *The Journal of Psychology,* 1969, *71,* 111-125.

Fisher, R. A. and Yates, F., *Statistical Tables for Biological, Agricultural and Medical Research,* 6th Ed., Edinburgh: Oliver and Boyd, 1963.

Fitts, P. M.,

— German applied psychology during World War II. *American Psychologist,* 1946, *1,* 151-161.

— Cognitive aspects of information processing: Set for speed versus accuracy. *Journal of Experimental Psychology,* 1966, *77,* 849-857.

Fitts, P., and Posner, M. I., *Human Performance.* Belmont, Cal.: Wadsworth, 1967.

Flanagan, J. L., Computers that talk and listen: man-machine communication by voice. *Proceedings of the IEEE.* 1976, *64,* 405-415.

Fleischman, J. S., Levin, E., Manley, W. W., Peavler, W. S., Yavelberg, I. S., Rebbin, T. J., Schwartz, B. K., Soth, M. W., and Vassallo, J. C., A collective view of human performance engineering, Private Communication, July 1979.

Fleishman, E. A., On the relation between abilities, learning, and human performance. *American Psychologist,* 1972, *27,* 1017-32.

Flesch, R., *The Art of Readable Writing.* New York: Collier Books, 1949.

Fletcher, H., *Speech and Hearing in Communication.* New York: D. Van Nostrand Company, Inc., 1953.

Flynn, M. A., *Analysis and Comparison of the Errors of a One-Year Typewriting Group with the Errors of a Three-Year Typewriting Group in Typing Selected Timed Writing Tests*. New York: Academic Press, 1964.

Foley, J. D. and Wallace, V. L., The art of natural graphic man-machine conversation. *Proceedings of the IEEE*, 1974, *62* (4), 462-471.

Ford, R. N., *Motivation Through the Work Itself*. American Management Association, Inc., 1969.

Fox, W. F.. Human performance in the cold, *Human Factors*, 1967, *9* (3), 203-220.

Franke, R. H. and Kaul, J. D., The Hawthorne experiments: first statistical interpretation. *American Sociological Review*, October 1978, *43* (5), 623-43

Frase, L. T., Computer aids for writing and text design. Presented at annual meeting of the American Educational Research Association, Boston, April 1980.

Frase, L. T., and Schwartz, B. J., Typographical cues that facilitate comprehension. *Journal of Educational Psychology*, 1979, *71*, 197-206.

Freedman, J., *Crowding and Behavior*. San Francisco: W. H. Freedman, 1975.

French, N. R. and Steinberg, J. C., Factors governing the intelligibility of speech sounds. *Journal of the Acoustical Society of America*, 1947, *19*, 90.

Freud, S., *The Psychopathology of Everyday Life*. London: The Hogarth Press, 1901.

Galanter, E., Contemporary psychophysics. In R. Brown, E. Galanter, E. H. Hess, and G. Mandler, *New Directions in Psychology 1*. New York: Holt, Rinehart and Winston, 1962.

Gallagher, C. C., The human use of numbering systems. *Applied Ergonomics*, 1974, *5* (4), 219-23.

Gallwey, W. T., *The Inner Game of Tennis*. Random House: New York, 1974.

Galton, F., *Inquiries into Human Faculty and its Development*. London: Macmillan, 1883.

Garner, W. R., The acquisition and application of knowledge: A symbiotic relation. *American Psychologist*, 1972, 941-46.

Garrett, J.W. and Kennedy, K.W., *A Collation of Anthropometry* (Volumes I and II), AMRL-TR-68-1, AD723630, 1971.

Geldard, F. A., Some neglected possibilities of communication. *Science*, 1960, *131*, 1583.

Gellerman, S. W., *Motivation and Productivity*. New York: American Management Association, 1963.

Gignoux, M., Martin, H., and Cajgfinger, H., troubles Cochles - Vestibulaires apres Tentative de suicide a laspirine. *Journal of Franc. Oto -rhinolaryng*, 1966, *15*, 631-635.

Gilb, T. and Weinberg, G. M., *Humanized Input: Techniques for Reliable Keyed Input*. Cambridge, Mass.: Winthrop Publishers, Inc., 1977.

Gilbert, T., *Human Competence: Engineering Worthy Performance*. New York: McGraw-Hill, 1978.

Gilbert, W., CAI notes. *USE Inc. Newsletter*, 1979, *79* 9.

Gilbreth, F. B., *Brick Laying System*. New York: Clark Publishing Company, 1911.

Gilbreth, F. B. and Gilbreth, L. M., *Applied Motion Study*. New York: The Macmillian Company, 1917.

Gilmartin, K. J., Newell, A. and Simon, H. A., A program modeling short-term memory under strategy control. In C. N. Cofer (ed.), *The Structure of Human Memory*. San Francisco: Freeman, 1975.

Gilmer, B. H., *Psychology*. New York: Harper and Row, 1970.

Gilson, E. Q. and Baddeley, A. D., Tactile short-term memory. *Quarterly Journal of Experimental Psychology*, 1969, *21*, 180-84.

Glass, A. L., Holyoak, K. J., and Santa, J. L., *Cognition*. Reading, Mass.: Addison-Wesley Publishing Company, 1979.

Glass, D. C., and Singer, J. E., *Urban Stress*. New York: Academic Press, 1972.

Glenn, J. W., Machines you can talk to. *Machine Design*, 1975, 72-75.

Glodman-Eisler, F. and Cohen, M., Is N, P, and PN difficulty a valid criterion of transformational operations? *Journal of Verbal Learning and Verbal Behavior*, 1970, *9*, 161-66.

Glucksberg, S., and Cowan, G. N., Memory for non-attended Auditory material. *Cognitive Psychology*, 1970, *1*, 149-156.

Glushko, R. J. and Bianchi, M. H., On-line documentation: Integrating development, distribution

and use, Private Communication, October 1980.
Goede, W. F. A digitally-addressed flat panel CRT review. *IEEE Intercon,* 1973, *5,* 33.
Goldschmidt, E. (ed.), *The Genetics of Migrant and Isolate Populations.* New York: Williams and Wilkens, 1963.
Goodman, L. A. Passive liquid displays: Liquid crystals, electrophoretics, and electrochromics. *IEEE Transactions on Consumer Electronics,* 1975, *CE-21,* 247-259.
Gordon, W. J. J., *Synetics.* New York: Harper and Row, 1961.
Gough, P. B., Grammatical transformations and speed of understanding. *Journal of Verbal Learning and Verbal Behavior,* 1965, *4,* 107-117.
Gould, J. D., Lewis, C., and Becker, C. A., Writing and following procedural, descriptive, and restrictive syntax language instructions. *IBM Research Report,* RC-5943, 1976.
Graham, C. H. (ed), *Vision and Visual Perception,* New York: John Wiley & Sons, Inc., 1965.
Grahn, D., Radiation. In P. Webb (ed.), *Bioastronautics Data Book.* Washington, D.C.: National Aeronautics and Space Administration, 1964.
Gray, M., Questionnaire typography and production. *Applied Ergonomics,* 1975, 81-89.
Green, H. J., and Spencer, J. P., *Drugs with Possible Ocular Side Effects.* Woburn, Mass.: Butterworths, 1969.
Greene, J. M., *Psycholinguistics: Chomsky and Psychology.* Harmondsworth, Middlesex U.K.: Penguin, 1972.
Gregory, R. L., *Eye and Brain: The Psychology of Seeing.* London: World University Library, 1966.
Griffith, R. T., The minimotion typewriter keyboard. *Journal of Franklin Institute,* 1949, 399-436.
Griffiths, I. D., and Boyce, P. R., Performance and thermal comfort. *Ergonomics,* 1971, *14,* 457-468.
Grignetti, M. C. and Miller, D., Modifying computer response characteristics to influence command choice. *Proceedings of IEEE Conference on Man-Computer Interaction.* London, September 1970, 201-205.
Grossberg, M., Wiesen, R. A., and Yntema, D. B., An experiment on problem solving with delayed computer responses. *IEEE Transactions on Systems, Man, and Cybernetics,* March 1976, 219-222.
Guignard, J. C., Noise and vibration. In J. A. Gillies (ed.), *A Textbook of Aviation Physiology.* Oxford: Pergamon, 1965.
Guion, R. M., *Personnel Testing.* New York: McGraw-Hill Book Company, 1965.
Gusinde, M., *Urwaldmenschen am Ituri.* Vienna: Springer - Verlag, 1948.
Guth, S., and Eastman, A., Lighting for the forgotten man. *American Journal of Optometry,* 1955, *32,* 413-421.
Haber, R. N., and Hershenson, M., *The Psychology of Visual Perception.* New York: Holt, Rinehart and Winston, 1973.
Hackman, J. R.,
— On the coming demise of job enrichment. In E. L. Cass and F. G. Zimmer (eds.), *Man and Work in Society.* New York: Van Nostrand Reinhold, 1975.
— Work redesign and motivation. *Professional Psychology,* June 1980, *11*(3), 445-455.
Hackman, J. R. and Oldham, G. R.,
— Motivation through the design of work: Test of a theory. *Organizational Behavior and Human Performance,* 1976, *16,* 250-279.
— *Work Redesign.* Reading, Mass.: Addison-Wesley, 1980.
Hakes, D. T. and Foss, D. J., Decision processes during sentence comprehension: effects of surface structure reconsidered. *Perception and Psychophysics,* 1970, 413-416.
Hall, E. T., *The Hidden Dimension.* New York: Doubleday, 1966.
Hall, J. F., Learning as a function of word frequency. *American Journal of Psychology,* 1954, 138-140.
Hammer, W., *Handbook of System and Product Safety.* Englewood Cliffs, N.J.: Prentice-Hall, Inc., 1972.

Hammerton, M., The use of same or different sensory modalities in information and instructions. AD-A026857, December 1974.

Hammond, N., Long, J., Clark, I., Barnard, P. and Morton, J., Documenting human-computer mismatch in interactive systems. *Proceedings of the Ninth International Symposium on Human Factors in Telecommunication,* 1980.

Haney, R. W., The effect of instructional format on functional testing performance. *Human Factors,* 1969, *11,* (2), 181-188.

Hansen, B. L., *Quality Control: Theory and Applications.* Englewood Cliffs, N.J.: Prentice-Hall, Inc., 1963.

Harless, J. H., *Analysis of Learning Problems.* Springfield, Va.: Guild V Publications, 1969.

Harris, C. S.,

— Perceptual adaptation to inverted, reversed, and displaced vision. *Psychological Review,* 1965, *72,* 419-444.

— Insight or out of sight?: Two examples of perceptual plasticity in the human adult. In C. S. Harris (ed.), *Visual Coding and Adaptability.* Hillsdale, N.J.: Lawrence Erlbaum Associates, 1980a.

—Human factors and graphics; Some examples. Private Communication, January 1980b.

Harris, C. S. and Harris, J. R., Rapid adaptation to right-left reversal of the visual field. Paper presented at Psychonomic Society, Chicago, October 1965.

Harris, J. R., and Harris, C. S. Through the looking glass; Rapid adaptation to right-left reversal of the visual field. *Psychological Research,* in press.

Harris, C. S., and Shoenberger, R. W., Combined effects of noise and vibration on psychomotor performance. AMRL-TR-70-14, Wright-Patterson Air Force Base, Ohio, 1970.

Harris, H. A., *Sport in Greece and Rome.* London: Thames and Hudson, 1972.

Hartley, J. and Burnhill, P.,

— Explorations in space. *Bulletin of the British Psychological Society,* 1975, 235.

— Fifty guide-lines for improving instructional text. *Programmed Learning and Educational Technology,* 1977, *14,* 65-73.

Hays, W. L.,

Statistics for the Social Sciences. 2nd Edition, New York: Holt, Rinehard and Winston, Inc., 1973.

Hebb, D. O., *The Organization of Behavior.* New York: John Wiley & Sons, Inc., 1949.

Heimstra, N. W. and Ellingstad, V. S., *Human Behavior:A Systems Approach.* Monterey, Cal.: Brooks/Cole Publishing Company, 1972.

Heron, W., Perception as a function of retinal locus and attention. *American Journal of Psychology,* 1957, *70,* 38-48.

Herriott, P., *An Introduction to the Psychology of Language.* London: Methuen, 1970.

Hershman, R. L. and Hillix, W. A., Data Processing in Typing, *Human Factors,* 1965, *7,* 483-492.

Hertzberg, F.,

Work and the Nature of Man. Cleveland, Illinois: World Publishing Company, 1966.

Herzberg, F., Mausner, B., and Snyderman, B., *The Motivation to Work* (2nd. Edition), New York: John Wiley & Sons, Inc., 1959.

Hertzberg, H.T.E., Daniels, G.S., and Churchill, E., Anthropometry of Flying Personnel--1950, WADC Technical Report 52-321, 1954.

Hertzberg, H.T.E., Emanuel, I. and Alexander, M., The Anthropometry of Working Positions: A Preliminary Study, WADC-TR-54-520, 1956.

Hilgard, E. R. and Bower, G. H., *Theories of Learning.* Engelwood Cliffs, N.J.: Prentice-Hall, Inc., 1975.

Hill, D. R., Man-machine interaction using speech. *Advances in Computers,* New York: Academic Press, 1971.

Hill, I. D., Wouldn't it be nice if we could write computer programs in ordinary English - or would it? *The Computer Bulletin,* June 1972, 306-312.

Hill, J. W., Applied problems of hot work in the glass industry. In C. N. Davies, P. R. Davis, and F. H. Tyrer (eds.). *The Effects of Abnormal Physical Conditions at Work,* London: E. and S. Livingstone, 1967.

Hill, W. D., An evaluation of five different abstract coding methods (Experiment IV). *Human Factors,* 1961, *3,* (2), 120-130.

Hirsch, R. S., Effects of standard versus alphabetical keyboard on typing performance, *Journal of Applied Psychology,* 1970, *54* (6), 484-490

Hodge, M. H., and Field, M. M., Human coding processes. *University of Georgia Report,* January 1970.

Hodge, M. H. and Pennington, F. M., Some studies of word abbreviation behavior. *Journal of Experimental Psychology,* 1973, *98,* 350-361.

Hornsby, T. G., Voice response systems. *Modern Data,* 1972, 46-50.

Howe, J. A., *Introduction to Human Memory.* New York: Harper and Row, 1970.

Howell, W. C. and Kreidler, D. L., Information processing under contradictory instructional sets. *Journal of Experimental Psychology,* 1963, *65,* 39-46.

Howes, D. H., On the relation between the intelligibility and frequency of occurrence of English words. *Journal of the Acoustical Society of America,* 1957, *29,* 296.

Hsia, Y. and Graham, C. H., *Color Blindness.* In C. H. Graham (ed.), *Vision and Visual Perception.* New York: John Wiley & Sons, Inc., 1965.

Hull, A. J.,

— A letter-digit matrix of auditory confusions. *British Journal of Psychology,* 1973, *64* (4), 579-585.

— Nine codes: a comparative evaluation of human performance with some numeric, alpha and alphanumeric coding systems. *Ergonomics,* 1975, *18,* 567-576.

— Reducing sources of human error in transmission of alphanumeric codes. *Applied Ergonomics,* June 1976, *7,* 75-78.

Hull, A. J., and Brown, I. D., Reduction of copying errors with selected alpha-numeric subsets. *Journal of Applied Psychology,* 1975, *60,* 231-237.

Hunt, J. McV., Intrinsic motivation and its role in psychological development. In D. Levine (ed.), *Nebraska Symposium on Motivation* (Vol. 13), Lincoln: University of Nebraska Press, 1965.

Hunter, I. M. L., *Memory: Facts and Fallacies.* Baltimore: Penguin, 1957.

Hyde, S. R., Automatic speech recognition: a critical survey and discussion of the literature. In E. E. David, Jr. and P. B. Denes (eds.), *Human Communication: A Unified View.* New York: McGraw-Hill Book Co., 1972.

Israelski, E. W., *Human Factors Handbook for Telecommunications Product Design.* Private Communication, 1977.

Jahnke, J., *Journal of Verbal Learning and Verbal Behavior.* New York: Academic Press, 1963.

James, W., *Habit.* New York: Henry Holt and Company, 1914.

Jamison, D., Suppes, P., and Wells, S., The effectiveness of alternative instructional media: A survey. *Review of Educational Research,* 1974, *44* (1), 1-67.

Jarvis, J. F., A case of unilateral permanent deafness following acetyl salicylic acid. *Journal of Laryng,* 1966, *80,* 318-320.

Jenkins, J. G. and Dallenbach, K. M., Obliviscence during sleep and waking. *American Journal of Psychology,* 1924, *35,* 605-612.

Jeste, R. E., Comprehension of connected meaningful discourse as a function of individual differences and rate of modality of presentation. Ph.D. dissertation, University of Utah, 1966.

John, B. E., Thoughts on system design, analysis, and human/machine allocation, Private Communication, September 1980.

Johnson, R. E., Recall of prose as a function of the structural importance of the linguistic units. *Journal of Verbal Learning and Verbal Behavior,* 1970, *9,* 12-20.

Johnston, J. C., and McClelland, J. L., Perception of files in words; seek not and ye shall find. *Science,* 1974, *184,* 1192-1194.

Jones, E. M. and Munger, S. J., The development of performance-based coding principles.

American Institutes for Research Reports, 1968 and 1969.

Jones, P. F., Four principles of man-computer dialogue. *Computer Aided Design*, 1978, *10*, 197-202.

Jones, S., *Design of Instructions*. London: Her Majesty's Stationary Office, 1968.

Jonides, J., Toward a model of attention shifts. In A. L. Glass, K. J. Holyoak, J. L. Santa, *Cognition*. Reading, Mass.: Addison-Wesley Publishing Co., 1979.

Joyce, R. P., Chenzoff, A. P., Mulligan, J. F. and Mallory, W. J., Fully proceduralized job performance aids (Volumes I, II and III), AFHRL-TR-73-43, December 1973.

Kammann, R., The comprehensibility of printed instructions and the flowchart alternative. *Human Factors*, 1975, *17* (2), 183-191.

Kaplan, R., Some methods and strategies in the prediction of preference. In E. H. Zube, J. G. Fabos, and R. O. Brush (eds.), *Landscape Assessment: Values, Perceptions, and Resources*. Stroudsburg, Pa.: Dowden, Hutchinson, and Ross, 1974.

Kaplan, S., An informal model for the prediction of preference. In E. H. Zube, J. G. Fabos, and R. O. Brush (eds.), *Landscape Assessment: Values, Perceptions, and Resources*. Stroudsburg, Pa.: Dowden, Hutchinson, and Ross, 1974.

Karlin, J. E. The changing and expanding role of human factors in telecommunication engineering at Bell Laboratories. In Proceedings of the eighth international symposium on human factors in telecommunications. Harlow, Essex, England. Standard Telecommunication Laboratories, 1977.

Katz, P., *Animals and Men*. Longmans Green, 1937.

Kaufman, J. E., *IES Lighting Handbook*, New York: Illuminating Engineering Society, 1972.

Kay, H., Learning of a serial task by different age groups. *Quarterly Journal of Experimental Psychology*, 1951, *3*, 166-183.

Kazan, B. Electroluminescent displays. *Proceedings of the S.I.D.*, 1976, *17*, 23-29.

Keele, S. W., Movement control in skilled motor performance. *Psychological Bulletin*, 1968, *70* (6), 387-403.

Keele, S. W., *Attention and Human Performance*. Pacific Palisades, Cal.: Goodyear Publishing Company, Inc., 1973.

Keele, S. W., and Chase, W. G., Short-term visual storage. *Perception and Psychophysics*, 1967, *2*, 383-386.

Kemeny, J. G., *The Accident at Three Mile Island*. Report of the Presidents Commission, October 1979.

Kennedy, K. W., The human body in equipment design: Reach capability of the USAF population, Phase I: The outer boundaries of grasping-reach envelopes for the shirt-sleeved, seated operator. AMRL-TDR-64-59, Aerospace Medical Research Laboratories, Wright-Patterson Air Force Base, Ohio, 1964.

Kennedy, T. C. S., The design of interaction procedures for man-machine communications. *International Journal of Man-Machine Studies*, 1974, *6*, 309-334.

Kherumian, R., and Pickford, R. W., *Heredite et Frequence des Anomalies Congenitales du Sens Chromatique*. Paris: Vigot Freres, 1959.

Kincaid, J. P., Fishburne, R. P., Rogers, R. L., and Chissom, B. S., Derivation of new readability formulas (Automated Readability Index, Fog Count, and Flesch Reading Ease Formula) for Navy enlisted personnel. Naval Training Command Research Branch Report 8-75. February 1975.

Kingsley, H. L., and Gerry, R., *The Nature and Conditions of Learning*. (2nd ed.), Englewood Cliffs, N.J.: Prentice-Hall, Inc., 1957.

Klare, G. R.,

— *The Measurement of Readability*. Ames, Iowa: Iowa State University Press, 1963.

— Assessing readability. *Reading Research Quarterly*, 1974-1975, *10*, 62-102.

— A second look at the validity of readability formulas. Invited address at the annual meeting of the National Reading Conference, St. Petersburg, Florida, December 1975.

Klare, G. R., Mabry, J. E., and Gustafson, L. M., The relationship of patterning (underlining) to immediate retention and to acceptability of technical material. *Journal of Applied Psychology*, 1955, *39*, 40-42.

Klare, G. R., Nichols, W. H., and Shuford, E. M., the relationship of typographical arrangement to the learning of technical material. *Journal of Applied Psychology*, 1957, *41* (1), 41-45.

Klare, G. R., Shuford, E. M., and Nichols, W. H., The relationship of style difficulty, practice and ability to efficiency of reading and to retention. *Journal of Applied Psychology*, 1957, *41*, (4), 222-226.

Klare, G. R., Shuford, E. H., and Nichols, W. H., The relation of format organization to learning. *Educational Research Bulletin*, 1958, 39-45.

Klemmer, E. T.,

— Grouping of printed digits for manual entry. *Human Factors*, 1969, *11* (4), 397-400.

— Keyboard entry. *Applied Ergonomics*, 1971, *2*, 2-6.

Klemmer, E. T. and Lockhead, G. R., Productivity and errors in two keying tasks: A field study. *Journal of Applied Psychology*, 1962, *46*, 401-408.

Knowlton, K., Computer displays optically superimposed on input devices. *Bell System Technical Journal*, 1977, 367-383.

Kohler, I.,

— Experiments with goggles. *Scientific American*, 1962, *206* (5), 62-72.

— The formation and transformation of the perceptual world, *Psychological Issues*, 1964, *3* (4), Monograph 12.

Korman, A. K.,

— Expectancies as determinents of performance. *Journal of Applied Psychology*, 1971, *55*, 218-222.

— Toward an hypothesis of work behavior. *Journal of Applied Psychology*, 1970, *54*, 31-41.

Kornhauser, A., *Mental Health of the Industrial Worker*. New York: John Wiley & Sons, Inc., 1965.

Kroemer, K. H. E.,

— Human engineering the keyboard. *Human Factors*, 1972, *14* (1), 51-63.

— Muscle strength as a criterion in control design for diverse populations. In A. Chapanis (ed.), *Ethnic Variables in Human Factors Engineering*, Baltimore: The Johns Hopkins University Press, 1975.

— Human strength: terminology, measurement and interpretation of data. *Human Factors*, 1970, *12*, 297-313.

— Designing for muscular strength of various populations. AMRL-TR-72-46, Wright-Patterson Air Force Base: Aerospace Medical Research Laboratory, 1974.

Krumm, R. L., and Allen, C. P. , Prediction of trainee errors: analysis of Sale reliability. Essex Corporation Report, June 1974.

Kryter, K. D.,

— *The Effects of Noise on Man*. New York: Academic Press, 1970.

— Speech communication. In H. P. Van Cott and R. G. Kinkade, *Human Engineering Guide to Equipment Design*. U. S. Government Printing Office, Washington, D.C., 1972.

Kryter, K. D. and Whitman, E. C., Some comparisons between Rhyme and PB - word intelligibility tests. *Journal of the Acoustical Society of America*, 1965, *39*, 1146.

Kubovy, M., and Howard, F. P., Persistence of a pitch segregating echoic memory. *Journal of Experimental Psychology: Human Perception and Performance*, 1976, *2*, 531-537.

Kucera, H. and Francis, W. N., *Computational Analysis of Present-Day American English*. Providence, R.I.: Brown University Press, 1967.

Kulp, M. J., The effects of position practice readability level on performance. Private Communication, February 1976.

Lachman, R., Mistler-Lachman, J., and Butterfield, E. C., *Cognitive Psychology and Information Processing: An Introduction*. Hillsdale, N.J.: Lawrence Erlbaum Associates, 1979.

Landauer, T. K. and Whiting, J. W. M., Correlates and consequences of stress in infancy, Private Communication, July 1979.

Lanning, E. D., *Peru Before the Incas*. Prentice-Hall, Inc., 1967.

Larkin, J., McDermott, J., Simon, D. P., and Simon, H. A., Expert and novice performance in solving physics problems. *Science 1980, 208,* 1335-1342.

Lashley, K. S., The problem of serial order in behavior. In L. A. Jeffress (ed.), *Cerebral Mechanisms in Behavior.* New York: John Wiley & Sons, Inc., 1951.

Lawton, R. W., Weightlessness. In P. Webb (ed.), *Bioastronautics Data Book.* Washington, D.C.: National Aeronautics and Space Administration, 1964.

Leigh, E., *On Religion and Learning,* 1656, In H. B. Wheatley, *Literary Blunders.* London: Elliot Stock, 1893. Reprinted by Gale Research Company, Detroit, 1969.

Lepper, M. R. and Greene, D., *The Hidden Costs of Reward.* Morristown, N. J.: Lawrence Erlbaum, 1979.

Lepper, M. R., Greene, D., and Nisbett, R. E., Undermining children's intrinsic interest with extrinsic rewards: A test of the overjustification hypothesis. *Journal of Personality and Social Psychology,* 1973, *28*, 129-137.

Levy, B. A., Role of articulation of auditory and visual short-term memory. *Journal of Verbal Learning and Verbal Behavior,* 1971, *10,* 123-132.

Lineberry, C. S., Jr., When to develop aids for on-the-job use and when to provide instruction. *Improving Human Performance: A Research Quarterly,* 1977, *6*, 87-92.

Linvill, J. G. and Bliss, J. C., A direct translation reading aid for the blind. *Proceedings of the IEEE,* 1966, *54,* 40-51.

Locke, J. L. and Locke, V. L., Deaf children's phonetic, visual, and dactylic coding in a grapheme recall task. *Journal of Experimental Psychology,* 1971, *89*, 142-146.

Loeb, M. and Holding, D. H., Backward interference by tones or noise in pitch perception as a function of practice. *Perception and Psychophysics,* 1975, *18*, 205-208.

Loftus, E. F. and Loftus, G. R., On the permanence of stored information in the human brain. *American Psychologist,* May 1980, 409-420.

Loftus, E. F., Freedman, J. L. and Loftus, G. R., Retrieval of words from subordinate and supraordinate categories in semantic hierarchies. *Psychonomic Science,* 1970, 235-236.

Lorayne, H., *Good memory—successful student!.* New York: Stein and Day Publishers, 1976.

Lovesey, E. S., The multi-axis vibration environment and man. *Ergonomics,* 1970, *1* (5), 258-261.

Lowe, D. G., Temporal aspects of selective masking. *Quarterly Journal of Experimental Psychology,* 1975, *27*, 375-385.

Luchins, A. S., Mechanization in problem solving - the effect of Einstellung. *Psychological Monographs,* 1942, *54* (6, Whole No. 248).

Lummis, R. C., Speaker verification: A step toward the "checkless" society. *Bell Laboratories Record,* 1972, 244-259.

Luria, A. R., *The Mind of a Mnemonist.* New York: Basic Books, 1968.

MacNeilage, P. F., Typing errors as clues to serial ordering mechanisms in language behavior. *Language and Speech,* 1964, *7,* 144-159.

MacNeilage, P. F., and MacNeilage, L. A., Central processes controlling speech production during sleep and waking. In F. J. McGuigan (ed.), *The Psychophysiology of Thinking.* New York: Academic Press, 1973.

Macdonald, N. H., Frase, L. T., Gingrich, P. S. and Keenan, S. A., Writer's Workbench: Computer programs for text editing and assessment, Private Communication, May 1980.

Macdonald, N. H., Frase, L. T., Gingrich, P. S. and Keenan, S. A., The Writer's Workbench: Computer aids for text analysis, *IEEE Transactions on Communications,* In Press.

Macdonald, N. H., Keenan, S. A., Gingrich, P. S., Fox, M. L., Frase, L. T., and Collymore, J. L., Writer's Workbench: Computer aids for writing, Private Communication, April, 1981.

Macdonald-Ross, M.,

— *Bibliography For Textual Communication.* Institute of Educational Technology, Open University, 1973.

— *Graphics in Text: A Bibliography.* Institute of Educational Technology, Open University, 1977.
Graphics in Texts. In L. S. Shulman (ed.), *Review of Research in Education* (Vol. 5), Hasca, Ill.:

Reacock, 1977.
Mackay, D. G., Aspects of the theory of comprehension memory and attention. *Quarterly Journal of Experimental Psychology,* 1973, 22-40.
Mackworth, J. F., The duration of the visual image. *Canadian Journal of Psychology,* 1963, *17*, 62-81.
Mager, R. F., On the other hand. *Improving Human Performance: A Research Quarterly,* 1973, *2,* 77-88.
Mager, R. F. and Beach, K. M., Jr., *Developing Vocational Instruction.* Belmont, Cal.: Fearon, 1967.
Malone, T. W., What makes things fun to learn? Palo Alto, *Xerox Research Center Technical Report,* August 1980.
Margolis, B. L. and Kroes, W. H., *The Human Side of Accident Prevention: Psychological Concepts and Principles Which Bear on Industrial Safety.* Springfield, Ill.: Charles C. Thomas, 1975.
Martin, C. R.,
— Contribution of position level testing to BISCUS/FACS system success. Private Communication, August 1974.
— The BISCUS/FACS position package testing program, Private Communication, April 1975.
— Human performance testing in the design of computer-based systems. Private Communication, September 1979.
Martin, T. B., Practical applications of voice input to machines. *Proceedings of the IEEE,* 1976, *64*(4), 487-501.
Maslow, A. H.,
— *Motivation and Personality.* New York: Harper, 1954.
— *Motivation and Personality* (2nd Ed.). New York: Harper and Row, 1970.
Massaro, D. W.,
— Preperceptual images, processing time and perceptual units in auditory perception. *Psychological Review,* 1972, *79*, 124-145.
— *Experimental Psychology and Information Processing.* Chicago: Rand McNally College Publishing Company, 1975.
Mayo, E.,
The Social Problems of an Industrial Civilization. New York: Macmillan, 1933.
The Social Problems of an Industrial Civilization. Boston: Harvard University Press, 1945.
Mayzner, M. S. and Tresselt, M. E., Tables of single-letter and digram frequency counts for various word-length and letter-position combinations. *Psychonomic Monograph Supplements,* 1965, *1*(2), 13-32.
Mayzner, M. S., Tresselt, M. E. and Wolin, B. R., Tables of trigram frequency counts for various word-length and letter-position combinations. *Psychonomic Monograph Supplements,* 1965, *1*(3), 33-78.
McArthur, B. N., Accuracy of source data: Human error in hand transcription, AD 623157, May 1965.
McClelland, D. C., *Assessing Human Motivation.* New York: General Learning Press, 1971.
McClelland, D. C., Atkinson, J. W., Clark, R., and Lowell, E., *The Achievement Motive.* New York: Appleton-Century-Crofts, 1953.
McClelland, L. A., Crowding and social stress. Unpublished Ph.D. dissertation, University of Michigan, 1974.
McCormick, E. J., *Human Factors Engineering.* 4rd ed., New York: McGraw-Hill Book Company, 1976.
McGregor, D., *The Human Side of Enterprise.* New York: McGraw-Hill, 1960.
Medsker, K. L., CAI/CMI, Private Communication, December 1979.
Meister, D., *Behavioral Foundations of System Development.* London: John Wiley & Sons, Inc., 1976.
Merikle, P. M., Coltheart, M. and Lowe, D. G. On the selective effects of a patterned masking stimulus. *Canadian Jornal of Psychology,* 1971, *25*, 264-79.
Merikle, P. M., and Coltheart, M. Selective forward masking. *Canadian Journal of Psychology,*

1972, *26*, 296-302.

Mewhort, D. J. K., and Cornett, S. Scanning and the familiarity effect in tachistoscopic recognition. *Canadian Journal of Psychology,* 1972, *26*, 181-189.

Michaels, S. E., Qwerty versus alphabetic keyboards as a function of typing skill. *Human Factors,* 1971, *13* (5), 419-426.

Milgram, S., Some conditions of obedience and disobedience to authority. *Human Relations,* 1965, *18,* 57-76.

Miller, G. A.,

— Some psychological studies of grammar. *American Psychologist,* 1962, *17*, 748-762.

— The Magical number seven, plus or minus two: Some limits on our capacity for processing information. *Psychological Review,* 1956, *63*, 81-97.

Miller G. A. and Nicely, P. E., An analysis of perceptual confusions among some English consonants. *Journal of the Acoustical Society of America,* 1955, *27,* 338.

Miller, G. A., Galanter, E., and Pribram, K. H., *Plans and the Structure of Behavior.* New York: Henry Holt and Company, 1960.

Miller R. B.,

— Response time in man-computer conversational transactions. *AFIPS Conference Proceedings* (Fall Joint Computer Conference, 1968). Washington, D.C.: Thompson Book Company, 1968.

— Survey of human engineering needs in maintenance of ground electronics equipment. Contract *AF* 30 (602) 24, February 1954, 274.

Miller, L. A. and Thomas, J. C., Behavioral issues in the use of interactive systems. *International Journal of Man-Machine Studies,* 1977, *9,* 509-536.

Milner, B. and Taylor. L., Right-hemisphere superiority in tactile pattern recognition after cerebral commisurotomy: evidence for non-verbal memory. *Neuropsychologia,* 1972, *10*, 1-15.

Montanelli, R. G. Using CAI to teach introductory computer programming. *ACM SEGCUE Bulletin,* 1977, *11* (1), 14-22.

Morehouse, L. E. and Gross, L., *Maximum Performance.* New York: Pocket Books, 1977.

Morgan, C. T., Cook, J. S., III, Chapanis, A., and Lund, M. W. (eds.), *Human Engineering Guide to Equipment Design.* New York: McGraw-Hill, 1963.

Morrison, P., and Morrison, E., *Charles Babbage and His Calculating Machines.* New York: Dover Publications, Inc., 1961.

Moser, H. and Fotheringham, W., *Number telling.* AD260457, December 1960.

Muller, G. E. and Pilzecker, A., An experimental contribution to teaching the mind. *Zeitscrift fur Psychology,* 1900, 212-216.

Munsterberg, H., *Psychology and Industrial Efficiency.* Boston: Houghton Mifflin, 1913.

Murdock, B. B., Jr., The retention of individual items. *Journal of Experimental Psychology,* 1961, *62*, 618-625.

Murray, A. A., *Explorations in Personality.* New York: Oxford University Press, 1938.

Murray, D. L., Articulation and acoustic confusability in short-term memory. *Journal of Experimental Psychology,* 1968, *78*, 679-684.

Murray, R. H., and McCally, M., Combined environmental stressors. In *Bioastronautics Data Book.* Washington, D.C.: National Aeronautics and Space Administration, 1972.

Murrell, K. F. H., *Ergonomics: Man in His Working Environment.* London: Chapman and Hall, 1965.

National Research Council, Committee on Undersea Warfare. *Human Factors in Undersea Warfare.* Washington, D.C., August 1949.

Nelson-Denny Reading Test Form D. Boston: Houghton Mifflin, 1973.

Neisser, U., *Cognition and Reality: Principles and Implications of Cognitive Psychology.* San Francisco: W. H. Freeman and Company, 1976.

Neisser, U. and Weene, P., A note on human recognition of hand-printed characters. *Information and Control,* 1963, *3* (2).

Newell, A., Speech understanding systems, Report AFOST-TR-72-0142. Carnegie-Mellon University, May 1971.

Newell, A. and Simon, H. A., *Human Problem Solving*. Englewood Cliffs, N.J.: Prentice-Hall, Inc., 1972.

Newman, R. W. and White, R. M., Reference Anthropometry of Army Men, Report No. 180. Lawrence, Mass.: Quartermaster Climatic Research Laboratory, 1951.

Nichols, R. G., Factors accounting for differences in comprehension of material presented orally in the classroom. State University of Iowa Ph.D. dissertation, 1948.

Nickerson, R. S., Auditory codability and the short-term retention of visual information. *Journal of Experimental Psychology*, 1972, *95*, 429-436.

Nickerson, R. S. and Pew, R. W., Oblique steps toward the human-factors engineering of interactive computer systems. In M. C. Grignetti, D. C. Miller, R. S. Nickerson, and R. W. Pew, *Information processing models and computer aids for human performance - final technical report*, AFOSR TR-72-1322, 1972.

Nievergelt, J. et al., An automated computer science education system. *Angewandte Informatik*, 1975, *4*, 135-142.

Norman, D. A., Post-Freudian slips. *Psychology Today*, April 1980, 42-50.

North, A. J., and Jenkins, L. B., Reading speed and comprehension as a function of typography. *Journal of Applied Psychology*, 1951, *35*(4), 225-228.

Olson, J. R., and Mackay, D. G., Completion and verification of ambiguous sentences. *Journal of Verbal Learning and Verbal Behavior*, 1974, 457-470.

Orr, D. B., Trainability of listening comprehension of speeded discourse. *Journal of Educational Psychology*, 1965, *56*, 148-156.

Osborn, A. F., *Applied Imagination*, 3rd ed. New York: Scribner, 1963.

Ota, I., Ohnishi, J., and Yoshiyama, M., Electrophoretic display panel--EPID. *Proceedings of the IEEE*, 1973, *61*, 832.

Owsowitz, S. and Sweetland, A., Factors affecting coding errors. *Rand Corporation Technical Report*, April 1965.

Palerno, D. S., More about less: a study of comprehension. *Journal of Verbal Learning and Verbal Behavior*, 1973, 211-221.

Parsons, H. M., What happened at Hawthorne? *Science*, 1974, *183*, 922-32.

Parsons, M. Mc I., Environmental design. *Human Factors*, 1972, *14*, 369-482.

Parsons, S. O., Eckert, S. K. and Seminara, J. L., Human Factors design practices for nuclear power plant control rooms. *Procedings of the Human Factors Society 22nd Annual Meeting*, 1978.

Penfield, W.,

— *The Excitable Cortex in Conscious Man*. Liverpool: Liverpool University Press, 1958.

— Consciousness, memory, and man's conditioned reflexes. In K. Pribram (ed.), *On the Biology of Learning*. New York: Harcourt, Brace and World, 1969.

Penfield, W. and Perot, P., The brain's record of auditory and visual experiences. *Brain*, 1963, *86*, 595-696.

Pepler, R., Thermal comfort of students in climate controlled and non-climate controlled schools. *ASHRAE Transactions*, 1972, *78*, 97-109.

Peterson, L. R. and Peterson, M., Short term retention of individual items. *Journal of Experimental Psychology*, 1959, *58*, 193-198.

Pew, R. W.,

— Acquisition of hierarchical control over the temporal organization of a skill. *Journal of Experimental Psychology*, 1966, *71*, 764-71.

— The speed-accuracy operating characteristic. *Acta Psychologica*, Publishing Company, Amsterdam, 1969, *30*, 16-26.

Pew, R. W. and Rollins, A. M., *Dialog Specification Procedures*, Cambridge, Mass.: Bolt, Beranek and Newman, Inc., September 1975.

Phillips, D. P., Deathday and birthday: An unexpected connection. In J. M. Tanur (ed.), *Statistics: A Guide to the Unknown*. San Francisco: Holden-Day, Inc.

Phillips, L. D., *Bayesian Statistics for Social Scientists*. New York: Thomas Y. Crowell Company,

1973.

Piaget, J., *The Origins of Intelligence in Children.* New York: International Universities Press, 1952.

Pierce, J. R. and Karlin, J. E., Reading rates and the information rate of a human channel. *Bell System Technical Journal,* 1957, *36,* 497-516.

Pierson, W. R., Body size and speed. *Research Quarterly,* 1961, *32,* 197-200.

Pollack, I., Message uncertainty and message reception. *Journal of the Acoustical Society of America,* 1959, *31,* 1500-1508.

Pollack, I. and Ficks, L., Information of multidimensional auditory displays. *Journal of the Acoustical Society of America,* 1954, *26,* 155-158.

Poller, M. F., Friend, E., Hegarty, J. A., Rubin, J. J., Dever, J. J., *Handbook for Writing Procedures.* Indianapolis, Indiana: Western Electric Publication Center, 1981.

Poller, M. F., Hegarty, J. A., Friend, J. A. and Rubin, J. J., TOP Format Study. Private Communication, 1979.

Porter, L. W. and Lawler, E. E., What job attitudes tell us about motivation. *Harvard Business Review,* 1968, *46* (1), 118-126.

Porter, L. W. and Lawler, E. E., *Managerial Attitudes and Performance.* Homewood, Ill.: Erwin, 1968.

Porter, L. W., Lawler, E. E. and Hackman, J. R., *Behavior in Organizations,* New York: McGraw-Hill, 1975.

Posner, M. I., *Cognition: An Introduction.* Glenview, Ill.: Scott, Foresman, 1973.

Potter, N. R., and Thomas, D. L., Evaluation of three types of technical data for troubleshooting: Results and project summary. AFHRL-TR-76-74(1), September 1976, 1-122.

Poulton, E. C.,

— Rate of comprehension of an existing teleprinter output and of possible alternatives. *Journal of Applied Psychology,* 1968, 16-21.

— Skimming lists of food ingredients printed in different sizes. *Journal of Applied Psychology,* 1969, 55-58.

— *Environment and Human Efficiency.* Springfield, Ill.: Charles C. Thomas Publisher, 1972.

Poulton, E. C., Hunt, J. C. R., Mumford, J. C., and Poulton, J., Mechanical disturbance produced by steady and gusty winds of moderate strength: Skilled performance and semantic assessments. *Ergonomics,* 1975, *18,* 651-673.

Provins, K. A. and Bell, C. R., Effects of heat stress on the performance of two tasks running concurrently. *Journal of Experimental Psychology,* 1970, *85,* 40-44.

Quetelet, A., *Anthropometrie.* Brussels: C. Munquardt, 1870.

Ramsey, H. R. and Atwood, M. E., *Human Factors in Computer Systems: A Review of the Literature.* ADA075679, 1979.

Reddy, D. R., *Speech Recognition.* New York: Academic Press, 1975.

Reder, L. M. and Anderson, J. R., A comparison of texts and their summaries: Memorial consequences. *Carnegie-Mellon University Technical Report,* October 1979.

Reicher, G. M., Perceptual recognition as a function of meaningfulness of stimulus material. *Journal of Experimental Psychology,* 1969, *81,* 275-280.

Restorff, H. Von, Uber die Wirkung von Bereichsbildungen im Spurenfeld. *Psychologische Forschung,* 1933, *18,* 299-342.

Riggs, L. A., Vision. In J. W. Kling and L. A. Riggs, *Experimental Psychology.* New York: Holt, Rinehart and Winston, Inc., 1971.

Roberts, D. F., Population differences in dimensions, their genetic basis and their relevance to practical problems of design. In A. Chapanis (ed.), *Ethnic Variables in Human Factors Engineering.* Baltimore: The Johns Hopkins University Press, 1975.

Roethliesberger, F. J. and Dickson, W. J., *Management and the Worker - An Account of a Research Program Conducted by the Western Electric Company, Hawthorne Works, Chicago.* Cambridge: Harvard University Press, 1939.

Ronco, P. G., Hanson, J. A., Raben, M. W., and Samuels, I. A., Characteristics of technical reports

that affect reader behavior: A review of the literature. National Science Foundation, 1966.
Rothkopf, E. Z.,
— Some observations on predicting instructional effectiveness by simple inspection. *The Journal of Programmed Instruction,* 1963, *2*, 18-20.
— The concept of mathemagenic activities. *Review of Educational Research,* 1970, *40,* 325-336.
— Adjunct aids, and the control of mathemagenic activities during purposeful reading. *Conference on Reading Expository Material,* University of Wisconsin, Madison, November 10-11, 1980.
Rotter, J. B., Generalized expectancies for internal versus external control of reinforcement. *Psychological Monographs,* 1966, *80*, 1-28.
Rubin, A. D. and Risley, J. F., The PROPHET system--An experiment in providing a computer resource to scientists. BBN Paper, 1976.
Ryan, J., Grouping and short-term memory: different means and patterns of grouping. *Quarterly Journal of Experimental Psychology,* 1969, *21*, 137-147.
Sackman, H., Outlook for man-computer symbiosis: Towards a general theory of man-computer problem solving. Paper presented at the *Advanced Study Institute of Man-Computer Interaction,* Mati, Greece, September 1976.
Safire, W., *On Language,* New York: Times Books, 1980
Sakitt, B., Locus of short-term visual storage. *Science,* 1975, *190*, 1318-1319.
Sanders, L., *The Tomorrow File.* New York: Berkley Publishing Corporation, 1976.
Schaffer, E., Writing space on forms. Private Communication, July 1980.
Schaffer, L. H. and Hardwick, J., Errors and error detection in typing. *Quarterly Journal of Experimental Psychology,* 1969, *21*, 209-213.
Scheibe, K. E., Shaver, P. R., and Carrier, S. C., Color association values and response interference on variants of the stroop test. *Acta Psychologica,* 1967, *28*, 286-295.
Seibel, R.,
— Data entry through chord parallel devices. *Human Factors,* 1964, *6,* 189-192.
— Data entry devices and procedures. In H. P. Van Cott, and R. G. Kinkade (eds.), *Human Engineering Guide to Equipment Design.* Washington, D. C.: U. S. Government Printing Office, 1972.
Seymour, P. J., A study of the relationship between the communication skills and a selected set of predicators and of the relationship among the communication skills. University of Minnesota, Ph.D. dissertation, 1965.
Shackel, B. (ed.),
— *Man/Computer Communication.* (Vol. 1), Maidenhead, Berkshire, England: Infotech International Limited, 1979.
— *Man/Computer Communication* (Vol. 2), Maidenhead, Berkshire, England: Infotech International Limited, 1979.
Shepard, R. N. and Sheenan, M. M., Immediate recall of numbers containing a familiar prefix or postfix. *Perceptual and Motor Skills,* 1965, *21,* 263-73.
Sherr, S., *Electronic Displays.* New York: Wiley, 1979.
Shneiderman, B., *Software Psychology.* Cambridge, Mass.: Winthrop Publishers, Inc., 1980.
Shoshkes, L., *Space Planning.* New York: Architectural Record, 1976.
Shulman, L. S., Loupe, M. J., and Piper, R. M., Studies of the inquiry process: Inquiry patterns of students in teacher-training programs. East Lansing: Educational Publications Services, Michigan State University, 1968.
Silbiger, H. R., Speech Communication. Private Communication, 1973.
Silvern, L. C.
— Systems engineering of education III: systems analysis and synthesis applied to occupational instruction in secondary schools. Los Angeles, Cal.: Education and Training Consultants, 1967.
— Training: man-man and man-machine communications. In K. B. DeGreene (ed.), *Systems Psychology.* New York: McGraw-Hill, 1970.
Sime, M. E., "So I said in the most natural way, if x=0 then begin...": The empirical study of

computer language. MRC Social and Applied Psychology Unit Report No. 132, 1976.
Simon, H. and Gilmartin, K., A simulation of memory for chess positions. *Cognitive Psychology*, 1973, *5*, 29-46.
Singleton, W. T., *Man-Machine Systems*. Baltimore: Penguin Books, 1974.
Skinner, B. F.,
— *The Behavior of Organisms*. New York: Appleton-Century-Crofts, 1938.
— *Science and Human Behavior*. New York: Free Press, 1953.
— *Beyond Freedom and Dignity*. New York: Alfred A. Knopf, 1971.
Slobin, D., Grammatical transformations and sentence comprehension in childhood and adulthood. *Journal of Verbal Learning and Verbal Behavior, 1966, 5*, 219-227.
Smith, E. A., and Kincaid, P., Derivation and validation of the automated readability index for use with technical materials. *Human Factors*, 1970, *12*, 457-464.
Smith, E. E., Choice reaction time: An analysis of the major theoretical positions. *Psychological Bulletin*, 1968, *69* (2), 77-110.
Smith, J., Deliver more - and maybe even better - training for less with microcomputers. *Training/HRD*, September 1979, 27-33.
Smith, K. U. and Sussman, H., Cybernetic theory and analysis of motor bearing memory. In E. A. Bilodeau (ed.), *Principles of Skill Acquisition*. New York: Academic Press, 1969, 103-138.
Smith, L. D., *Cryptography: the Science of Secret Writing*. New York: W. W. Norton and Company, 1943.
Smith, S. L., Requirements definition and design guidelines for the man-machine interface in C^3 system acquisition, MITRE Technical Report, MTR-3888, April 1980.
Soble, A. "Gas-Discharge Displays: The State of the Art," *Conference Record of 1976 Biennial Display Conference*, Institute of Electrical and Electronic Engineers, 1976, 99-103.
Snyder, H. L., Human visual performance and Flat panel display image quality, AD A092685, July, 1980.
Solomon, R. L. and Postman, L., Frequency of usage as a determinant of recognition threshold for words. *Journal of Experimental Psychology*, 1952, 195-201.
Sonntag, L., Designing human-oriented codes. *Bell Laboratories Record*, February 1971, 43-49.
Soth, M. W., The human work module: A structural entity for personnel subsystem design, Private Communication, November 1976.
Spence, R., Human factors in interactive graphics. *Computer Aided Design*, 1976, *8*, 49-53.
Sperling, G., The information available in brief visual presentations. *Psychological Monographs*, 1960, 74.
Starch, D., *Advertising*. Chicago: Scott, Foresman, 1914.
Starkieweiz, W. and Kuliszewski, T., In L. Clark (ed.), *Proceedings of the International Congress on Technology and Blindness*. 1962, 157-166.
Starkieweiz, W., Kuprinoowicz, W., and Petruczenko, F., In T. D. Sterling (ed.), *Visual Prosthesis - The Interdisciplinary Dialogue*. New York: Academic Press, 1971, 295-299.
Steinmetz, C. S., The Evolution of Training. *Training and Development Handbook*, New York: McGraw-Hill, 1967.
Sternberg, S.,
— Estimating the distribution of additive reaction-time components. Paper presented at a meeting of the Psychometric Society, Niagara Falls, Ontario, October 1964.
— Memory scanning: Mental process revealed by reaction-time experiments. *American Scientist*, 1969a, *57*, 421-457.
— The discovery of processing stages: Extensions of Donders' method. In W. G. Koster (ed.), *Attention and Performance II, Acta Psychologica*, 1969b, *30*, 276-315.
— Character delimiters are a liability in data-entry forms. Private Communication, May 1978.
Stewart, T. F. M., Damodaran, L. and Eason, K. D., Interface problems in man-computer interaction. In E. Humford and H. Sackman (eds.), *Human Choice and Computers*. New York: American Elsevier Publishing Co., 1976.

Stewart, T. F. M.,
— The software interface and displays. HUSAT Memo No. 95, Department of Human Sciences, Univ. of Technology, Loughborough, 1974.
— The specialist user. *Proceedings of NATO Advanced Study Institute,* Mati, Greece, September 1976.
Sticht, T. G. (ed.), *Reading For Working: A Functional Literacy Anthology.* Alexandria, Va.: Human Resources Research Organization, 1975.
Stokols, D., On the distinction between density and crowding: Some implications for future research. *Psychological Review,* 1972, *79* (3), 275-278.
Strachey, J., *The Standard edition of the Complete Psychological Works of Sigmund Freud.* (Vol. 6), London: The Hogarth Press, 1960.
Streeter, L. A., Ackroff, J. M. and Taylor, G. A., On abbreviating command names. Private Communication, August 1980.
Strong, E. P., A comparative experiment in simplified keyboard retraining and standard keyboard supplementary training. Washington, D.C.: General Services Administration, 1956.
Strunk, W. and White, E. B., *The Elements of Style.* New York: Macmillan Publishing Company, Inc., 1972.
Swenson, J. S., A review of some issues relating to programming language design. Private Communication, 1979.
Swensson, R. G., The elusive tradeoff: Speed vs. accuracy in visual discrimination tasks. *Perception and Psychophysics,* 1972, *12*, 16-32.
Swink, J. R., Intersensory comparisons of reaction time using an electropulse tactile stimulus. *Human Factors,* 1966, *8* (2), 143-145.
Tannehill, R. E., *Job Enrichment.* Chicago, Ill.: The Dartnell Corporation, 1974.
Tannenbaum, J. J. The MAP process. Private Communication, 1977.
Taylor, D. A., Stage analysis of reaction time. *Psychological Bulletin,* 1976, *83* (2), 161-181.
Taylor, F. W.,
— *The Principles of Scientific Management.* New York: Harper, 1911.
— *Principles of Scientific Management.* New York: Harper and Brothers, 1947.
— Psychology and the design of machines. *American Psychologist,* 1957, *12,* 249-258.
Terrace, H. S., Errorless transfer of a discrimination across two continua. *Journal of the Experimental Analysis of Behavior,* 1963, *6* (2), 223-232.
Thompson, E. T., How to write clearly. *Rolling Stone,* November 15, 1979.
Thorpe, C. E. and Rowland, G. E., The effect of natural grouping of numerals on short-term memory. *Human Factors,* 1965, *7,* 38-44.
Tichauer, E. R.,
— The effects of climate on working efficiency. *Impetus,* Australia, 1962, *1* (5), 24-31.
— *The Mechanical Basis of Ergonomics.* New York: John Wiley & Sons, Inc., 1978.
Tinker, M. A.,
— Illumination intensities for reading newspaper type. *Journal of Education Psychology,* 1943, *34,* 247-250.
Legibility of Print. Ames, Iowa: Iowa State University Press, 1963.
— *Bases for effective reading.* Minneapolis: University of Minnesota Press, 1965, 136-141.
Tinker, M. A. and Paterson, D. G., Influence of type form on speed reading. *Journal of Applied Psychology,* 1928, 359-368.
Triplett, N., The dynamogenic factors in pacemaking and competition. *American Journal of Psychology,* 1897, *9,* 507-533.
Trollip, S. R. The evaluation of a complex computer-based flight procedures trainer, *Human Factors,* 1979, *21* (1), 47-54.
Tsemel, G. I., Recognition of spoken signals. Moscow, USSR; English translation, Springfield, Va.: Joint publications Research Service, National Technical Information Service, 1972.
Tulving, E. and Thomson, D. M., Encoding specificity and retrieval processes in episodic memory.

Psychological Review, 1973, *80*, 352-373.
Turn, R., The use of speech for man-computer communication, Report R-1386-ARPA, Santa Monica, California: Rand Corporation, January 1974.
Turner, A. N. and Lawrence, P. R., *Industrial Jobs and the Worker*, Boston: Harvard Graduate School of Business Administration, 1965.
Turoff, M., Knowns and unknowns of immediate access time-shared systems. Paper presented at the American Management Association Conference on the Computer Utility, 1967.
U.S. Environmental Protection Agency, *Noise on Wheels*. February 1977.
VanCott, H. P. and Kinkade, R. G. (eds.), *Human Engineering Guide to Equipment Design*. Washington, D.C.: U.S. Government Printing Office, 1972.
VanCott, H. P. and Chapanis, A., Human engineering tests and evaluation, In. H. P. VanCott and R. G. Kinkade (eds.), *Human Engineering Guide to Equipment Design*. Washington, D.C.: U. S. Government Printing Office, 1972.
Vanek, M. and Cratty, B. J., *Psychology and the Superior Athlete*. London: The Macmillian Company/Collier - Macmillian Limited, 1970.
Velichko, V. M. and Zagoruyko, N. G., Automatic recognition of 200 words. *International Journal of Man-Machine Studies*, 1970, *2*, 223-234.
Viteles, M. S., Postlude: The past and future of industrial psychology. *Journal of Consulting Psychology*, 1944, *8*, 182-185.
Von Wright, J. M., A note on the role of "guidance" in learning. *British Journal of Psychology*, 1957, *48*, 133-137.
Vroom, V. H.,
— *Source Personality Determinants of the Effects of Participation*. Englewood Cliffs, N.J.: Prentice-Hall, 1960.
— *Work and Motivation*. New York: John Wiley & Sons, Inc., 1964.
Wallas, G., *The Art of Thought*. New York: Harcourt, Brace and World, 1926.
Wargo, M. J., Human operator response speed, frequency, and flexibility: A review and analysis. *Human Factors*, 1967, *9* (3), 221-238.
Warner, H. D., Effects of intermittent noise on human target detection. *Human Factors*, 1969, *11* (3), 245-250.
Warren, R. E.,
— Norms of restricted color association. *Bulletin of the Psychonomic Society*, 1974, *4*, 37-38.
— Color associations of alarm and status labels. Private Communication, March 1980.
Wason, P. C., Response to affirmative and negative binary statements. *British Journal of Psychology*, 1961, 133-142.
Wason, P. C. and Johnson-Laird, P. N., *Psychology of Reasoning: Structure and Content*, London: Batsford, 1972.
Weidman, T. G., and Ireland, F. M., A new look at procedures manuals. *Human Factors*, 1965, *7*, 371-377.
Weiner, B. and Kukla, A., An attributional analysis of achievement motivation. *Journal of Personality and Social Psychology*, 1970, *15*, 1-20.
Weiss, D. S., The effects of text seqmentation on reading. University of Toronto, Ph.D. dissertation, 1980.
Weisstein, N. and Harris, C. S., Visual detection of line segments: an object-superiority effect, *Science*, 1974, *186*, 752-755.
Welch, J. R., Automatic speech recognition—putting it to work in industry. *Computer*, May 1980.
Welford, A. T.,
— The measurement of sensory-motor performance: Survey and reappraisal of twelve years' progress. *Ergonomics*, 1960, *3*, 189-230.
— *Skilled Performance: Perceptual and Motor Skills*. Glenview, Ill.: Scott, Foresman and Company, 1976.
West, L. J., Vision and kinethesis in the acquisition of typewriting skill. *Journal of Applied*

Psychology, 1967, *51*, 161-166.
Weston, G. F. Plasma panel displays. *Journal of Physics E: Schientific Instruments*, December 1975, *8*, 981-991.
Weston, H. C., *Slight Light and Efficiency*. London: Lewis, 1949.
Wheatley, D. M. and Unwin, A. W., *The Algorithm Writer's Guide*. London: Longman Group Limited, 1972.
Wheatley, H. B., *Literary Blunders*. London: Elliot Stock, 1983, Reprinted by Gale Research Company, Detroit, 1969.
White, R. W., Motivation recountered: The concept of competence. *Psychological Review*, 1959, *66*, 297-333.
Wickelgren, W. A.,
— Acoustic similarity and intrusion errors in short-term memory. *Journal of Experimental Psychology*, 1965, *70*, 102-108.
— Associative intrusions in short-term recall. *Journal of Experimental Psychology*, 1966, *72* (6), 853-858.
— Auditory or articulatory coding in verbal short-term memory. *Psychological Review*, 1969, *76*, 232-235.
Williams, C. M., System response time: A study of users' tolerance. *IBM Technical Report*, 1973, 217-232.
Williams, J. D., Swenson, J. S., Hegarty, J. A., and Tullis, T. S., The effects of mean CSS response time and task type on operator performance in an interactive computer system. Private Communication, 1977.
Winkler, R. L., *Introduction to Bayesian Inference and Decision*. New York: Holt, Rinehart and Winston, Inc., 1972.
Wohlwill, J. F.,
— Human adaption to levels of environmental stimulation. *Human Ecology*, 1974, *2*, 127-147.
— Environmental aesthetics: The environment as a source of affect. In I. Altman and J. F. Wohlwill (eds.), *Human Behavior and Environment: Advances in Theory and Research* (Vol. 1). New York: Plenum, 1976.
Woodson, W.E. and Conover, D.W., *Human Engineering Guide for Equipment Designers* (2nd Edition). Berkeley, Cal.: University of California Press, 1964.
Woodworth, R. S.,
— The accuracy of voluntary movement. *Psychological Monographs*, 1899, 3 (Whole No. 13).
— *Experimental Psychology*. New York: Holt, 1938.
Wooldridge, S., *Computer Input Design*. New York: Petrocelli Books, 1974.
Work in America, Report of a Special Task Force to the Secretary of Health, Education, and Welfare. Cambridge: The MIT Press, 1974.
Wright, P.,
— Transformations and the understanding of sentences. *Language and Speech*, 1969, *12*, 156-166.
— Presenting technical information: a survey of research findings. *Instructional Science*, 1977, *6*, 93-134.
Wright, P. and Barnard, P., Just fill in this form—a review for designers. *Applied Ergonomics*, 1975, *6* (4), 213-220.
Wright, P. and Reid, F., Written information: Some alternatives to prose for expressing the outcomes of complex contingencies. *Journal of Applied Psychology*, 1973, *57*(2), 160-166.
Wulff, J. J., and Berry, P. C. Aids to job performance. In R. M. Gagne and A. W. Melton, *Psychological Principles in System Development*. New York: Holt, Rinehart and Winston, 1962.
Wyatt, S., and Langdon, J. N., *Inspection Process in Industry*. London: HMSO, 1932. Industrial Health Board, Report 3.
Wyndham, C., Adaption to heat and cold. *Environmental Research*, 1969, *2*, 442-469.
Wyon, D. P., Studies of children under imposed noise and heat stress. *Ergonomics*, 1970, *13*, 598-612.

Yerkes, R. M. and Dodson, J. D., The relation of strength of stimulus to rapidity of habit formation. *Journal of Comparative Neurology and Psychology,* 1908, *18*, 459-482.

Zimmerman, R., Telephone as terminal dialogue between inexperienced users and computer-controlled information systems, 1977.

Zinke, R. A., A study of the application of personnel subsystem development technology in the design of the BISCUS/FACS Conversion System. Private Communication, April 1977.

Zipf, G., *The Psycho-biology of Language*. Boston: Hougton Mifflin, 1935.

INDEX

A

C

D

E

F

G

I

J

K

L

M

N

O

P

Q

R

S

T

U

V

W

X

Y

Z